Remote Sensing of Atmospheric Pollution

Special Issue Editors

Yang Liu
Jun Wang
Omar Torres

MDPI • Basel • Beijing • Wuhan • Barcelona • Belgrade

MDPI

Special Issue Editors
Yang Liu
Emory University
USA

Jun Wang
University of Iowa
USA

Omar Torres
NASA Goddard Space Flight Center
USA

Editorial Office
MDPI AG
St. Alban-Anlage 66
Basel, Switzerland

This edition is a reprint of the Special Issue published online in the open access journal *Remote Sensing* (ISSN 2072-4292) from 2016–2017 (available at: http://www.mdpi.com/journal/remotesensing/special_issues/pollution).

For citation purposes, cite each article independently as indicated on the article page online and as indicated below:

Author 1; Author 2. Article title. *Journal Name* **Year**. Article number, page range.

First Edition 2017

ISBN 978-3-03842-640-0 (Pbk)
ISBN 978-3-03842-641-7 (PDF)

Table of Contents

About the Special Issue Editors vii

Preface to "Remote Sensing of Atmospheric Pollution" ix

J. H. Belle and Yang Liu
Evaluation of Aqua MODIS Collection 6 AOD Parameters for Air Quality Research over the
Continental United States
doi: 10.3390/rs8100815 1

Wei Chen, Aiping Fan and Lei Yan
Performance of MODIS C6 Aerosol Product during Frequent Haze-Fog Events: A Case
Study of Beijing
doi: 10.3390/rs9050496 15

Yi Wang, Jun Wang, Robert C. Levy, Xiaoguang Xu and Jeffrey S. Reid
MODIS Retrieval of Aerosol Optical Depth over Turbid Coastal Water
doi: 10.3390/rs9060595 34

Minghui Tao, Zifeng Wang, Jinhua Tao, Liangfu Chen, Jun Wang, Can Hou, Lunche Wang, Xiaoguang Xu and Hao Zhu
How Do Aerosol Properties Affect the Temporal Variation of MODIS AOD Bias in
Eastern China?
doi: 10.3390/rs9080800 47

Wei Wang, Feiyue Mao, Zengxin Pan, Lin Du and Wei Gong
Validation of VIIRS AOD through a Comparison with a Sun Photometer and MODIS AODs
over Wuhan
doi: 10.3390/rs9050403 62

Jun Zhu, Xiangao Xia, Jun Wang, Huizheng Che, Hongbin Chen, Jinqiang Zhang, Xiaoguang Xu, Robert C. Levy, Min Oo, Robert Holz and Mohammed Ayoub
Evaluation of Aerosol Optical Depth and Aerosol Models from VIIRS Retrieval Algorithms
over North China Plain
doi: 10.3390/rs9050432 75

Kun Sun, Xiaoling Chen, Zhongmin Zhu and Tianhao Zhang
High Resolution Aerosol Optical Depth Retrieval Using Gaofen-1 WFV Camera Data
doi: 10.3390/rs9010089 93

Yang Wang, Liangfu Chen, Shenshen Li, Xinhui Wang, Chao Yu, Yidan Si and Zili Zhang
Interference of Heavy Aerosol Loading on the VIIRS Aerosol Optical Depth (AOD)
Retrieval Algorithm
doi: 10.3390/rs9040397 112

Xi Chen, Dongxu Yang, Zhaonan Cai, Yi Liu and Robert J. D. Spurr
Aerosol Retrieval Sensitivity and Error Analysis for the Cloud and Aerosol Polarimetric
Imager on Board TanSat: The Effect of Multi-Angle Measurement
doi: 10.3390/rs9020183 128

Man Jiang, Weiwei Sun, Gang Yang and Dianfa Zhang
Modelling Seasonal GWR of Daily PM$_{2.5}$ with Proper Auxiliary Variables for the Yangtze River Delta
doi: 10.3390/rs9040346 . . . . **146**

Wei Wang, Feiyue Mao, Lin Du, Zengxin Pan, Wei Gong and Shenghui Fang
Deriving Hourly PM2.5 Concentrations from Himawari-8 AODs over Beijing–Tianjin–Hebei in China
doi: 10.3390/rs9080858 . . . . **166**

Yawei Qu, Yong Han, Yonghua Wu, Peng Gao and Tijian Wang
Study of PBLH and Its Correlation with Particulate Matter from One-Year Observation over Nanjing, Southeast China
doi: 10.3390/rs9070668 . . . . **183**

Sergey I. Dolgii, Alexey A. Nevzorov, Alexey V. Nevzorov, Oleg A. Romanovskii and Olga V. Kharchenko
Intercomparison of Ozone Vertical Profile Measurements by Differential Absorption Lidar and IASI/MetOp Satellite in the Upper Troposphere–Lower Stratosphere
doi: 10.3390/rs9050447 . . . . **197**

Lei Liu, Xiuying Zhang, Wen Xu, Xuejun Liu, Xuehe Lu, Shanqian Wang, Wuting Zhang and Limin Zhao
Ground Ammonia Concentrations over China Derived from Satellite and Atmospheric Transport Modeling
doi: 10.3390/rs9050467 . . . . **212**

Jianbin Gu, Liangfu Chen, Chao Yu, Shenshen Li, Jinhua Tao, Meng Fan, Xiaozhen Xiong, Zifeng Wang, Huazhe Shang and Lin Su
Ground-Level NO$_2$ Concentrations over China Inferred from the Satellite OMI and CMAQ Model Simulations
doi: 10.3390/rs9060519 . . . . **231**

Daewon Kim, Hanlim Lee, Hyunkee Hong, Wonei Choi, Yun Gon Lee and Junsung Park
Estimation of Surface NO$_2$ Volume Mixing Ratio in Four Metropolitan Cities in Korea Using Multiple Regression Models with OMI and AIRS Data
doi: 10.3390/rs9060627 . . . . **248**

Kai Qin, Lanlan Rao, Jian Xu, Yang Bai, Jiaheng Zou, Nan Hao, Shenshen Li and Chao Yu
Estimating Ground Level NO$_2$ Concentrations over Central-Eastern China Using a Satellite-Based Geographically and Temporally Weighted Regression Model
doi: 10.3390/rs9090950 . . . . **264**

Adrin Yuchechen, Susan Gabriela Lakkis and Pablo Canziani
Linear and Non-Linear Trends for Seasonal NO$_2$ and SO$_2$ Concentrations in the Southern Hemisphere (2004–2016)
doi: 10.3390/rs9090891 . . . . **284**

Matas Osorio, Nicols Casaballe, Gastn Belsterli, Miguel Barreto, lvaro Gmez, Jos A. Ferrari and Erna Frins
Plume Segmentation from UV Camera Images for SO$_2$ Emission Rate Quantification on Cloud Days
doi: 10.3390/rs9060517 . . . . **308**

Mika G. Tosca, James Campbell, Michael Garay, Simone Lolli, Felix C. Seidel, Jared Marquis
and Olga Kalashnikova
Attributing Accelerated Summertime Warming in the Southeast United States to Recent
Reductions in Aerosol Burden: Indications from Vertically-Resolved Observations
doi: 10.3390/rs9070674 325

About the Special Issue Editors

Yang Liu received his Ph.D. in environmental science and engineering from Harvard University in 2004 and completed his postdoctoral training in Harvard T.H. Chan School of Public Health. He is an Associate Professor at the Rollins School of Public Health of Emory University. For 18 years, his research and teaching have focused on the applications of satellite remote sensing in air pollution and health effects research at the urban to national scales using data from MODIS, MISR, MAIAC, GOES, OMI and VIIRS. He has led and participated in multiple projects funded by NASA, CDC, EPA, NIH, and HEI to apply satellite data in air pollution exposure modeling and health effects research. He has authored over 120 peer-reviewed articles, and is a science team member of the NASA EVI-3 MAIA instrument, the Terra MISR instrument, the Aura satellite, and a PI member of the NASA AQAST and HAQAST team. From 2012 to 2014, He was an instructor of NASA Applied Remote Sensing Education and Training (ARSET) team with which he provided trainings to policy makers, regulatory agencies, NGOs, and other applied science professionals on the benefits and applications of NASA Earth observations.

Jun Wang is a Professor in the University of Iowa, with joint appointments in Department of Chemical and Biochemical Engineering and Iowa Informatics Initiative, and secondary affiliation in Center for Global and Regional Environmental Studies and Department of Civil and Environmental Engineering. Prior to joining the University of Iowa, he worked in University of Nebraska—Lincoln first as Assistant Professor and then Associate Professor (with Rosowski professorship). His research focuses on the integration of satellite remote sensing and chemistry transport model to study air quality, wildfires, aerosol-cloud-radiation interaction, and land-atmospheric interaction. He has authored 100+ peer-reviewed articles, and has been a science team member of several NASA satellite missions, focusing on development and improvement of aerosol retrieval algorithms. He received his Ph.D. from University of Alabama—Huntsville in 2005, and was a postdoctoral researcher in Harvard Uni-versity in 2005–2007. More about his research can be found at https://arroma.uiowa.edu.

Omar Torres received a B.S. degree in Geodetic Engineering from Bogota District FJC University, in Bogota, Colombia. Ho got an M.S. degree in Meteorology from the University of the Philippines, in Quezon City, Philippines, and a Ph.D degree in Atmospheric Sciences at The Georgia Institute of Technology, Atlanta, and GA in 1990. Between 1989 and 1999, he was involved in a variety of NASA atmospheric remote sensing projects involving the use of TOMS satellite UV observations. From 1999 to 2010 he was an Associate Professor at the University of Maryland Baltimore County and Hampton University. In 2011, he joined NASA Goddard Space Flight Center, Greenbelt, MD, where he has worked in UV atmospheric remote sensing applications with special emphasis in the characterization of aerosol absorption properties from satellite UV observations by OMI, OMPS, and EPIC instruments. He has authored and co-authored over 100 peer-reviewed publications.

Preface to "Remote Sensing of Atmospheric Pollution"

This book recompiles the most recent articles published in Remote Sensing as a special issue focusing on the latest research development and applications in the field of remote sensing of air pollution. In brief, these twenty articles cover a wide range of topics as follows.

Geospatial and statistical modeling to estimate hourly surface PM2.5 from Himawari-8, surface NO_2 from OMI and AIRS, southern hemispheric NO_2 and SO_2 trend from OMI, and surface NH3 from IASI.

Evaluation and error characterization of satellite products such as aerosol optical depth products (AOD) retrieved from MODIS Dark-Target algorithm (at 3 km), MODIS Deep-Blue algorithm and VIIRS over Asia during heavy aerosol loading conditions, and O_3 profile from IASI with ground-based lidar data.

Development of new algorithms for retrieving costal AOD from MODIS, AOD at high spatial resolution from GaoFen satellite, for estimating potential improvement in AOD from using polarization on TanSat, and for deriving SO_2 emission rate using ground-based UV cameras.

Process studies of aerosol impact on surface temperature trend using satellite and ground-based aerosol data, and PBL vs. surface PM2.5 correlations.

To the best of our knowledge, there has not been a book like this one that collectively shows the fast and exciting progresses in remote sensing of nearly all key air pollutants such as aerosols, NO_2, SO_2, O_3 and NH_3, not only from mature satellite sensors (e.g., MODIS, AIRS, and IASI), but also from newer sensors (such as VIIRS, GaoFen and TanSat) and ground-based lidar and cameras. We hope this book serves as a frequent reference for graduate students, faculty, researchers, and professionals in the field of remote sensing of aerosols and trace gases, especially those with a high interest in air quality and applied sciences.

Yang Liu, Jun Wang , Omar Torres

Special Issue Editors

Article

Evaluation of Aqua MODIS Collection 6 AOD Parameters for Air Quality Research over the Continental United States

J. H. Belle and Yang Liu *

Department of Environmental Health, Rollins School of Public Health, Emory University, Atlanta, GA 30322, USA; jessica.hartmann.belle@emory.edu
* Correspondence: yang.liu@emory.edu; Tel.: +1-404-727-8744

Academic Editors: Jun Wang, Omar Torres, Alexander A. Kokhanovsky, Richard Müller and Prasad S. Thenkabail
Received: 18 July 2016; Accepted: 26 September 2016; Published: 1 October 2016

Abstract: Satellite-retrieved aerosol optical depth (AOD) has become an important predictor of ground-level particulate matter (PM) and greatly empowered air pollution research worldwide. We evaluated the AOD parameters included in the Collection 6 aerosol product of the Moderate Resolution Imaging Spectroradiometer (MODIS) for two key factors affecting their applications in air quality research—coverage and accuracy—over the continental US. For the high confidence retrievals (QAC 3), the 10 km DB-DT combined AOD has the best coverage nationwide (29.7% of the days in a year in any given 12 km grid cell). While the Eastern US generally had more successful AOD retrievals, the highest spatial coverage of AOD parameters were found in California (>55%) and other vegetated parts of the Western US. If lower QAC retrievals were included, the coverage of the 10 km DB AOD was dramatically increased to 49.6%. In the Eastern US, the QAC 3 retrievals of all four AOD parameters are highly correlated with AERONET observations (correlation coefficients between 0.80 and 0.92). In the Western US, positive retrieval errors existed in all MODIS AOD parameters, resulting in lower correlations with AERONET. AOD retrieval errors showed significant dependence on flight geometry, land cover type, and weather conditions. To ensure appropriate use of these AOD values, air quality researchers should carefully balance the needs for coverage and accuracy, and develop additional data screening criteria based on their study design.

Keywords: MODIS; AOD; remote sensing; United States; retrieval accuracy; satellite coverage

---

1. Introduction

Aerosol optical depth (AOD) is 'the single most comprehensive variable to remotely assess the aerosol burden in the atmosphere' [1]. It is used to characterize ambient aerosols, either for land-based remote sensing applications where it is used to remove atmospheric influences, or directly, to assess atmospheric pollution, primarily fine particulate matter, and its impacts on the climate, ecosystems, and human populations. Exposure to fine particulate matter ($PM_{2.5}$, airborne particles with an aerodynamic diameter of 2.5 micrometers or less) was identified as a leading risk factor for global disease burden with an estimated 2.9 million attributable deaths in the year 2013 [2]. Historically, the estimation of population exposure to $PM_{2.5}$ depends on filter-based ground monitors. However, because of its high operation and maintenance costs, these ground-based monitoring networks do not achieve comprehensive spatial coverage. With its comprehensive spatial coverage, spatial models driven by MODIS AOD are able to estimate the $PM_{2.5}$ exposure levels in many parts of the world where ground observations are sparse or nonexistent [3]. The MODerate Resolution Imaging Spectroradiometer (MODIS) instruments on board the Aqua and Terra satellite platforms have been providing daily, near-global satellite coverage since 2000 and 2002, respectively [4]. MODIS-retrieved aerosol optical depth (AOD) has been used extensively in estimating ground-level fine particulate matter (PM$_{2.5}$) concentrations [5]. Over the past decade, various MODIS-driven PM$_{2.5}$ exposure models have been developed, from relatively simple linear regressions [6] to complex multi-level spatial models [7] and Bayesian hierarchical models [8]. Because PM$_{2.5}$ is linked to adverse health outcomes even at the low concentrations commonly observed in the cities of North America [9], PM$_{2.5}$ models based on MODIS retrievals have been used to extend ground air quality monitoring networks to cover the suburban and rural populations in the U.S. [10] and Canada [11].

Accuracy and coverage are the most important factors affecting the application of satellite AOD in air quality research. The retrieval error in AOD has a major influence in the PM$_{2.5}$ prediction error, as AOD is often used as the primary predictor in various PM$_{2.5}$ exposure models. If the AOD retrieval error varies by season or with land use types, the PM$_{2.5}$ prediction error will also display spatiotemporal patterns. This is especially true at the low AOD levels, typically below 0.2, commonly observed in developed countries [12]. On the one hand, availability of AOD data coverage determines whether satellite-driven models are feasible for a given study region. On the other hand, it plays an important role in determining the design of PM$_{2.5}$ health effect studies [13]. For example, the health effect of short-term PM$_{2.5}$ exposure such as asthma exacerbation is often evaluated in a time series model where temporal missingness of exposure estimates can substantially limit model performance [10]. Cohort studies designed for associating long-term PM$_{2.5}$ exposure with cardiovascular morbidity and mortality would benefit from complete spatial coverage [14].

The most recent MODIS collection 6 (C6) aerosol products include enhanced Dark-Target (10 km DT) and Deep-Blue (10 km DB) AOD present in collection 5 (C5), a 'merged' DB-DT parameter (10 km DB-DT) and a 3 km AOD based off of the 10 km DT retrieval algorithm (3 km DT) [15,16]. The MODIS science team has conducted a few global validation studies to document the collective impact of these changes and differences between the various parameters [12,16–18]. These studies mainly focused on estimating the AOD retrieval errors by comparing with collocated measurements from the Aerosol Robotic Network (AERONET) at the global scale. Because of the large spatial differences in aerosol loading, global performance metrics such as regression slopes and correlation coefficients are often driven by regions of high AOD values. To date, only a handful of evaluation studies were reported in North America, none of which had both accuracy and coverage as their primary research objectives [19,20]. Therefore, there remains a need for detailed validation studies in dominantly low-AOD regions to investigate issues related to surface reflectance treatment and extreme events [12]. In addition, the accuracy and potential usability of lower quality retrievals needs to be better characterized, and could have important implications on the coverage issue in air quality applications of MODIS data.

In the current analysis, we focused on characterizing the accuracy and coverage of various MODIS AOD parameters in the continental US, a dominantly low-AOD area. We focused on examining the degree to which changes in surface properties and retrieval conditions, such as viewing angle and land use, affect AOD retrieval error. In addition to the highest quality AOD data, we evaluate the impact of including lower quality AOD values on the spatial and temporal coverage statistics. Additionally, we use a case study to demonstrate the practical implications of including lower-quality retrievals and accounting for major sources of bias on the ability of each AOD parameter to accurately estimate ground-level PM$_{2.5}$ concentrations. Finally, we summarize the strengths and weaknesses of these AOD parameters in the context of air quality research.

2. Materials and Methods

2.1. Satellite and Ground Datasets

We collected Aqua MODIS level 2 AOD data [21] between 1 January 2004 and 31 December 2013 in the Continental US. Level 2 cloud-screened and quality assured AOD retrievals from 120 permanent AERONET stations and 73 temporary stations from the Distributed Regional Aerosol Gridded

Observation Networks (DRAGON) were collected to validate MODIS retrievals. Out of the 120 permanent AERONET stations, 48% had been in operation for less than one year (Figure 1). Total column precipitable water estimates were also collected from these stations to evaluate their impact on MODIS AOD retrieval error. Since AERONET does not directly measure AOD at the 0.55 μm wavelength reported by MODIS, values were interpolated to this wavelength with a quadratic fit in log-log space based on valid AOD values at a minimum of 4 of any of the 15 wavelengths potentially reported by AERONET [22]. Ancillary datasets were collected for identifying surface properties and retrieval conditions that could have affected MODIS retrieval accuracy. The MODIS 16-day gridded NDVI parameter at 1 km spatial resolution [23] was used to calculate NDVI values at individual MODIS level 2 AOD pixels. The National Land Cover Database (NLCD) with a 30 m spatial resolution was used for land cover type calculation at individual MODIS level 2 AOD pixels [24]. The 2006 NLCD was used for collocations occurring prior to 2009 and the 2011 NLCD was used for collocations occurring after 2009. Information on scattering, viewing, and solar angles for each AOD retrieval was obtained for each MODIS pixel from the MODIS AQUA level 2 Aerosol product [21].

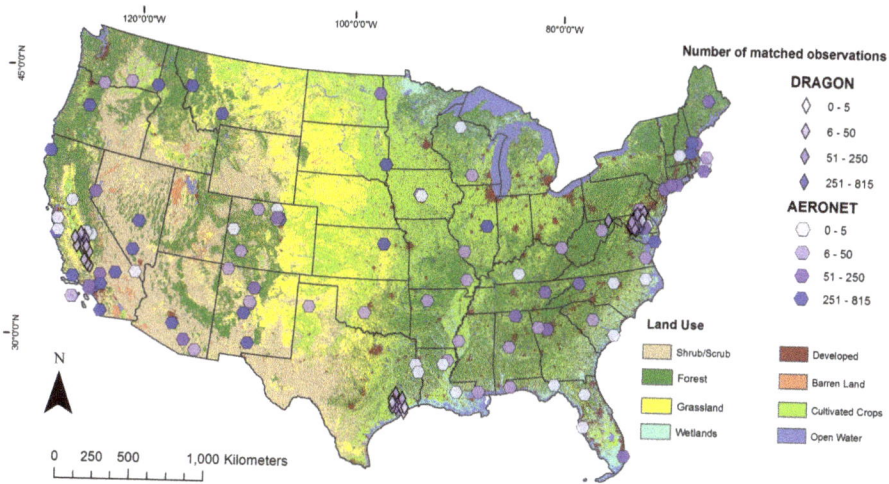

Figure 1. Spatial distribution of AERONET (hexagons) and DRAGON (diamonds) sites over the study period from 1 January 2004 to 31 December 2013. The color of the symbols represents the number of collocations at each site.

2.2. Coverage

Since MODIS pixels are created relative to each satellite view and the MODIS instrument exhibits a fish-eye effect, the size and location of individual pixels is not constant from one day to the next. To compensate for this, a 12 km grid commonly used in the Community Multi-scale Air Quality (CMAQ) modeling system was created for our coverage calculation (a total of 55,031 cells). The grid size roughly corresponds to the nadir resolution of the MODIS 10 km AOD parameters and represents an important application of the MODIS data, where AOD observations are assimilated into air quality models to improve model performance [25]. MODIS pixels were determined to be within a grid cell if, for the 10 km DT, 10 km DB, and 10 km DB-DT AOD, the polygon representing the pixel area, reconstructed from the pixel centroids using a Voronoi tessellation algorithm [26], lay at least partially within the grid cell. Pixels from the 3 km DT parameter were determined to be within a grid cell if the centroid of the 3 km pixel fell within the grid cell, allowing the increased resolution of the 3 km DT parameter relative to the grid to compensate for the lack of smoothing between cells that the Voronoi tessellation would have provided. The percentage of days with a valid, QAC 3 retrieval were calculated for each grid cell and parameter. Results were interpreted relative to averages over the continental U.S. (CONUS) for each parameter. Coverage statistics were also calculated for QAC 1, 2, and 3 retrievals together to provide an accounting of the gains in coverage by including lower confidence retrievals in an analysis.

2.3. Accuracy

AERONET observations were collocated with each MODIS AOD parameter respectively, so that a temporal average of AERONET observations within ±30 min of the MODIS pass was compared to the spatial average of MODIS pixels within a ~25 km radius for the 10 km DT, DB and DB-DT AOD, and a ~7.5 km radius for the 3 km DT AOD [17]. Following previous work, a collocation was only considered valid if a minimum of three MODIS pixels, two AERONET measurements, and at least 20% of the total number of MODIS pixels included in the 25/7.5 km radius had valid values with a QAC code of 3 assigned to the pixel [27]. AERONET stations were categorized as either being in the East or the West, using the 100° W longitude line [16]. The east/west division was necessary because previous work had found large differences in MODIS performance between the two regions [25]. Retrieval error, or the difference between MODIS and AERONET AOD at each collocation ($\tau_M - \tau_A$) and the percentage of MODIS observations within the 10 km DT expected error envelope (EE$_{DT}$)—defined as ±(0.05 + 0.15)τ—[16] were calculated and linear regression models were used to quantify retrieval errors. In order to evaluate the QAC code assignments as indicators of retrieval errors, independent collocations were created for QAC 1 and QAC 2 retrievals with AERONET, using the same criteria as for the QAC 3 collocations. Finally, for each AOD parameter, retrieval error in MODIS AOD relative to AERONET was examined within quintiles of the surface and retrieval parameters. These parameters include median NDVI, total column precipitable water from AERONET, land-cover type mode, mean solar zenith, sensor zenith, and scattering angles. Linear regression models were used to identify any significant linear trends in retrieval error for each surface and retrieval parameter.

3. Results

During our study period, 193 ground stations reported a total of 286,055 observations that could be interpolated to AOD at 550 nm. Of these, 262,491 originated from a permanent AERONET station and 23,564 were recorded during a DRAGON campaign. In the Eastern US, the number of valid collocations at the 127 stations with high confidence MODIS retrievals ranges from 5616 for 3 km DT to 6617 for 10 km DB observations. AERONET AOD ranged from 0.0005 to 1.26, with mean values of 0.12, 0.12, 0.12, and 0.10 for collocations with the high confidence 3 km DT, 10 km DT, 10 km DB-DT and 10 km DB parameters, respectively. MODIS AOD ranged from −0.05 to 2.77, with mean values of 0.13, 0.13, 0.13, and 0.11 for these four AOD parameters, respectively. In the Western US, the number of valid collocations at 66 AERONET stations with high confidence MODIS retrievals ranges from 6251 for 3 km DT to 11,590 for 10 km DB-DT AOD. AERONET AOD values ranged from 0.0003 to 1.43, with mean values of 0.09, 0.09, 0.08, and 0.08 for collocations with the high confidence 3 km DT, 10 km DT, 10 km DB-DT, and 10 km DB parameters, respectively. MODIS AOD values ranged from −0.05 to 2.35, with mean values of 0.09, 0.12, 0.10, and 0.08 for these parameters, respectively. In both regions, and for all four products, the majority of collocations occurred in the fall and summer, while the fewest occurred in winter months.

3.1. Coverage of High-Confidence Retrievals

Table 1 shows that on average, a valid AOD retrieval was available on 25%–30% of days in any given grid cell. However, there is considerable spatial heterogeneity in coverage rates for each parameter (Figure 1). The highest rates of coverage are found on the western coast, near the large cities of Los Angeles and San Francisco, and in California's central valley. In these areas, all four AOD parameters achieve coverage rates of over 55%, and in the area right around Los Angeles, coverage rates are above 70%. Similarly high coverage rates, between 50% and 60%, are also observed over the national forests north of Phoenix in Arizona and rates of 40%–50% are observed over the south-central plains covering the areas of central Texas, Oklahoma, and Kansas. A north-to-south and elevation gradient in coverage rates can also be observed in Figure 2. The lowest coverage rates were observed over the Great Salt Lake desert, where a few locations had no valid retrieval. Outside of the Rockies, average coverage rates in the northern parts of the CONUS—an area that includes the large cities of Chicago and New York—were typically only 10%–20%. Coverage rates further south were 30%–40%, slightly higher than the CONUS-wide average. This north-south and elevation-based gradient in coverage rates can be linked to seasonal snow-cover occurring primarily at higher latitudes and elevations.

Table 1. Coverage statistics for both QAC 3 retrievals only, and for all AOD retrievals. Coverage is calculated as the percentage of days with a valid Aqua retrieval for each AOD parameter.

AOD Parameter	Coverage % (QAC 3 Only)	Coverage % (QAC 1, 2, 3)
3 km DT	28.2	28.9
10 km DT	24.3	32.8
10 km DB-DT	29.7	31.1
10 km DB	28.9	49.6

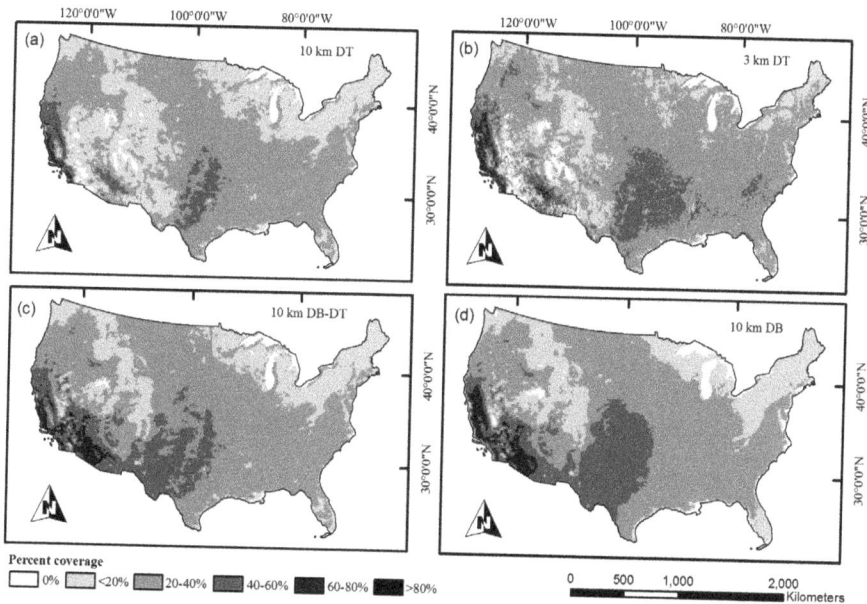

Figure 2. Mean coverage statistics for high confidence AOD retrievals in the CONUS for 10 km DT (**a**); 3 km DT (**b**); 10 km DB-DT (**c**); and 10 km DB (**d**). Coverage is calculated as the percentage of days between 1 January 2004 and 31 December 2013 with a valid MODIS retrieval from Aqua.

The 3 km DT AOD, with a CONUS-wide coverage rate of 28.2%, is comparable to the 10 km AOD parameters in terms of coverage. In contrast to the 10 km products, the 3 km DT AOD excels over areas where the surface is more complex but not arid, such as the Pacific Northwest, and over the Carolinas. It achieves slightly higher coverage rates on the eastern coast than the 10 km parameters, and retrieves at higher rates at high to moderate latitudes and elevations than the 10 km AOD parameters (Figure 2a). The most likely explanation for these higher coverage rates at higher elevations and latitudes would be an increased ability, on the part of the higher resolution parameter, to retrieve aerosols over patchy snow-cover. However, while it has been previously noted that the higher resolution parameter is able to retrieve aerosol information at higher rates over complex landscapes, coastlines, and between clouds, the extension of this ability to complex snow-cover has not been investigated [17,19]. The 10 km merged AOD has the highest overall coverage, averaging 29.7% for the CONUS. This parameter aims to maximize the number of high-confidence AOD retrievals by using AOD values from the 10 km DT algorithm over locations where the NDVI is higher and to use 10 km DB AOD values over locations where NDVI is lower and the 10 km DT algorithm is less likely to accurately retrieve AOD. The result is that the spatial patterns of coverage for the merged AOD are similar to 10 km DT AOD, but without the gaps in coverage over the arid southwest. The coverage of 10 km DB AOD along the east coast is lower than the other three AOD parameters especially in the summer months (see Supplementary Figure S1) and over Florida, and balances out the additional coverage gained in the west and south-central plains (Figure 2). The reasons for this are currently unknown, but slight differences exist between the DT and DB retrieval processes in the tests used to identify cloud-cover and distinguish it from aerosols and this could explain summertime differences in coverage over the highly vegetated Eastern CONUS [15,16].

3.2. Coverage Gains with Lower-Quality Retrievals

When lower-quality retrievals were included, coverage rates increased for all AOD parameters. However, there were large differences in relative increases between products, reflective of differences in QAC code assignments among AOD parameters. On the one hand, coverage rates for the 3 km DT AOD increased only slightly from 28.2% to 28.9%, similar to that observed for the 10 km DB-DT AOD. Coverage rates for the 10 km DT parameter increased substantially from 24.3% to 32.8%. On the other hand, coverage of the 10 km DB AOD increased dramatically from 28.9% to 49.6% when lower-confidence retrievals were included. Of the four AOD parameters, the 10 km merged AOD provides the highest overall coverage over the CONUS if only high-confidence retrievals are considered. Coverage patterns are similar to those observed for high confidence observations (Supplementary Figure S2). Previous studies have examined how AOD missingness impacts the representativeness of the sample within the CONUS, with mixed results [28,29]. For some applications, the typical coverage rates from high quality retrievals of ~30% may be too low to preserve study power, and investigators may seek to boost coverage rates through the use of noisier, lower-confidence observations. In this instance, the 10 km DB AOD offers the greatest potential gains in coverage.

3.3. Accuracy of High-Confidence Retrievals

Error statistics for all four AOD parameters, broken down by region, are presented in Table 2. In the eastern region, 76% of QAC 3 3 km DT AOD retrievals were within pre-launch expectations with a mean retrieval error of 0.01, and the correlation coefficient between AERONET and MODIS values was 0.89. However, the slope of a regression line fit between the two was 1.24, indicating over-prediction. In the western region, 49% of 3 km DT observations were within the EE and the slope of a line relating AERONET AOD values to MODIS observations was 1.41. The QAC 3 10 km DT AOD retrievals in the East had a slightly positive, but near 0 retrieval error, and its correlation coefficient with AERONET observations (0.92) was the highest of the four parameters. Over-prediction was a problem in the western region, but was more problematic at low AOD levels. The performance of the 10 km DB-DT AOD across accuracy metrics was nearly identical to the 10 km DT AOD in the Eastern US, but it had an additional 341 QAC 3 collocations. In the western region, it was the most highly correlated with AERONET (0.73). The QAC 3 10 km DB AOD had somewhat uneven performance across accuracy metrics relative to findings from global validations [12]. It had the highest percentage of observations within EE_{DT} (87% in the East, 83% in the West). Correlation coefficients, however, were lower than the other three parameters (0.80 in the East, 0.63 in the West) and global estimates [12]. Median retrieval error estimates in both regions were low (0.01 in the East, −0.00 in the West), but intercepts from regression modeling were higher (0.03 in the East, 0.02 in the West) and slopes were below 1 (0.79 in the East, 0.75 in the West). When assessed together, these metrics indicated over-prediction at lower AOD values and under-prediction at higher AOD values. Previous global validations have found similar patterns for 10 km DB retrievals, but over-prediction was more severe in the Western US than was noted in the global studies.

Table 2. Performance statistics for each AOD parameter.

Region	QAC	Parameter	N	Error	Intercept	Slope	R	% Above EE	% Below EE	% Inside EE
West	1	3 km DT	546	0.02	0.02	1.03	0.86	9.0	0.9	90.1
		10 km DT	1487	0.04	0.05	1.46	0.58	43.4	1.6	55.0
		10 km DB-DT	482	0.01	0.01	1.06	0.85	7.3	1.5	91.3
		10 km DB	10603	0.04	0.07	1.01	0.52	42.0	2.0	55.9
	2	10 km DT	2577	0.08	0.08	1.30	0.58	58.1	2.4	39.6
		10 km DB-DT	1321	0.01	0.03	0.95	0.47	27.0	3.3	69.6
		10 km DB	2316	0.01	0.04	0.90	0.62	27.8	3.4	68.9
	3	3 km DT	6251	0.06	0.06	1.41	0.64	48.9	2.5	48.6
		10 km DT	7814	0.02	0.03	0.98	0.71	25.3	5.6	69.1
		10 km DB-DT	11590	0.01	0.01	1.01	0.73	17.3	5.1	77.6
		10 km DB	10662	−0.00	0.02	0.75	0.63	11.2	5.9	82.9
East	1	3 km DT	883	0.02	0.04	0.80	0.79	12.6	4.6	82.8
		10 km DT	1322	0.02	0.02	1.19	0.82	26.6	5.6	67.8
		10 km DB-DT	440	0.00	0.04	0.50	0.71	3.9	8.4	87.7
		10 km DB	7019	0.05	0.08	0.84	0.76	36.4	1.0	62.7
	2	10 km DT	1538	0.04	0.02	1.26	0.88	37.1	6.1	56.8
		10 km DB-DT	16	0.13	0.17	0.42	0.21	62.5	0.0	37.5
		10 km DB	368	0.05	0.09	0.70	0.73	39.1	1.4	59.5
	3	3 km DT	5616	0.01	−0.01	1.24	0.89	16.0	7.6	76.4
		10 km DT	6409	0.00	−0.01	1.18	0.92	10.8	8.4	80.9
		10 km DB-DT	6750	0.00	−0.01	1.17	0.91	11.7	7.5	80.8
		10 km DB	6617	0.01	0.03	0.79	0.80	10.5	2.6	86.9

N: number of collocations. Error: $\tau_M - \tau_A$ The intercept, slope, and correlation coefficient (r) are calculated using a linear regression model relating MODIS to AERONET AOD values. The error envelope is defined as $\pm(0.05 + 0.15)\tau_A$.

3.4. Accuracy Assessment of Lower-Confidence Retrievals

The more mature 10 km DB and 10 km DT AOD had 20,306 and 6924 valid lower-confidence collocations, respectively. In contrast, the less mature 3 km DT and 10 km DB-DT products had only 1429 and 2259 valid lower-confidence collocations, respectively. As expected, lower confidence collocations for the 10 km DB and 10 km DT AOD were noisier and had larger retrieval errors when compared to QAC 3 observations (Table 2). For both parameters, a lower proportion of low-confidence observations were within EE$_{DT}$, ranging from 40% to 68% of retrievals, median retrieval error estimates were higher, ranging from 0.02 to 0.09, and correlation coefficients were lower, ranging from 0.52 to 0.88. There were a substantial number of valid QAC 1 10 km DB collocations with a positive retrieval error at lower AOD values. This is typically attributable to cloud contamination, and so it may be possible to use some of these observations in an analysis with caution and additional cloud screening procedures [25]. Accuracy statistics for lower-confidence 3 km DT and 10 km merged AOD, on the other hand, were comparable with high confidence retrievals (Table 2). The 3 km DT AOD had a high percentage of observations within EE$_{DT}$, 83 and 90% for the eastern and western regions, respectively, and strong correlations (0.79 in the East, 0.85 in the West). All low-confidence 10 km DB-DT AOD except QAC 2 retrievals in the eastern region (only 16 valid collocations) met pre-launch expectations, having between 70% and 90% of retrievals within EE$_{DT}$. The QAC code assignments for the two new AOD parameters do not seem to accurately reflect retrieval errors in the same way as for the more mature AOD parameters.

3.5. Dependence of Retrieval Errors on Flight Geometry and Land Cover Type

Figure 3 illustrates the dependence of AOD errors on the scattering, solar zenith, and viewing angles for QAC 3 retrievals. Scattering angle was associated with a statistically significant, positive trend in retrieval error in all four parameters in both regions. This trend is most pronounced for 3 km DT AOD in the western region with a median retrieval error of 0.12 for the highest quintile, and 0.04

for the lowest quintile of scattering angle. The median retrieval errors of 10 km DT, DB, and DB-DT AOD in both regions, and 3 km DT observations in the eastern region increase slightly with scattering angle (but remain below 0.04). This type of dependence could be related to issues with accounting for anisotropy in the surface reflectance over the CONUS [30]. Our findings in the CONUS disagree with those presented in global evaluations, which found tendencies of median retrieval error with scattering angle to be small and negative [12]. Our findings on the association between retrieval errors and solar zenith angles are only partially consistent with Sayer et al. [12] which found solar zenith angles below 20 degrees to have positive retrieval errors for the 10 km DT parameter and negative retrieval errors for the 10 km DB parameter, but retrievals at angles greater than 20 to be relatively unbiased. In the Eastern US, our results show a similar pattern to those present in Sayer et al. [12] for 10 km DB retrievals relative to 10 km DT retrievals, but with a slight positive retrieval error for 10 km DB. Additionally, we observed fairly substantial retrieval errors at solar zenith angles greater than 20 degrees. In the Western US, we observed negative retrieval errors in 10 km DB observations spanning solar zenith angles from 25 to 43 degrees, while the first quintile, containing observations with solar zenith angles less than 25 degrees, was relatively unbiased. This finding runs contrary to previous observations which have suggested that it is primarily low solar zenith angles that are problematic [12,25]. AOD retrieval error shows a small negative trend with sensor zenith angle for the 10 km DB AOD in both regions and for the 3 km DT in the West, and a small positive trend for the 10 km DT, 10 km DB-DT, and 3 km DT AOD in the East. The largest change was for 10 km DB observations in the East, where the median retrieval error estimate in AOD within the first quintile of sensor zenith angle, near the nadir, was 0.024 and the median retrieval error in the highest quintile was −0.005.

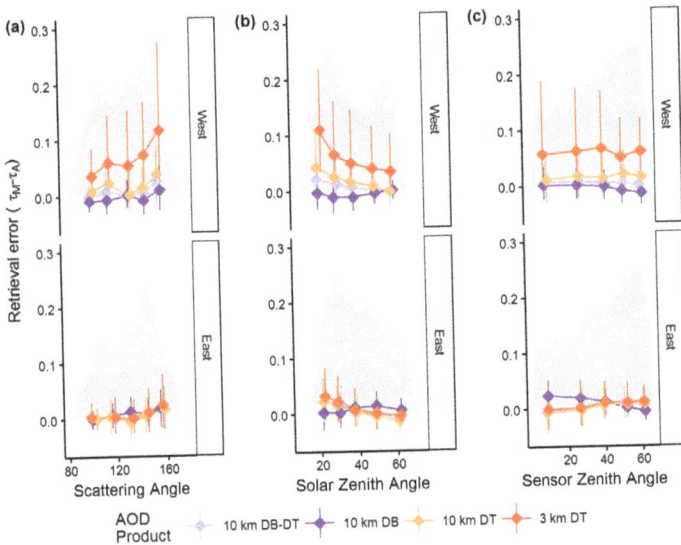

Figure 3. The dependence of AOD retrieval error and distributions of values for scattering angle (**a**); solar zenith angle (**b**); and sensor zenith angle (**c**). Median error (points) and the IQR (vertical line ranges from 25th to 75th percentile) is shown within quintiles. The distribution of values is shown in the background in gray and represents proportional frequency, where 0.25 on the y-axis represents the most frequent value in the category.

We assessed AOD retrieval errors by six land cover types, i.e., developed, forest, shrub, grass, cultivated, and wetland (see Supplementary Figure S3). All AOD parameters showed positive retrieval errors over developed areas, particularly in the Western US (0.03 for 10 km DB, and 0.21 for 3 km

DT AOD). Small but consistent positive errors were also observed over wetlands in the Eastern US (mean retrieval errors of 0.04 for 3 km DT, 0.03 for 10 km DT, 0.03 for 10 km DB-DT, and 0.02 for 10 km DB). The 3 km DT and, to a lesser degree, 10 km DT AOD also showed significant positive errors over shrub lands in the Western US (mean retrieval errors were 0.10 and 0.06 for 3 km DT and 10 km DT, respectively). The best agreement between MODIS and AERONET was over forests, grasslands, and cultivated lands. Overall, the 10 DB AOD had the least retrieval errors across all land cover types (<0.03), followed by the 10 km DB-DT AOD (greatest mean retrieval error of 0.10 over developed areas in the Western US). Previous studies have identified high retrieval error in AOD retrievals over developed areas, and the retrieval error in DT products over poorly vegetated surfaces to which 10 km DB retrievals are more robust [16,20,31].

3.6. Dependence of Retrieval Errors on Season and Weather Conditions

Figure 4 summarizes monthly retrieval errors from each AOD parameter. Median retrieval errors in the 10 km DT, 10 km DB-DT, and 3 km DT parameters were the highest in the summer months. The 3 km DT AOD had the widest fluctuation of retrieval errors over the course of the year (0.009 in December to 0.056 in May). The 10 km DB product had a more even distribution over time, from −0.01 in August to 0.01 in February. The reasons for the increased positive retrieval error in the DT-based AOD parameters in the summer months is unclear, and has not been well-characterized in previous work on this collection. However, despite the fact that collocations in summer months are associated with increased mean NDVI values, which typically result in better accuracy statistics for DT products, collocations in these months also have higher scattering angles, lower solar zenith angles, and higher values of total column precipitable water, all factors that result in positive retrieval errors over the CONUS.

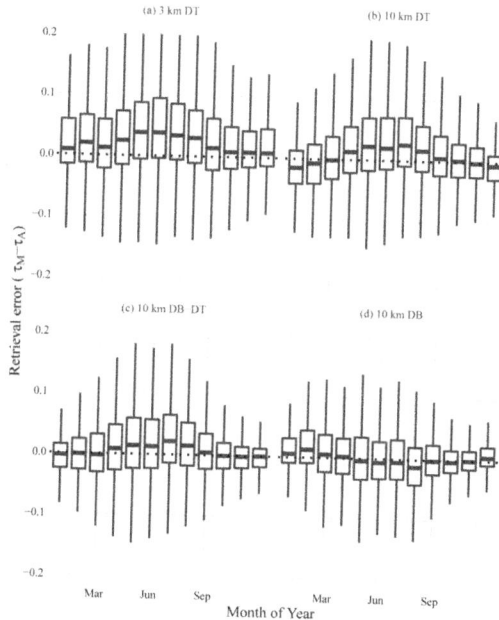

Figure 4. Boxplot showing the distribution of retrieval errors in MODIS AOD relative to AERONET, for each month of the year, for 3 km DT (**a**); 10 km DT (**b**); 10 km DB-DT (**c**); and 10 km DB (**d**). For each box, the midline represents the median, upper, and lower hinges represent the 25th and 75th percentiles, whiskers extend out to 5th and 95th percentiles.

As mentioned above, lower NDVI has been associated with increased retrieval error and noise in MODIS AOD retrievals, particularly for DT-based products, in previous works [16]. In the West, this is clearly shown in the 3 km DT AOD, and to a lesser degree in 10 km DT AOD (Figure 5). The 10 km DB AOD in the Western US was unbiased in the lowest three quintiles of NDVI, but was negatively biased in the upper two quintiles (up to −0.03 in the highest quintile). This pattern was observed in the global validations as well [12] and it likely points to an overestimation of the surface reflectance over vegetated areas in the eastern US. In the East, AOD retrieval errors are less dependent on NDVI, and the negative retrieval error observed for DB at higher NDVI values was not observed. Both humidity and potential cloud contamination have been shown to bias MODIS observations, and total column precipitable water (TCPW) can be a marker for both factors [32]. Figure 5 shows a complex relationship between AOD retrieval errors and TCPW. In the Western US, TCPW has little impact on the retrieval errors of 10 km DB and DB-DT AOD, but both very high or very low TCPW values are associated with positive retrieval errors in the 3 km and 10 km DT AOD. In the Eastern US, the 10 km DB AOD is negatively associated with TCPW. However, the impact of TCPW is generally small for all AOD parameters, except at very high levels where both the 3 km and 10 km DT AOD showed a small positive retrieval error. At higher TCPW values, this bias is likely indicative of cloud contamination, and the lack of retrieval error in 10 km DB product under these conditions fits with our coverage results, which suggests more conservative cloud screening procedures for this product.

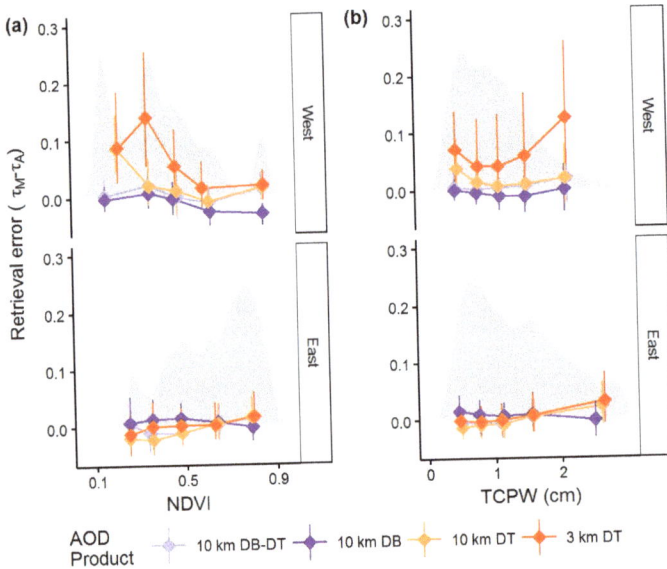

Figure 5. Dependence of AOD retrieval errors and distributions of values in QAC 3 MODIS AOD for NDVI (**a**) and TCPW (**b**). Median AOD retrieval error (dots) and the IQR (vertical line ranges from 25th percentile to 75th) is shown within quintiles of NDVI, and total column precipitable water. The density distribution of values is shown in the background in gray and represents proportional frequency, where 0.25 on the y-axis represents the most frequent value in the category.

4. Case Study

We conducted a case study over the Atlanta Metropolitan Area from 1 January 2004 to 31 December 2013. The study area stretched from 32°N to 36°N latitude and from 83°W to 86°W longitude, and included 23 ground-level $PM_{2.5}$ monitors in 19 distinct grid cells from the same ~12 km × 12 km grid used in the coverage analysis. This case study compared the ability of each of the four MODIS AOD products to predict ground-level $PM_{2.5}$ in a widely used linear mixed effect (LME) model framework [33]. Three AOD datasets were generated for each of the four MODIS AOD products: (1) AOD values with only QAC = 3; (2) AOD values with the highest available QAC (1, 2 or 3); and (3) filtered and corrected AOD values using the relationships examined in Sections 3.5 and 3.6 prior to inclusion in the model. To produce the filtered and corrected dataset, AOD values from dataset #2 with scattering angles over 165° or solar zenith angles less than 15° were first eliminated. This removed ~9% of observations while some were additionally lost in the matching process. This filtered dataset was then corrected, using a linear regression model fit to the dataset of matched AERONET and MODIS AOD observations in the Eastern CONUS, used in the accuracy analysis, for QAC value, land use type, sensor zenith angle, total column precipitable water, and NDVI. All 12 combinations of AOD were fit using the LME model, of the form: $PM_{2.5,s,t} = (b_0 + b_{0,t}) + (b_1 + b_{1,t})AOD_{s,t}$, where b_0 is the fixed intercept, $b_{0,t}$ the random intercept for each day, b_1 the fixed slope, and $b_{1,t}$ the random slope for each day [33]. These results are presented in Table 3.

Table 3. Performance statistics for each AOD parameter in the Atlanta case study.

Parameter	QAC 3 Only			Best of QAC 1, 2, and 3			Filtered & Corrected AOD		
	N	R^2	Fixed Slope	N	R^2	Fixed Slope	N	R^2	Fixed Slope
3 km DT	10,438	0.76	14.4	10,438	0.76	14.4	9680	0.75	20.2
10 km DT	8994	0.72	20.4	11,511	0.77	16.1	10,593	0.75	23.4
10 km DB-DT	8994	0.72	20.4	8994	0.72	20.4	8457	0.71	25.7
10 km DB	8560	0.80	31.2	14,706	0.83	11.7	13,448	0.83	16.7

For the parameters where the number of observations increased with the addition of lower QAC valued observations, 10 km DB and 10 km DT, R^2 values for a model relating ground-level $PM_{2.5}$ concentrations to AOD actually increased slightly. Increasing from 0.80 to 0.83 for 10 km DB and from 0.72 to 0.75 for 10 km DT. When AOD values were filtered and corrected to remove potentially biased observations, model fits for the 10 km DT product decreased slightly, from 0.77 to 0.75 and remained the same for the 10 km DB product. For the parameters with relatively few lower quality observations, 10 km DB-DT and 3 km DT, neither the number of observations included in the model nor the resulting R^2 values changed when lower confidence observations were included in the model. When filtering and correction was applied, model fits, as measured by the R^2 values, actually decreased by 0.01 relative to the 'best of' models. These results run counter to what would be expected: that R^2 values for all four parameters would decrease slightly with the inclusion of lower-confidence retrievals, given the fact that the lower confidence observations for the 10 km DB and 10 km DT parameters are noisier. However, in this case study, the additional number of observations appears to have offset the additional noise introduced via these observations in the model and resulted in better prediction of ground-level $PM_{2.5}$ via this simple model. Despite the smaller sample sizes, the models using the corrected and filtered AOD values achieved similar R^2 values as the uncorrected AOD models.

These results illustrate some of the key points made in this paper, namely that coverage is an often-overlooked but important factor, when considering AOD accuracy statistics, and that, because of the role played by coverage, the inclusion of lower-quality AOD observations in a model can provide some benefit. These results additionally highlight our observation that the lower confidence designations for the newer products, 10 km DB-DT and 3 km DT, are very few in number. The utility of correcting for major sources of bias or error in the AOD values was demonstrated in this limited example by greater fixed effect regression slopes, indicating greater sensitivity of the corrected and filtered AOD values to $PM_{2.5}$ concentrations.

5. Conclusions

We conducted a detailed analysis on the coverage and accuracy of Collection 6 MODIS AOD parameters in the CONUS. With their applications in air quality research in mind, we examined the benefits and risks of including lower QAC retrievals in order to improve data coverage, as well as how

AOD retrieval errors depend on various factors. Our recommendation is that, for inexperienced users who are beginning to explore MODIS AOD data for air quality research, the QAC 3 10 km DB-DT AOD is their best choice. For more experienced users, the ideal AOD parameter could depend on the purpose as well as domain of their study. The coverage of QAC 3 retrievals is comparable among all four AOD parameters, ranging from 25% to 30%. The Eastern US in general had higher and more consistent data coverage. However, much higher coverage rates were found in highly developed Southern California and over the south-central plains with limited ground-level air pollution monitoring, a surprising fact since these areas are traditionally regarded as having too high of surface brightness for DT to retrieve reliably. These findings are promising to researchers interested in conducting regional air quality assessments in these regions. Including lower QAC retrievals marginally improved coverage for the 3 km DT and 10 km DB-DT AOD. However, since these QAC assignments do not appear to reflect retrieval accuracy and are few in number, including lower QAC retrievals of these two parameters is probably beneficial, and unlikely to be harmful. On the other hand, lower QAC retrievals could increase the coverage of 10 km DT AOD by ~20% and that of 10 km DB AOD by ~70% on average. Caution must be given when including them to enhance coverage as these retrievals are often noisier, as shown in Table 2. However, as demonstrated in the case study, sufficient daily sample sizes can sometimes be more important than retaining only the high quality AOD values for the purposes of improving prediction errors with ground-level $PM_{2.5}$. To take advantage of the dramatic coverage gain offered by these lower QAC retrievals, retrieval error correction steps using local AERONET observations could be valuable [7].

In terms of data accuracy, the 10 km DB-DT AOD had the best performance in terms of correlation and linear model fit statistics, although QAC 3 retrievals for all but the 3 km DT AOD over the Western US met pre-launch expectations for the percentage of collocations within EE_{DT}. However, the 10 km DB product performs well in the context of a prediction model and may be an understudied AOD parameter in the US, where the 10 km DT product is currently used more frequently. The robustness of this product to major sources of bias additionally makes it an attractive option in the Western United States. The noisier 3-km DT AOD, however, can be valuable over dark targets in the Eastern US, particularly over areas where it tends to retrieve at higher rates than the lower-resolution products, such as in the Northeast and Northern Midwest, over the South Central plains in Texas, and over Southern Florida. The errors in MODIS AOD parameters vary in time and space, and are dependent on various retrieval conditions. Additional data screening and retrieval error correction steps should be considered other than simply relying on the QAC values, particularly in the Western United states, where these biases tend to have a larger impact. For example, AOD retrievals associated with high scattering angles and lower solar zenith angles may be excluded to avoid data contamination. Such parameters can be found in the operational MODIS aerosol product. In addition, categorical variables of land cover types as well as time trends can be introduced in $PM_{2.5}$ exposure models to control for the systematic retrieval errors in DT-based AOD retrievals. NDVI and TCPW had statistically significant, distinct impacts on all AOD parameters in the Western US, and therefore are probably worth considering when analyzing AOD data. Since they must be extracted from separate MODIS data products, users would need to consider the nontrivial time and computational demands associated with dealing with these large datasets.

Supplementary Materials: The following are available online at www.mdpi.com/2072-4292/8/10/815/s1, Figure S1: Seasonal coverage for high confidence retrievals, Figure S2: Seasonal coverage for all-confidence retrievals, Figure S3: AOD retrieval errors by land cover type.

Acknowledgments: This work was partially supported by the NASA Applied Sciences Program (grant No. NNX11AI53G, PI: Liu). In addition, this publication was made possible by the AERONET principal investigators and their staff for establishing and maintaining the 193 sites used in this investigation. The original MODIS AOD, AERONET AOD, land use, and MODIS NDVI data used in this paper are available free through the links referenced in Section 2 of the paper. We have additionally made our processing and analysis code, gridded coverage estimates, and MODIS and AERONET collocations produced for this analysis available on GitHub (https://github.com/jhbelle/MODIS_C6_Evaluation_over_the_CONUS).

Author Contributions: J.H.B. and Y.L. conceived and designed the experiments; J.H.B. performed the experiments; J.H.B. analyzed the data; J.H.B. and Y.L. contributed reagents/materials/analysis tools; J.H.B. and Y.L. wrote the paper.

Conflicts of Interest: The authors declare no conflict of interest.

References

1. Holben, B.; Tanre, D.; Smirnov, A.; Eck, T.; Slutsker, I.; Abuhassan, N.; Newcomb, W.; Schafer, J.; Chatenet, B.; Lavenu, F. An emerging ground-based aerosol climatology: Aerosol optical depth from aeronet. *J. Geophys. Res. Atmos.* **2001**, *106*, 12067–12097. [CrossRef]
2. Brauer, M.; Freedman, G.; Frostad, J.; Van Donkelaar, A.; Martin, R.V.; Dentener, F.; Dingenen, R.V.; Estep, K.; Amini, H.; Apte, J.S. Ambient air pollution exposure estimation for the global burden of disease 2013. *Environ. Sci. Technol.* **2015**, *50*, 79–88. [CrossRef] [PubMed]
3. Van Donkelaar, A.; Martin, R.V.; Brauer, M.; Boys, B.L. Use of satellite observations for long-term exposure assessment of global concentrations of fine particulate matter. *Environ. Health Perspect.* **2015**, *123*, 135–143. [CrossRef] [PubMed]
4. Remer, L.A.; Kaufman, Y.; Tanré, D.; Mattoo, S.; Chu, D.; Martins, J.V.; Li, R.-R.; Ichoku, C.; Levy, R.; Kleidman, R. The MODIS aerosol algorithm, products, and validation. *J. Atmos. Sci.* **2005**, *62*, 947–973. [CrossRef]
5. Hoff, R.M.; Christopher, S.A. Remote sensing of particulate pollution from space: Have we reached the promised land? *J. Air Waste Manag. Assoc.* **2009**, *59*, 645–675. [PubMed]
6. Gupta, P.; Christopher, S.A.; Wang, J.; Gehrig, R.; Lee, Y.; Kumar, N. Satellite remote sensing of particulate matter and air quality assessment over global cities. *Atmos. Environ.* **2006**, *40*, 5880–5892. [CrossRef]
7. Ma, Z.; Hu, X.; Sayer, A.M.; Levy, R.; Zhang, Q.; Xue, Y.; Tong, S.; Bi, J.; Huang, L.; Liu, Y. Satellite-based spatiotemporal trends in PM$_{2.5}$ concentrations: China, 2004–2013. *Environ. Health Perspect.* **2015**, *124*, 184–192. [CrossRef] [PubMed]
8. Chang, H.H.; Hu, X.; Liu, Y. Calibrating MODIS aerosol optical depth for predicting daily PM$_{2.5}$ concentrations via statistical downscaling. *J. Expo. Sci. Environ. Epidemiol.* **2014**, *24*, 398–404. [CrossRef] [PubMed]
9. Smith, K.R.; Jantunen, M. Why particles? *Chemosphere* **2002**, *49*, 867–871. [CrossRef]
10. Strickland, M.; Hao, H.; Hu, X.; Chang, H.; Darrow, L.; Liu, Y. Pediatric emergency visits and short-term changes in PM$_{2.5}$ concentrations in the US State of Georgia. *Environ. Health Perspect.* **2015**, *124*, 690–696. [CrossRef] [PubMed]
11. Crouse, D.L.; Peters, P.A.; van Donkelaar, A.; Goldberg, M.S.; Villeneuve, P.J.; Brion, O.; Khan, S.; Atari, D.O.; Jerrett, M.; Pope, C.A., III. Risk of nonaccidental and cardiovascular mortality in relation to long-term exposure to low concentrations of fine particulate matter: A Canadian national-level cohort study. *Environ. Health Perspect.* **2012**, *120*, 708–714. [CrossRef] [PubMed]
12. Sayer, A.; Munchak, L.; Hsu, N.; Levy, R.; Bettenhausen, C.; Jeong, M.J. MODIS Collection 6 aerosol products: Comparison between Aqua's e-deep blue, dark target, and "merged" data sets, and usage recommendations. *J. Geophys. Res. Atmos.* **2014**, *119*. [CrossRef]
13. Liu, Y. New directions: Satellite driven PM$_{2.5}$ exposure models to support targeted particle pollution health effects research. *Atmos. Environ.* **2013**, *68*, 52–53. [CrossRef]
14. Kloog, I.; Ridgway, B.; Koutrakis, P.; Coull, B.A.; Schwartz, J.D. Long-and short-term exposure to PM$_{2.5}$ and mortality: Using novel exposure models. *Epidemiology* **2013**, *24*, 555–561. [CrossRef] [PubMed]
15. Hsu, N.; Jeong, M.J.; Bettenhausen, C.; Sayer, A.; Hansell, R.; Seftor, C.; Huang, J.; Tsay, S.C. Enhanced deep blue aerosol retrieval algorithm: The second generation. *J. Geophys. Res. Atmos.* **2013**, *118*, 9296–9315. [CrossRef]
16. Levy, R.; Mattoo, S.; Munchak, L.; Remer, L.; Sayer, A.; Hsu, N. The collection 6 MODIS aerosol products over land and ocean. *Atmos. Meas. Tech.* **2013**, *6*, 2989–3034. [CrossRef]
17. Remer, L.; Mattoo, S.; Levy, R.; Munchak, L. MODIS 3 km aerosol product: Algorithm and global perspective. *Atmos. Meas. Tech. Discuss.* **2013**, *6*, 69–112. [CrossRef]
18. Sayer, A.; Hsu, N.; Bettenhausen, C.; Jeong, M.J. Validation and uncertainty estimates for MODIS collection 6 "deep blue" aerosol data. *J. Geophys. Res. Atmos.* **2013**, *118*, 7864–7872. [CrossRef]

19. Livingston, J.; Redemann, J.; Shinozuka, Y.; Johnson, R.; Russell, P.; Zhang, Q.; Mattoo, S.; Remer, L.; Levy, R.; Munchak, L. Comparison of MODIS 3 km and 10 km resolution aerosol optical depth retrievals over land with airborne sunphotometer measurements during arctas summer 2008. *Atmos. Chem. Phys.* **2014**, *14*, 2015–2038. [CrossRef]

20. Munchak, L.; Levy, R.; Mattoo, S.; Remer, L.; Holben, B.; Schafer, J.; Hostetler, C.; Ferrare, R. MODIS 3 km aerosol product: Applications over land in an urban/suburban region. *Atmos. Meas. Tech.* **2013**, *6*, 1747–1759. [CrossRef]

21. Levy, R. Collection 006 (C6) MODIS Atmosphere l2 Aerosol Product, 6th ed.LAADS Web, 2014. Available online: https://ladsweb.nascom.nasa.gov/data/search.html (accessed on 15 July 2014).

22. Slutsker, I.; Kinne, S. Wavelength dependence of the optical depth of biomass burning, urban, and desert dust aerosols. *J. Geophys. Res.* **1999**, *104*, D24.

23. Carroll, M.; DiMiceli, R.; Sohlberg, R.; Townshend, J. *1 km MODIS Normalized Difference Vegetation Index*, 5th ed.; University of Maryland, LAADS Web: College Park, MD, USA, 2015.

24. Fry, J.A.; Xian, G.; Jin, S.; Dewitz, J.A.; Homer, C.G.; Limin, Y.; Barnes, C.A.; Herold, N.D.; Wickham, J.D. Completion of the 2006 national land cover database for the conterminous United States. *Photogramm. Eng. Remote Sens.* **2011**, *77*, 858–864.

25. Hyer, E.; Reid, J.; Zhang, J. An over-land aerosol optical depth data set for data assimilation by filtering, correction, and aggregation of MODIS collection 5 optical depth retrievals. *Atmos. Meas. Tech.* **2011**, *4*, 379–408. [CrossRef]

26. Turner, R. Deldir: Delaunay Triangulation and Dirichlet (Voronoi) Tessellation. 2009. R Package Version 0.0-8. Available online: http://cran.r-project.org/web/packages/deldir (accessed on 31 July 2014).

27. Petrenko, M.; Ichoku, C.; Leptoukh, G. Multi-sensor aerosol products sampling system (MAPSS). *Atmos. Meas. Tech.* **2012**, *5*, 913–926. [CrossRef]

28. Yu, C.; Di Girolamo, L.; Chen, L.; Zhang, X.; Liu, Y. Statistical evaluation of the feasibility of satellite-retrieved cloud parameters as indicators of $PM_{2.5}$ levels. *J. Exposure Sci. Environ. Epidemiol.* **2015**, *25*, 457–466. [CrossRef] [PubMed]

29. Christopher, S.A.; Gupta, P. Satellite remote sensing of particulate matter air quality: The cloud-cover problem. *J. Air Waste Manag. Assoc.* **2010**, *60*, 596–602. [CrossRef] [PubMed]

30. Levy, R.C.; Remer, L.A.; Mattoo, S.; Vermote, E.F.; Kaufman, Y.J. Second-generation operational algorithm: Retrieval of aerosol properties over land from inversion of moderate resolution imaging spectroradiometer spectral reflectance. *J. Geophys. Res. Atmos.* **2007**, *112*, D13. [CrossRef]

31. Tao, M.; Chen, L.; Wang, Z.; Tao, J.; Che, H.; Wang, X.; Wang, Y. Comparison and evaluation of the MODIS collection 6 aerosol data in China. *J. Geophys. Res. Atmos.* **2015**, *120*, 6992–7005. [CrossRef]

32. Ford, B.; Heald, C.L. Aerosol loading in the southeastern united states: Reconciling surface and satellite observations. *Atmos. Chem. Phys.* **2013**, *13*, 9269–9283. [CrossRef]

33. Lee, H.; Liu, Y.; Coull, B.; Schwartz, J.; Koutrakis, P. A novel calibration approach of MODIS aod data to predict $PM_{2.5}$ concentrations. *Atmos. Chem. Phys.* **2011**, *11*, 7991–8002. [CrossRef]

remote sensing

MDPI

Technical Note

Performance of MODIS C6 Aerosol Product during Frequent Haze-Fog Events: A Case Study of Beijing

Wei Chen [1,*], Aiping Fan [1] and Lei Yan [2,*]

[1] College of Geoscience and Surveying Engineering, China University of Mining & Technology, Beijing 100083, China; fap_fanaiping@163.com
[2] Beijing Key Lab of Spatial Information Integration & Its Applications, Peking University, Beijing 100871, China
* Correspondence: chenw@cumtb.edu.cn (W.C.); lyan@pku.edu.cn (L.Y.); Tel.: +86-10-6233-9335 (W.C.); +86-10-6275-9765 (L.Y.)

Academic Editors: Yang Liu, Jun Wang, Omar Torres, Richard Müller and Prasad S. Thenkabail
Received: 22 March 2017; Accepted: 14 May 2017; Published: 18 May 2017

Abstract: The newly released MODIS Collection 6 aerosol products have been widely used to evaluate fine particulate matter with a 10 km Dark Target aerosol optic depth (DT AOD) product, a new 3 km DT AOD product and an enhanced Deep Blue (DB) AOD product. However, the representativeness of MODIS AOD products under different air quality conditions remains unclear. In this study, we obtained all three types of MODIS Terra AOD from 2001 to 2015 and Aqua AOD from 2003 to 2015 for the Beijing region to study the performance of the different AOD products (Collection 6) under different air quality situations. The validation of three MODIS AOD products suggests that DB AOD has the highest accuracy with an expected error (EE) envelope (containing at least 67% of the matchups on a scatter plot) of $0.05 + 0.15\tau$, followed by 10 km DT AOD ($0.08 + 0.2\tau$) and 3 km DT AOD ($0.35 + 0.15\tau$), specifically for Beijing. Near-surface $PM_{2.5}$ concentrations during the passage of MODIS from 2013 to 2015 were also obtained to categorize air quality as unpolluted, moderately, and heavily polluted, as well as to analyze the performance of the different AOD products under different air quality conditions. Very few MODIS 3 km DT retrievals appeared on heavily polluted days, making it almost impossible to play an effective role in air quality applications in Beijing. While the DB AOD allowed for considerable retrievals under all air quality conditions, it had a coarse spatial resolution. These results demonstrate that the MODIS 3 km DT AOD product may not be the appropriate proxy to be used in the satellite retrieval of surface $PM_{2.5}$, especially for those areas with frequent haze-fog events like Beijing.

Keywords: AOD; MODIS; dark target; deep blue; air quality

1. Introduction

Aerosols are important components of the atmosphere that can affect atmospheric environment [1], weather [2], climate [3], and human health [4]. Aerosol particles in the atmosphere absorb and scatter incident solar radiation, and affect the Earth's radiation budget [5]. Aerosol particles can act as condensation nuclei to form clouds and affect precipitation, while excessive aerosol loading can change cloud properties (e.g., albedo, etc.) by means of the indirect radiative forcing effects of the aerosol [6]. Therefore, the analysis of aerosol properties has been the focus of atmospheric research via both ground-based sun-photometers and satellite-based observations [7]. Satellite observation of aerosols utilizes the reflected signals of the atmosphere to retrieve the optical properties of the aerosol (primarily the aerosol optical depth, AOD), and is very powerful for monitoring the global distribution of aerosol loadings due to its instantaneous coverage of the Earth's surface. Various satellite sensors have been launched, such as the Moderate Resolution Imaging Spectroradiometer

(MODIS), Multi-angle Imaging Spectroradiometer (MISR), and POLarization and Directionality of the Earth's Reflectances (POLDER) [8–10]. The MODIS aerosol products employ different aerosol retrieval algorithms over land and ocean. Over land, AOD over vegetation and other dark objects is retrieved with the Dark Target (DT) algorithm, which primarily utilizes the relationship between two visible bands (red and blue) and a shortwave infrared band (2.1 μm). The Deep Blue (DB) algorithm was developed to retrieve the AOD over bright surfaces [11], and was later extended to cover vegetated surfaces. MISR utilizes multi-angular observation to retrieve aerosol properties with nine discrete view angles in four bands [12]. POLDER is also able to retrieve aerosol properties with multi-angular observations with additional polarization signals, but the land aerosol properties retrieved by POLDER are currently limited to fine mode aerosols, as it assumes that the polarization signal is sensitive to fine mode aerosols over land [13]. Among these satellite aerosol products, the MODIS products have been the most widely used because of their high accuracy; global expected error envelope of ±(0.05 + 15%) over land and +(0.04 + 10%) to −(0.2 + 10%) over ocean [14]; wide-coverage (a swath of 2330 km); long-term dataset (beginning in 2000); and high temporal resolution (Terra and Aqua in the morning and afternoon) [15]. However, satellite-based sensors retrieve aerosol properties by utilizing relatively weak reflected signals from the atmosphere, and are easily contaminated by surface reflective signals if the surface reflectance is too high [16]. Therefore, over a bright surface such as a desert or city, surface-reflected signals dominate and the accuracy of the satellite-retrieved aerosol properties is limited [17]. Although the accuracy of AOD retrieval is relatively high, other aerosol properties—such as the single scattering albedo (SSA), size distribution, and refractive index—cannot be effectively obtained via satellite remote sensing [18]. Furthermore, the retrieval of aerosol properties relies on the radiative transfer calculation for reflected signals, involving both surface and atmospheric properties. If the surface is "masked" by clouds or high concentrations of particulate matter, the radiative transfer models may not be able to decouple surface and atmosphere contributions, making aerosol retrieval inaccurate. For example, the Second Simulation of the Satellite Signal in the Solar Spectrum (6S) suggests that visibility should be larger than 5 km; otherwise, the accuracy is doubtful [19]. This might be acceptable for regions such as the United States and Western Europe that have good air quality resulting from strict environmental policies; however, for areas such as northern China that have frequent haze-fog events when the particulate matter concentration could be as high as 300 μg/m^3 [20], the retrieval of aerosol properties has a high fail rate, as the high concentration of particulate matter acts as a particle "wall" that stops radiative transfer from the land surface to the satellite.

Ground-based sun-photometer observations retrieve AOD by aiming directly towards the sun, thus avoiding surface reflectance contamination [21,22]. Considering the high accuracy of ground-based sun-photometer observations and the limitations imposed by a fixed location, it is necessary to construct a network to observe the aerosol with sun-photometers with comparable accuracy and retrieval methods, such as Aerosol RObotic NETwork (AERONET). China has also established its own ground-based sun-photometer networks, such as the China Aerosol Remote Sensing Network (CARSNET) [23] and Chinese Sun Hazemeter Network (CSHNET) [24], to monitor the properties of aerosols in several key regions. Due to the high accuracy of sun-photometer observations, they have been regarded as benchmarks for the validation of satellite aerosol retrievals [25]. Although many sun-photometer sites exist worldwide, they do not cover a wide spatial range, so their application for local monitoring and satellite result validation is limited. Sun-photometer observations—especially AERONET retrievals—provide a unique tool to investigate the comprehensive properties of aerosols for local issues [26].

Beijing is the capital of China, is located in Northern China, and has a population greater than 20 million [27]. During the rapid economic development that occurred during the past three decades, air pollution in Beijing has become an issue of comprehensive public concern because of the frequent haze-fog events in late autumn, winter, and early spring [28]. To investigate the properties of aerosols in Beijing, four AERONET sites were deployed in and around Beijing (i.e., the Beijing site, the Beijing-CAMS site, the PKU_PEK site, and the Yufa_PEK site), and these provide relatively long-term monitoring data of aerosol properties in Beijing. In addition to the AERONET observations, 35 air quality monitoring stations have been deployed in Beijing since 2013 to obtain real-time concentrations of six key atmospheric pollutants ($PM_{2.5}$, PM_{10}, SO_2, NO_2, O_3, and CO). Therefore, the properties of aerosols during different air quality situations can be obtained with the aid of air quality monitoring data.

In previous studies, the performance of MODIS aerosol products has been validated globally by means of AERONET observations [29,30]. However, these validations have been primarily focused on successfully retrieving pairs of both effective MODIS and AERONET results. The results may be correct for areas that have few occurrences of heavy particulate matter pollution, as aerosol retrieval is likely to be successful. However, for areas such as Beijing that have frequent particulate matter pollution episodes, MODIS retrieval is highly possible to fail, while AERONET—which is aimed directly at the sun—could produce sufficiently accurate results [31,32]. As MODIS AOD products—including 3 km DT, 10 km DT, and 10 km DB—have been widely used to estimate the surface $PM_{2.5}$ concentrations over various situations [33,34], it raises a question of what the representativeness of different MODIS AOD products under different air quality conditions is. By failing to retrieve effective AOD under heavily polluted days, retrieving $PM_{2.5}$ from MODIS AOD may be biased due to the lack of sufficient samples under heavy polluted areas such as Northern China, with frequent haze-fog events.

In this study, we documented AOD data in Beijing collected from two AERONET (Beijing station and Beijing-CAMS station) and three MODIS AOD products from 2001 to 2015 to obtain successful retrieval pairs to evaluate the performance and representativeness of MODIS aerosol retrievals over Beijing. Furthermore, detailed $PM_{2.5}$ concentrations measured by two air quality monitoring stations close to the two AERONET stations were also collected to characterize the air quality. The major objectives of this study are to: (1) investigate the performance of the MODIS AOD product under different air quality situations; and (2) to figure out whether these AOD products are suitable for $PM_{2.5}$ retrieval in frequent haze-fog areas such as Beijing.

2. Study Area and Data Sets

2.1. Study Area

Beijing (40°N, 116°E) is the capital city of the People's Republic of China, and has over 20 million inhabitants and an area of 16,800 km^2. It lies on the northwest border of the Northern China Plain and is surrounded by the Yansan Mountains in the north and west. The climate in Beijing is semi-humid continental with an annual precipitation of 644 mm. Affected by the East Asia monsoon, approximately 80% of annual precipitation occurs in the summer months, causing wet, hot summers, and dry, cold winters. As the political and cultural center of China, air quality in Beijing has raised both domestic and foreign concerns. In the late 20th century and the beginning of the 21st century, frequent dust storms in the winter and early spring caused major atmospheric pollution issues [35]. Since 2012, serious haze events with high concentrations of $PM_{2.5}$ have aroused public concern. To mitigate serious atmospheric problems, the Beijing municipal government launched the "Defending Blue Sky" campaign to report the "Blue Sky Index" to delineate the "Blue Sky Days" each year since 1998 [36]. The official data show that "Blue Sky Days" have increased from 100 per year in 1998 to 286 per year in 2012, but the public has remained confused by the deteriorating air quality conditions [37]. The reason for the confusion is that the previous ambient air quality assessments only considered PM_{10}, NO_2, and SO_2, but $PM_{2.5}$ has become an increasingly important issue regarding deteriorating air quality. Partly due to public pressure, the new ambient air quality standards (National Ambient Air Quality Standards, GB 3095-2012) have included $PM_{2.5}$ as a key pollutant since 2013. Furthermore, air quality control guidelines such as the "Air Pollution Prevention and Control Action Plan" released by the State Council of China has requested a reduction of 25% in $PM_{2.5}$ concentration in the Beijing-Tianjin-Hebei region by 2017 compared with that in 2012. To achieve these goals, a number of air quality monitoring stations were deployed to provide real-time data for the evaluation of air quality.

2.2. AERONET Data

AERONET is a global aerosol property observation network that has more than 500 stations, and some of the monitoring at AERONET began prior to 2000 [38]. AODs of several bands (typically, 340 nm, 380 nm, 440 nm, 500 nm, 670 nm, 870 nm, and 1020 nm) can be obtained through direct observation aimed at the sun. Through sky radiance measurements in the almucantar plane at 440 nm, 670 nm, 870 nm, and 1020 nm combined with the directly-obtained direct sun measured AOD at these four bands, the microphysical properties of the aerosol particles (i.e., the size distribution, the refractive index, the single scattering albedo, etc.) are retrieved via radiative transfer calculations. Furthermore, the radiative forcing and forcing efficiency are also integrated into the AERONET inversion scheme by utilizing the AOD, surface albedo, size distribution, refractive indices, particle size parameters, and so on [39]. Three levels of aerosol retrieval data are provided by AERONET: Level 1.0, Level 1.5, and Level 2.0. The Level 1.0 product provides the direct inversion of the aerosol properties without a strict quality check. The Level 1.5 product is the cloud screen product that provides a series of quality checks: data quality checks, triplet stability criterion checks, a diurnal stability check, and three standard deviation criteria checks to exclude the clouds [40]. The Level 2.0 product has more rigorous quality checks: an instrument performance check, a temperature sensor check, a calibration check, an aerosol optical depth spectral dependency check, a cloud contamination check, a consistency check, and a historical data impact check [41]. Although the Level 2.0 product has the highest quality, the strict quality checks screen out a large amount of data. Moreover, some AERONET stations do not have enough Level 2.0 data, thus limiting the scientific applications of the Level 2.0 product. Therefore, the Level 1.5 product was deemed to be of sufficient quality and quantity, and was selected and analyzed in this study.

Four AERONET stations have been deployed in Beijing, and two stations' data with nearby air quality monitoring stations and with more than three years of data were included in this study: the Beijing station (39°58′37″N, 116°22′51″E since March 2001) and the Beijing-CAMS station (39°55′58″N, 116°19′01″E since August 2012). Both AERONET stations both have a nearby air quality monitoring station to obtain AOD–$PM_{2.5}$ data pairs.

2.3. MODIS Data

Two MODIS sensors on the Terra and Aqua platforms have provided extensive aerosol data sets since 2000 and 2002, respectively. The aerosol products provided by MODIS have been updated to Collection 6 (C6) and primarily use three algorithms: (1) the Dark Target (DT) land algorithm (dark vegetated/soil lands); (2) the Dark Target ocean algorithm; and (3) the enhanced Deep Blue (DB) algorithm (originally over desert or arid lands in C5.1 and expanded to cover vegetated land surfaces in C6). Over land, the main principle of the DT algorithm relies on a consistent relationship between the 0.47, 0.66, and 2.1 µm bands in dense vegetated regions [42,43]. Therefore, the DT algorithm only provides aerosol retrievals over vegetated land and other dark land surfaces, which restricts its application. The DB algorithm was developed to obtain aerosol properties over desert and other arid surfaces, and later extended to cover vegetated land surfaces [44]. The DB algorithm makes use of the relatively low reflectance of the desert and other surfaces in the deep blue bands to retrieve aerosol properties and then extrapolate them to the 0.55 µm band. Compared to the Collection 5.1 data, the accuracy and extension of the DB results have largely increased. In Collection 5.1, the MODIS aerosol products have a spatial resolution of 10 km at the nadir, which has been retained in Collection 6. In addition to the 10 km product, Collection 6 has provided a 3 km DT product for air quality applications. In this study, both the 10 km and 3 km level 2 MODIS Collection 6 Terra and Aqua aerosol products were obtained and analyzed to provide a comprehensive analysis of the performance of MODIS aerosol products over Beijing with frequent haze-fog events.

2.4. Air Quality Monitoring Data

AERONET and MODIS aerosol retrievals provide the column-integrated effects of aerosol particles, but air quality monitoring data—including $PM_{2.5}$ and PM_{10} measurements—are another method to evaluate the dry aerosol concentrations near the surface. Before 2013, the daily air quality for main cities only considered PM_{10}, SO_2, and NO_2. Since 2013, the new national ambient air quality standards required 74 key cities to monitor six key pollutants at one hour intervals. From 2015, 367 cities with more than 1400 air quality monitoring stations published real-time air pollutant concentration data on the web platform of Ministry of Environmental Protection of China [45]. A total of 12 state-controlled stations and 23 municipal stations were deployed in Beijing. In this study, two air quality stations in Beijing (the West Park Officials station that matches the Beijing-CAMS station and the Olympic Sports Center station that matches the Beijing station) were selected to match the AERONET monitoring observations. The air quality monitoring data were used to characterize air quality during the overpassing of MODIS and the AERONET retrieval as follows: unpolluted ($PM_{2.5}$ concentration $\leq 75\ \mu g/m^3$), moderately polluted ($75\ \mu g/m^3 < PM_{2.5}$ concentration $\leq 150\ \mu g/m^3$), and heavily polluted ($PM_{2.5}$ concentration $> 150\ \mu g/m^3$). The distribution of the two AERONET stations and the two air quality monitoring stations are shown in Figure 1.

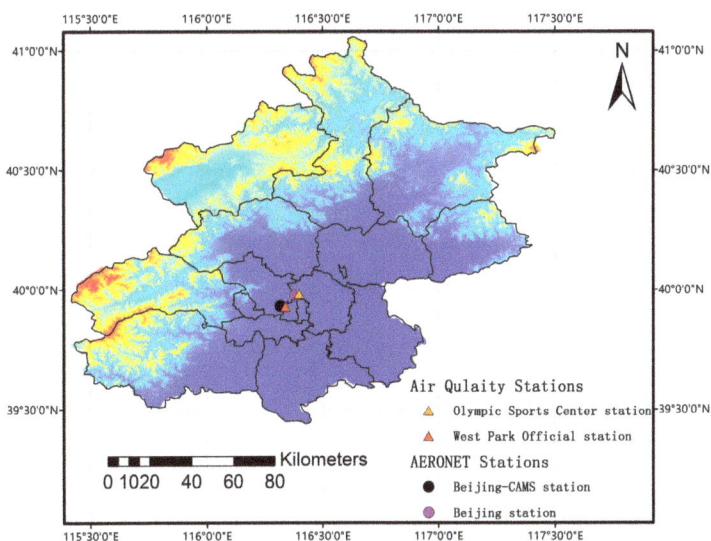

Figure 1. Distributions of the two Aerosol RObotic NETwork (AERONET) stations and two air quality stations used in this study. The two AERONET stations are: Beijing-CAMS station (39.93°N, 116.32°E) and Beijing station (39.97°N, 116.38°E). The two air quality stations are Olympic Sports Center station (39.98°N, 116.40°E) and West Park Official station (39.93°N, 116.34°E).

3. Analytical Methods

AERONET observations during MODIS overpasses have been used as benchmarks in the evaluation and application of MODIS aerosol products. However, such comparisons require both effective AERONET and high quality MODIS retrievals. If the MODIS retrieval fails or is less reliable, the comparison cannot be made; this may be due to a variety of sources, including cloud contamination, bright surface contamination, haze-fog masking, and so on. As MODIS aerosol retrievals rely on the separation of surface and atmospheric contributions, the retrieval scheme is highly likely to fail when the particulate matter concentrations are too high (moderately polluted and heavily polluted). Urban monitoring stations in Beijing indicate that the average annual $PM_{2.5}$ concentration is approximately

78 μg/m^3 (5.5–475.0 μg/m^3), and particulate matter pollution is present for one-third of a year. Therefore, it is highly possible that MODIS aerosol products may only retrieve aerosol optical properties when the atmosphere is not heavily polluted, causing errors in the estimates of the long-term aerosol properties from MODIS. In this study, only those MODIS retrievals with sufficient quality (Quality Assurance Confidence, QAC = 3) were included: SDS Optical_Depth_Land_And_Ocean in the 3 km DT and 10 km DT products and SDS Deep_Blue_Aerosol_Optical_Depth_550_Land selected with "QAC = 3" in the 10 km DB product.

In this study, we selected MODIS pixels within ±25 km of the two AERONET stations and averaged the all the effective AOD retrievals as effective MODIS AOD retrievals for 10 km DT and DB products [46]. The AERONET retrievals within a temporal interval of ±30 min of the MODIS overpass time were averaged to obtain effective AERONET retrievals. Because 25 km is around the 2.5 pixel resolution of MODIS 10 km DT and DB products, we restricted the spatial searching range of MODIS 3 km DT products to ±7.5 km proportionally. Both effective data pairs (with both effective MODIS and AERONET data) and ineffective data pairs (with either the MODIS or AERONET data missing) were collected and analyzed for different air quality conditions (unpolluted, moderately polluted, and heavily polluted) to evaluate the performance of MODIS aerosol products.

4. Results

4.1. Performance of the MODIS 10 km DT, 10 km DB, and 3 km DT Products

AOD data retrieved from the MODIS 10 km DT, 10 km DB, and 3 km DT products are shown in Figures 2–4. In this study, Terra AOD observations from 2001 to 2015 and Aqua AOD observations from 2003 to 2015 were acquired from the MODIS Level-1 and Atmosphere Archive and Distribution System (http://ladsweb.nascom.nasa.gov) for the Beijing station, as the AERONET retrievals for this station began in 2001. For the Beijing-CAMS station (which began operation in 2012), both the MODIS Terra and Aqua data that were used to match the AERONET data were from 2012. In this study, only the highest-quality MODIS data over land (QAC = 3) were selected to match the AERONET data.

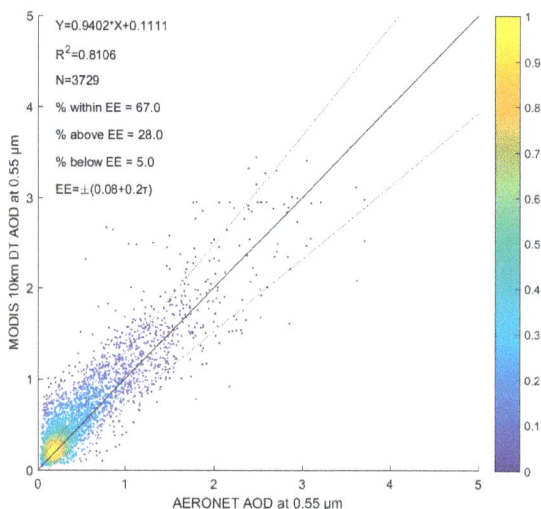

Figure 2. Validation of the Moderate Resolution Imaging Spectroradiometer (MODIS; both Terra and Aqua) 10 km Dark Target aerosol optical depth (DT AOD) against AERONET measurements at the Beijing station and Beijing-CAMS station. EE: expected error.

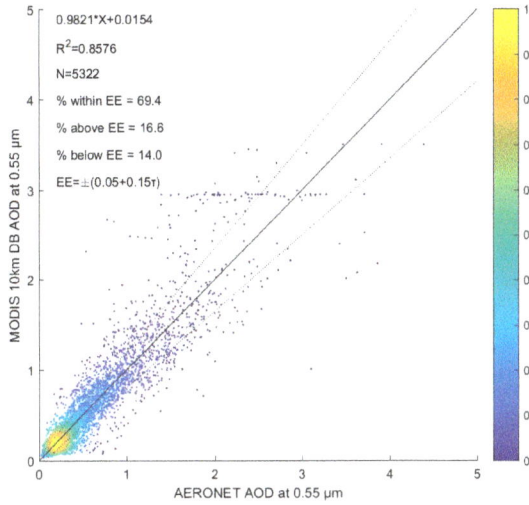

Figure 3. Validation of the MODIS (both Terra and Aqua) 10 km Deep Blue (DB) AOD against AERONET measurements at the Beijing station, and Beijing-CAMS station.

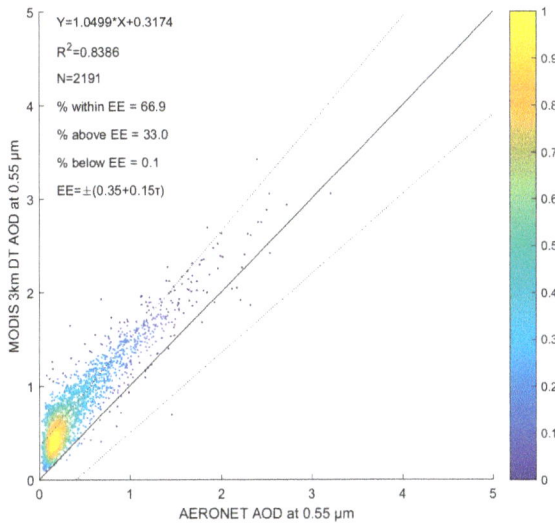

Figure 4. Validation of the MODIS (both Terra and Aqua) 3 km Dark Target AOD against AERONET measurements at the Beijing station, and Beijing-CAMS station.

As shown in Figure 2, a total of 3729 pairs of MODIS 10 km DT AOD and AERONET retrievals were obtained for the two stations. Most of the retrieval pairs ranged from 0 to 1.0. A linear relationship between the MODIS 10 km DT AOD and AERONET AOD with a very high determination coefficient ($R^2 = 0.8106$) was found. The expected error (EE, with at least 67% of data falling within the EE) envelope was approximately $\pm(0.08 + 0.2\tau)$, which is slightly larger than the global average accuracy of the MODIS AOD over land ($\pm(0.05 + 0.15\tau)$). As shown in Figure 3, a total 5322 pairs of MODIS 10 km DB AOD and AERONET retrievals were collected, as the DB algorithm was developed to overcome the restriction of the DT algorithm only covering densely vegetated areas or water bodies and failing for retrieval over bright surfaces. Although the DB algorithm is relatively less accurate than DT globally; for the three stations in Beijing, the DB AOD in Beijing was relatively highly accurate ($\pm(0.05 + 0.15\tau)$) [47] and had more successful pairs ($N = 5322$). The accuracy of the DB AOD was comparable to the global average accuracy of the DT AOD. The 3 km MODIS aerosol products (shown in Figure 4) was designed to provide refined spatial resolution of aerosol retrieval for regional air quality applications [48]. However, in our study, the accuracy of the 3 km MODIS aerosol product was much lower ($\pm(0.35 + 0.15\tau)$) at Beijing, suggesting that despite a higher spatial resolution, the MODIS 3 km products were less reliable, which is consistent with a previous validation of the Aqua 3 km products over Asia [49]. However, our result was slightly different because we selected a 7.5×7.5 km square rather than a 9×9 km square. The validation results of the 3 km MODIS AOD suggested that in a heavily-polluted area such as Beijing, the MODIS 3 km AOD was neither highly accurate nor had a wide coverage, and only had a high spatial resolution. Most previous studies have compared AERONET AOD in a range of 0–2.0, and values larger than 2.0 seldom occurred. In this study, we found many retrievals larger than 2.0, with some retrievals as high as approximately 3.5, suggesting that the aerosol load in Beijing is very high.

Beijing station began monitoring aerosols in March 2001, and has continuously released data since 2003, while the Beijing-CAMS station started in 2012. We have collected the monthly retrieval counts for MODIS 10 km DT, 10 km DB, 3 km DT, and AERONET retrievals during the MODIS overpass since 2005 at the Beijing stations (Figure 5). Very few successful retrievals were obtained in winter (December, January, and February) from the two DT algorithms (10 km and 3 km), and more retrievals were obtained in the middle of the year (April–October), in the shape of an "M". However, the successful DB algorithms and AERONET retrievals had an opposite pattern, with the largest values occurring in winter and fewer values in summer. The successful DB and AERONET retrievals were comparable. The fewer successful AERONET retrievals in summer may have been due to the rainy season, while in autumn and winter, less precipitation and fewer clouds allowed more AOD retrievals.

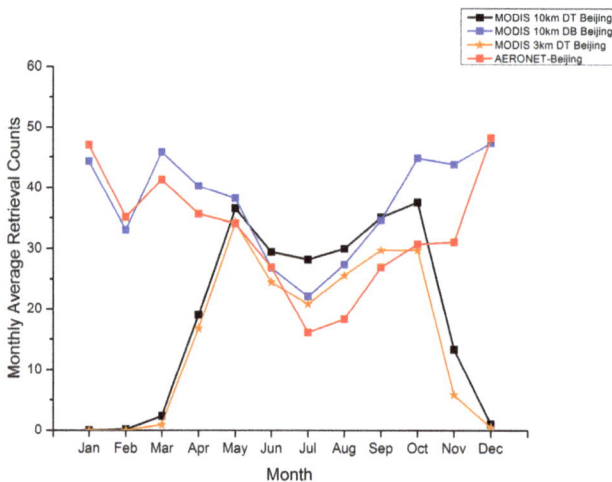

Figure 5. Monthly successfully retrieval counts of the MODIS 10 km Dark Target AOD, 10 km Deep Blue AOD, 3 km Dark Target AOD, and AERONET retrievals during the overpass of Terra and Aqua for the Beijing station.

In Beijing, air quality is good in the middle of the year and poor in autumn and winter. Particulate matter ($PM_{2.5}$ and PM_{10}) is the principal reason for the deterioration of air quality in Beijing. Although the MODIS 3 km DT products were developed for the applications by the air quality community, the low number of successful retrievals in winter in a heavily polluted region such as Beijing, restricts its effective application. Therefore, in this study we analyzed the retrievals of various MODIS AOD products and AERONET AOD retrievals under different air quality conditions to estimate the performance of different MODIS AOD products.

4.2. Retrieval Statistics under Different Air Quality Situations

Beijing began to publish real-time air quality data—including $PM_{2.5}$—at one hour intervals on its official website in January 2013, providing a powerful data source for atmospheric pollution research. Two stations (West Park Official and Olympic Sports Center stations) out of the 35 air quality monitoring stations are very close to two AERONET stations (the Beijing and Beijing-CAMS stations). The successful retrieval counts of the MODIS 3 km DT, 10 km DT, 10 km DB, and AERONET during different $PM_{2.5}$ concentrations were obtained for the West Park Official–Beijing-CAMS and Olympic Sports Center–Beijing station pairs. In total, 2028 and 1920 successful $PM_{2.5}$ concentration data pairs were collected during the overpass of the twin MODIS sensors for the Beijing-CAMS station and the Beijing station, respectively. For the Beijing-CAMS station, 1203 $PM_{2.5}$ data indicated less than 75 $\mu g/m^3$ (unpolluted), 479 $PM_{2.5}$ data indicated a range of 75–150 $\mu g/m^3$ (moderately polluted), and 346 $PM_{2.5}$ data were larger than 150 $\mu g/m^3$ (heavily polluted). The average $PM_{2.5}$ concentration was 82.1 $\mu g/m^3$. For the Beijing station, the corresponding values were 1151 (unpolluted), 455 (moderately polluted), and 314 (heavily polluted), with an average $PM_{2.5}$ concentration of 78.9 $\mu g/m^3$. It should be mentioned that the average $PM_{2.5}$ concentration only considered data collected during the MODIS fly-over time, which was different from the actual annual average. Additionally, the West Park Official station data started in January 2013, while the Olympic Sports Center station started in March 2013. Histograms of the retrievals from the different MODIS products, AERONET retrievals, and total $PM_{2.5}$ data are shown in Figure 6. The $PM_{2.5}$ data had the largest effective number of retrievals due to the automatic $PM_{2.5}$ monitoring and was affected by weather conditions, while AOD retrievals were restricted to cloud-free conditions. For both stations, the MODIS 10 km DB products had the largest number of successful retrievals, followed by the AERONET retrievals, retrievals of the MODIS 10 km DT products, and retrievals of the MODIS 3 km DT products. The number of successful MODIS 10 km DB products was even greater than the AERONET retrievals. Furthermore, when the $PM_{2.5}$ concentration increased to more than 150 $\mu g/m^3$ (heavily polluted), very few successful retrievals resulted from the MODIS 3 km DT products, but a considerable number resulted from the MODIS 10 km DB products and AERONET. In this study, we categorized air qualities into three levels: unpolluted, moderately polluted, and heavily polluted, as described in Section 2 and summarized the air quality situations regarding effective MODIS or AERONET retrievals as shown in Table 1. By simply analyzing the $PM_{2.5}$ monitoring data during the overpass of MODIS, approximately 17% of these data were from heavily polluted times, 23% from moderately polluted times, and 60% from unpolluted times, which suggests that the air quality in Beijing was polluted 40% of the time. When considering rainy or cloudy conditions, the number of aerosol retrievals was less than the $PM_{2.5}$ monitoring data. The results in Table 1 showed that three MODIS AOD products and AERONET retrievals all have comparable successful retrieval percentages on moderately polluted days (75 $\mu g/m^3$ < $PM_{2.5}$ concentration $\leq$ 150 $\mu g/m^3$) of approximately 20%, which was slightly less than the 23% directly derived from the $PM_{2.5}$ monitoring data. All four AOD products had far lower successful retrieval rates during heavily polluted days than the actual situation. For the MODIS 3 km DT AOD, only 3–4% of the successful retrievals occurred during heavily polluted conditions, and for the MODIS 10 km DT AOD, the rate was approximately 6.5%—far less than the 17% demonstrated by the monitoring data. The successful retrieval rate on heavily polluted days for the MODIS 10 km DB AOD was slightly less than 10%. The successful retrieval rates on heavily polluted days for the two AERONET stations were very different: 11.9% from Beijing station, and 3.1% from the Beijing-CAMS station. These results suggest that when considering air quality in Beijing, the MODIS 3 km DT products may not be able to reflect the actual air quality when the air is moderately or heavily polluted. Although developed for air quality monitoring, the MODIS 3 km DT products may be unsuitable for air quality remote sensing retrieval due their low success rate.

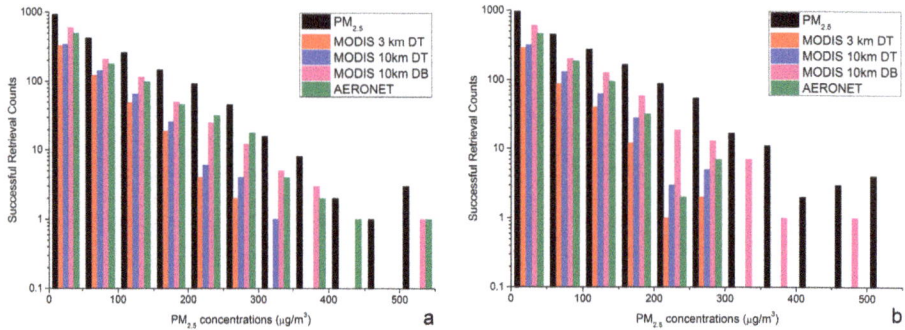

Figure 6. Retrieval number histograms of the MODIS 10 km Dark Target AOD, 10 km Deep Blue AOD, 3 km Dark Target AOD, AERONET retrievals, and total $PM_{2.5}$ data during the overpass of Terra and Aqua for (**a**) Beijing and (**b**) Beijing-CAMS station.

Table 1. Air qualities statistics of successful retrievals of different AOD products.

Station	AOD Products/PM$_{2.5}$	PM$_{2.5}$ Concentrations	Retrieval Counts (Percentage)
Beijing	MODIS 3 km DT	<75 µg/m^3	396 (76.6%)
		75–150 µg/m^3	97 (18.6%)
		>150 µg/m^3	25 (4.8%)
	MODIS 10 km DT	<75 µg/m^3	422 (72.5%)
		75–150 µg/m^3	123 (21.1%)
		>150 µg/m^3	37 (6.4%)
	MODIS 10 km DB	<75 µg/m^3	716 (70.5%)
		75–150 µg/m^3	203 (20.0%)
		>150 µg/m^3	96 (9.5%)
	AERONET	<75 µg/m^3	601 (68.6%)
		75–150 µg/m^3	171 (19.5%)
		>150 µg/m^3	104 (11.9%)
	PM$_{2.5}$	<75 µg/m^3	1151 (60.0%)
		75–150 µg/m^3	455 (23.7%)
		>150 µg/m^3	314 (16.3%)
Beijing-CAMS	MODIS 3 km DT	<75 µg/m^3	352 (82.2%)
		75–150 µg/m^3	61 (14.3%)
		>150 µg/m^3	15 (3.5%)
	MODIS 10 km DT	<75 µg/m^3	400 (73.8%)
		75–150 µg/m^3	106 (19.6%)
		>150 µg/m^3	36 (6.6%)
	MODIS 10 km DB	<75 µg/m^3	728 (70.7%)
		75–150 µg/m^3	203 (19.6%)
		>150 µg/m^3	100 (9.7%)
	AERONET	<75 µg/m^3	556 (76.0%)
		75–150 µg/m^3	153 (20.9%)
		>150 µg/m^3	23 (3.1%)
	PM$_{2.5}$	<75 µg/m^3	1203 (59.3%)
		75–150 µg/m^3	479 (23.6%)
		>150 µg/m^3	346 (17.1%)

5. Discussion

Satellite-based remote sensing techniques are powerful tools for monitoring aerosol properties and distribution, of which MODIS aerosol products have been widely used during the past decade due to their ability to provide continuous monitoring with relatively high accuracy [50,51]. The major method of validating MODIS data is through surface-deployed sun-photometers [52–54]. MODIS C5.1 aerosol products (mainly the 10 km DT product) have already been validated over China. The inter-comparison of MODIS, MISR, and GOCART (Goddard Global Ozone Chemistry Aerosol Radiation and Transport) aerosol products over Beijing from 2001 to 2010 suggested that only 40.95% of MODIS C5.1 retrievals fell within the EE envelope, while the corresponding values of MISR and GOCART were 70.90% and 32.40%, respectively [55]. However, the validation results may vary from place to place. The validation over Beijing and Xianghe (a small county near Beijing) show that MISR retrievals in Beijing were higher, but lower in Xianghe than MODIS retrievals when considering the correlation coefficients and root-mean-square errors [56]. The validations of MODIS C5.1 all demonstrated that the accuracy of MODIS AOD in China was lower than the global average accuracy of MODIS products. The validation of MODIS C5.1 also showed that MODIS had poor performance in extreme aerosol conditions—especially under dust events or heavy haze [57]. Since the release of MODIS C6 aerosol products, validation and evaluation of these products have been conducted. The validation of 3 km DT MODIS/Aqua data in 18 Asia AERONET stations suggested that only 55% of MODIS retrievals fell within the nominal EE ($0.05 + 0.15\tau$) of the 3 km DT product. For Beijing, only 6.5% of MODIS retrievals fell within the nominal EE, suggesting that the 3 km DT product was less reliable than the 10 km DT product [49]. Furthermore, another validation work in Northern China from 2013 to 2015 suggested that only 53% of the 3 km DT AODS and 66% of the 10 km DT AODS were within the error range [51]. The overall evaluation of both MODIS DT and DB products showed that DT results tended to overestimate the aerosol loading over Northern China, while the DB product exhibited better performance [58]. The results given in Figures 2–4 showed that the 10 km DB AODs had the highest accuracy ($\pm(0.05 + 0.15\tau)$) of the three types of MODIS AOD products, while the 3 km DT AOD had the lowest accuracy ($\pm(0.35 + 0.15\tau)$) in Beijing. The result may be surprising, but it is realistic and suggests that the DB AODs may be more accurate when the air quality is relatively poor, or for urban situations. One suggestion to overcome this problem is to identify a "clearest" day in a certain temporal window to obtain the surface reflectance and retrieve AOD during this temporal window [59].

The update from C5.1 to C6 products brought many improvements to the MODIS aerosol products: the DB algorithm was enhanced to cover more land regions with a higher accuracy, and a 3 km DT product was developed for air quality applications. Validation of the 10 km C6 DB product suggested the number within nominal EE ($0.05 + 0.15\tau$) has increased from 56% to 76% in Beijing when compared with the C5.1 DB product [60]. Two years of validation for the C5.1 and C6 products over Beijing from 2012 to 2013 showed that the accuracy of the 10 km DB and 10 km DT have improved from 46% to 80% and 10% to 18%, respectively [61]. Furthermore, the new C6 DB product was expanded to cover regions like Northern China [62]. The validation of both C5.1 and C6 AOD in Pakistan over AERONET suggested that MODIS C6 DT AOD had significantly improved, but the DB products retained similar accuracies [63]. While the validation of MODIS data over a mountain sun–sky radiometer site suggested that DT products had a high accuracy, with more than 70% of retrievals falling within the nominal EE [52]. C6 AODs were systematically higher than C5.1 AODs over the Mediterranean ocean regions [64]. Despite these improvements, the performance of the C6 products over heavily polluted regions has yet to be deeply investigated. Three questions are germane: first, which dataset is more accurate in a frequently polluted region such as Beijing; second, which data can be more effectively retrieved at different particulate matter concentration conditions; and third, which dataset is more useful for air quality applications. For air quality monitoring applications with aerosol products on a large scale, sufficient successful retrievals must be made for different air quality situations. If one AOD algorithm cannot obtain enough successful retrievals on heavily polluted days, the satellite-retrieved $PM_{2.5}$ concentrations may be biased and underestimate the actual situation.

Most validations of MODIS AOD are primarily focused on relatively low aerosol loads (e.g., an AOD of less than 2.0). These results may not be biased for an area such as Western Europe or North America, where air qualities are good. For example, the number of successful retrievals of 3 km DT was far more than those of 10 km DT and 10 km DB over Atlanta, USA [65]. In China, however—especially Northern China, which has frequent haze-fog events when $PM_{2.5}$ concentrations are very high—most satellite-based aerosol retrieval schemes have difficulties in separating the surface and atmospheric contribution. Despite these shortcomings, the MODIS products have been widely used for monitoring particulate matter in Northern China using various algorithms that take surface particulate matter monitoring data, meteorological data, and satellite data into account [66]. To cover more regions, the DB products are also included in some studies as supplementary data, as the DT algorithm—despite its claims of high accuracy—has relatively low spatial coverage for urban areas with a large population [67].

The results in Figure 6 and Table 1 show that the least number of MODIS 3 km DT AODs was retrieved on heavily polluted days, while the largest number was retrieved with the MODIS 10 km DB AODs. For the Beijing station (Figure 6a), most $PM_{2.5}$ monitoring values (1346 of a total 1920 monitoring values) during the overpass of MODIS were less than $100 \ \mu g/m^3$. The monitoring counts gradually reduced as the $PM_{2.5}$ concentrations increased. There were only 168 $PM_{2.5}$ monitoring values greater than $200 \ \mu g/m^3$. For the 3 km DT AODs, 444 effective values were obtained when $PM_{2.5}$ concentrations were less than $100 \ \mu g/m^3$, only six retrievals were obtained when $PM_{2.5}$ concentrations were greater than $200 \ \mu g/m^3$, and no retrievals were obtained when $PM_{2.5}$ concentrations were greater than $300 \ \mu g/m^3$. This result suggested that the satellite retrieval of $PM_{2.5}$ from MODIS 3 km DT AOD may not be able to gain sufficiently high $PM_{2.5}$ results [68]. Meanwhile, for the 10 km DB (DT) products, the corresponding numbers were 804 (480), 46 (11), and 9 (1). In the Beijing-CAMS station, only 37 $PM_{2.5}$ values (in total 2028 effective $PM_{2.5}$ monitoring data were obtained) greater than $300 \ \mu g/m^3$, and 1408 $PM_{2.5}$ values less than $100 \ \mu g/m^3$ were observed. For the 3 km DT AODs, 393 effective values were obtained when $PM_{2.5}$ concentrations were less than $100 \ \mu g/m^3$, and only eight values were greater than $200 \ \mu g/m^3$. None of the effective 3 km DT AODs were retrieved when $PM_{2.5}$ concentrations were larger than $300 \ \mu g/m^3$. Only the 10 km DB product in the Beijing-CAMS station could obtain effective retrievals when $PM_{2.5}$ concentrations were greater than $300 \ \mu g/m^3$. Therefore, some studies have utilized the DB product to retrieve surface $PM_{2.5}$ concentrations. Nevertheless, these results demonstrate that all MODIS AOD products have a relatively low successful retrieval ratio on heavily polluted days, suggesting that satellite retrieval of $PM_{2.5}$ concentration may underestimate the actual situation by failing to obtain sufficient successful retrievals when the air is heavily polluted. The correlation between the AODs and $PM_{2.5}$ concentrations must be considered along with the number of retrievals and their accuracy against AERONET observations when deciding which dataset is more useful for air quality applications. Figures 7 and 8 show the correlations between the $PM_{2.5}$ concentrations during a MODIS fly-over and different AOD products for the Beijing-CAMS and Beijing stations. The MODIS 10 km DB AODs had higher determination coefficients (R^2) and number of successful retrievals that correlated with the $PM_{2.5}$ concentration in both stations. The correlation of the MODIS 3 km DT AODs with the $PM_{2.5}$ concentrations was in second place, but had the lowest number of retrievals. Although the actual retrieval of the $PM_{2.5}$ concentration requires more data than the AOD (e.g., pressure and relative humidity), the planetary boundary layer, and so on, the original correlation between the AOD and $PM_{2.5}$ plays an important role in the final accuracy of the retrieval scheme. Therefore, despite being originally designed for air quality applications, the MODIS 3 km DT AOD did not seem to be accurate enough or have sufficient spatial coverage for air quality monitoring. In contrast, the MODIS 10 km DB AOD had surprisingly high accuracy and spatial coverage for highly polluted areas such as Beijing. The only limitation was that its spatial resolution is 10 km, which may restrict its application for air quality monitoring on a citywide scale.

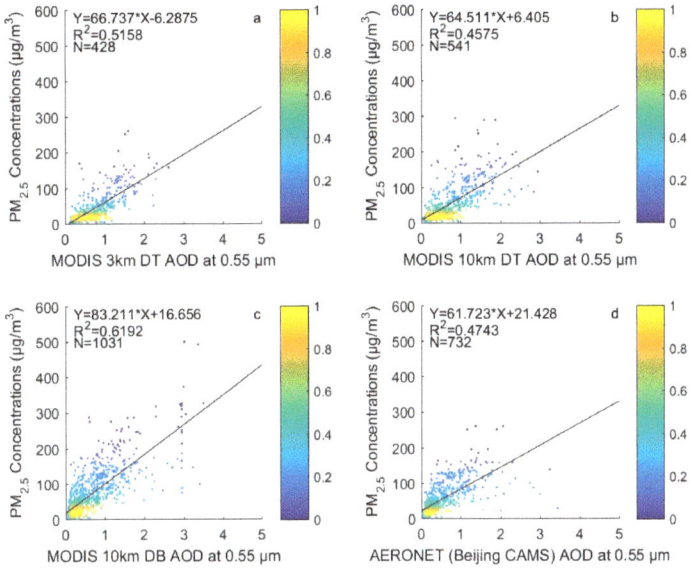

Figure 7. Correlation of PM$_{2.5}$ concentrations against the (**a**) MODIS 3 km DT AOD; (**b**) MODIS 10 km DT AOD; (**c**) MODIS 10 km DB AOD; and (**d**) AERONET AOD for the Beijing-CAMS station.

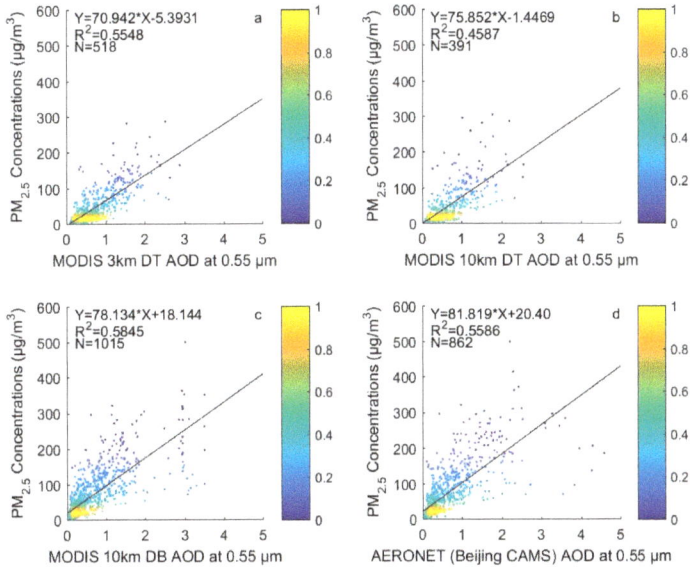

Figure 8. Correlation of PM$_{2.5}$ concentrations against the (**a**) MODIS 3 km DT AOD; (**b**) MODIS 10 km DT AOD; (**c**) MODIS 10 km DB AOD; and (**d**) AERONET AOD for the Beijing station.

In addition to air quality monitoring, we calculated the monthly average AODs of the MODIS 3 km DT, MODIS 10 km DT, and MODIS 10 km DB, and compared them with the AERONET AODs for the Beijing and Beijing-CAMS stations from 2005. Figure 9 shows these monthly average AODs grouped by seasons. The monthly MODIS 3 km DT AODs had the lowest correlation with the AERONET monthly AODs, while the monthly MODIS 10 km DB AODs had the highest correlation. For the two DT products, the winter and spring seasons had the smallest number of successful retrievals and the lowest correlation with the AERONET retrievals. For the DB product, the difference between different seasons was as large as those of the two DT products. Furthermore, the correlation between the AERONET and MODIS products was not high, suggesting that a large difference existed between the satellite-retrieved AODs and the sun-photometer retrievals. Beijing is seriously affected by particulate matter, especially during the autumn and winter months [69], meaning that the satellite retrieval of $PM_{2.5}$ concentrations in autumn and winter was more important than those in summer. However, the newly released 3 km MODIS DT products did not have enough retrievals during the heavily polluted seasons, making it unsuitable to efficiently evaluate the air quality in Beijing.

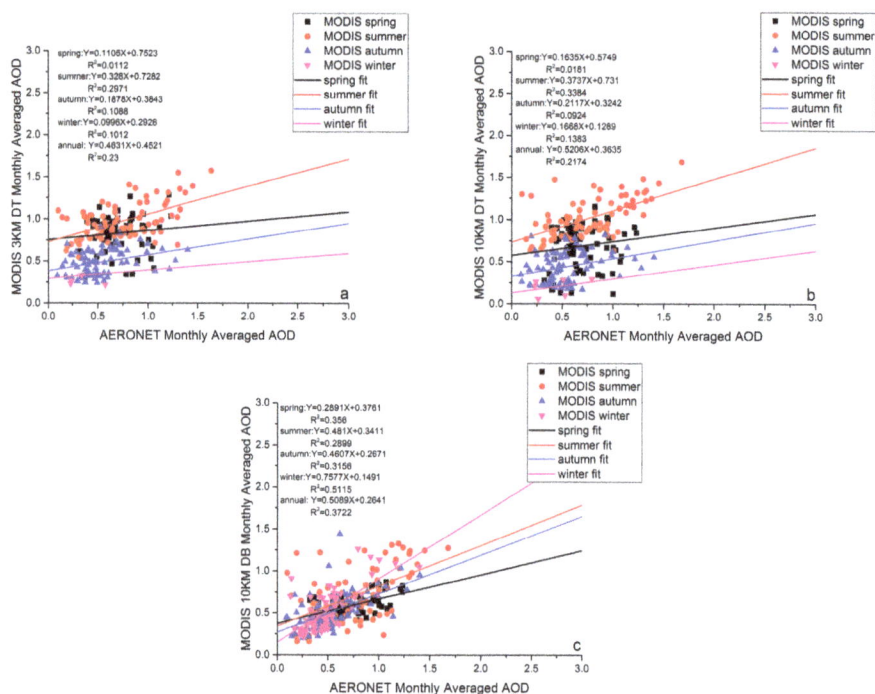

Figure 9. Comparisons of the monthly average AOD of (**a**) MODIS 3 km DT; (**b**) MODIS 10 km DT; and (**c**) MODIS 10 km DB with the AERONET retrieved monthly average AOD for the Beijing and Beijing-CAMS stations.

6. Conclusions

The air quality in Beijing varied greatly with $PM_{2.5}$ concentrations ranging from 5.5 to 475.0 µg/m^3, and over 40% of days had $PM_{2.5}$ concentrations larger than 75 µg/m^3. As the capital of China, station-based monitoring and satellite retrieval of $PM_{2.5}$ concentrations have attracted much attention. Among various other satellite aerosol products, the MODIS aerosol products have been widely used to retrieve surface $PM_{2.5}$ concentrations. The newly released MODIS C6 AOD products have improved the algorithms for the 10 km DT and 10 km DB products, and included a new 3 km DT product with a finer spatial resolution for the air quality community. One challenge in the satellite retrieval of $PM_{2.5}$ concentration is to get sufficient samples when the real atmospheric environment has high $PM_{2.5}$ concentrations. However, satellites often fail to obtain successful AOD retrievals when the aerosol loading is too high, or when surface reflectance is too high. Therefore, it poses the question of how well the MODIS C6 AOD products perform under different air quality situations, given that many studies focus on retrieving $PM_{2.5}$ concentrations with satellite signals. The objective of this study was to evaluate the performances of different MODIS C6 AOD products under different air quality situations and understand whether these AOD products were suitable to retrieve surface $PM_{2.5}$ concentrations in Beijing, given the frequent haze-fog events during winter and autumn. In this study, three MODIS AOD products (3 km DT, 10 km DT, and 10 km DB) were collected at two AERONET stations to validate these three types of AOD products. The results suggested that the 10 km DB AODs had the largest number of effective retrievals as well as the highest retrieval accuracy, followed by the 10 km DT and 3 km DT products. Although it is designed to provide a high spatial resolution aerosol product for air quality monitoring, the 3 km DT product did not perform well in the region of Beijing.

Due to the bright surface problem in urban Beijing and the high concentrations of particulate matter, the DT algorithm often failed to obtain effective results in the winter and autumn when frequent haze-fog events occur. Thus, the successful retrievals by the MODIS DT products mainly happened in the middle of the year, which poses a large challenge for the application of MODIS DT AODs for air quality monitoring, as they fail to obtain effective retrievals when the $PM_{2.5}$ concentrations were high during autumn and winter. Therefore, it is highly possible that the satellite retrieval of $PM_{2.5}$ concentrations MODIS DT AODs would underestimate real situations, especially during autumn and winter. The DB products—initially developed to provide a low accuracy AOD product for desert areas—were further enhanced in the C6 release and contained both high accuracy ($\pm(0.05 + 0.15\tau)$) and spatial-temporal coverage for Beijing. The 10 km DB product also provided sufficient retrievals at high $PM_{2.5}$ concentrations across all four seasons. These results suggest that the MODIS 10 km DB AOD product has high accuracy and temporal coverage for heavily polluted areas such as Beijing, and is a more suitable proxy for the estimation of aerosol loads in urban regions than the other two MODIS aerosol products. For regions like Beijing with frequent haze-fog events in autumn and winter, the MODIS 3 km DT and 10 km DT products may not be able to obtain effective and sufficient retrievals. Therefore, the satellite-based retrieval of $PM_{2.5}$ concentrations with DT products may not reflect the real aerosol pollution situations in Beijing. Thus, for regions with frequent haze-fog events when $PM_{2.5}$ concentrations are high, the newly developed 3 km DT product is recommended as a proxy for air quality monitoring. Furthermore, it is possible that a DB product with a finer space resolution (e.g., 3 km or even 1 km) may be more suitable for air quality monitoring applications.

Acknowledgments: This research was supported by the Fundamental Research Funds for the Central Universities under Grant 2014QD02, National Natural Science Foundation of China under Grant 41671383 and 61405204, the Open Fund of State Key Laboratory of Remote Sensing Science under Grant OFSLRSS201623, Open Fund of State Key Laboratory of Information Engineering in Surveying, Mapping and Remote Sensing under Grant 16R01, and the outstanding talent training project of Beijing under Grant 2014000020124G0441. We acknowledge the PIs of the Beijing Station (Hong-Bin Chen and Philippe Goloub) and the Beijing-CAMS station (Huizheng Che) for providing AERONET observation data.

Author Contributions: All authors have made significant contributions to the paper. Wei Chen designed and wrote the paper; Aiping Fan processed the data; and Lei Yan provided key data.

Conflicts of Interest: The authors declare no conflict of interest.

References

1. Lynch, P.; Reid, J.S.; Westphal, D.L.; Zhang, J.L.; Hogan, T.F.; Hyer, E.J.; Curtis, C.A.; Hegg, D.A.; Shi, Y.X.; Campbell, J.R.; et al. An 11-year global gridded aerosol optical thickness reanalysis (v1.0) for atmospheric and climate sciences. *Geosci. Model Dev.* **2016**, *9*, 1489–1522. [CrossRef]
2. Remer, L.A.; Kaufman, Y.J.; Tanre, D.; Mattoo, S.; Chu, D.A.; Martins, J.V.; Li, R.R.; Ichoku, C.; Levy, R.C.; Kleidman, R.G.; et al. The MODIS aerosol algorithm, products, and validation. *J. Atmos. Sci.* **2005**, *62*, 947–973. [CrossRef]
3. Zhang, Z.B.; Meyer, K.; Yu, H.B.; Platnick, S.; Colarco, P.; Liu, Z.Y.; Oreopoulos, L. Shortwave direct radiative effects of above-cloud aerosols over global oceans derived from 8 years of CALIOP and MODIS observations. *Atmos. Chem. Phys.* **2016**, *16*, 2877–2900. [CrossRef]

4. Zhou, M.G.; Liu, Y.N.; Wang, L.J.; Kuang, X.Y.; Xu, X.H.; Kan, H.D. Particulate air pollution and mortality in a cohort of Chinese men. *Environ. Pollut.* **2014**, *186*, 1–6. [CrossRef] [PubMed]

5. Kahn, R.A.; Gaitley, B.J. An analysis of global aerosol type as retrieved by MISR. *J. Geophys. Res.* **2015**, *120*, 4248–4281. [CrossRef]

6. Ghan, S.J.; Liu, X.; Easter, R.C.; Zaveri, R.; Rasch, P.J.; Yoon, J.H.; Eaton, B. Toward a Minimal Representation of Aerosols in Climate Models: Comparative Decomposition of Aerosol Direct, Semidirect, and Indirect Radiative Forcing. *J. Clim.* **2012**, *25*, 6461–6476. [CrossRef]

7. Hsu, N.C.; Tsay, S.C.; King, M.D.; Herman, J.R. Aerosol properties over bright-reflecting source regions. *IEEE Trans. Geosci. Remote Sens.* **2004**, *42*, 557–569. [CrossRef]

8. Levy, R.C.; Remer, L.A.; Kaufman, Y.J. Effects of neglecting polarization on the MODIS aerosol retrieval over land. *IEEE Trans. Geosci. Remote Sens.* **2004**, *42*, 2576–2583. [CrossRef]

9. Hasekamp, O.P.; Litvinov, P.; Butz, A. Aerosol properties over the ocean from PARASOL multiangle photopolarimetric measurements. *J. Geophys. Res.* **2011**, *116*. [CrossRef]

10. Diner, D.J.; Martonchik, J.V.; Kahn, R.A.; Pinty, B.; Gobron, N.; Nelson, D.L.; Holben, B.N. Using angular and spectral shape similarity constraints to improve MISR aerosol and surface retrievals over land. *Remote Sens. Environ.* **2005**, *94*, 155–171. [CrossRef]

11. Sayer, A.M.; Munchak, L.A.; Hsu, N.C.; Levy, R.C.; Bettenhausen, C.; Jeong, M.J. MODIS Collection 6 aerosol products: Comparison between Aqua's e-Deep Blue, Dark Target, and "merged" data sets, and usage recommendations. *J. Geophys. Res.* **2014**, *119*, 13965–13989. [CrossRef]

12. Diner, D.J.; Abdou, W.A.; Bruegge, C.J.; Conel, J.E.; Crean, K.A.; Gaitley, B.J.; Helmlinger, M.C.; Kahn, R.A.; Martonchik, J.V.; Pilorz, S.H.; et al. MISR aerosol optical depth retrievals over southern Africa during the SAFARI-2000 dry season campaign. *Geophys. Res. Lett.* **2001**, *28*, 3127–3130. [CrossRef]

13. Su, X.; Goloub, P.; Chiapello, I.; Chen, H.; Ducos, F.; Li, Z. Aerosol variability over East Asia as seen by POLDER space-borne sensors. *J. Geophys. Res.* **2010**, *115*, D24215. [CrossRef]

14. Levy, R.C.; Mattoo, S.; Munchak, L.A.; Remer, L.A.; Sayer, A.M.; Patadia, F.; Hsu, N.C. The Collection 6 MODIS aerosol products over land and ocean. *Atmos. Meas. Tech.* **2013**, *6*, 2989–3034. [CrossRef]

15. Yu, L.; Liu, T.; Cai, H.; Tang, J.; Bu, K.; Yan, F.; Yang, C.; Yang, J.; Zhang, S. Estimating land surface radiation balance using MODIS in northeastern China. *J. Appl. Remote Sens.* **2014**, *8*, 083523. [CrossRef]

16. Payra, S.; Soni, M.; Kumar, A.; Prakash, D.; Verma, S. Intercomparison of Aerosol Optical Thickness Derived from MODIS and in Situ Ground Datasets over Jaipur, a Semi-arid Zone in India. *Environ. Sci. Technol.* **2015**, *49*, 9237–9246. [CrossRef] [PubMed]

17. Misra, A.; Jayaraman, A.; Ganguly, D. Validation of Version 5.1 MODIS Aerosol Optical Depth (Deep Blue Algorithm and Dark Target Approach) over a Semi-Arid Location in Western India. *Aerosol Air Qual. Res.* **2015**, *15*, 252–262. [CrossRef]

18. Xie, Y.; Li, Z.; Li, D.; Xu, H.; Li, K. Aerosol Optical and Microphysical Properties of Four Typical Sites of SONET in China Based on Remote Sensing Measurements. *Remote Sens.* **2015**, *7*, 9928. [CrossRef]

19. Vermote, E.F.; Tanre, D.; Deuze, J.L.; Herman, M.; Morcrette, J.J. Second Simulation of the Satellite Signal in the Solar Spectrum, 6S: An overview. *IEEE Trans. Geosci. Remote Sens.* **1997**, *35*, 675–686. [CrossRef]

20. Chen, W.; Tang, H.; Zhao, H. Diurnal, weekly and monthly spatial variations of air pollutants and air quality of Beijing. *Atmos. Environ.* **2015**, *119*, 21–34. [CrossRef]

21. Che, H.; Shi, G.; Uchiyama, A.; Yamazaki, A.; Chen, H.; Goloub, P.; Zhang, X. Intercomparison between aerosol optical properties by a PREDE skyradiometer and CIMEL sunphotometer over Beijing, China. *Atmos. Chem. Phys.* **2008**, *8*, 3199–3214. [CrossRef]

22. Eck, T.F.; Holben, B.N.; Dubovik, O.; Smirnov, A.; Goloub, P.; Chen, H.B.; Chatenet, B.; Gomes, L.; Zhang, X.Y.; Tsay, S.C.; et al. Columnar aerosol optical properties at AERONET sites in central eastern Asia and aerosol transport to the tropical mid-Pacific. *J. Geophys. Res.* **2005**, *110*. [CrossRef]

23. Zhu, J.; Xia, X.; Che, H.; Wang, J.; Zhang, J.; Duan, Y. Study of aerosol optical properties at Kunming in southwest China and long-range transport of biomass burning aerosols from North Burma. *Atmos. Res.* **2016**, *169*, 237–247. [CrossRef]

24. Xin, J.Y.; Wang, Y.S.; Li, Z.Q.; Wang, P.C.; Hao, W.M.; Nordgren, B.L.; Wang, S.G.; Liu, G.R.; Wang, L.L.; Wen, T.X.; et al. Aerosol optical depth (AOD) and Angstrom exponent of aerosols observed by the Chinese Sun Hazemeter Network from August 2004 to September 2005. *J. Geophys. Res.* **2007**, *112*, D05203. [CrossRef]

25. Lee, K.H.; Kim, Y.J. Satellite remote sensing of Asian aerosols: A case study of clean, polluted, and Asian dust storm days. *Atmos. Meas. Tech.* **2010**, *3*, 1771–1784. [CrossRef]

26. Xia, X.A.; Chen, H.B.; Wang, P.C.; Zong, X.M.; Qiu, J.H.; Gouloub, P. Aerosol properties and their spatial and temporal variations over North China in spring 2001. *Tellus B* **2005**, *57*, 28–39. [CrossRef]

27. Zhang, A.; Qi, Q.; Jiang, L.; Zhou, F.; Wang, J. Population Exposure to $PM_{2.5}$ in the Urban Area of Beijing. *PLoS ONE* **2013**, *8*, e63486. [CrossRef] [PubMed]

28. Li, R.; Li, Z.; Gao, W.; Ding, W.; Xu, Q.; Song, X. Diurnal, seasonal, and spatial variation of $PM_{2.5}$ in Beijing. *Sci. Bull.* **2015**, *60*, 387–395. [CrossRef]

29. Kumar, K.R.; Yin, Y.; Sivakumar, V.; Kang, N.; Yu, X.; Diao, Y.; Adesina, A.J.; Reddy, R.R. Aerosol climatology and discrimination of aerosol types retrieved from MODIS, MISR and OMI over Durban (29.88°S, 31.02°E), South Africa. *Atmos. Environ.* **2015**, *117*, 9–18. [CrossRef]

30. He, J.L.; Zha, Y.; Zhang, J.H.; Gao, J. Aerosol Indices Derived from MODIS Data for Indicating Aerosol-Induced Air Pollution. *Remote Sens.* **2014**, *6*, 1587–1604. [CrossRef]

31. Wei, J.; Sun, L. Comparison and Evaluation of Different MODIS Aerosol Optical Depth Products Over the Beijing-Tianjin-Hebei Region in China. *IEEE J. Sel. Top. Appl. Earth Obs.* **2017**, *10*, 835–844. [CrossRef]

32. Zhang, Y.; Li, Z.Q.; Zhang, Y.H.; Chen, Y.; Cuesta, J.; Ma, Y. Multi-peak accumulation and coarse modes observed from AERONET retrieved aerosol volume size distribution in Beijing. *Meteorol. Atmos. Phys.* **2016**, *128*, 537–544. [CrossRef]

33. Hu, X.F.; Waller, L.A.; Al-Hamdan, M.Z.; Crosson, W.L.; Estes, M.G.; Estes, S.M.; Quattrochi, D.A.; Sarnat, J.A.; Liu, Y. Estimating ground-level $PM_{2.5}$ concentrations in the southeastern US using geographically weighted regression. *Environ. Res.* **2013**, *121*, 1–10. [CrossRef] [PubMed]

34. Lee, H.J.; Chatfield, R.B.; Strawa, A.W. Enhancing the Applicability of Satellite Remote Sensing for $PM_{2.5}$ Estimation Using MODIS Deep Blue AOD and Land Use Regression in California, United States. *Environ. Sci. Technol.* **2016**, *50*, 6546–6555. [CrossRef] [PubMed]

35. Song, Y.; Xu, D.D.; Chai, Z.F.; Ouyang, H.; Feng, W.Y.; Mao, X.Y. INAA study for characterization of PM_{10} and $PM_{2.5}$ in Beijing and influence of dust storm. *J. Radioanal. Nucl. Chem.* **2006**, *270*, 29–33. [CrossRef]

36. Chen, W.; Wang, F.; Xiao, G.; Wu, K.; Zhang, S. Air Quality of Beijing and Impacts of the New Ambient Air Quality Standard. *Atmosphere* **2015**, *6*, 1243. [CrossRef]

37. Zhang, R.; Jing, J.; Tao, J.; Hsu, S.C.; Wang, G.; Cao, J.; Lee, C.S.L.; Zhu, L.; Chen, Z.; Zhao, Y.; et al. Chemical characterization and source apportionment of $PM_{2.5}$ in Beijing: Seasonal perspective. *Atmos. Chem. Phys.* **2013**, *13*, 7053–7074. [CrossRef]

38. Xia, X.G. Variability of aerosol optical depth and Angstrom wavelength exponent derived from AERONET observations in recent decades. *Environ. Res. Lett.* **2011**, *6*, 044011. [CrossRef]

39. Che, H.; Xia, X.; Zhu, J.; Li, Z.; Dubovik, O.; Holben, B.; Goloub, P.; Chen, H.; Estelles, V.; Cuevas-Agullo, E.; et al. Column aerosol optical properties and aerosol radiative forcing during a serious haze-fog month over North China Plain in 2013 based on ground-based sunphotometer measurements. *Atmos. Chem. Phys.* **2014**, *14*, 2125–2138. [CrossRef]

40. Smirnov, A.; Holben, B.N.; Eck, T.F.; Dubovik, O.; Slutsker, I. Cloud-screening and quality control algorithms for the AERONET database. *Remote Sens. Environ.* **2000**, *73*, 337–349. [CrossRef]

41. Dubovik, O.; Smirnov, A.; Holben, B.N.; King, M.D.; Kaufman, Y.J.; Eck, T.F.; Slutsker, I. Accuracy assessments of aerosol optical properties retrieved from Aerosol Robotic Network (AERONET) Sun and sky radiance measurements. *J. Geophys. Res.* **2000**, *105*, 9791–9806. [CrossRef]

42. Kaufman, Y.J.; Tanré, D.; Gordon, H.R.; Nakajima, T.; Lenoble, J.; Frouin, R.; Grassl, H.; Herman, B.M.; King, M.D.; Teillet, P.M. Passive remote sensing of tropospheric aerosol and atmospheric correction for the aerosol effect. *J. Geophys. Res.* **1997**, *102*, 16815–16830. [CrossRef]

43. Levy, R.C.; Remer, L.A.; Mattoo, S.; Vermote, E.F.; Kaufman, Y.J. Second-generation operational algorithm: Retrieval of aerosol properties over land from inversion of Moderate Resolution Imaging Spectroradiometer spectral reflectance. *J. Geophys. Res.* **2007**, *112*. [CrossRef]

44. Di Tomaso, E.; Schutgens, N.A.J.; Jorba, O.; Garcia-Pando, C.P. Assimilation of MODIS Dark Target and Deep Blue observations in the dust aerosol component of NMMB-MONARCH version 1.0. *Geosci. Model Dev.* **2017**, *10*, 1107–1129. [CrossRef]

45. Distribution of Real Time Air Qulaity of China. Available online: http://113.108.142.147:20035/emcpublish/ (accessed on 15 May 2017).

46. Petrenko, M.; Ichoku, C.; Leptoukh, G. Multi-sensor Aerosol Products Sampling System (MAPSS). *Atmos. Meas. Tech.* **2012**, *5*, 913–926. [CrossRef]
47. Kuang, Y.; Zhao, C.S.; Tao, J.C.; Ma, N. Diurnal variations of aerosol optical properties in the North China Plain and their influences on the estimates of direct aerosol radiative effect. *Atmos. Chem. Phys.* **2015**, *15*, 5761–5772. [CrossRef]
48. Remer, L.A.; Mattoo, S.; Levy, R.C.; Munchak, L.A. MODIS 3 km aerosol product: Algorithm and global perspective. *Atmos. Meas. Tech.* **2013**, *6*, 1829–1844. [CrossRef]
49. Nichol, J.; Bilal, M. Validation of MODIS 3 km Resolution Aerosol Optical Depth Retrievals over Asia. *Remote Sens.* **2016**, *8*, 328. [CrossRef]
50. Sanchez-Romero, A.; Gonzalez, J.A.; Calbo, J.; Sanchez-Lorenzo, A.; Michalsky, J. Aerosol optical depth in a western Mediterranean site: An assessment of different methods. *Atmos. Res.* **2016**, *174*, 70–84. [CrossRef]
51. Yan, X.; Shi, W.; Luo, N.; Zhao, W. A new method of satellite-based haze aerosol monitoring over the North China Plain and a comparison with MODIS Collection 6 aerosol products. *Atmos. Res.* **2016**, *171*, 31–40. [CrossRef]
52. Lee, H.J.; Son, Y.-S. Spatial Variability of AERONET Aerosol Optical Properties and Satellite Data in South Korea during NASA DRAGON-Asia Campaign. *Environ. Sci. Technol.* **2016**, *50*, 3954–3964. [CrossRef] [PubMed]
53. Ma, Y.; Li, Z.; Li, Z.; Xie, Y.; Fu, Q.; Li, D.; Zhang, Y.; Xu, H.; Li, K. Validation of MODIS Aerosol Optical Depth Retrieval over Mountains in Central China Based on a Sun-Sky Radiometer Site of SONET. *Remote Sens.* **2016**, *8*, 111. [CrossRef]
54. Witek, M.L.; Diner, D.J.; Garay, M.J. Satellite assessment of sea spray aerosol productivity: Southern Ocean case study. *J. Geophys. Res.* **2016**, *121*, 872–894. [CrossRef]
55. Cheng, T.; Chen, H.; Gu, X.; Yu, T.; Guo, J.; Guo, H. The inter-comparison of MODIS, MISR and GOCART aerosol products against AERONET data over China. *J. Quant. Spectrosc. Radiat.* **2012**, *113*, 2135–2145. [CrossRef]
56. Qi, Y.L.; Ge, J.M.; Huang, J.P. Spatial and temporal distribution of MODIS and MISR aerosol optical depth over northern China and comparison with AERONET. *Sci. Bull.* **2013**, *58*, 2497–2506. [CrossRef]
57. Li, B.G.; Yuan, H.S.; Feng, N.; Tao, S. Comparing MODIS and AERONET aerosol optical depth over China. *Int. J. Remote Sens.* **2009**, *30*, 6519–6529. [CrossRef]
58. Tao, M.; Chen, L.; Wang, Z.; Tao, J.; Che, H.; Wang, X.; Wang, Y. Comparison and evaluation of the MODIS Collection 6 aerosol data in China. *J. Geophys. Res.* **2015**, *120*, 6992–7005. [CrossRef]
59. Huang, G.H.; Huang, C.L.; Li, Z.Q.; Chen, H. Development and Validation of a Robust Algorithm for Retrieving Aerosol Optical Depth over Land From MODIS Data. *IEEE J. Sel. Top. Appl. Earth Obs.* **2015**, *8*, 1152–1166. [CrossRef]
60. Sayer, A.M.; Hsu, N.C.; Bettenhausen, C.; Jeong, M.J. Validation and uncertainty estimates for MODIS Collection 6 "Deep Blue" aerosol data. *J. Geophys. Res.* **2013**, *118*, 7864–7872. [CrossRef]
61. Bilal, M.; Nichol, J.E. Evaluation of MODIS aerosol retrieval algorithms over the Beijing-Tianjin-Hebei region during low to very high pollution events. *J. Geophys. Res.* **2015**, *120*, 7941–7957. [CrossRef]
62. Zhang, Q.; Xin, J.Y.; Yin, Y.; Wang, L.L.; Wang, Y.S. The Variations and Trends of MODIS C5 & C6 Products' Errors in the Recent Decade over the Background and Urban Areas of North China. *Remote Sens.* **2016**, *8*, 754. [CrossRef]
63. Bilal, M.; Nichol, J.E.; Nazeer, M. Validation of Aqua-MODIS C051 and C006 Operational Aerosol Products Using AERONET Measurements over Pakistan. *IEEE J. Sel. Top. Appl. Earth Obs.* **2016**, *9*, 2074–2080. [CrossRef]
64. Georgoulias, A.K.; Alexandri, G.; Kourtidis, K.A.; Lelieveld, J.; Zanis, P.; Amiridis, V. Differences between the MODIS Collection 6 and 5.1 aerosol datasets over the greater Mediterranean region. *Atmos. Environ.* **2016**, *147*, 310–319. [CrossRef]
65. Belle, J.H.; Liu, Y. Evaluation of Aqua MODIS Collection 6 AOD Parameters for Air Quality Research over the Continental United States. *Remote Sens.* **2016**, *8*, 815. [CrossRef]
66. Ma, Z.; Hu, X.; Sayer, A.M.; Levy, R.; Zhang, Q.; Xue, Y.; Tong, S.; Bi, J.; Huang, L.; Liu, Y. Satellite-Based Spatiotemporal Trends in PM$_{2.5}$ Concentrations: China, 2004–2013. *Environ. Health Perspect.* **2016**, *124*, 184–192. [CrossRef] [PubMed]

67. Sayer, A.M.; Hsu, N.C.; Bettenhausen, C.; Jeong, M.J.; Meister, G. Effect of MODIS Terra radiometric calibration improvements on Collection 6 Deep Blue aerosol products: Validation and Terra/Aqua consistency. *J. Geophys. Res.* **2015**, *120*, 12157–12174. [CrossRef]
68. Xie, Y.; Wang, Y.; Zhang, K.; Dong, W.; Lv, B.; Bai, Y. Daily Estimation of Ground-Level $PM_{2.5}$ Concentrations over Beijing Using 3 km Resolution MODIS AOD. *Environ. Sci. Technol.* **2015**, *49*, 12280–12288. [CrossRef] [PubMed]
69. Zhou, Y.; Cheng, S.; Chen, D.; Lang, J.; Wang, G.; Xu, T.; Wang, X.; Yao, S. Temporal and Spatial Characteristics of Ambient Air Quality in Beijing, China. *Aerosol Air Qual. Res.* **2015**, *15*, 1868–1880. [CrossRef]

remote sensing

MDPI

Article

MODIS Retrieval of Aerosol Optical Depth over Turbid Coastal Water

Yi Wang [1,2,3], Jun Wang [1,2,3,*], Robert C. Levy [4], Xiaoguang Xu [1,2] and Jeffrey S. Reid [5]

1   Department of Chemical and Biochemical Engineering, The University of Iowa, Iowa City, IA 52242, USA;
     yi-wang-4@uiowa.edu (Y.W.); xiaoguang-xu@uiowa.edu (X.X.)
2   Center of Global and Regional Environmental Research, The University of Iowa, Iowa City, IA 52242, USA
3   Interdisciplinary Graduate Program in Informatics, The University of Iowa, Iowa City, IA 52242, USA
4   NASA Goddard Space Flight Center, Greenbelt, MD 20771, USA; robert.c.levy@nasa.gov
5   Marine Meteorology Division, Naval Research Laboratory, Monterey, CA 93943, USA;
     jeffrey.reid@nrlmry.navy.mil
*   Correspondence: jun-wang-1@uiowa.edu; Tel.: +1-319-353-4483

Academic Editors: Yang Liu, Omar Torres, Alexander A. Kokhanovsky and Prasad S. Thenkabail
Received: 23 March 2017; Accepted: 7 June 2017; Published: 12 June 2017

Abstract: We present a new approach to retrieve Aerosol Optical Depth (AOD) using the Moderate
Resolution Imaging Spectroradiometer (MODIS) over the turbid coastal water. This approach
supplements the operational Dark Target (DT) aerosol retrieval algorithm that currently does not
conduct AOD retrieval in shallow waters that have visible sediments or sea-floor (i.e., Class 2 waters).
Over the global coastal water regions in cloud-free conditions, coastal screening leads to ~20%
unavailability of AOD retrievals. Here, we refine the MODIS DT algorithm by considering that
water-leaving radiance at 2.1 μm to be negligible regardless of water turbidity, and therefore the 2.1 μm
reflectance at the top of the atmosphere is sensitive to both change of fine-mode and coarse-mode
AODs. By assuming that the aerosol single scattering properties over coastal turbid water are similar
to those over the adjacent open-ocean pixels, the new algorithm can derive AOD over these shallow
waters. The test algorithm yields ~18% more MODIS-AERONET collocated pairs for six AERONET
stations in the coastal water regions. Furthermore, comparison of the new retrieval with these
AERONET observations show that the new AOD retrievals have equivalent or better accuracy than
those retrieved by the MODIS operational algorithm's over coastal land and non-turbid coastal water
product. Combining the new retrievals with the existing MODIS operational retrievals yields an
overall improvement of AOD over those coastal water regions. Most importantly, this refinement
extends the spatial and temporal coverage of MODIS AOD retrievals over the coastal regions where
60% of human population resides. This expanded coverage is crucial for better understanding of
impact of anthropogenic aerosol particles on coastal air quality and climate.

Keywords: AOD; coastal water; MODIS; retrieval

---

1. Introduction

Aerosols are a colloidal system of particles suspended in the atmosphere, and have significant
impacts on weather, climate, and human health [1–3]. Because of its global observational coverage,
satellite remote sensing has a critical role in quantifying these impacts. Such satellite remote sensing
is being used to retrieve aerosol properties such as the aerosol optical depth (AOD), along with a
well-characterized uncertainty envelope, at high spatial resolution across the globe. Indeed, since the
launch of the Moderate Resolution Imaging Spectroradiometer (MODIS) on the NASA Terra satellite
in 1999, AOD retrievals derived by Dark Target (DT) algorithm [4] or Deep Blue (DB) algorithm
(land only) [5,6] have been used widely in the research community [7].

One persistent challenge to aerosol remote sensing is retrieval of environmental properties over coastal (or littoral) waters. Variable ocean color, along with the presence of visible sediments and sea floor along coastlines, has strong spectral and spatial variability leading to poorly constrained lower boundary conditions for aerosol retrievals. Yet, retrieving AOD in coastal waters is a much-desired part of new systems characterizing air quality and aerosol radiative effects. As ~60% of the human population lives in coastal areas [8], it is crucial to expand the satellite-remote sensing datasets to include these areas. This study aims to address this observational gap by refining the MODIS DT-ocean algorithm to retrieve AOD over turbid coastal waters.

The MODIS DT-ocean algorithm uses top of atmosphere (TOA) reflectance in six wavelength bands, ranging from 0.55 to 2.1 μm, to simultaneously retrieve AOD and Fine Mode Fraction (FMF) based on a lookup table approach. The lookup table is constructed by assuming aerosol optical properties of four fine aerosol modes and five coarse aerosol modes, coupled with the boundary conditions of molecular scattering, and a rough ocean surface (glitter, whitecaps, foam). Additionally, the lookup table assumes zero water-leaving radiance at all bands except 0.55 μm (at which a fixed water-leaving reflectance of 0.005 is used) [4,9]. The retrieval process searches the lookup table to find the best combination of fine mode and coarse mode aerosol type (out of possibly 20 combinations) such that the AOD and FMF retrievals render the best match between observed and simulated (e.g., those from the lookup table) radiances. However, while the assumption of nearly-zero water-leaving radiance may be suitable for open ocean, it is clearly not applicable to turbid coastal waters. There, the water-leaving radiances can be contributed by reflection of shallow-water sea floor (particularly in blue wavelengths) and either suspended or dissolved particulate matter in the water, especially in green to red wavelengths such as 0.55, 0.66, and 0.86 μm [10]. Hence, as part of the DT-ocean algorithm, the turbid water pixels are masked and not considered for retrieval. The method used for such masking compares the TOA reflectance at 0.55 μm with the expected counterpart from the power-law fitting using the TOA reflectance at 0.47, 1.2, 1.6, and 2.1 μm; if a significant difference (larger than 0.01) is found, the corresponding pixel is masked out for AOD retrieval [8].

We analyze the MODIS AOD retrieval unavailability over the cloud-free conditions at both global and regional scale to reveal how often the AOD is not retrieved only because of water turbidity (and not other factors such as cloud cover). As shown in Figure 1, this data availability is near total over all open ocean and decreases dramatically (by 90–100%) toward coastlines. In a global average, the unavailability of AOD is ~20% over coastal water region which is labelled as shallow ocean (within 5 km of coastline or with water depth less than 50 m) in the MODIS geolocation product [11]. In other words, in these 20% of cloud-free cases, the AOD should have been retrieved if the water was not turbid.

Based on the principle that liquid water absorption increases in shortwave infrared (SWIR), we present a new approach that uses MODIS-measured radiance at 2.1 μm to retrieve AOD over turbid coastal water no matter whether the aerosol is fine or coarse mode dominated. In the spectral range of 0.55 to 2.1 μm, the transparency of pure water to sunlight decreases rapidly with the increase of wavelength; the penetration depth (at which light attenuation is 90%) is 41 m at 0.55 μm and drops to 0.001 m at 2.1 μm [8]. Hence, unless the water is shallower than 1 mm, the water-leaving radiance contributed from the sea floor, sediments, and other contaminants in the water at 2.1 μm is nearly negligible; this vastly simplifies the lower boundary condition of the retrieval.

We present the data and the new approach in Sections 2 and 3, respectively, followed by description of retrieval results and the validation against measurements from six Aerosol Robotic Network (AERONET) coastal sites in Section 4. Discussions and conclusions are in Sections 5 and 6, respectively.

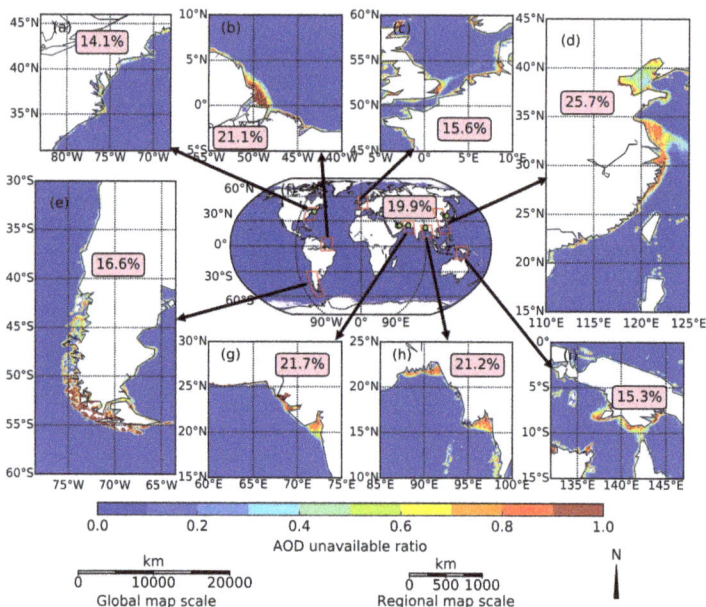

Figure 1. Regional (**a–e** and **g–i**) and global (**f**) distribution of AOD unavailable ratio which is defined as the ratio between number of non-retrieval pixels and all pixels in cloud-free conditions. MYD04 product in 2016 are used to conduct the statistics. Zoom in on the red-box regions of (**f**) are shown in (**a–e** and **g–i**). Lime points in (**f**) are AERONET sites used for validation and more information is in Table 1. Numbers in the pink boxes are AOD unavailable ratio over coastal water in clear sky condition.

2. Data

2.1. MODIS Data

MODIS is an earth-viewing sensor on board the Terra and Aqua polar-orbiting satellites, launched in December 1999 and May 2002, respectively. MODIS has 36 channels spanning the spectral range from 0.41 to 15 µm with spatial resolutions of 250 m (2 channels), 500 m (5 channels), and 1 km (29 channels). Its 2330-km swath width enables it to provide near-global converge daily. Terra and Aqua cross the equator from north to south (descending node) at approximately 10:30 a.m. local time and from south to north (ascending node) at approximately 1:30 p.m. local time, respectively.

Here, all MODIS data products are labeled as MxDNN, where x is substituted by O for Terra and Y for Aqua, respectively, and NN is the serial number of a specific product. King et al. [12] presented a general description of MODIS atmosphere data processing architecture and products. In this study, the MODIS-calibrated TOA reflectance product (MxD02) (http://mcst.gsfc.nasa.gov/content/l1b-documents), atmospheric profile product (MxD07), geolocation product (MxD03), and aerosol product (MxD04) [4] are used for retrieving aerosol optical depth over turbid coastal water. TOA reflectance of 2.1 µm with spatial resolution of 500 m from MxD02 is used to retrieve AOD through a lookup table (LUT) method. Gas absorption by water vapor and ozone column are corrected from the TOA reflectance with spatial resolution of 5 km from MxD07 product [4]. The MxD03 product has a spatial resolution of 1 km and is used to distinguish between water and land pixels. All other ancillary information from MxD04 aerosol product for each valid AOD retrieval (at spatial resolution of 10 km at nadir) are also used as input to the retrieval algorithm in this study; the information includes aerosol mode index, reflectance weighting parameter, and National Centers for Environmental Prediction

(NCEP) analysis of wind speed (2 m above the surface). In addition, a cloud mask with a spatial resolution of 500 m from MxD04 aerosol product is also used to ensure only AOD during cloud-free conditions over the turbid coastal region is retrieved by this study.

2.2. AERONET Data

AOD measurements from the ground-based AERONET sun photometers are commonly used for validating MODIS retrievals. Here we use data from six coastal water sites for this study (Table 1), and their locations are marked with green dots in global map in Figure 1f. All AERONET sun photometers (SP) measure direct solar radiation at 0.44 μm, 0.67 μm, 0.87 μm, 0.94 μm and 1.02 μm, and these measurements are used to infer AOD through Beer-Lambert-Bouguer law with quality at Level 1.0 (unscreened), Level 1.5 (cloud screened), and Level 2.0 (cloud-screened and quality-assured) [13,14]. We evaluate MODIS AOD (both MxD04 product and retrieval of this study) at 0.55 μm against an AERONET counterpart that is derived through linearly interpolating AERONET AOD at 0.44 and 0.67 μm in the logarithm domain.

Table 1. Information of AERONET sites used for validation.

Site	Location *	Period	Data Level	Monthly Mean 0.44–0.87 μm Ångström Exponent
MVCO, New England	41.3°N 70.6°W	August 2015	2.0	1.842
Bhola, Bangladesh	22.2°N 90.8°E	December 2015	1.5	1.206
Anmyon, S. Korea	36.5°N 126.3°E	May 2016	1.5	1.076
Dalma, UAE	24.5°N 52.3°E	August 2004	2.0	0.711
Karachi, Pakistan	24.9°N 67.0°E	March 2014	2.0	0.701
MAARCO, UAE	24.7°N 54.7°E	September 2004	2.0	0.597

* All the locations are shown as lime points in Figure 1f.

Dalma and MVCO are two sites over the ocean, with distance to coastline being 48 km and 29 km, respectively, while the rest of the sites are over land within 6 km from the coastline. According to our analysis, most AERONET sites are more than 10 km away from the coast lines, and lack dedicated long-term continuous measurements of AOD over the turbid coastal water. Here, we use the SP AOD data from August 2015, December 2015, May 2016, August 2004, March 2014, and September 2004 at MVCO Bhola, Anmyon, Dalma, Karachi, and MAARCO sites, respectively, because these time periods have the most measurements available at their corresponding sites. AERONET Level 2.0 data are used for most sites except Bhola and Anmyon where only Level 1.5 data are available and utilized here. The monthly mean AERONET 0.44–0.87 μm Ångström exponent ranges from 0.597 to 1.842 over these six sites. Hence, these sites represent a wide range of atmospheric conditions, ranging from coarse mode dominated to fine mode dominated cases [15].

2.3. Data Extraction Procedure

MxD04 [4] AOD and the AOD retrieval from the new algorithm are evaluated against AERONET measurements. The spatio-temporal matching approach by Ichoku et al. [16] is applied to collocate AERONET AOD measurements and MODIS retrievals (MxD04 and/or new algorithm) for comparison. AERONET measurements within ±30 min of the MODIS overpass time are averaged and compared against MODIS retrievals averaged within a 50-km diameter circular region centered over the AERONET sites. MxD04 products include DT, DB, and DT/DB merged retrievals, corresponding to 0.55 μm AOD from Dark Target algorithms (DT-land and DT-ocean algorithms), Deep Blue algorithm, and combination of the two retrievals, respectively [4]. DT retrievals cover both vegetated land and ocean while DB is limited to land. As the DT-land algorithm is not designed to retrieve AOD over bright desert, there are significant retrieval gaps over land. The DB algorithm was originally designed to complement the gaps, based on the principal that desert pixels are relatively darker in deep-blue bands [5]. However, now DB has been extended to vegetated land surface as well. Thus, for this study, we use the DB/DT merged product (Quality flag = 1, 2, 3 over ocean, and Quality flag = 3 over land) from MxD04. Specifically, DB/DT merged product over ocean is indeed DT.

3. Retrieval Algorithm

3.1. Retrieval Principal and Sensitivity Analysis

Like the existing DT algorithm, our new algorithm is based on a lookup table (LUT) approach, meaning that it is attempting to match simulated TOA reflectance to the observed TOA reflectance. The best-match solution represents the AOD and other properties of the aerosol. Over the open ocean, the atmospheric signal tends to dominate that of clear water. Near coastlines, however, sediments and turbid waters can dominate the signal, making aerosol retrieval impossible. However, in longer wavelengths, such as 2.1 μm (at which the imaginary part of refractive index for liquid water is several of magnitude larger than that in the visible), the penetration depth is so small, that contribution from the water should be nearly negligible.

To test this assumption, we compare the sensor sensitivity at 2.1 μm at which water-leaving radiance is negligible regardless of water turbidity with that at shorter wavelength which has large water-leaving radiance over turbid coastal water (taking 0.65 μm as an example). Figure 2 shows 2.1 μm TOA reflectance has better sensitivity to aerosol than 0.65 μm in both fine and coarse mode aerosol situations when 0.65 μm surface reflectance is large (turbid coastal water). When AOD is small (less than 0.15 at 0.55 μm, Figure 2b), the TOA reflectance (Figure 2c,d) is nearly the surface reflectance. Surface reflectance of turbid coastal water is up to 0.4 at 0.65 μm (Figure 2c) while it is very small (less than 0.0035) at 2.1 μm (Figure 2d). Figure 2e–h present a simulation of TOA reflectance through UNL-VRTM model [17] with various surface reflectances in fine aerosol (average TOA reflectance of the four fine aerosol modes defined in the DT-ocean LUT) and coarse aerosol (average TOA reflectance of the coarse aerosol modes defined in the DT-ocean LUT) situations. In a fine aerosol situation, 0.65 μm TOA reflectance is an increasing function of AOD when surface reflectance is small (0.03, clear water) (Figure 2e). However, it first decreases and then increases slowly when surface reflectance is large (0.3 or 0.4, turbid coastal water) (Figure 2e). The gradient of 2.1 μm MODIS digital signal (defined as change of MODIS digital count, $dn**$, with respect to 0.55 μm AOD, or $\partial(dn**)_{2.1}/\partial(AOD)_{0.55}$) is ~25 regardless of AOD values (Figure 2j) while $\partial(dn**)_{0.65}/\partial(AOD)_{0.55}$ is smaller than 25 when AOD is less than 0.5 (1.2) and surface reflectance is 0.3 (0.4). This means that it is better to use TOA reflectance at 2.1 μm rather than 0.65 μm to retrieve fine AOD in turbid coastal water situation. Additionally, while the sensitivity of digital signal at 2.1 μm to the AOD change is a factor of 3–8 smaller than the counterpart at 0.65 μm (e.g., contrasts of blue lines between Figure 2i,j) in the clear water situation, 2.1 μm still has significant sensitivity to the change of fine-mode AOD, and its detection limit for fine-mode AOD is ~0.04 (i.e., the inverse of $\partial(dn**)_{2.1}/\partial(AOD)_{0.55}$). The reason that 2.1 μm still has reasonably good sensitivity for fine-mode aerosols is because the ocean surface is nearly black at 2.1 μm, albeit fine aerosol extinction decrease significantly.

In a coarse aerosol situation, $\partial(dn**)_{0.65}/\partial(AOD)_{0.55}$ is similar to that in a fine aerosol situation and is a little smaller than $\partial(dn**)_{2.1}/\partial(AOD)_{0.55}$ when 0.65 μm surface reflectance is small (Figure 2k,l). However, $\partial(dn**)_{0.65}/\partial(AOD)_{0.55}$ is a factor of 6–10 smaller than $\partial(dn**)_{2.1}/\partial(AOD)_{0.55}$ when water is turbid (e.g., contrast of red/green curves between Figure 2k,l). Overall, over turbid water, it is better to use TOA reflectance at 2.1 μm rather than 0.65 μm to retrieve AOD, regardless of whether fine or coarse mode dominates. In other words, 2.1 μm should be used to retrieve coarse-mode AOD regardless of water turbidity.

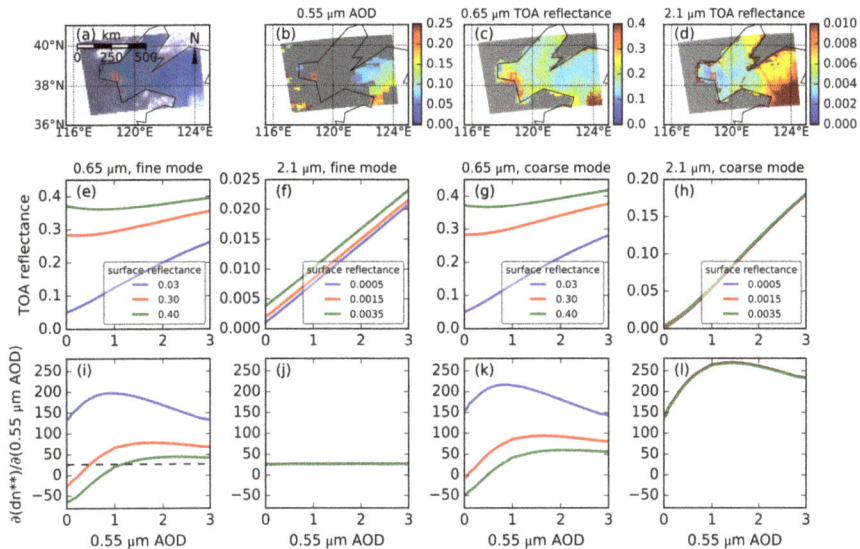

Figure 2. (a–d) are MODIS Aqua true color image, 0.55 μm AOD, 0.65 μm TOA reflectance, and 2.1 μm TOA reflectance at Bohai Sea, China on 23 February 2017, respectively. Red box is the region where water is turbid and atmosphere is clean. (e,f) are simulation (solar zenith angle is 24°, view zenith angle 54°, and relative azimuth angle 60°) of 0.65 μm and 2.1 μm TOA reflectance with fine aerosol model, respectively. (g,h) are similar to (e,f) but for coarse aerosol model. (i,j) are gradient of 0.65 μm and 2.1 μm MODIS digital signal (dn**) with respect to 0.55 μm AOD with fine aerosol model. (k,l) are similar to (i,j) but for coarse aerosol model. Values in the legends are surface reflectance used to simulate TOA reflectance at corresponding wavelength.

3.2. Algorithm Implementation and Steps

The retrieval algorithm is designed to supplement the existing DT algorithm. Hence, the lookup table (LUT) from the MODIS DT-ocean AOD retrieval algorithm is also used here [4,9]. That LUT is created by using Ahmad and Fraser's [18] radiative transfer code, assuming aerosol optical properties from four "fine" (effective radius < 1 μm) and five "coarse" (effective radius > 1 μm) lognormal mode. The fine modes include two water-insoluble and two soluble modes, whereas the coarse modes are separated into three soluble sea-salt like, and two insoluble dust-like modes (http://darktarget.gsfc. nasa.gov/algorithm/ocean/aerosol-models). This LUT is defined at various 0.55 μm AOD values in the range of 0 to 3 and at different Sun-Earth-satellite geometries [4,9]. Furthermore, as water surface reflectance depends on surface wind speed, the LUT is constructed at four wind speeds, specifically $2\,\mathrm{m\,s^{-1}}, 6\,\mathrm{m\,s^{-1}}, 10\,\mathrm{m\,s^{-1}}$, and $14\,\mathrm{m\,s^{-1}}$.

Like the DT algorithm, combination of a fine mode and a coarse mode is required for AOD retrieval. In the retrieval procedure, MxD02 TOA reflectance at 2.1 μm is used to fit the corresponding LUT value $\rho_{2.1}^{LUT}\left(\tau_{0.55}^{tot}\right)$ which is a weighted sum of pure fine mode LUT value $\rho_{2.1}^{f}\left(\tau_{0.55}^{tot}\right)$ and pure coarse mode LUT $\rho_{2.1}^{c}\left(\tau_{0.55}^{tot}\right)$ at a given 0.55 μm AOD value $\tau_{0.55}^{tot}$.

$$\rho_{2.1}^{LUT}\left(\tau_{0.55}^{tot}\right) = \eta\rho_{2.1}^{f}\left(\tau_{0.55}^{tot}\right) + (1-\eta)\rho_{2.1}^{c}\left(\tau_{0.55}^{tot}\right) \tag{1}$$

The value of reflectance weighting parameter η equals the fraction of total AOD at 0.55 μm contributed by the fine mode [9]. In addition, the LUT is calculated under the assumption of "gas free" (gas absorption is not included), thus correction [4] of MxD02 TOA reflectance is required to match the LUT before retrieving AOD.

Figure 3 shows the steps of our algorithm to retrieve AOD over coastal water, and hereafter, we call this algorithm the Coastal Water (CW) algorithm. Each step of CW is described below with a note that these steps essentially follow DT-Ocean algorithm except CW uses the 2.1 μm to retrieve AOD for those cloud-free turbid coastal water scenes. Hence, for each valid cloud-free turbid coastal water scene (i.e., a box of 20 × 20 pixels at 500 m resolution or at 10 km resolution at nadir, as available in the standard MODIS aerosol product), the following steps are implemented.

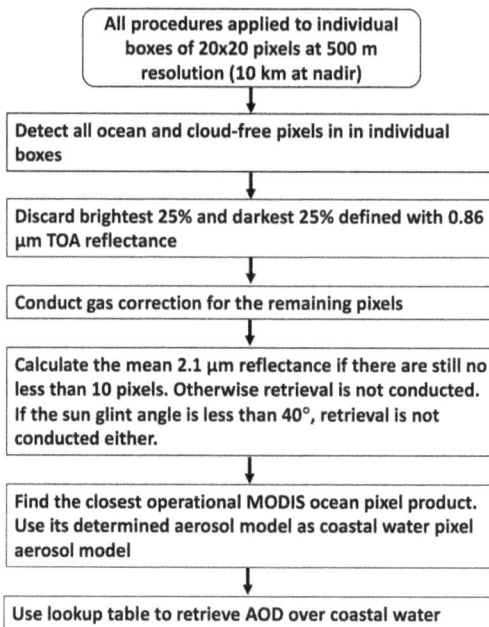

All procedures applied to individual boxes of 20x20 pixels at 500 m resolution (10 km at nadir)

↓

Detect all ocean and cloud-free pixels in in individual boxes

↓

Discard brightest 25% and darkest 25% defined with 0.86 μm TOA reflectance

↓

Conduct gas correction for the remaining pixels

↓

Calculate the mean 2.1 μm reflectance if there are still no less than 10 pixels. Otherwise retrieval is not conducted. If the sun glint angle is less than 40°, retrieval is not conducted either.

↓

Find the closest operational MODIS ocean pixel product. Use its determined aerosol model as coastal water pixel aerosol model

↓

Use lookup table to retrieve AOD over coastal water

Figure 3. Flowchart of retrieving AOD over coastal water.

1. Collect and organize 20 × 20 pixels at 500 m resolution, remove pixels that are defined by land/sea mask as "land", designated by ice/snow mask to be "ice", designated by the cloud mask to be "cloud", or removed by other tests.
2. Discard the brightest 25% and darkest 25% pixels defined with 0.86 μm reflectance.
3. Conduct gas (H_2O, CO_2, and O_3) correction [4] for the remaining pixels.
4. Calculate the mean 2.1 μm TOA reflectance if there are still no less than 10 pixels. Otherwise retrieval is not conducted. Calculate the sun glint angle [9]. If the sun glint angle is less than 40°, the retrieval is not conducted.
5. Prescribe single scattering properties of the aerosol. By only using 2.1 μm reflectance to retrieve AOD, there is no sensitivity to aerosol optical properties. Figure 4 shows that AOD can differ up to 0.2 in 100 km from the coast, but FMF differs only by 0.08 in 100 km from the coast. Thus, we assume the single scattering properties (including FMF) and surface wind speed for a turbid coastal water pixel is the same as those used for the AOD retrieval by the standard MODIS algorithm over its closest open-ocean pixel (within 100 km radius). The assumption that aerosol type does not change over moderate spatial scale is reasonable and was used in the atmospheric correction of SeaWiFS imagery over turbid coastal waters [19].
6. Use the mean 2.1 μm TOA reflectance and lookup table determined by Equation (1) to retrieve AOD over the turbid coastal water where MxD04 product is unavailable. In application of Equation (1), all ancillary information (aerosol mode selection and FMF) is obtained from step 5.

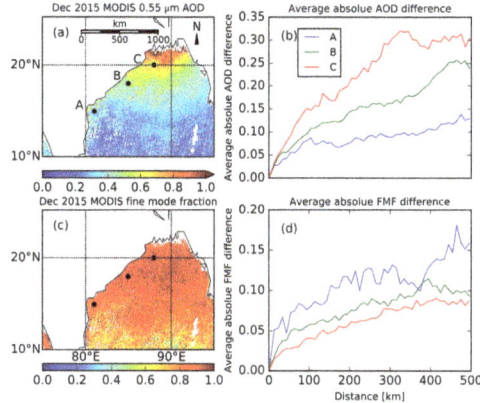

Figure 4. (**a**) is monthly mean 0.55 μm AOD and fine mode fraction (FMF) from MOD04 (Terra) in December, 2015. (**b**) is average absolute AOD difference as a function a distance with respect to reference points (black solid circles in (**a**)). (**c**) is monthly mean 0.55 μm AOD and fine mode fraction (FMF) from MOD04 (Terra) in December, 2015. (**d**) is similar to (**b**), but for FMF.

4. Results

Figure 5 shows the example of CW applied to an Aqua-observed scene (10 December 2015) over the Bay of Bengal. According to the standard DT retrieval (MYD04), AOD reaches 1.6 in the center of the granule and decreases gradually southward. MYD04 does not provide retrievals over the coastal water at north part of the Bay of Bengal (Figure 5c). This non-retrieval region is in cloud-free conditions (Figure 5a), so AOD was not retrieved due to turbid water or underlying sea-floor. Figure 5b shows non-retrievals are in 36.4% of this cloud-free, coastal area (within 5 km of coastline or with water depth less than 50 m).

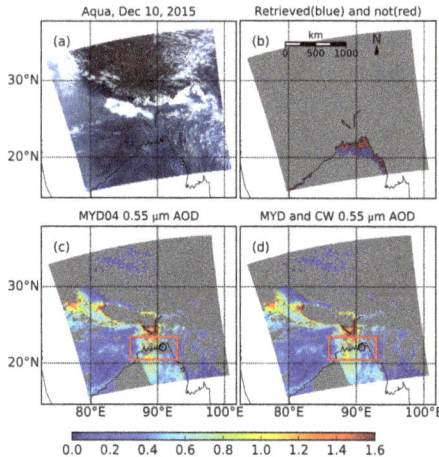

Figure 5. (**a**) Aqua MODIS true color image on 10 December 2015. (**b**) Blue and red represents coastal water where MYD AOD retrievals are available (blue) and unavailable (red), respectively in clear sky condition. (**c**) MYD04 0.55 μm DT/DB merged AOD retrievals. (**d**) is similar to (**c**), but the coastal water gaps where MYD04 AOD retrievals are not available in clear sky condition (red box) are filled by AOD retrievals from CW algorithm. AERONET AOD measurement overlaps on (**c,d**).

The CW algorithm enhances DT-ocean, by retrieving AOD over those turbid water conditions (Figure 5d). These "new" retrievals are consistent, in that there appears to be smooth transition in retrieved AOD (e.g., land → CW → ocean). In addition, the new CW AOD retrievals compare well with corresponding values observed by AERONET (overlaid as filled circles in Figure 5c,d).

To validate, we follow the standard protocol, in that AERONET measurements within ±30 min of the MODIS overpass time are averaged and compared against MODIS retrievals averaged within a 25-km radius circular region centered over the AERONET site. This means that there are three situations: MxD04 retrievals only, CW retrievals only, and cases where both are available within the 25-km of the AERONET site.

AOD (at 0.55 μm) retrievals from MxD04 product and CW algorithm are evaluated against AERONET data (as shown in Figure 6). We use a Venn diagram in Figure 6 to represent how MODIS AOD retrievals (MxD04 and/or CW) and AERONET observations are collocated. Figure 6a is a scatter plot of MxD04 retrievals versus AERONET measurements in the cases that have retrievals only from MxD04, and these cases can be divided into two categories: (a) all cloud-free pixels within 25-km radius proximity of AERONET site are retrieved through MODIS operational algorithm or (b) only part of cloud-free pixels are retrieved through MODIS DT algorithm, but the AOD of the rest of the cloud-free pixels cannot be retrieved with the CW algorithm because these pixels are over turbid water and no open-ocean pixel is close enough (i.e., within 100 km) to provide aerosol single scattering property for CW algorithm. Figure 6b is a scatter plot of CW retrievals versus AERONET measurements in the cases where all cloud-free pixels within 25-km radius proximity of AERONET site cannot be retrieved through MODIS DT algorithm but some or all of these cloud-free pixels can be retrieved through CW algorithm. Comparison of Figure 6a,b shows that normalized mean bias (NMB, $NMB = \sum_{i=1}^{N} ((\tau_i^{MODIS} - \tau_i^{AEROENT})/\tau_i^{AEROENT})/N$) and root mean square error (RMSE) of CW algorithm (12.0% and 0.141) are smaller than that of MxD04 (15.9% and 0.213). The percentage of collocated pairs within expected error envelope (+(0.04 + 10%), −(0.02 + 10%), asymmetric) and correlation coefficient increase from 35.5% and 0.72 in MxD04 only situation (Figure 6a) to 39.3% and 0.94 in CW only situation (Figure 6b), respectively.

In addition to the situations where only MxD04 or only CW retrievals are available within the 25-km radius proximity of AERONET site, there is a third one where some of the cloud-free pixels are retrieved through the operational MODIS algorithm while some are retrieved through the CW algorithm. Figure 6c,d are similar to Figure 6a,b, respectively, but both include the AOD retrievals from their respective counterparts of this third situation. Overall, for all possible retrievals by each algorithm, CW retrievals are comparable to MxD04 in quality, although the MODIS operational algorithm has slightly more samples.

The basis of MODIS-AERONET collocation is that air masses are always in motion and the average of MODIS AOD retrievals in a certain area which encompass an AERONET site should be comparable to the temporal statistics of the AERONET measurements [16]. In the situation that both MxD04 and CW retrievals are available within the 25-km radius proximity of AERONET site, if only MxD04 or CW retrievals are collocated with AERONET observation, it will be biased comparison. Therefore, we further show the inter-comparison between either MxD04 or CW AOD with AERONET AOD at this situation (Figure 6f–h). The percentage of collocated pairs in the EE for (MxD04 + CW) combined vs. AERONET (53.1%, Figure 6h) is larger than that of MxD04-AERONET (49.2%, Figure 6f) and CW-AERONET (40.6%, Figure 6g). The RMSE of (MxD04 + CW)-AERONET (0.110) is smaller than that of MxD04-AERONET (0.114) and CW-AERONET (0.148).

Inter-comparison of MxD04 and CW merged AOD with AERONET AOD is shown in Figure 6e; such merged AOD include all data points in Figure 6a,b,h. The number of collocated pairs increases from 190 (62 in MxD04 only situation plus 128 in the situation that both MxD04 and CW retrievals are available) to 218 (28 in CW only situation are added). In addition, the inter-comparison statistics (Figure 6e) is still comparable to or better than those retrieved from MxD04 (Figure 6a,c) in quality.

Overall, CW retrievals supplement MODIS DT, and improve the AOD retrievals both spatially and temporally without degrading (and sometimes increasing) the DT-Ocean AOD retrieval quality.

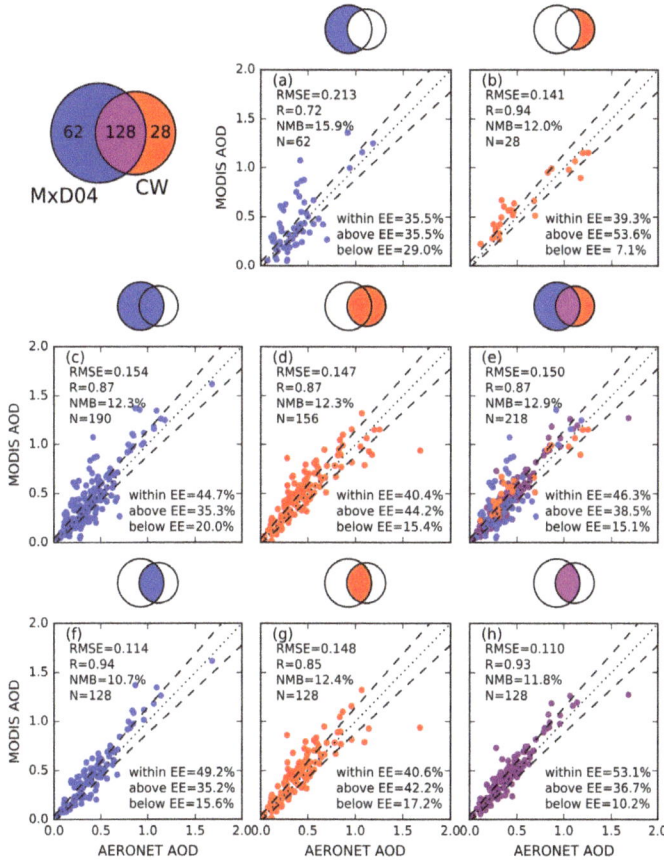

Figure 6. Scatter plots of 0.55 μm AOD retrievals from MODIS (MxD04 and/or CW) versus AERONET observations of 6 sites in one month. Venn diagram on the upper left represents number of collocated data sets for MxD04-AERONET (blue), CW-AERONET (red), (MxD04 + CW)-AERONET (purple). Color part of Venn diagram on the top of each scatter plot (**a–h**) represents which collocated data sets are plotted; see details in the text for further description of the Venn diagram. (**a**) is MxD04-AERONET in the situation that only MxD04 retrievals are available within the 25-km of AERONET site. (**b**) is CW-AERONET in the situation that only CW retrievals are available. (**c,d**) are similar to (**a,b**), respectively, but both additionally include AOD retrievals from their respective counterparts of the situation that both MxD04 and CW retrievals are available. (**e**) is (MxD04 + CW)-AERONET in all the three situations. (**f–h**) are MxD04-AERONET, CW-AERONET, (MxD04 + CW)-AERONET, respectively, in the situation that both MxD04 and CW retrievals are available. 1:1 lines and expected error (EE) envelopes (+(0.04 + 10%), −(0.02 + 10%), asymmetric) are plotted as dot and dashed lines. The number of collocated pairs (N), root mean square error (RMSE), normalized mean bias (NMB), and linear correlation coefficients are also shown.

In addition to evaluation of MODIS AOD diagnostic error in Figure 6, prognostic error is presented in Figure 7. Two thirds of AERONET-MODIS collocations are within the expected error envelope (y = ±(0.05 + 0.15x)) (Figure 7c), thus 0.05 + 0.15 (MODIS AOD) can be considered as its prognostic error. Figure 7d–f show RMSE as a function of MODIS; RMSE goes up as MODIS AOD increases and combined retrievals (Figure 7f) are comparable to MxD04 (Figure 7d).

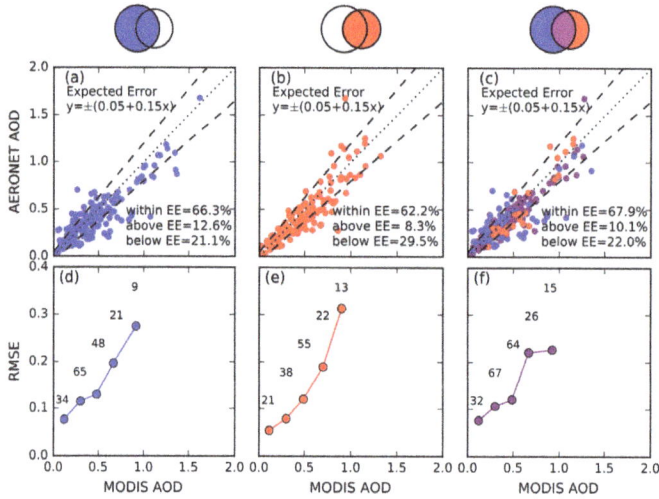

Figure 7. (**a**–**c**) are scatter plots of AERONET 0.55 µm AOD observations versus retrievals from MODIS (MxD04 and/or CW) for the 6 monthly cases. Sampling method is same as in Figure 6. 1:1 lines and expected error (EE) envelopes (+(0.05 + 15%), −(0.05 + 15%), symmetric) are plotted as dotted and dashed lines. (**d**–**f**) are root mean square error (RMSE) as a function of MODIS AOD in the 0.2 interval. Numbers above the averaged dot show how many AERONET-MODIS collocations are available to calculate RMSE.

5. Discussion

The CW algorithm uses 2.1 µm to retrieve AOD over turbid coastal water as water-leaving radiance is negligible at that band. As only one band is used in the algorithm, aerosol single scattering properties need to be prescribed. We assume the single scattering properties for a turbid coastal water pixel are the same as those used for the AOD retrieval by the standard MODIS algorithm over its closest open-ocean pixel (within 100 km radius). When no pixel is within 100 km, we do not conduct retrieval. Further evaluation can be targeted at the assumptions regarding the use of aerosol single scattering properties from adjacent non-turbid water pixels within certain threshold distance.

The 6 AERONET sites used for this analysis were located in polluted regions. We expect to evaluate how the algorithm performs in an unpolluted zone if there is a new AERONET site or other field campaign that satisfies all the conditions: (1) it is close to coast line; (2) coastal water is turbid; and (3) the region is unpolluted.

The CW algorithm potentially can also be used for the other aerosol retrieval algorithm over the ocean such as the SeaWiFS Ocean Aerosol Retrieval (SOAR) algorithm developed by Sayer et al. [20]. In SOAR, TOA reflectance measured by the SeaWiFS bands centered near 0.51 µm, 0.67 µm, and 0.86 µm are used to retrieve AOD regardless of water turbidity. Hence, as pointed out by Sayer et al. [20], the performance of SOAR retrieval is expected to be poorer in turbidity cases and a double of TOA reflectance as a result of water turbidity causes a positive error in retrieved 0.55 µm AOD of 0.25 [20]. Indeed, in the evaluation of SOAR algorithm, AOD retrievals that are proximity to coast are not included [20].

As MODIS sensors aboard on Terra and Aqua have been providing data for more than one decade and will be decommissioned by the early 2020s, its aerosol record is expected to be continued by the Visible Infrared Imaging Radiometer Suite (VIIRS) on board Suomi-NPP (S-NPP) which was launched in late 2011. Thus, future investigation will also include the application of the approach in this study to VIIRS. VIIRS has 2.26 µm band at which water leaving radiance is also negligible like 2.1 µm band on MODIS and thus the method can be likewise applied.

6. Conclusions

The MODIS Dark Target (DT) algorithm has been applied to retrieve AOD over land and ocean since early 2000. Although there have been significant improvements to the algorithm in the past decade, it has not yet been able to retrieve AOD over turbid coastal water due to its high water-leaving radiance from the ocean bottom or water color. We designed the Coastal Water AOD retrieval algorithm (CW algorithm) for these regions to supplement the current DT algorithm. The AOD retrieval algorithm for turbid coastal water takes advantage of the fact that water-leaving radiance is negligible at 2.1 µm, hence this band is only used to retrieve while other auxiliary information such as aerosol single scattering properties are obtained from DT retrievals over the nearby non-turbid water surfaces. In other words, the aerosol modes and reflectance weighting parameter of the pixel to be retrieved is substituted by the closest counterpart from the MxD04 ocean AOD product.

The CW algorithm not only fills the gaps of MxD04 AOD retrievals over turbid coastal water but also improves their comparison with AERONET measurements. CW AOD retrievals are validated against measurements of six AERONET sites that are located at coastal regions. The new algorithm yields ~18% more of MODIS-AERONET collocated pairs, and CW AOD retrievals are comparable or better than MxD04 in quality; the RMSE of MxD04 and CW product is 0.154 and 0.147, respectively. In the situation that both AOD retrievals from MxD04 and CW are available within a 50 km diameter circular regions centered at AERONET sites, the merged AOD retrievals show better agreement with AERONET AOD either of them alone, in terms of the percentage of collocated pairs in the EE (49.2%, 40.6%, and 53.1% for MxD04, CW, and merged product, respectively) and RMSE (0.114, 0.148, 0.110 for MxD04, CW, and merged product, respectively). In addition, 0.05 + 0.15(MODIS AOD) can be considered as prognostic error of the merged product.

Acknowledgments: Funding for this study was provided by the NASA Earth Science Division as part of NASA's Terra/Aqua and VIIRS science team program (grant numbers: NNX16AT63G and NNX17AC94A) and NASA's GEO-CAPE and KORUS-AQ program (grant number: NNX16AT82G), and was also supported by Office of Naval Research (ONR's) Multidisciplinary University Research Initiatives (MURI) Program under the award N00014-16-1-2040.

Author Contributions: Yi Wang, Jun Wang and Robert C. Levy conceived and designed the experiments; Yi Wang and Xiaoguang Xu performed the experiments and analyzed the data; Yi Wang and Jun Wang wrote the paper; Jeffrey S. Reid provided guidance to the project, reviewed and edited the paper.

Conflicts of Interest: The authors declare no conflict of interest.

References

1. Haywood, J.; Boucher, O. Estimates of the direct and indirect radiative forcing due to tropospheric aerosols: A review. *Rev. Geophys.* **2000**, *38*, 513–543. [CrossRef]
2. Rosenfeld, D.; Lohmann, U.; Raga, G.B.; Dowd, C.D.; Kulmala, M.; Fuzzi, S.; Reissell, A.; Andreae, M.O. Flood or drought: How do aerosols affect precipitation? *Science* **2008**, *321*, 1309. [CrossRef] [PubMed]
3. Lim, S.S.; Vos, T.; Flaxman, A.D.; Danaei, G.; Shibuya, K.; Adair-Rohani, H.; AlMazroa, M.A.; Amann, M.; Anderson, H.R.; Andrews, K.G.; et al. A comparative risk assessment of burden of disease and injury attributable to 67 risk factors and risk factor clusters in 21 regions, 1990–2010: A systematic analysis for the global burden of disease study 2010. *Lancet* **2012**, *380*, 2224–2260. [CrossRef]
4. Levy, R.C.; Mattoo, S.; Munchak, L.A.; Remer, L.A.; Sayer, A.M.; Patadia, F.; Hsu, N.C. The collection 6 modis aerosol products over land and ocean. *Atmos. Meas. Tech.* **2013**, *6*, 2989–3034. [CrossRef]
5. Hsu, N.C.; Jeong, M.J.; Bettenhausen, C.; Sayer, A.M.; Hansell, R.; Seftor, C.S.; Huang, J.; Tsay, S.C. Enhanced deep blue aerosol retrieval algorithm: The second generation. *J. Geophys. Res.* **2013**, *118*, 9296–9315. [CrossRef]

6. Sayer, A.M.; Hsu, N.C.; Bettenhausen, C.; Jeong, M.J. Validation and uncertainty estimates for modis collection 6 "deep blue" aerosol data. *J. Geophys. Res.* **2013**, *118*, 7864–7872. [CrossRef]
7. IPCC. Contribution of working group I to the fifth assessment report of the intergovernmental panel on climate change. In *Climate Change 2013: The Physical Science Basis*; Cambridge University Press: Cambridge, UK; New York, NY, USA, 2013; p. 1535.
8. Li, R.-R.; Kaufman, Y.J.; Gao, B.-C.; Davis, C.O. Remote sensing of suspended sediments and shallow coastal waters. *IEEE Trans. Geosci. Remote Sens.* **2003**, *41*, 559–566.
9. Remer, L.A.; Kaufman, Y.J.; Tanré, D.; Mattoo, S.; Chu, D.A.; Martins, J.V.; Li, R.R.; Ichoku, C.; Levy, R.C.; Kleidman, R.G.; et al. The modis aerosol algorithm, products, and validation. *J. Atmos. Sci.* **2005**, *62*, 947–973. [CrossRef]
10. Anderson, J.C.; Wang, J.; Zeng, J.; Leptoukh, G.; Petrenko, M.; Ichoku, C.; Hu, C. Long-term statistical assessment of aqua-modis aerosol optical depth over coastal regions: Bias characteristics and uncertainty sources. *Tellus B* **2013**, *65*. [CrossRef]
11. Wolfe, R.E.; Nishihama, M.; Fleig, A.J.; Kuyper, J.A.; Roy, D.P.; Storey, J.C.; Patt, F.S. Achieving sub-pixel geolocation accuracy in support of modis land science. *Remote Sens. Environ.* **2002**, *83*, 31–49. [CrossRef]
12. King, M.D.; Menzel, W.P.; Kaufman, Y.J.; Tanre, D.; Bo-Cai, G.; Platnick, S.; Ackerman, S.A.; Remer, L.A.; Pincus, R.; Hubanks, P.A. Cloud and aerosol properties, precipitable water, and profiles of temperature and water vapor from modis. *IEEE Trans. Geosci. Remote Sens.* **2003**, *41*, 442–458. [CrossRef]
13. Smirnov, A.; Holben, B.N.; Eck, T.F.; Dubovik, O.; Slutsker, I. Cloud-screening and quality control algorithms for the aeronet database. *Remote Sens. Environ.* **2000**, *73*, 337–349. [CrossRef]
14. Holben, B.N.; Eck, T.F.; Slutsker, I.; Tanré, D.; Buis, J.P.; Setzer, A.; Vermote, E.; Reagan, J.A.; Kaufman, Y.J.; Nakajima, T.; et al. Aeronet—A federated instrument network and data archive for aerosol characterization. *Remote Sens. Environ.* **1998**, *66*, 1–16. [CrossRef]
15. O'Neill, N.T.; Eck, T.F.; Smirnov, A.; Holben, B.N.; Thulasiraman, S. Spectral discrimination of coarse and fine mode optical depth. *J. Geophys. Res.* **2003**, *108*, 4559. [CrossRef]
16. Ichoku, C.; Chu, D.A.; Mattoo, S.; Kaufman, Y.J.; Remer, L.A.; Tanré, D.; Slutsker, I.; Holben, B.N. A spatio-temporal approach for global validation and analysis of modis aerosol products. *Geophys. Res. Lett.* **2002**, *29*. [CrossRef]
17. Wang, J.; Xu, X.; Ding, S.; Zeng, J.; Spurr, R.; Liu, X.; Chance, K.; Mishchenko, M. A numerical testbed for remote sensing of aerosols, and its demonstration for evaluating retrieval synergy from a geostationary satellite constellation of geo-cape and goes-r. *J. Quant. Spectrosc. Radiat. Transf.* **2014**, *146*, 510–528. [CrossRef]
18. Ahmad, Z.; Fraser, R.S. An iterative radiative transfer code for ocean-atmosphere systems. *J. Atmos. Sci.* **1982**, *39*, 656–665. [CrossRef]
19. Hu, C.; Carder, K.L.; Muller-Karger, F.E. Atmospheric correction of seawifs imagery over turbid coastal waters: A practical method. *Remote Sens. Environ.* **2000**, *74*, 195–206. [CrossRef]
20. Sayer, A.M.; Hsu, N.C.; Bettenhausen, C.; Ahmad, Z.; Holben, B.N.; Smirnov, A.; Thomas, G.E.; Zhang, J. Seawifs ocean aerosol retrieval (soar): Algorithm, validation, and comparison with other data sets. *J. Geophys. Res. Atmos.* **2012**, *117*. [CrossRef]

remote sensing

MDPI

Article

How Do Aerosol Properties Affect the Temporal Variation of MODIS AOD Bias in Eastern China?

Minghui Tao [1], Zifeng Wang [1,*], Jinhua Tao [1,*], Liangfu Chen [1], Jun Wang [2], Can Hou [1], Lunche Wang [3], Xiaoguang Xu [2] and Hao Zhu [1]

[1]   State Key Laboratory of Remote Sensing Science, Institute of Remote Sensing and Digital Earth, Chinese Academy of Sciences, Beijing 100101, China; taomh@radi.ac.cn (M.T.); chenlf@radi.ac.cn (L.C.); wumagua@163.com (C.H.); zhuhao453271826@163.com (H.Z.)

[2]   Department of Chemical and Biochemical Engineering, Center of Global and Regional Environmental Research, University of Iowa, Iowa City, IA 52242, USA; jun-wang-1@uiowa.edu (J.W.); xiaoguang-xu@uiowa.edu (X.X.)

[3]   Laboratory of Critical Zone Evolution, School of Earth Sciences, China University of Geosciences, Wuhan 430074, China; wang@cug.edu.cn

*   Correspondence: wangzf@radia.ac.cn (Z.W.); taojh@radi.ac.cn (J.T.); Tel.: +86-010-6488-9567 (Z.W. & J.T.)

Academic Editors: Yang Liu and Omar Torres

Received: 21 June 2017; Accepted: 1 August 2017; Published: 3 August 2017

Abstract: The rapid changes of aerosol sources in eastern China during recent decades could bring considerable uncertainties for satellite retrieval algorithms that assume little spatiotemporal variation in aerosol single scattering properties (such as single scattering albedo (SSA) and the size distribution for fine-mode and coarse mode aerosols) in East Asia. Here, using ground-based observations in six AERONET sites, we characterize typical aerosol optical properties (including their spatiotemporal variation) in eastern China, and evaluate their impacts on Moderate Resolution Imaging Spectroradiometer (MODIS) Collection 6 aerosol retrieval bias. Both the SSA and fine-mode particle sizes increase from northern to southern China in winter, reflecting the effect of relative humidity on particle size. The SSA is ~0.95 in summer regardless of the AEROENT stations in eastern China, but decreases to 0.85 in polluted winter in northern China. The dominance of larger and highly scattering fine-mode particles in summer also leads to the weakest phase function in the backscattering direction. By focusing on the analysis of high aerosol optical depth (AOD) (>0.4) conditions, we find that the overestimation of the AOD in Dark Target (DT) retrieval is prevalent throughout the whole year, with the bias decreasing from northern China, characterized by a mixture of fine and coarse (dust) particles, to southern China, which is dominated by fine particles. In contrast, Deep Blue (DB) retrieval tends to overestimate the AOD only in fall and winter, and underestimates it in spring and summer. While the retrievals from both the DT and DB algorithms show a reasonable estimation of the fine-mode fraction of AOD, the retrieval bias cannot be attributed to the bias in the prescribed SSA alone, and is more due to the bias in the prescribed scattering phase function (or aerosol size distribution) in both algorithms. In addition, a large yearly change in aerosol single scattering properties leads to correspondingly obvious variations in the time series of MODIS AOD bias. Our results reveal that the aerosol single scattering properties in the MODIS algorithm are insufficient to describe a large variation of aerosol properties in eastern China (especially change of particle size), and can be further improved by using newer AERONET data.

Keywords: aerosol; MODIS; retrieval bias; eastern China; AERONET

1. Introduction

Atmospheric aerosols play a vital role in regional climate by redistributing solar radiation in the Earth-atmosphere system and modifying cloud properties [1]. Furthermore, fine particles near the surface can cause air pollution, which has robust relation with epidemic diseases affecting human health [2]. Unlike long-lived greenhouse gases such as CO_2, the amount and properties of aerosol particles vary largely over space and time due to diverse emission sources and short lifetime, which make it a challenge to quantify the magnitude of aerosols and their climate effects [3]. Since the 1990s, a ground-based remote sensing network has been established to explore aerosol optical and microphysical properties [4]. Meanwhile, a variety of sophisticated sensors, such as Moderate Resolution Imaging Spectroradiometer (MODIS), Multi-angle Imaging SpectroRadiometer (MISR), Ozone Monitoring Instrument (OMI), CALIPSO, and other sensors have been launched to obtain aerosol information with global coverage [5–8].

Aerosol optical properties, including aerosol optical depth (AOD) and aerosol single scattering properties (single scattering albedo (SSA) and phase function) regulate the role of aerosols in radiative transfer calculations; these properties in turn change with mass, size, shape, and the composition of the particles [9]. Since satellite spectral radiances at the top of the atmosphere (TOA) are affected by the radiative interactions between surface reflectance and aerosol scattering, not all aerosol properties can be fully constrained and retrieved reliably from satellite measurements at the TOA. Hence, aerosol single scattering properties (such as SSA and size distribution for fine or coarse particles that affect phase function) are often derived from a cluster analysis of ground observations, and are subsequently used in the algorithms for the satellite remote sensing of aerosols. For example, the MODIS aerosol algorithm only retrieves the AOD and the fraction between fine-mode and coarse mode AODs [10,11], while aerosol single scattering properties are fixed with only consideration of seasonal and continental-scale variation.

The aerosol properties in eastern China are characterized by large spatiotemporal variations [12,13], which are not considered in the MODIS retrieval algorithm, and hence the impact of such variations on MODIS AOD biases deserve a dedicated investigation. Widespread haze pollution usually occurs over eastern China, with the mixing of anthropogenic emissions, natural dust, and biomass burning smoke [14,15]. Ground observations show a distinct temporal variation and area-dependent change of optical and microphysical properties of aerosol particles [16–19], which resonates with our recent evaluation of MODIS Collection (C) 6 aerosol retrievals in China that show considerable bias with obvious geographic difference and temporal trend [20]. However, previous ground-based validations of satellite retrievals usually focus on an evaluation of the AOD, with less attention paid to understanding error sources from the variation of aerosol single scattering properties in real atmosphere [21,22]. It was found that the temporal variations of aerosol scattering properties in regions such as southern Africa caused a seasonal shift of MODIS AOD biases [23]. To date, how the change of aerosol properties in China influences MODIS AOD bias at different spatial and temporal scales remains unclear, and is the focus of this study.

We conduct a comprehensive analysis of aerosol single scattering properties in eastern China, and examine their impacts on biases in MODIS aerosol retrievals. Section 2 introduces the Aerosol Robotic Network (AERONET) and MODIS aerosol data sets used. Monthly variations of typical aerosol properties in six sites of eastern China are described in Section 3.1. Section 3.2 investigates the monthly bias of MODIS AOD and its connection with aerosol properties. The influence of decadal variations in aerosol properties on satellite retrievals is discussed specially in Section 3.3. The main purpose of this work is to present a reference for aerosol model assumptions in satellite retrievals over China.

2. Data and Methods

2.1. AERONET Aerosol Data Sets

The AERONET is a worldwide network of sun photometers that provides continuous observations and inversions of aerosol optical and microphysical properties [4]. By measuring the direct solar irradiance, AOD can be acquired with a high accuracy of 0.01–0.02, which is usually taken as the "true" value in ground validation. With skylight observations, microphysical and single scattering parameters such as volume size distribution, SSA, and phase function can be retrieved [9]. The retrieval error of the size distribution is within 10% of the maximum value in the median particle size range (0.1–7 μm). To obtain sufficient information, the SSA has to be retrieved in high aerosol loading (AOD 440 nm >0.4) and solar zenith angle (>50°) with an accuracy of ~0.03.

Aerosol data from six AERONET sites in typical regions were selected to investigate variations in aerosol optical properties in eastern China (Figure 1). There are nearly 15 years of continuous observations from the Beijing site since 2001, and 10 years of observations from the Xianghe, Taihu, and Hongkong sites from 2005 at Level 2.0 (cloud-screened and quality-assured), respectively. Observation over 4 years is available in Xuzhou from 2013 at Level 1.5 (cloud-screened). More than one year's observation exists in Qiandaohu at Level 1.5 from 2007 to 2009, but the inversions are much fewer, which is mainly used for reference here. The monthly mean values of aerosol microphysical and optical parameters are analyzed to show the temporal variations of the aerosol scattering properties. Outliers caused by few available inversions during the cloudy season are examined and removed. To match with the satellite data, the spatial average of MODIS AOD in 5 × 5 pixel around the ground site is compared with the temporal mean values of AERONET inversions within ±30 min of the satellite's passing time [24]. Since there is no 550 nm band in the sun photometers, we interpolate the AOD from the nearest bands on the two sides of 550 nm with their Ångström exponent.

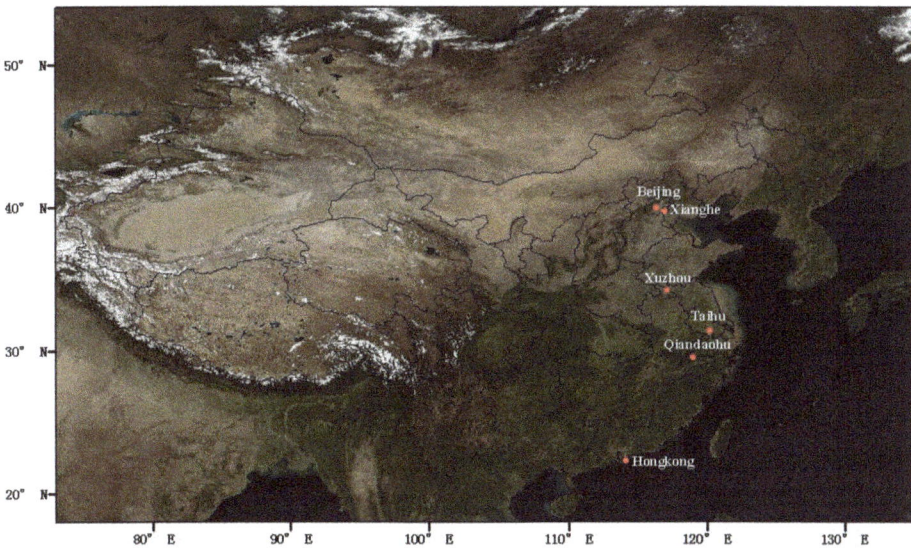

Figure 1. Geographic location of the six typical Aerosol Robotic Network (AERONET) sites in eastern China. Beijing: megacity site, Xianghe: background site among megacities, Xuzhou: industrial city site in hinterland of northern China, Taihu: background site of the Yangtze River Delta, Qiandaohu: background site of southern China, Hongkong: coastal megacity site in southern China.

2.2. MODIS Aerosol Data Sets

The MODIS sensor onboard the Terra satellite from 2000 and the Aqua satellite from 2002 provides daily global detection of reflected and emitted radiance from the Earth-atmospheric system with a broad spectrum range of 0.4–14.4 μm, fine spatial resolution at 250–1000 m, and swath width of ~2330 km. MODIS aerosol retrieval over land employs the Dark Target (DT) and Deep Blue (DB) algorithms over dense vegetation and bright surfaces, such as urban and deserts, respectively. In recent MODIS C6 aerosol products, DB retrieval has been expanded to all cloud-free and snow/ice-free regions, and a merged AOD data product combining DT and DB retrievals developed over dense vegetation regions [25]. To raise the signal-to-noise ratio and minimize cloud contamination, DT retrieval is performed over an area of 20 × 20 500 m pixels or at a nominal spatial resolution of AOD at 10 × 10 km. By contrast, DB aerosol properties are retrieved at 1 km resolution and then averaged to 10 × 10 km scale. Since the sensor degradation of the MODIS aboard Terra is much larger than that of Aqua, only aerosol data from the Aqua MODIS is used here [26].

The DT algorithm retrieves aerosol properties over dense vegetation utilizing a linear relationship in surface reflectance between visible and shortwave infrared bands [11]. With weighted fine and coarse aerosol models (that describe aerosol single scattering properties), the AOD and fine-mode AOD fraction are retrieved by matching the calculated apparent reflectance with satellite TOA spectral reflectance in 0.47, 0.67, and 2.1 μm. The surface reflectance relationship between 0.67 and 2.1 μm varies with view geometry and surface type. One coarse aerosol model and one of three fine aerosol models are employed in DT retrieval (Table 1), and the fine aerosol model is prescribed by location and season [27]. As a by-product, surface reflectance in 2.12 μm is also retrieved. Global validation shows that more than 66% of the DT retrievals are within the error envelope of ±(0.05 + 15% AOD) [25].

To retrieve the AOD over a bright surface, the DB algorithm utilizes a pre-calculated surface reflectance database in blue channels, where the surface reflectance is much lower than in longer channels [10]. The surface reflectance database in the C6 DB algorithm has been improved as a function of season, normalized difference vegetation index (NDVI), and scattering angle [28]. Similarly to DT, the DB algorithm divides the globe into several regions, and assumes fixed aerosol types in certain areas such as dust and smoke models in early retrievals over East Asia (Table 1). In cases of mixed aerosol types, two unknowns, the AOD and the Ångström exponent (and hence, the AOD fraction of aerosol types) are retrieved from TOA radiance in 0.41 and 0.47 μm. For heavy dust loading conditions, the radiance in the red band at 0.67 μm is added to further retrieve the SSA in the blue bands. The global bias of DB AOD is approximately within ±(0.03 + 0.2 AOD_{MODIS}) [29].

Table 1. List of typical optical parameters in aerosol models used in MODIS Dark Target (DT) and Deep Blue (DB) algorithms [10,25]. The DT aerosol model varies with aerosol loading and the case of AOD_{550} = 0.5 is presented here; R denotes the effective radius in the DT model, and for DB it is the mean radius; since information of Collection 6 DB aerosol model is not available, the original version is used for reference. SSA, single scattering albedo.

Aerosol Model	Algorithm	SSA, 412 nm	SSA, 470 nm	SSA, 660 nm	R, μm	Standard Deviation, μm
Dust/Spheroid	DT	-	0.94	0.96	0.68	-
Absorbing/Smoke	DT	-	0.88	0.85	0.256	-
Moderately Absorbing	DT	-	0.93	0.91	0.261	-
Nonabsorbing/Urban-Industrial	DT	-	0.95	0.94	0.207	-
Dust/Spheroid	DB	0.91	0.96	-	1.0	1.45
Smoke	DB	0.90	0.89	-	0.14	1.45
"whiter" Dust/Spheroid	DB	0.98	0.99	≈1.0	-	-
"redder" Dust/Spheroid	DB	0.91	0.94	≈1.0	-	-

3. Results and Analysis

3.1. Aerosol Optical Properties in Eastern China

The angular distribution of scattered light largely depends on particle size, which associates closely with emission sources [9]. Figure 2 shows the variations of monthly volume size distribution in typical regions of eastern China. There is a notable bimodal distribution in all the sites, with a large spatial and temporal difference. Coarse particles are dominant in northern China (Beijing, Xianghe, and Xuzhou) and the Yangtze River Delta (YRD, Taihu) during spring. Although there are few dust storm events in eastern China every year, floating dust transported from the deserts of East Asia is prevalent over northern China [14]. The vertical distribution of the floating dust particles is very inhomogeneous [30]. Hence, the fraction of coarse particles may not be considerable in ground sampling, but can be significant in column observations. The contribution of coarse particles also shows large monthly variations even in Qiandaohu in the north part of southern China, and in some months, fine and coarse particles have a nearly equivalent volume. By contrast, fine particles play a major role in July and August, modulated mainly by anthropogenic emissions and photochemical processes. It should be noted that the volume of coarse particles is similar to that of fine aerosols during fall and winter in northern China, which indicates the common mixing of dust and anthropogenic pollutants. While Hongkong is dominated by fine aerosol all year round, the fraction of coarse aerosols is still significant (~30%), and has much less monthly or seasonal variation, suggesting the influence of sea salt. Indeed, the fraction of coarse particles in Hongkong is higher than its counterpart in the continental background site of Qiandaohu during September. Similar observations also exist in other coastal sites, such as Zhoushan in the YRD [17].

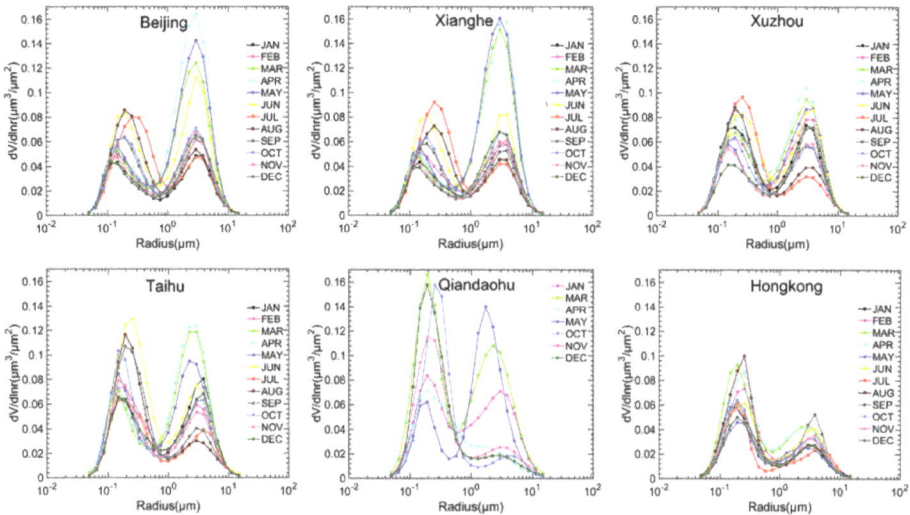

Figure 2. Monthly mean value of volume size distribution in AERONET sites of Beijing, Xianghe, Xuzhou, Taihu, Qiandaohu, and Hongkong, respectively.

Aerosol SSA exhibits large temporal variations in eastern China (Figure 3). The SSA values at 440 nm in northern China range from ~0.85 in the winter season (that uses coal-burning for heating) to ~0.95 in humid summer. The SSA increases gradually from January, reaches its maximum value in July, and then begins to decrease. In southern China (Qiandaohu and Hongkong), where no coal-burning is needed for heating in winter, the temporal change of the SSA is smaller (~0.90–0.95), but there is also an obvious increase in the SSA during summer associated with enhanced hygroscopic growth in humid conditions. It is worth noting that the SSA in Qiandaohu may be not typical due to only few available inversion data from AERONET in most months. Overall, for most sites, the variations of the SSA show a larger dependence on wavelength in winter and spring but a smaller dependence in summer and fall, which can caused by a mixture of anthropogenic aerosols and coarse dust in winter and spring [31]. Large daily variations in the SSA also exist in Beijing and Hongkong, where regional transport usually superposes on local pollution.

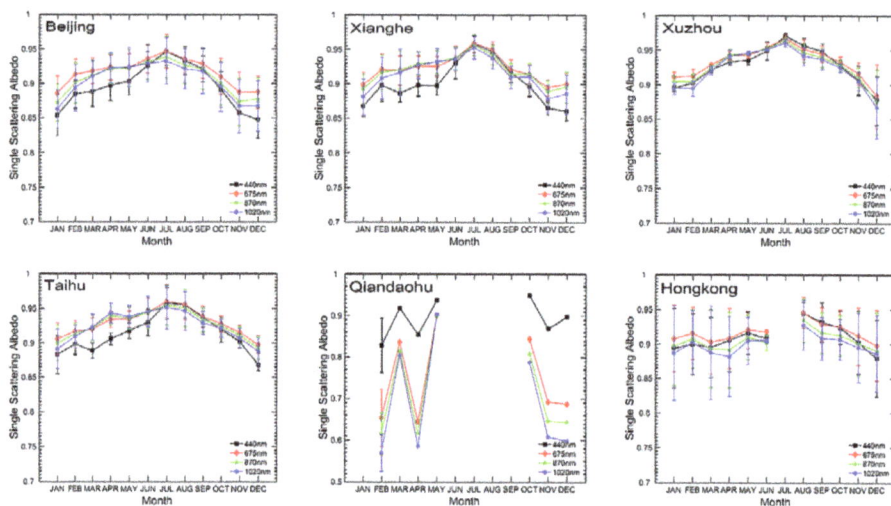

Figure 3. Monthly mean values of single scattering albedos in eastern China. The monthly values with available daily values of no more than 3 days every year were deleted.

Corresponding to the temporal variation of size distribution and SSA, and relevant to the satellite retrieval algorithm that uses backscattered solar radiation, Figure 4 shows the variation of phase function at backward direction over different AEROENT sites. In northern China, the monthly phase function exhibits approximately 20–30% of its variation at around 90–150° in the back-scattering directions. It can be seen that the phase function is lowest in summer (red lines) with fine particles dominant and highest in spring (blue lines) with prevalent dust, reflecting the change of particle size. Indeed, the composite phase function for fine-mode and coarse-mode aerosols in an atmospheric column depends not only on the size distributions for both modes, but also on the scattering optical depth of both modes and the relative weighting. Fine-mode particles have a relatively larger size (due to efficient hygroscopic growth, Figure 2) in summer than in winter, and hence a relatively smaller phase function in the backscattering direction. This size effect, together with more efficient scattering (again due to hygroscopic growth) and a more dominant contribution of fine mode aerosol to the total aerosol optical depth, makes the columnar phase function as whole smaller in summer. Indeed, with a reduction in the fraction of coarse mode particles, the phase function in summer gets lower than 0.1 at Xuzhou and other sites in the south to Xuzhou. Notably, the phase function is much lower at the Qiandaohu site in most scattering angles during October and December, when fine particles are absolutely dominant with very low volume in coarse mode (Figure 2). However, temporal variations of the phase function in Hongkong are much smaller, which is consistent with the particle size distribution. Thus, it is important to make a proper assumption of particle size in aerosol models for satellite retrievals.

To synthesize, the aerosol properties in eastern China exhibit considerable variability over space and time. Spatially, and for the same season, the fraction of the coarse mode deceases from north to south, which leads to weaker backscattering; the SSA in northern China is much lower than that in southern China during winter (0.85 vs. 0.90), but is nearly the same in summer (~0.95). Hence, the aerosol optical properties in northern China are weaker in scattering during winter with more absorption, but the backscattering ability is stronger than in southern China due to the higher phase function associated with the smaller size of fine-mode aerosols (because of less hygroscopic growth, as seen in Figure 2). Temporally, the volume of coarse particles accounts for the higher fraction in northern China and the YRD during winter and spring, and fine particles are dominant in summer and fall. There is a notable increase and decease in oscillation for the SSA in eastern China in the transition of dry winter (~0.85–0.90) to humid summer (~0.95). Corresponding with variations of the particle size, the scattering phase function of the columnar aerosols is high in winter and spring, and more than 20–30% lower in summer and fall in the backscattering direction.

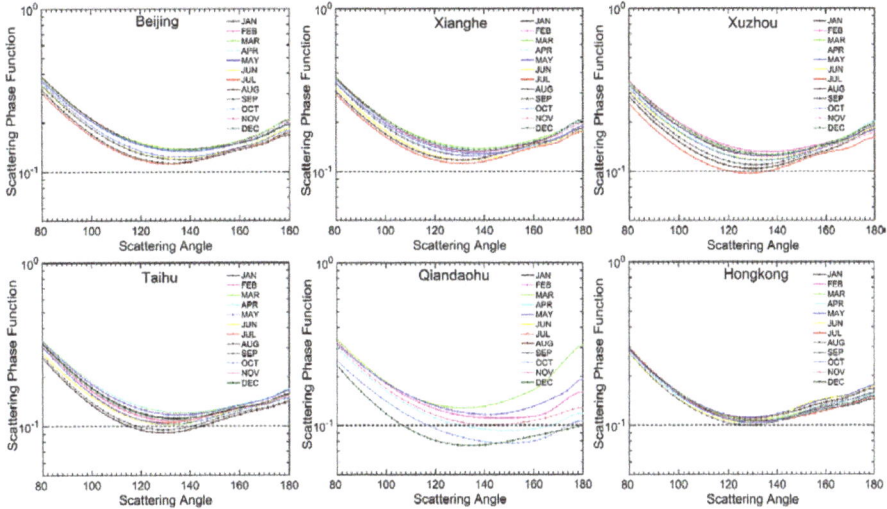

Figure 4. Monthly mean values of scattering phase function at 440 nm in eastern China.

3.2. Temporal Characteristics of MODIS Aerosol Retrieval Bias

Figure 5 shows the daily variations of MODIS DB and DT AOD biases compared with AERONET AOD in northern China (Bejing, Xianghe, and Xuzhou). To reduce the influence from uncertainties of surface reflectance estimation, the bias of satellite retrievals was considered to be robust and analyzed only when the AERONET AOD_{550} is >0.4. With only a few exceptions, MODIS DB retrieval of the AOD overall has a distinct positive bias in winter (in the range of 0.13–0.24) and slight overestimation in fall, but small negative bias in other seasons (except in Xianghe, Table 2). In particular, a clear switch between positive and negative bias can be found at all of the three sites (Bejing, Xianghe, and Xuzhou) with the same time nodes. In contrast, the AOD in MODIS DT retrievals exhibit obvious overestimation during the whole year, with few values in winter due mostly to the bright surface that is not suitable for retrieval from the DT algorithm (red line Figure 5). Considering that the aforementioned biases are systematic and periodic, it is likely that the difference between the actual and assumed aerosol properties by season can play a dominant role.

Table 2. List of seasonal mean aerosol optical depth (AOD) biases for MODIS DT and DB retrievals when AERONET AOD_{550} >0.4 of eastern China. Samples with available values of no more than five were removed to be representative.

Site	Winter DT	Spring DT	Summer DT	Fall DT	Winter DB	Spring DB	Summer DB	Fall DB
Beijing	-	0.3	0.38	0.04	0.24	−0.11	−0.095	−0.003
Xianghe	-	0.353	0.42	0.201	0.153	−0.04	0.1	0.064
Xuzhou	0.299	0.31	0.26	0.273	0.137	0.006	−0.176	0.088
Taihu	0.219	0.483	0.366	0.266	0.08	−0.021	−0.229	0.021
Qiandaohu	−0.192	0.108	0.082	−0.027	−0.293	−0.21	−0.347	−0.251
Hongkong	0.09	0.179	-	0.137	−0.113	−0.274	−0.5	−0.218

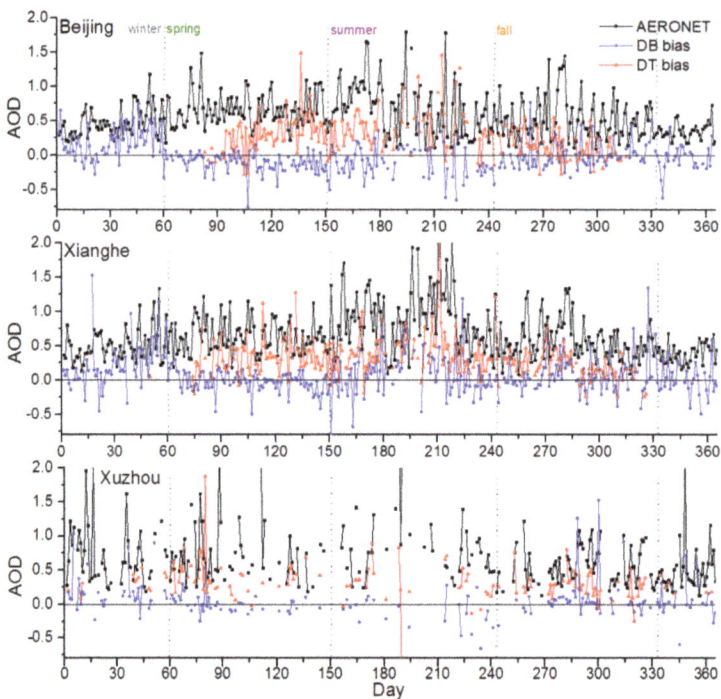

Figure 5. Daily mean values of AERONET AOD (black curve) and bias of MODIS DB and DT AOD (red and blue curves, respectively) when AERONET $AOD_{550} > 0.4$ in northern China.

As shown in Figure 6, the temporal variations of MODIS AOD bias by season in the YRD (Taihu) is similar to their counterparts in northern China. However, there is a notable change in MODIS DB bias in southern China. Differently from the overestimation in northern China and YRD during winter, DB retrievals exhibit a slight negative bias during the whole year in Qiandaohu and Hongkong (Table 2). If the aerosol model was the same as that in northern China, the increase of SSA in southern China would lead to more overestimation in DB retrievals, but the decrease in backscattering can offset the overestimation. Thus, the negative switch in DB AOD bias indicates that the weak backscattering in a phase function can make a more important contribution in southern China. It should be noted the assumed aerosol model can be different in southern China. Therefore, the systematic switch of MODIS DB AOD biases from positive to negative can be caused by changes of aerosol properties or aerosol models in southern China. Correspondingly, when the same aerosol model is used in DT retrieval, the DT bias becomes much smaller with the increase in SSA but decrease in phase function in southern China, and underestimation even appears in Qiandaohu during winter. In addition, significant errors in surface reflectance estimation can also influence the bias in DT retrieval [20].

Since both of the MODIS DB and DT algorithms employ mixed coarse and fine models in aerosol retrieval, the systematic bias of satellite AOD is determined by the combined uncertainties of the mixed models and their respective fractions. Figure 7 shows that MODIS retrieval can generally make a reasonable discrimination of the contribution between fine and coarse aerosol particles except some obvious deviation. The fraction of fine AOD_{500} in Beijing is around 0.7, with lower values in spring and a higher contribution in summer. The fine aerosol model used for DT and DB retrieval can be inferred from "Aerosol_Type_Land" and "Quality_Assurance_Land", respectively. It is found that a moderately absorbing fine model is assumed in winter and spring in northern China and a non-absorbing one during summer and fall in DT algorithm. While the moderately absorbing model is replaced by a non-absorbing one in DT retrieval during winter in Taihu and Qiandaohu, the moderately absorbing model is changed back in Hongkong. A dust and smoke model is employed in DB retrieval in eastern China with aerosol phase function and SSA selected by season and location [28], but detailed information is not available in the current literature.

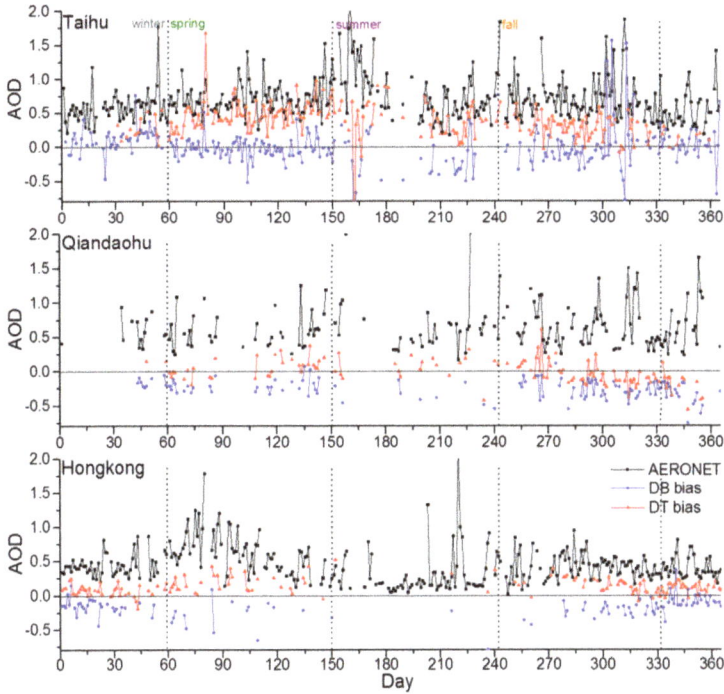

Figure 6. Daily mean values of AERONET AOD (black curve) and bias of MODIS DB and DT AOD (red and blue curves, respectively) when AERONET $AOD_{550} > 0.4$ in Taihu, Qiandaohu, and Hongkong.

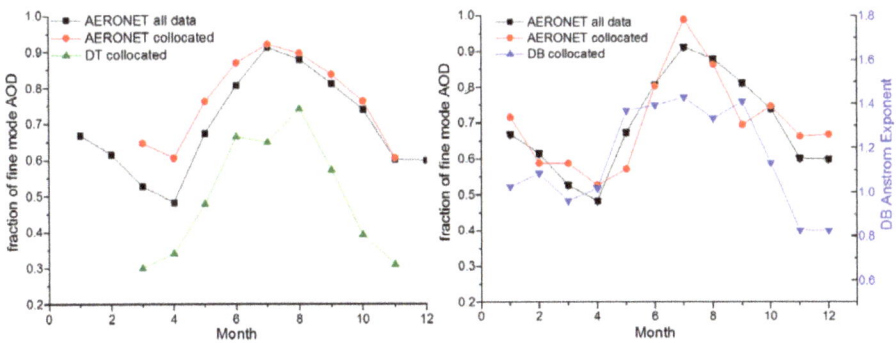

Figure 7. (**Left**) Comparison of AERONET fine AOD fraction at 500 nm (black), collocated AERONET (red) and MODIS DT fine mode aerosol fraction at 550 nm (green), (**right**) and DB Ångström exponent at 412 and 470 nm (blue) in Beijing.

Differently from the high spatial dependence in uncertainties induced by surface reflectance estimation in low-AOD conditions [20], MODIS retrieval bias in moderate and high loading exhibits clear periodic variations in regional scale, which is associated with the large temporal variations of aerosol properties. Compared with the assumed moderately absorbing fine aerosol model (SSA = 0.90 at 670 nm) in DT retrieval during winter and spring, the SSA values in northern China vary from 0.88 to 0.94, indicating that a large deviation exists in calculating scattering contribution. However, overestimation remains prevalent in DT retrievals when a non-absorbing aerosol model with SSA = 0.95 is used (Figure 8a), demonstrating that the aerosol scattering is generally underestimated. Since the magnitude of aerosol scattering is determined by SSA and phase function besides aerosol loading, improper assumptions of aerosol phase function play a major role in this systematic positive AOD bias. The effective radius in the DT algorithm is around 0.25 µm for fine mode and 0.68 µm for coarse particles (Table 1), which is larger than the fine mode (~0.13 µm) in AERONET inversion but much smaller than the coarse mode (~2.0 µm) (Figure 8b). As shown in Figure 4, the backscattering ability is much weaker due to the dominance of fine-mode aerosols with larger size particles in humid summer. The large variations in particle size distribution in eastern China were not well-considered in current MODIS retrievals. Furthermore, at 2.12 µm, where the fine-mode aerosol effect is much less, the phase function used in the DT algorithm is likely to have a large low bias because the DT algorithm assumes a much smaller coarse-mode particle size (than that which AERONET retrieves). Hence, the underestimation of the phase function at 2.12 µm can lead to the spectral slope of retrieved AOD skewing upward, thereby causing an overestimation of AOD in the visible. According to the connections between aerosol properties and the MODIS AOD bias, a preliminary suggestion for the modification of the aerosol models in MODIS retrievals is presented in Table 3.

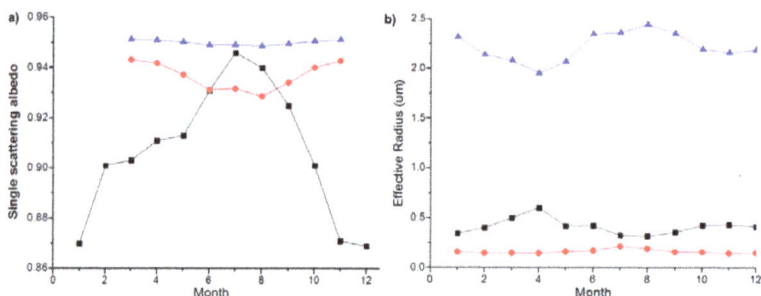

Figure 8. (a) Comparison of monthly mean SSA (black) and MODIS DT SSA from fine (blue for non-absorbing and red for moderately absorbing) and coarse aerosol models at 0.55 µm; (b) Monthly mean AERONET effective radius (µm) for total column (black), fine mode (red), and coarse mode aerosols (blue) in Beijing.

Table 3. Proposed modification for aerosol models of MODIS DT and DB retrievals in eastern China. S and P denote SSA and Particle size of fine aerosol models of DT; T signifies the total scattering properties of DB aerosol models (SSA and particle size used unknown); symbols of '+', '−', and '*' mean the parameter should be increased, deceased, or unchanged, respectively.

Site	Winter DT	Spring DT	Summer DT	Fall DT	Winter DB	Spring DB	Summer DB	Fall DB
Beijing	S−P−	S*P−	S*P−	S−P−	T−	T+	T*	T−
Xianghe	S−P−	S*P−	S*P−	S−P−	T−	T*	T−	T*
Xuzhou	S*P−	S+P−	S*P−	S−P−	T−	T*	T+	T*
Taihu	S-P−	S*P−	S*P−	S−P−	T*	T*	T+	T*
Qiandaohu	S−P*	S*P*	S+P*	S−P*	T+	T*	T+	T+
Hongkong	S*P*	S*P*	S−P*	S−P*	T+	T−	T+	T+

3.3. Decadal Variations of the Aerosol Optical Properties and MODIS AOD Bias

Besides the large seasonal variations caused by dust transport, winter coal-burning heating, and meteorological conditions, the concurrent economic development and government regulation for improving air quality have brought dramatic changes in anthropogenic emissions in eastern China during the last decades [19,32]. Figure 9 shows an obvious decreasing trend (−0.11 per decade) of AOD in Beijing, where great efforts has been made to reduce local and surrounding emissions. Another prominent feature is that the total effective radius in Beijing becomes smaller with control measures on primary emissions. Moreover, there is a large increase (~0.05) in the SSA of the aerosol particles, which is consistent with the considerable reduction (~38%) of black carbon [33]. These striking changes in aerosol scattering ability and angular distribution can to lead to serious deviations from those of the fixed fine aerosol model for satellite retrievals.

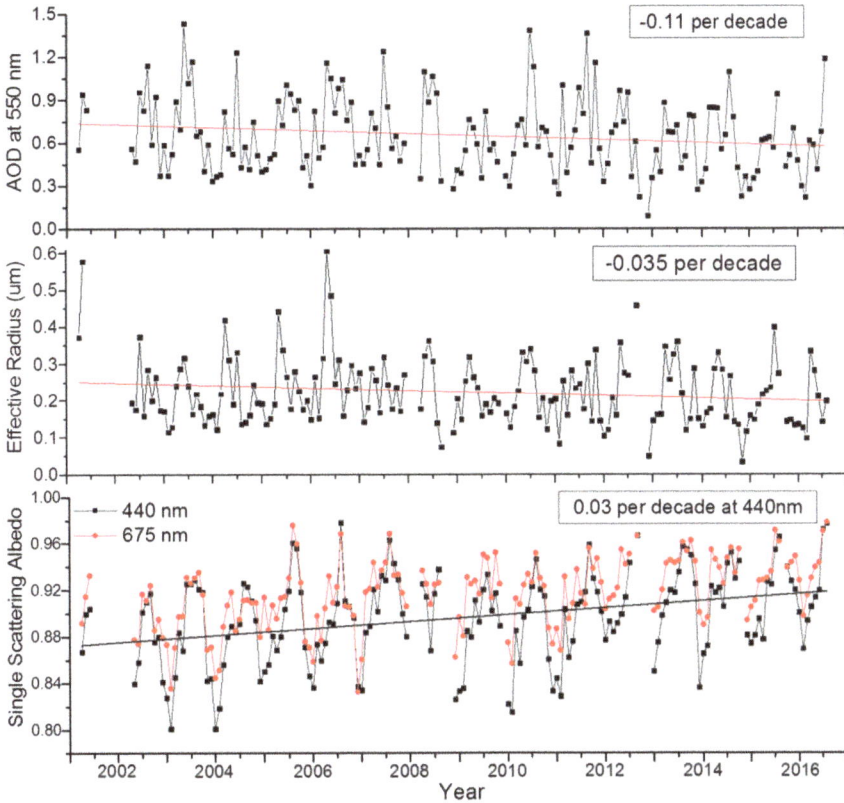

Figure 9. Variations of monthly AOD at 550 nm, total effective radius, and single scattering albedo in Beijing from 2001 to 2016.

Figure 10 displays the trends of MODIS retrieval bias in moderate and high AOD conditions from 2003 to 2016. It is surprising that the overestimation of DT retrievals becomes larger with time. The SSA values in Beijing become higher than both of these in the moderately absorbing and weakly absorbing fine models used in the DT algorithm (Figure 9), leading to more overestimation of aerosol loading by underestimating the aerosol scattering contribution. By comparison, variations of DB retrieval bias are relatively complicated, with negative bias getting much smaller and positive bias becoming slightly larger. While the obvious decrease in the underestimation of the DB retrievals during spring and summer can be associated with the increasing SSA in Beijing, the overestimation in fall and winter can be enlarged due to further difference with the underestimated scattering ability in the DB aerosol model. Such variations of aerosol optical properties can cause a difference of >0.1–0.2 in annual mean AOD, which exerts non-negligible uncertainties in analyzing the trends of aerosol loading with satellite retrievals.

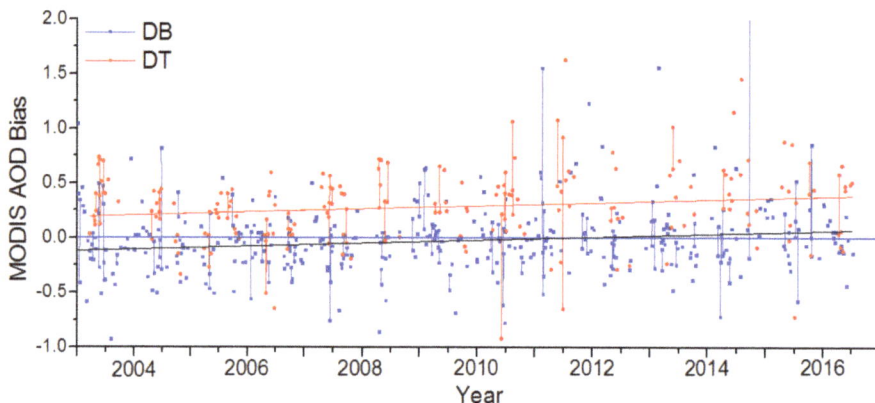

Figure 10. Daily bias of Aqua MODIS C6 DB and DT AOD (AOD_{550} > 0.4) in Beijing from 2003 to 2016.

4. Discussions

As shown above, the complicated aerosol properties in eastern China, with both striking seasonal and yearly changes, have caused a large bias in MODIS aerosol retrievals. The several aerosol models in the MODIS DT and DB algorithms can represent most situations in the world [25,28], but obviously miss some unique properties of the aerosol particles, such as the prevalent coarse dust in eastern China. Different assumptions of the aerosol properties lead to distinct deviations between the DB and DT retrievals. Although detailed information of MODIS aerosol models, especially for the DB algorithm, is not available, their overall performance from the combined contribution of the assumed SSA and the scattering phase function can be inferred from the AOD bias of the satellite retrievals. A clear switch in MODIS DB retrieval bias from spring can be caused by the seasonal change of assumed aerosol models (Figure 5). However, to identify the specific error sources, detailed information, including both the SSA and phase function (or refractive index and particle size), is necessary to compare with ground observations.

The lookup table of aerosol properties is significant for the efficiency of MODIS operational global daily retrieval. Despite the large spatial and temporal variations of aerosol properties in eastern China, reasonable assumptions can constrain the typical properties with mixed and fixed aerosol models. In the other hand, the yearly variations of aerosol properties have to be considered by time or a more flexible combination of aerosol models. More available information, such as multi-angle and polarization observations, can enable investigators to better capture the aerosol properties in satellite retrievals. Ground observations in a sufficient number of sites are needed to characterize the geographic difference of aerosol properties. Luckily, there have been many observations, such as the China Aerosol Remote Sensing Network (CARSNET) [16,18], which can substantially improve MODIS retrieval bias in China.

5. Conclusions

MODIS aerosol products have been widely used in climate and air pollution research due to their near daily global observations and coverage since 2000, and at the same time have received much concern about their uncertainties. Despite reasonable accuracy at the global scale, the performance of MODIS retrievals in China has suffered deeply from the complicated and dramatic aerosol properties. Differently from previous validations for general evaluation, here we present a special study on aerosol scattering properties in typical regions of eastern China and the dependence of MODIS C6 aerosol retrieval bias on ground-based observations over six AERONET sites. Striking systematic deviation is found in both the MODIS DT and DB retrievals, with distinct biases caused by their respective aerosol models. However, the fraction of the weighted fine and coarse models can be roughly estimated with the MODIS retrievals.

The scattering ability of the aerosol models is constrained by both SSA and scattering phase function. Despite the high SSA (~0.95) assumed in summer and fall, DT retrievals in eastern China exhibit obvious overestimation rather than underestimation during the whole year. The effective radius of fine aerosol models used in DT retrievals (~0.26 μm when AOD = 0.5) is much larger than the actual size (~0.13 μm) in ground observations, which leads to a serious overestimation of aerosol loading due to the assumed low scattering ability. Meanwhile, the smaller particle size (~0.68 μm when AOD = 0.5) assumed in the DT coarse model is much smaller than that (~2.0 μm) used in ground inversions. Considering the negligible scattering contribution of fine aerosols in shortwave infrared bands, the underestimation of the phase function at 2.1 μm can further aggravate the overestimation of aerosol loading. Correspondingly, DT bias gets much smaller in southern China, which is dominated by fine particles with larger size in humid conditions. By comparison, positive bias is dominant in DB retrieval during fall and winter with smaller negative bias in spring and summer, indicating the clear difference caused by the seasonal aerosol models. On the other hand, the yearly variations of aerosol properties lead to considerable changes in the magnitude of MODIS AOD bias. While the DT bias gets larger with increasing SSA from 2003 to 2016, the negative bias in DB retrieval becomes much smaller in spring and summer.

Our results demonstrate that the unique aerosol properties in eastern China should be included in satellite retrievals. Since MODIS DT aerosol models were fixed within half a year in eastern China, the notable variations of monthly aerosol scattering properties have caused considerable retrieval bias, since even deviations of SSA can partly cancel out the errors from an assumed phase function. On the other hand, the distinct spatial difference in particle size and SSA demonstrates that the assumed aerosol models should constrain such geographic variations rather than employ one fixed fine aerosol model in the whole of eastern China. In addition, the obvious changes in yearly aerosol scattering properties have to be taken into account. The DT algorithm utilizes a combination of one fine and one coarse aerosol model to characterize real situations. Despite the substantial variations, aerosol properties in eastern China exhibit clear change patterns in spatial and temporal scales. To establish proper and flexible aerosol models, the division of geographic sectors (e.g., northern and southern China), time range (e.g., seasonal), and typicality of the combined aerosol models should be considered based on the aerosol scattering properties and radiative transfer simulations.

Acknowledgments: This study was supported by the National Science Foundation of China (Grant No. 41401482, 41571347). Z. Wang was supported by National High Technology Research and Development Program of China (Grant No. 2014AA06A511). J. Tao was supported by the public welfare science and technology project of Beijing (Z161100001116013). J. Wang and X. Xu are supported by the research funds from the University of Iowa. The authors are grateful for the MODIS aerosol team, AERONET team, and site PIs (H. B. Chen, P. Goloub, P. Wang, X. Xia, L. Wu, R. Ma, J. Hong, and J. E. Nichol) for the data used in this investigation.

Author Contributions: Minghui Tao, Zifeng Wang, Jinhua Tao, and Liangfu Chen conceived and designed the experiments; Minghui Tao, Can Hou, Lunche Wang, and Hao Zhu performed the experiments and analyzed the data; Minghui Tao wrote the paper; Jun Wang and Xiaoguang Xu provided guidance to the project, reviewed and edited the paper.

Conflicts of Interest: The authors declare no conflict of interest.

References

1. Kaufman, Y.J.; Tanre, D.; Boucher, O. A satellite view of aerosols in the climate system. *Nature* **2002**, *419*, 215–223. [CrossRef] [PubMed]
2. Pope, C.A.; Dockery, D.W. Health Effects of Fine Particulate Air Pollution: Lines that Connect. *J. Air Waste Manag. Assoc.* **2006**, *56*, 709–742. [CrossRef] [PubMed]
3. IPCC. *Climate Change 2013: The Physical Science Basis. Contribution of Working Group I to the Fifth Assessment Report of the Intergovernmental Panel on Climate Change*; Stocker, T.F., Qin, D., Plattner, G.K., Tignor, M.M.B., Allen, S.K., Boschung, J., Nauels, A., Xia, Y., Bex, V., Midgley, P.M., Eds.; Cambridge University Press: Cambridge, UK; New York, NY, USA, 2013; 1535 p.
4. Holben, B.N.; Eck, T.F.; Slutsker, I.; Tanre, D.; Buis, J.P.; Setzer, A.; Lavenu, F. AERONET—A Federated Instrument Network and Data Archive for Aerosol Characterization. *Remote Sens. Environ.* **1998**, *66*, 1–16. [CrossRef]
5. Kaufman, Y.J.; Tanré, D.; Remer, L.A.; Vermote, E.F.; Chu, A.; Holben, B.N. Operational remote sensing of tropospheric aerosol over land from EOS moderate resolution imaging spectroradiometer. *J. Geophys. Res.* **1997**, *102*, 17051–17068. [CrossRef]
6. Diner, D.J.; Martonchik, J.V.; Kahn, R.A.; Pinty, B.; Gobron, N.; Nelson, D.L.; Holben, B.N. Using angular and spectral shape similarity constraints to improve MISR aerosol and surface retrievals over land. *Remote Sens. Environ.* **2005**, *94*, 155–171. [CrossRef]
7. Torres, O.; Tanskanen, A.; Veihelmann, B.; Ahn, C.; Braak, R.; Bhartia, P.K.; Veefkind, P.; Levelt, P. Aerosols and surface UV products from Ozone Monitoring Instrument observations: An overview. *J. Geophys. Res.* **2007**, *112*. [CrossRef]
8. Winker, D.M.; Vaughan, M.A.; Omar, A.; Hu, Y.; Powell, K.A.; Liu, Z.; Hunt, W.H.; Young, S.A. Overview of the CALIPSO Mission and CALIOP Data Processing Algorithms. *J. Atmos. Ocean. Technol.* **2009**, *26*, 2310–2323. [CrossRef]
9. Dubovik, O.; Holben, B.; Eck, T.F.; Smirnov, A.; Kaufman, Y.J.; King, M.D.; Tanre, D.; Slutsker, I. Variability of absorption and optical properties of key aerosol types observed in worldwide locations. *J. Atmos. Sci.* **2002**, *59*, 590–608. [CrossRef]
10. Hsu, N.C.; Tsay, S.C.; King, M.D.; Herman, J.R. Aerosol properties over bright-reflecting source regions. *IEEE Trans. Geosci. Remote Sens.* **2004**, *42*, 557–569. [CrossRef]
11. Levy, R.C.; Remer, L.A.; Mattoo, S.; Vermote, E.F.; Kaufman, Y.J. Second generation operational algorithm: Retrieval of aerosol properties over land from inversion of Moderate Resolution Imaging Spectroradiometer spectral reflectance. *J. Geophys. Res.* **2007**, *112*, 3710–3711. [CrossRef]
12. Li, Z.; Li, C.; Chen, H.; Tsay, S.C.; Holben, B.; Huang, J.; Xia, X. East Asian Studies of Tropospheric Aerosols and their Impact on Regional Climate (EAST-AIRC): An overview. *J. Geophys. Res. Atmos.* **2011**, *116*, D00K34. [CrossRef]
13. Zhang, X.Y.; Wang, Y.Q.; Niu, T.; Zhang, X.C.; Gong, S.L.; Zhang, Y.M.; Sun, J.Y. Atmospheric aerosol compositions in China: Spatial/temporal variability, chemical signature, regional haze distribution and comparisons with global aerosols. *Atmos. Chem. Phys.* **2012**, *11*, 779–799. [CrossRef]
14. Tao, M.H.; Chen, L.F.; Su, L.; Tao, J.H. Satellite observation of regional haze pollution over the North China Plain. *J. Geophys. Res.* **2012**, *117*, D12203. [CrossRef]
15. Tao, M.H.; Chen, L.F.; Wang, Z.F.; Tao, J.H.; Su, L. Satellite observation of abnormal yellow haze clouds over East China during summer agricultural burning season. *Atmos. Environ.* **2013**, *79*, 632–640. [CrossRef]
16. Che, H.; Zhang, X.Y.; Xia, X.; Goloub, P.; Holben, B.; Zhao, H.; Damiri, B. Ground-based aerosol climatology of China: Aerosol optical depths from the China Aerosol Remote Sensing Network (CARSNET) 2002–2013. *Atmos. Chem. Phys.* **2015**, *15*, 7619–7652. [CrossRef]
17. Xie, Y.; Li, Z.; Li, D.; Xu, H.; Li, K. Aerosol Optical and Microphysical Properties of Four Typical Sites of SONET in China Based on Remote Sensing Measurements. *Remote Sens.* **2015**, *7*, 9928–9953. [CrossRef]
18. Xin, J.; Wang, Y.; Pan, Y.; Ji, D.; Liu, Z.; Wen, T.; Wang, P. The Campaign on Atmospheric Aerosol Research Network of China: CARE-China. *Bull. Am. Meteorol. Soc.* **2015**, *96*, 1137–1155. [CrossRef]
19. Lyapustin, A.; Smirnov, A.; Hoben, B.; Chin, M.; Streets, D.G.; Lu, Z.; Tanre, D. Reduction of aerosol absorption in Beijing since 2007 from MODIS and AERONET. *Geophys. Res. Lett.* **2011**, *38*, L10803. [CrossRef]

20. Tao, M.H.; Chen, L.F.; Wang, Z.F.; Tao, J.H.; Che, H.Z.; Wang, X.H.; Wang, Y. Comparison and evaluation of MODIS Collection 6 aerosol data in China. *J. Geophys. Res. Atmos.* **2015**, *120*, 6992–7005. [CrossRef]
21. Zhang, Q.; Xin, J.Y.; Yin, Y.; Wang, L.L.; Wang, Y.S. The Variations and Trends of MODIS C5 & C6 Products' Errors in the Recent Decade over the Background and Urban Areas of North China. *Remote Sens.* **2016**, *8*, 754.
22. Zhu, J.; Xia, X.A.; Wang, J.; Che, H.; Chen, H.; Zhang, J.; Ayoub, M. Evaluation of Aerosol Optical Depth and Aerosol Models from VIIRS Retrieval Algorithms over North China Plain. *Remote Sens.* **2017**, *9*, 432. [CrossRef]
23. Eck, T.; Holben, B.; Reid, J.; Mukelabai, M.; Piketh, S.; Torres, O.; Jethva, H.; Hyer, E.; Ward, D.; Dubovik, O. A seasonal trend of single scattering albedo in southern African biomass-burning particles: Implications for satellite products and estimates of emissions for the world's largest biomass-burning source. *J. Geophys. Res. Atmos.* **2013**, *118*, 6414–6432. [CrossRef]
24. Ichoku, C.; Chu, D.A.; Mattoo, S.; Kaufman, Y.J.; Remer, L.A.; Tanre, D.; Slutsker, I.; Holben, B.N. A spatio-temporal approach for global validation and analysis of MODIS aerosol products. *Geophys. Res. Lett.* **2002**, *29*, 8006. [CrossRef]
25. Levy, R.C.; Mattoo, S.; Munchak, L.A.; Remer, L.A.; Sayer, A.M.; Patadia, F.; Hsu, N.C. The Collection 6 MODIS aerosol products over land and ocean. *Atmos. Meas. Tech.* **2013**, *6*, 2989–3034. [CrossRef]
26. Lyapustin, A.; Wang, Y.; Xiong, X.; Meister, G.; Platnick, S.; Levy, R.; Hall, F. Science impact of modis C5 calibration degradation and C6+ improvements. *Atmos. Meas. Tech.* **2014**, *7*, 4353–4365. [CrossRef]
27. Levy, R.C.; Remer, L.A.; Dubovik, O. Global aerosol optical properties and application to Moderate Resolution Imaging Spectroradiometer aerosol retrieval over land. *J. Geophys. Res.* **2007**, *112*, D13210. [CrossRef]
28. Hsu, N.C.; Jeong, M.J.; Bettenhausen, C.; Sayer, A.M.; Hansell, R.; Seftor, C.S.; Huang, J.; Tsay, S.C. Enhanced Deep Blue aerosol retrieval algorithm: The second generation. *J. Geophys. Res. Atmos.* **2013**, *118*, 9296–9315. [CrossRef]
29. Sayer, A.M.; Hsu, N.C.; Bettenhausen, C.; Jeong, M.J. Validation and uncertainty estimates for MODIS Collection 6 "Deep Blue" aerosol data. *J. Geophys. Res. Atmos.* **2013**, *118*, 7864–7872. [CrossRef]
30. Tao, M.H.; Chen, L.; Wang, L.; Ma, P.; Tao, J.; Jia, S. A study of urban pollution and haze clouds over northern China during the dusty season based on satellite and surface observations. *Atmos. Environ.* **2014**, *82*, 183–192. [CrossRef]
31. Li, J.; Carlson, B.E.; Lacis, A.A. Using Single Scattering Albedo Spectral Curvature to Characterize East Asian Aerosol Mixtures. *J. Geophys. Res. Atmos.* **2015**, *120*, 2037–2052. [CrossRef]
32. Lu, Z.; Zhang, Q.; Streets, D.G. Sulfur dioxide and primary carbonaceous aerosol emissions in China and India, 1996–2010. *Atmos. Chem. Phys.* **2011**, *11*, 9839–9864. [CrossRef]
33. Chen, Y.; Schleicher, N.; Fricker, M.; Cen, K.; Liu, X.L.; Kaminski, U.; Yu, Y.; Wu, X.F.; Norra, S. Long-term variation of black carbon and PM2.5 in Beijing, China with respect to meteorological conditions and governmental measures. *Environ. Pollut.* **2016**, *212*, 269–278. [CrossRef] [PubMed]

remote sensing

MDPI

Article

Validation of VIIRS AOD through a Comparison with a Sun Photometer and MODIS AODs over Wuhan

Wei Wang [1], Feiyue Mao [1,2,3,*], Zengxin Pan [1], Lin Du [1,4] and Wei Gong [1,3,5]

[1] State Key Laboratory of Information Engineering in Surveying, Mapping and Remote Sensing (LIESMARS), Wuhan University, Wuhan 430079, China; wangweicn@whu.edu.cn (W.W.); pzx@whu.edu.cn (Z.P.); linyufocus@foxmail.com (L.D.); weigong@whu.edu.cn (W.G.)
[2] School of Remote Sensing and Information Engineering, Wuhan University, Wuhan 430079, China
[3] Collaborative Innovation Center for Geospatial Technology, Wuhan 430079, China
[4] School of Physics and Technology, Wuhan University, 299, Bayi Road, Wuhan 430071, China
[5] Hubei Collaborative Innovation Center for High-Efficiency Utilization of Solar Energy, Wuhan 430068, China
*   Correspondence: maofeiyue@whu.edu.cn

Academic Editors: Yang Liu, Jun Wang, Omar Torres, Richard Müller and Prasad S. Thenkabail
Received: 22 February 2017; Accepted: 21 April 2017; Published: 25 April 2017

Abstract: Visible Infrared Imaging Radiometer Suite (VIIRS) is a next-generation polar-orbiting operational environmental sensor with a capability for global aerosol observations. A comprehensive validation of VIIRS products is significant for improving product quality, assessing environment quality for human life, and studying regional climate change. In this study, three-year (from 1 January 2014 to 31 December 2016) records of VIIRS Intermediate Product (IP) data and Moderate Resolution Imaging Spectroradiometer (MODIS) retrievals on aerosol optical depth (AOD) at 550 nm were evaluated by comparing them to ground sun photometer measurements over Wuhan. Results indicated that VIIRS IP retrievals were underestimated by 5% for the city. A comparison of VIIRS IP retrievals and ground sun photometer measurements showed a lower R^2 of 0.55 (0.79 for Terra-MODIS and 0.76 for Aqua-MODIS), with only 52% of retrievals falling within the expected error range established by MODIS over land (i.e., $\pm(0.05 + 0.15AOD)$). Bias analyses with different Ångström exponents (AE) demonstrated that land aerosol model selection of the VIIRS retrieval over Wuhan was appropriate. However, the larger standard deviations (i.e., uncertainty) of VIIRS AODs than MODIS AODs could be attributed to the less robust retrieval algorithm. Monthly variations displayed largely underestimated AODs of VIIRS in winter, which could be caused by a large positive bias in surface reflectance estimation due to the sparse vegetation and greater surface brightness of Wuhan in this season. The spatial distribution of VIIRS and MODIS AOD observations revealed that the VIIRS IP AODs over high-pollution areas (AOD > 0.8) with sparse vegetation were underestimated by more than 20% in Wuhan, and 40% in several regions. Analysis of several clear rural areas (AOD < 0.2) with native vegetation indicated an overestimation of about 20% in the northeastern region of the city. These findings showed that the VIIRS IP AOD at 550 nm can provide a solid dataset with a high resolution (750 m) for quantitative scientific investigations and environmental monitoring over Wuhan. However, the performance of dark target algorithms in VIIRS was associated with aerosol types and ground vegetation conditions.

Keywords: aerosol; VIIRS; MODIS; sun photometer; AERONET

1. Introduction

Atmospheric aerosols such as those from biomass burning, dust minerals, volcanic ash, smoke, sea salt, and particulate pollution, are emitted from various natural and anthropogenic activities [1,2]. Atmospheric aerosols significantly influence the radiation budget of the Earth by affecting the lifetime and microphysical properties of clouds, as well as precipitation rates and tropospheric photochemistry; therefore, they are significant in climate change studies [3–5]. However, aerosol sources, transport, and sinks possess a relatively short lifetime of one to two weeks in the atmosphere, and this characteristic restricts the understanding of their chemical and physical properties, as well as their spatiotemporal distribution characteristics [6,7]. A global network of ground-based sun photometers, such as the aerosol robotic network (AERONET) [8], provides regular measurements of aerosol optical properties, such as aerosol optical depth (AOD), at high temporal and spectral resolutions to better understand aerosol distributions in the atmosphere. However, these measurements are limited over space. This spatial limitation is addressed by satellite remote sensing, which provides systematic near-real-time AOD observations at low to high spatial resolutions [2,9,10]. Satellite remote sensing is recognized as an ideal method for monitoring the spatiotemporal distribution of AOD at regional and global scales. Aerosol retrieval algorithms are developed for global distributions of AOD by using different satellite sensors, such as the advanced very-high-resolution radiometer (AVHRR) [11], sea-viewing wide field of view sensor [12], total ozone-mapping spectroradiometer [13], ozone-monitoring instrument [14], multi-angle imaging spectroradiometer [15], moderate-resolution imaging spectroradiometer (MODIS) [2,9,10], and Visible Infrared Imaging Radiometer Suite (VIIRS) [16,17].

In October 2011, VIIRS was launched aboard the Suomi National Polar-orbiting Partnership (S-NPP) satellite; VIIRS is a new generation of operational satellite sensors for the characterization aerosol [16]. The VIIRS instrument is designed using many of the features of the National Aeronautics and Space Administration's Earth-Observing System MODIS, which has produced near-real-time aerosol data products for over a dozen years [18,19]. Given the similarity in the design of the two instruments, VIIRS is expected to produce aerosol products that are similar in scope and capability as those of MODIS [16]. The VIIRS aerosol calibration/validation team continuously monitors, evaluates, and improves the performance of VIIRS aerosol retrievals [16,17]. However, the accuracy and consistency of retrieving aerosol products via VIIRS remain worse than that of MODIS [16], due to uncertainties from cloud screening, radiance calibration, and aerosol optical property modeling. Thus, further studies that find uncertainty sources or reduce uncertainties are significant for the successful accomplishment of the VIIRS mission.

With the rapid growth of the Chinese economy in recent years, air pollution has reached a critical level, resulting in the uncertain aerosol climate on Earth due to rapid urbanization and increased industrial activity [20]. However, spatial and temporal variations in aerosols in China are poorly understood because of the sparse network of observations, or limited satellite observations with high precision and resolution [21]. VIIRS is expected to serve as a powerful tool for large-scale aerosol observations with a high spatial resolution (750 m). Therefore, the performance of VIIRS should be compared and validated against that of ground- and space-based sensors before VIIRS aerosol products are applied in scientific research in China. The VIIRS AOD Intermediate Product (IP) was evaluated with ground-measured AOD from over 12 selected AERONET sites and compared with MODIS aerosol data over China in 2013 [22]. The spatiotemporal variations in AOD retrieved from VIIRS in eastern China was also investigated [23]. Emerging aerosol products from VIIRS, MODIS (Collection 6), and Geostationary Ocean Color Imager in East Asia were evaluated in 2012 and 2013 by using ground AOD observations from AERONET and handheld sun photometers [24]. However, these upfront validation studies were conducted in the coastal areas of China. VIIRS AOD validation with AERONET is limited spatially, because most regions in Asia are empty, especially in central China [16]. Furthermore, the VIIRS data of urban areas exhibit more discrepancy than the ground measurements [23]. Therefore, additional validation study in urban region in central China is necessary to determine the performance of VIIRS.

The objective of this study is to investigate the performances of VIIRS and MODIS in the AOD retrievals over Wuhan, China by comparing data recorder in three years (from 1 January 2014 to 31 December 2016) with those from a ground-based sun photometer. The qualities of VIIRS IP and

MODIS Terra/Aqua AODs are evaluated in Section 4.1. The monthly AOD characteristics are studied with VIIRS IP, MODIS, and sun photometer data in Section 4.2. The spatial distribution of VIIRS and MODIS AOD observations is shown in Section 4.3. The reasons for the AOD difference between the three tools over Wuhan are also discussed.

2. Study Area and Datasets

2.1. Study Area

The rapid economic growth and population expansion in China in the past three decades has resulted in drastic increases in energy consumption, which has significantly increased AOD over a large part of China [22]. Wuhan is the largest city in central China, with a dense population and heavy industrialization on the Yangtze River Basin (indicated by the white point in Figure 1). Wuhan experiences a typical north subtropical humid monsoon climate, with an annual average temperature of 15.8 °C to 17.5 °C and an annual average rainfall of 1050 mm to 2000 mm [25,26]. Most areas in Wuhan are 50 m above sea level. Figure 1 shows the location and terrain of Wuhan.

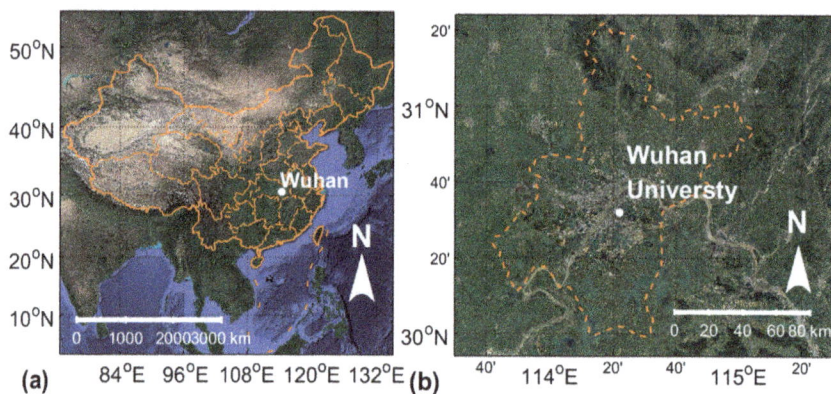

Figure 1. (a) Location of Wuhan in China and (b) location of the sun photometers at Wuhan University (white point).

2.2. Datasets

2.2.1. Sun Photometer Data

A Cimel sun photometer, CE-318, was placed and operated on top of the LIESMARS building (30°32′N, 114°21′E, and 30 m above sea level) at Wuhan University (WHU), Wuhan, Hubei Province, China, in July 2007 (Figure 1b). The CE-318 sun photometer manufactured by Cimel Electronique Company (France), is a multi-channel, automatic, sun-and-sky-scanning radiometer that measures direct solar irradiance and sky radiance. The instrument performs direct spectral solar radiation measurements within a 1.2 full field of view every 15 min at eight normal bands of 340, 380, 440, 500, 675, 870, 940, and 1020 nm [27]. The total uncertainty of AOD is approximately 0.01 to 0.02 [28]. The observations from August 2007 can be used to investigate the aerosol optical properties in central China. Several problems occurred in the equipment from April to November 2013. CE-318 is annually calibrated with China Meteorological Administration Aerosol Remote Sensing Network (CARSNET) reference instruments to ensure data accuracy and reliability; the detailed calibration procedures are described in [29]. The AOD data were calculated using ASTPwin software (Cimel Co., Ltd., Paris, Phalsbourg, France) for level 1.5 AOD. The retrieval method is found in [30].

2.2.2. VIIRS Data

VIIRS is one of the key environmental remote-sensing instruments onboard the Suomi NPP satellite. This instrument is a scanning radiometer that can extend and improve upon the heritage of AVHRR and MODIS [16]. VIIRS aerosol retrieval is performed at the pixel level and produces aerosol products with a spatial resolution of 0.75 km [31]. The product of this process, known as IP, is then aggregated and designated as an Environmental Data Record (EDR) reported at 6 km (8 × 8 pixels) resolution at nadir [16]. In this work, we evaluated VIIRS AOD550s at IP level. AOD550s are the most important aerosol parameters used by models and other community-wide applications. Quality assurance is applied at IP levels, with the resulting flags indicating the confidence of the retrievals as described in detail in [16]. For the current study, a three-year (from 1 January 2014 to 31 December 2016) dataset of AOD from VIIRS IP with quality flag = 0 (good), was analyzed to evaluate the performance of VIIRS in Wuhan.

2.2.3. MODIS Data

The detailed retrieval principle of the MODIS dark target (DT) algorithm over land can be found in [9,18,32]. The DT AOD product at 3 km is developed from spectral reflectance using a similar look-up-table and inversion based on the ratio of visible and shortwave infrared as the 10-km product [19,33]. MODIS AOD products with high resolutions (3 km) are expected to address aerosol gradients and pollution sources missed at 10 km. The quality of MODIS aerosol retrievals generally depends on the accuracy of the surface reflectance and the aerosol model, and over- or underestimation under clear and polluted conditions are normally caused by an error in these two factors [34,35]. The MODIS C6 DT AOD product significantly and systematically overestimates the AOD of Asian cities [35,36]. In this study, MOD04 and MYD04 C006 DT aerosol products were obtained over Wuhan, and only the highest-quality-flag (QF = 3) AOD observations were considered for analysis. To distinguish the different MODIS DT C006 datasets in this study, we marked them as "MOD04_3K" and "MYD04_3K" for Terra-MODIS and Aqua-MODIS AOD at 3 km, respectively.

3. Comparison and Verification of Methods

According to suggestions from other research groups [16,22,24], data matching was performed with the following rules. First, the mean AOD was averaged when at least 20% of the pixels fell within the sampling box. Second, observed AODs with a standard deviation greater than 0.5 were excluded, whereas satellite data (covering an area of 20 km × 20 km) near each WHU site were selected to reduce validation uncertainty due to the atmospheric variability imposed by atmospheric motion. Sun photometer data acquired within 30 min of the satellite overpass times were collected from 2014 to 2016. The data provided by a sun photometer did not have the 550 nm channel, and AODs at 550 nm were calculated by linear interpolation at a log scale from two measurements with adjacent wavelengths [16]. To show how accurately the satellite AOD matched the evaluation datasets, the following metrics were applied. A regression technique was used to estimate the slope and intercept of the datasets, and the uncertainty in the aerosol algorithms was evaluated using the expected error (EE, Equation (1)) over land, the relative mean bias (RMB, Equation (2)) that indicates the average overestimation (RMB > 1.0) or underestimation (RMB < 1.0) for the retrieval AODs, the root-mean-square error (RMSE, Equation (3)), the mean absolute error (MAE, Equation (4)), and the Pearson correlation coefficient (R).

$$EE = \pm(0.05 + 0.15 AOD_{ground}) \tag{1}$$

The MODIS AOD expected errors (EE) were ±(0.05 + 0.15AOD) over land [16,19,33].

$$RMB = (\overline{AOD}_{(satellite)} / \overline{AOD}_{(ground)}) \tag{2}$$

$$RMSE = \sqrt{\frac{1}{n}\sum_{i=1}^{n}\left(AOD_{(satellite)i} - AOD_{(ground)i}\right)^2} \qquad (3)$$

$$MAE = \frac{1}{n}\sum_{i=1}^{n}\left|AOD_{(satellite)i} - AOD_{(ground)i}\right| \qquad (4)$$

4. Results and Analysis

4.1. Validation of VIIRS IP and MODIS C006 AOD

Figure 2a–c show the comparisons of satellite AOD retrievals with ground observations over the Wuhan region. The total times that the satellite passed over the WHU site within 20 km and 30 min for VIIRS, Terra, and Aqua were about 1092, 1084, and 1097, respectively; by contrast, the numbers of valid matchups with the ground sun photometer were 284, 242, and 202, respectively. The relatively few valid matchups over the three-year period were caused by cloud cover and sun photometer maintenance.

VIIRS IP indicated a linear regression slope of 0.69 and a positive intercept of 0.18 on average against ground observations. The comparison of VIIRS IP retrievals with the ground sun photometer measurements showed a low R^2 of 0.55, with RMB being equal to 0.95, and only 52% of retrievals falling within EE (Figure 2a). RMB = 0.95 suggested that VIIRS IP underestimated (5%) the retrieval AODs. The scatter plot illustrates that the IP retrievals varied substantially, especially given high AOD values. A global study from 23 January 2013 to 31 December 2014 revealed similarly underestimated results with sample numbers (20269), accuracy (0.0415), precision (0.155), uncertainty (0.160), slope (0.730), intercept (0.089), and R^2 (0.549) in the land AOD IP versus AERONET [17]. However, this result was inconsistent with a previous validation study, in which 32% of VIIRS IP retrievals fell into the EE, with a slightly low R^2 of 0.63, and a relatively large positive bias (0.25) in Beijing within 3 km around AERONET [24]. The different aerosol types may have caused this difference between Beijing and Wuhan. Beijing is easily affected by downwind dust (weakly absorbing and coarse mode aerosol) from large northern deserts, but the aerosol model over this region in VIIRS has too much absorption [2]. Therefore, weakly absorbing aerosol model may be considered when dust is prominent [19].

MODIS C006 3 km products showed a large RMB at 1.13 for Terra-MODIS and 1.07 for Aqua-MODIS, which suggests a mean overestimation of 13% for MOD04_3K and 7% for MYD04_3K. Of the Terra-MODIS C6 3 km retrievals and the Aqua-MODIS C6 3 km retrievals, 66% and 71% fell within EE, respectively. MODIS with high spatial retrievals (3 km) was highly correlated with the ground AOD with R^2 of 0.79 and 0.76, for MOD04_3K and MYD04_3K, respectively. The linear regressions of MODIS retrievals and ground AOD were close to the 1:1 line. MODIS valid matchups were smaller than those of VIIRS over the same sun photometer sites, but the MODIS retrievals were better correlated with ground measurements than the VIIRS data (Figure 2 and Table 1). A previous global validation study of the 3 km MODIS AOD data reported similar retrieval errors (R^2 for Aqua and Terra were 0.68 and 0.85 respectively, and the intercepts for Aqua and Terra were 0.22 and 0.30 respectively) in urban areas [33]. Moreover, a recent study reported that MODIS C6 3 km product produced a higher bias (0.21 for Aqua and 0.29 for Terra) in a comparison with AERONET in Beijing, as well as the lowest within EE (44% for Aqua and 25% for Terra) [24]. The higher bias may be attributed to the lower average area (within 9×9 km^2) around AERONET. Similar results were reported by other evaluation studies on MODIS C6 3 km aerosol retrieval algorithms over bright urban surfaces of Beijing during low and high aerosol loadings [32]. Similarly, a recent evaluation study for the MYD04_3K over Asian countries with severe pollution showed that a large, significant overestimation was observed at urban sites dominated by coarse aerosols, including Beijing, Karachi, and Osaka, at 93.20%, 94.55%, and 75.76% of observations above the EE, respectively [2]. These results are similar to the C6 DT algorithm at 10 km, which was also found to be overestimated over cities in China and Pakistan against AERONET [1,37]. These overestimations in MODIS C006 3 km products may be attributed to a large underestimation in surface reflectance, because these study regions are highly urbanized with bright surfaces, which posed a challenge to the DT algorithm [2,24]. In addition, mixing aerosols with non-absorbing and absorbing fine mode aerosols over urban regions causes over-prediction for absorption, which results in AOD overestimation.

Comparisons over the Wuhan region showed more MODIS retrievals falling within EE and larger R^2 than VIIRS in the WHU site. From preliminary global verification over land from 23 January 2013 to 1 September 2013 by the VIIRS aerosol validation team, compared with AERONET, VIIRS retrievals showed comparable accuracy (−0.009 versus −0.005), larger uncertainty (0.130 versus 0.106), and lower correlation (*R*: 0.773 versus 0.886) than MODIS [16]. The fine-resolution aerosol products showed greater noise than the low-resolution products, which may explain the better performance of MODIS.

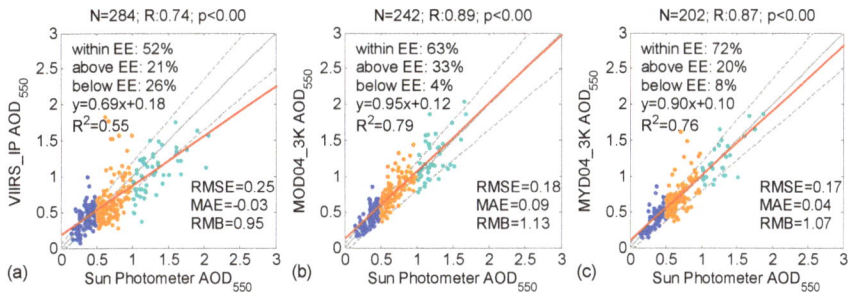

Figure 2. Validation of Visible Infrared Imaging Radiometer Suite (VIIRS) (**a**), and Moderate Resolution Imaging Spectroradiometer (MODIS) C006 aerosol optical depth (AOD) (**b**,**c**) observations (QF = high) against the Wuhan sun photometer AOD at 550 nm measurements from 2014 to 2016. The red line is the regression line, the gray solid line is the 1:1 line, and the gray dashed lines are the expected errors (EE) envelopes.

Figure 3 shows the box plots indicating the difference between satellite AOD retrievals and ground observations. The box plot in Figure 3a presents the VIIRS IP retrievals underestimated (overestimated) AOD under AOD > 1.0 (AOD < 0.3). The AOD, ranging between 0.3 and 1.0, agreed well with the ground sun photometer observations. This finding was inconsistent with a previous validation study, wherein the VIIRS product tended to overestimate AOD at low (AOD < 0.3) and high (AOD > 1.0) AOD values in East Asia [24]. Figure 3b,c present box plots showing the difference between MODIS AOD retrievals and ground observations. The bias between MODIS and sun photometer AODs was small across the entire AOD range, and was within EE when the AOD was above 0.3. However, several overestimations in AOD retrievals were also observed when AOD was low (AOD < 0.3) (Figure 3b). This finding was consistent with a previous global evaluation study, in which MODIS C6 3 km products tended to overestimate AOD [33]. The results indicated that MODIS has better accuracy than VIIRS in terms of AOD retrievals over Wuhan. The statistical results of the temporal comparisons between satellite retrievals and ground AOD measurements at 550 nm over Wuhan from 2014 to 2016 are shown in Table 1.

Figure 3d–f plot the differences between VIIRS (MODIS) and sun photometer AOD against the AE measured by sun photometer from 440 nm to 870 nm, which depicts the relationship between AOD biases and aerosol particle sizes [16,17]. AE can reflect aerosol particle sizes and their corresponding aerosol model selections. The present study found that average positive biases (the middle circle in each figure) of MODIS AOD in Figure 3e,f are larger than VIIRS IP retrievals (Figure 3d) when AE is less than 0.6. The aerosol types in Wuhan may be influenced by downwind dust (low AE) transported by prevailing north winds from the large northern deserts at winter monsoon period [26]. The smaller biases of VIIRS IP AODs confirmed the use of a proper aerosol model in VIIRS. However, the larger standard deviations (i.e., uncertainty in AODs) of VIIRS AODs could be attributed to the less robust retrieval algorithm. The biases of VIIRS IP AODs arose with increases in AE (Figure 3d), indicating more negative biases for fine particles, especially where AE > 1.3, whereas MODIS AOD showed less biases against sun photometer AODs. The negative bias at AE > 1.3 influenced most of the systematic underestimation of VIIRS IP AODs (Figure 2a), and the underestimation is explained in the next subsection analyses about monthly variations. The fine mode aerosols with strong absorption and large AE over Wuhan often have a dominant function due to automobile exhaust and the use of coal for domestic cooking, heating, and industrial processes [25]. Moreover, most matchups over a broad range of particle sizes (0.6–1.3 of AE), show smaller retrieval bias but larger uncertainty than MODIS. The results demonstrate that the aerosol model selection of the VIIRS retrieval is appropriate in this evaluation region, but the robustness of the retrieval algorithm needs improvement.

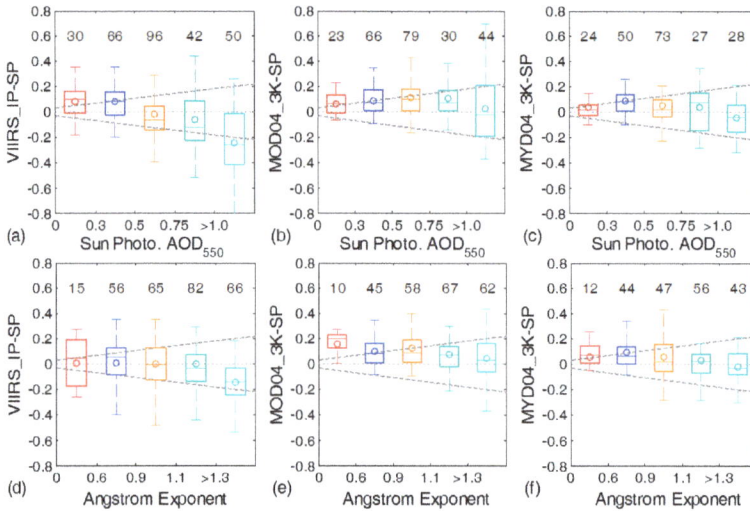

Figure 3. Box plots of AOD$_{550}$ differences (satellite—sun photometer) versus sun photometer AOD at 550 nm (**a–c**) and Ångström Exponent (**d–f**) over the Wuhan region. The number above each box refers to the corresponding statistical collocations. The y = 0 line (zero error) is shown as a fine dashed line, and the boundary lines of the expected error are depicted as gray coarse dashed lines. The properties and statistics representing each box whisker include the following: the solid lines in each box indicate the 25th and 75th percentiles of the AOD error, the whiskers are the maximum and minimum of the AOD error, the middle line is the median value of the AOD error, and the middle circle is the mean value of the AOD error.

Table 1. Statistics of the comparisons between satellite retrievals and ground AOD measurements at 550 nm over Wuhan from 2014 to 2016.

	N	R^2	Slope	Intercept	% Above/Within/Below EE
VIIRS IP 0.75 km	284	0.55	0.69	0.18	21/52/26
Terra MODIS C6 3 km	242	0.79	0.95	0.12	33/63/4
Aqua MODIS C6 3 km	202	0.76	0.90	0.10	20/72/8

4.2. Monthly Variations in VIIRS, MODIS, and Sun Photometer AOD Observations

The monthly AOD observations derived from VIIRS, MODIS, and the sun photometer over the WHU sites during the study period were analyzed (Figure 4). The annual average AODs were 0.66 ± 0.33 for 10:30 local time (LT) (Figure 4a) and 0.63 ± 0.33 for 13:30 LT (Figure 4b). Compared to the previous study, the result was smaller than the multi-year average of AOD at 550 nm measured by the sun photometer at 1.05 ± 0.66 in urban Wuhan [38]. The bias attributes to the average AOD of the latter were from a whole day. Moreover, the result was similar to those of suburban and background stations in the literature; for example, yearly mean AODs are equal to 0.82 at 500 nm at Xianghe [39], about 0.6 at Shenyang [40] and about 0.50 ± 0.06 in the Bohai Rim economic zone [41].

The monthly AODs varied with a bimodal curve. The peak values at 10:30 LT appeared in April and August (Figure 4a), and two peak values at 13:30 LT appeared in spring and autumn. The seasonal cycles of both the VIIRS IP, MODIS derived and sun photometer-measured AODs showed the same trends, which are consistent with previous studies, such as the work of [23,42]. The monthly mean AODs were related to the Asian dust and anthropogenic emission patterns in March–May, but they were modified by precipitation in June–July. A large amount of straw is burned in farmlands from nearby provinces during the harvest seasons (summer and autumn), thereby leading to the frequent haze conditions in recent years [43]. Moreover, increasing industrial and human activities, such as cement processing, smelting, coal combustion, and automobile emissions, generally lead to severe air pollution [25]. Climatologically, Wuhan is located in the East Asian monsoon area. The Meiyu period in the Yangtze River Delta during mid-June and late July [44] corresponds to the low AODs during the rainy period from June to July.

Overall, the VIIRS and MODIS AODs presented the same trend, but were over- or underestimated in different months against the ground-measured results in the WHU site. The comparison of monthly Terra-MODIS and sun photometer AODs measured at about 10:30 LT indicated a significant overestimation by MOD04_3k, especially from March to June (Figure 4a). Although the Aqua-MODIS product tended to overestimate AOD from February to June, the data agreed well with ground observations from June to November (Figure 4b). The MODIS overestimation may have been related to the difference in surface reflection caused by surface vegetation in various months. The VIIRS IP products and ground sun photometer measurements indicated similar monthly mean AOD observations over the Wuhan sites from June to November, and larger underestimations in winter (January, February, and December). Liu et al. pointed out that the seasonal variability in the biases relatively depended on the seasonally variation of vegetation growth and senescence [16]. Therefore, the larger underestimated AODs in winter could be attributed to the large positive bias in surface reflectance estimation, due to sparse vegetation with larger surface brightness in Wuhan during this season.

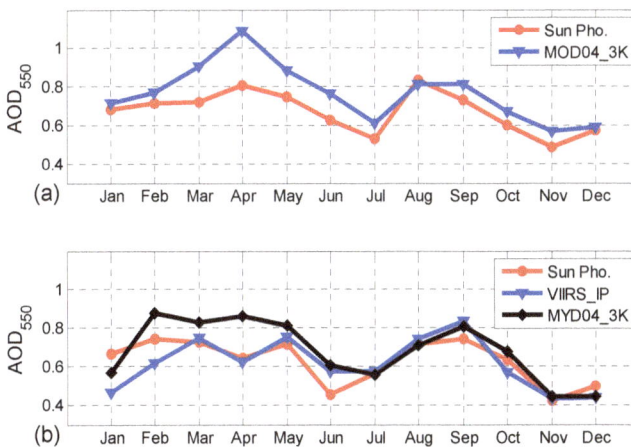

Figure 4. Monthly variations using different collocated satellite remote-sensing AODs and ground sun photometer measurements (AOD at 550 nm) over Wuhan from 1 January 2014 to 31 December 2016: (**a**) 10:30 local time (LT) and (**b**) 13:30 LT.

4.3. Spatial Distribution of VIIRS and MODIS AOD Observations

Figure 5 shows the three-year mean spatial distribution of VIIRS IP, MOD04 3 km, and MYD04 3 km (QF = high) over the Wuhan region and its surroundings used for comparisons with VIIRS and MODIS AOD products. Similar aerosol load distributions are found in Figure 5a–c. The aerosol loads around the Yangtze River and at the city center of Wuhan were significantly higher than nearby northeast rural areas, possibly because of intense anthropogenic activity and industrialized pollution. The comparison results showed that VIIRS and MODIS could realize AOD retrievals over Wuhan, but VIIRS was able to describe aerosol distribution and variability in greater detail at a higher spatial resolution (750 m) than the current MOD04 AOD products, which could produce 3 km spatial resolution can under both low and high aerosol loads (Figure 5a). Although the three collections showed a similar spatial distribution pattern, the underestimated AOD retrieved by VIIRS over the city center of Wuhan was significant compared to MODIS AOD.

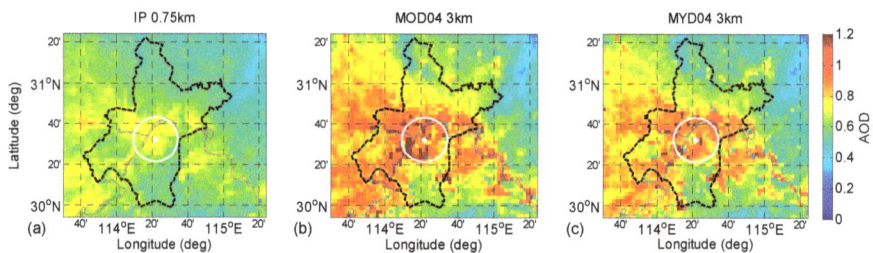

Figure 5. Mean AOD spatial distribution of (a) VIIRS IP 0.75 km, (b) MOD04 3 km, and (c) MYD04 3 km from 2014 to 2016. The white point denotes the Wuhan University (WHU) site; the white circle represents the city center of Wuhan and the average area for satellite data.

As stated in Section 4.1, the MODIS AOD retrievals had higher accuracy than VIIRS (Figure 2 and Table 1); therefore, the spatial distribution of VIIRS IP AOD could be evaluated by MODIS AOD retrievals. VIIRS IP AODs at 0.75 km resolution were resampled to 3 km spatial resolution, similar to the MODIS AOD retrievals. The AOD difference (%) in Figure 6 is defined as "100 × (VIIRS−MODIS)/MODIS", which describes the average difference between VIIRS IP and MODIS retrievals. The greater differences between VIIRS IP and Terra-MODIS (Figure 6a) than those between VIIRS IP and Aqua-MODIS (Figure 6b) may contribute to Terra-MODIS overpassing Wuhan at approximately 10:30 LT, and VIIRS IP at approximately 13:30 LT. Unlike Terra-MODIS, Aqua-MODIS is quasi-synchronous with VIIRS.

The VIIRS IP AODs over the high-pollution areas (AOD > 0.8) with sparse vegetation were underestimated by more than 20% and 40% in Wuhan and several regions, respectively. By contrast, several clear areas (AOD < 0.2) presented an overestimation of about 20% in the northeastern region, as depicted in Figure 6. The smallest AOD differences (−10% to 10%) were observed in the transition regions (0.2 < AOD < 0.8). These conclusions are similar to those of the VIIRS and ground measurement comparisons in Section 4.1. As discussed in [16], the biases between the VIIRS and MODIS AODs are potentially related to surface conditions such as surface brightness and vegetation coverage, because the aerosol retrieval algorithm does not work well in urban areas with bright reflective regions [19,23]. The biases are due to the limitations of the DT algorithm over sparse vegetated surfaces, because Wuhan is dominated by built-up and bare land surfaces (Figure 1), and the retrieval quality depends on the surface reflectance and aerosol model schemes used in the look-up table. These results suggest that the aerosol model and surface reflectance applied in the DT algorithm should be treated separately according to different aerosol types and land cover characteristics. These analyses could help reduce the uncertainty in AOD products using the DT retrieval algorithm of VIIRS and MODIS. Overall, the DT algorithm in VIIRS still needs improvement and modifications to achieve good accuracy, similar to that of MODIS AOD retrievals.

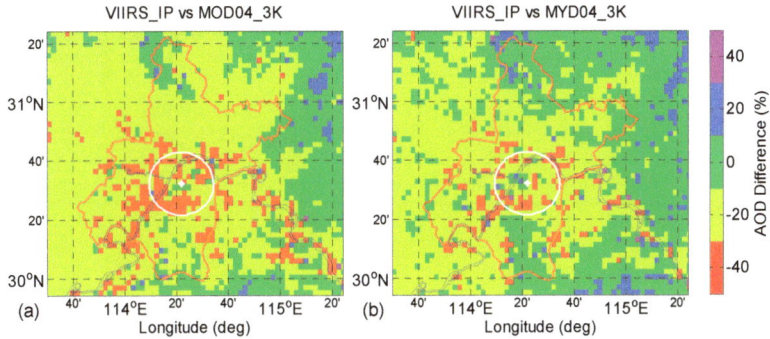

Figure 6. Mean spatial distribution of the AOD difference from 2014 to 2016: (**a**) VIIRS IP vs. MOD04 3 km and (**b**) VIIRS IP vs. MYD04 3 km. The negative and positive signs indicate under- and overestimations, respectively.

5. Conclusions

The VIIRS sensor, which is a next-generation polar-orbiting operational environmental sensor with a capability for global aerosol observations, provided the multi-year global aerosol data used in this study. In this work, VIIRS and MODIS high-quality AODs at 550 nm over Wuhan were validated against ground sun photometer measurements. The comparison spanned from 1 January 2014 to 31 December 2016. Compared with the ground sun photometer measurements over Wuhan, the VIIRS AOD IP exhibited an underestimation by 5% for cities. On the contrary, the MODIS C006 AOD retrievals were significantly overestimated over the Wuhan sites, with mean overestimations of 13% and 7% for Terra and Aqua, respectively. The evaluation and comparison of the results showed that the VIIRS IP retrievals and ground sun photometer measurements had a low R^2 of 0.55, with only 52% of retrievals falling within the expected error range established by MODIS over land. The MODIS AODs indicated higher correlations (0.79 for Terra and 0.76 for Aqua) with the ground sun photometer measurements and lower RMSEs (0.18 for Terra and 0.17 for Aqua) and MAE values (0.09 for Terra and 0.04 for Aqua) than the VIIRS IP AOD products. Bias analyses demonstrated an appropriate aerosol model selection of the VIIRS retrieval over Wuhan, but the larger standard deviations (i.e., uncertainty) of VIIRS AODs than that of MODIS AODs were attributed to the less robust retrieval algorithm. Monthly variations displayed larger underestimated AODs of VIIRS in winter, which was attributed to the large positive bias in surface reflectance estimation, due to sparse vegetation and the larger surface brightness of Wuhan in this season. The spatial distribution of VIIRS and MODIS AOD observation revealed that the VIIRS IP AODs over the high-pollution areas (AOD > 0.8) with sparse vegetation were underestimated by more than 20% and 40% in Wuhan and several regions, respectively. By contrast, several clear areas (AOD < 0.2) indicated an overestimation of about 20% in the northeastern region. In summary, the VIIRS IP AOD at 550 nm can provide a solid dataset with a high resolution (750 m) for quantitative scientific investigations and environmental monitoring over Wuhan. However, the performance of dark target algorithm in VIIRS is associated with aerosol types and ground vegetation conditions, and it needs to be improved and modified to achieve good accuracy, similar to that of MODIS AOD retrievals. This study and its results are indispensable for achieving a better, more accurate evaluation of VIIRS AODs with high spatial resolution in urban cities, which will play an important role in the assessment of environment quality for human life, and research on regional climate change.

Acknowledgments: This research is supported by the National Key Research and Development Program of China (2016YFC0200900), the National Natural Science Foundation of China (41627804), the Program for Innovative Research Team in University of the Ministry of Education of China (IRT1278), the National Science Foundation of Hubei Province (2015CFA002), and the China Postdoctoral Science Foundation (2016T90731).

Author Contributions: The work presented here was conducted in collaboration with all authors. Feiyue Mao, Zengxin Pan, Lin Du, Wei Gong and Wei Wang defined the research theme. All authors performed the experiments. This manuscript was written by Wei Wang. Feiyue Mao and Zengxin Pan checked the experimental results. All authors agreed to the submission of the manuscript.

Conflicts of Interest: The authors declare no conflict of interest.

References

1. Bilal, M.; Nichol, J.E.; Nazeer, M. Validation of Aqua-MODIS C051 and C006 operational aerosol products using AERONET measurements over Pakistan. *IEEE J. Sel. Top. Appl. Earth Obs. Remote Sens.* **2016**, *9*, 2074–2080. [CrossRef]
2. Nichol, J.; Bilal, M. Validation of MODIS 3 km resolution aerosol optical depth retrievals over Asia. *Remote Sens.* **2016**, *8*, 328. [CrossRef]
3. Twomey, S. Aerosols, clouds and radiation. *Atmos. Environ. Part A Gen. Top.* **1991**, *25*, 2435–2442. [CrossRef]
4. Pan, Z.; Gong, W.; Mao, F.; Li, J.; Wang, W.; Li, C.; Min, Q. Macrophysical and optical properties of clouds over east Asia measured by CALIPSO. *J. Geophys. Res. Atmos.* **2015**, *120*. [CrossRef]
5. Pan, Z.; Mao, F.; Gong, W.; Wang, W.; Yang, J. Observation of clouds macrophysical characteristics in China by CALIPSO. *J. Appl. Remote Sens.* **2016**, *10*, 036028. [CrossRef]
6. He, Q.; Li, C.; Mao, J.; Lau, A.; Li, P. A study on the aerosol extinction-to-backscatter ratio with combination of micro-pulse lidar and MODIS over Hong Kong. *Atmos. Chem. Phys.* **2006**, *6*, 3243–3256. [CrossRef]
7. Logan, T.; Xi, B.; Dong, X.; Li, Z.; Cribb, M. Classification and investigation of Asian aerosol absorptive properties. *Atmos. Chem. Phys.* **2013**, *13*, 2253–2265. [CrossRef]
8. Holben, B.; Eck, T.; Slutsker, I.; Tanre, D.; Buis, J.; Setzer, A.; Vermote, E.; Reagan, J.; Kaufman, Y.; Nakajima, T. Aeronet—A federated instrument network and data archive for aerosol characterization. *Remote Sens. Environ.* **1998**, *66*, 1–16. [CrossRef]
9. Remer, L.A.; Kaufman, Y.; Tanré, D.; Mattoo, S.; Chu, D.; Martins, J.V.; Li, R.-R.; Ichoku, C.; Levy, R.; Kleidman, R. The MODIS aerosol algorithm, products, and validation. *J. Atmos. Sci.* **2005**, *62*, 947–973. [CrossRef]
10. Kittaka, C.; Winker, D.; Vaughan, M.; Omar, A.; Remer, L. Intercomparison of column aerosol optical depths from CALIPSO and MODIS-Aqua. *Atmos. Meas. Tech.* **2011**, *4*, 131–141. [CrossRef]
11. Hauser, A.; Oesch, D.; Foppa, N.; Wunderle, S. NOAA AVHRR derived aerosol optical depth over land. *J. Geophys. Res. Atmos.* **2005**, *110*. [CrossRef]
12. Sayer, A.M.; Hsu, N.C.; Bettenhausen, C.; Jeong, M.J. Global and regional evaluation of over-land spectral aerosol optical depth retrievals from SeaWIFS. *Atmos. Meas. Tech.* **2012**, *5*, 2169–2220. [CrossRef]
13. Torres, O.; Bhartia, P.K.; Herman, J.R.; Sinyuk, A.; Ginoux, P.; Holben, B. A long-term record of aerosol optical depth from toms observations and comparison to AERONET measurements. *J. Atmos. Sci.* **2002**, *59*, 398–413. [CrossRef]
14. Torres, O.; Tanskanen, A.; Veihelmann, B.; Ahn, C.; Braak, R.; Bhartia, P.K.; Veefkind, P.; Levelt, P. Aerosols and surface uv products from ozone monitoring instrument observations: An overview. *J. Geophys. Res. Atmos.* **2007**, *112*, 1–14. [CrossRef]
15. Kahn, R.A.; Gaitley, B.J.; Garay, M.J.; Diner, D.J.; Eck, T.F.; Smirnov, A.; Holben, B.N. Multiangle imaging spectroradiometer global aerosol product assessment by comparison with the aerosol robotic network. *J. Geophys. Res. Atmos.* **2010**, *115*, D23209. [CrossRef]
16. Liu, H.; Remer, L.A.; Huang, J.; Huang, H.-C.; Kondragunta, S.; Laszlo, I.; Oo, M.; Jackson, J.M. Preliminary evaluation of s-NPP VIIRS aerosol optical thickness. *J. Geophys. Res. Atmos.* **2014**, *119*, 3942–3962. [CrossRef]
17. Huang, J.; Kondragunta, S.; Laszlo, I.; Liu, H.; Remer, L.A.; Zhang, H.; Superczynski, S.; Ciren, P.; Holben, B.N.; Petrenko, M. Validation and expected error estimation of Suomi-NPP VIIRS aerosol optical thickness and angström exponent with AERONET. *J. Geophys. Res. Atmos.* **2016**, *121*, 7139–7160. [CrossRef]

18. Levy, R.C.; Remer, L.A.; Mattoo, S.; Vermote, E.F.; Kaufman, Y.J. Second-generation operational algorithm: Retrieval of aerosol properties over land from inversion of moderate resolution imaging spectroradiometer spectral reflectance. *J. Geophys. Res. Atmos.* **2007**, *112*. [CrossRef]

19. Levy, R.C.; Mattoo, S.; Munchak, L.A.; Remer, L.A.; Sayer, A.M.; Patadia, F.; Hsu, N.C. The collection 6 MODIS aerosol products over land and ocean. *Atmos. Meas. Tech.* **2013**, *6*, 2989–3034. [CrossRef]

20. Zhang, Y.-L.; Cao, F. Fine particulate matter (PM2.5) in China at a city level. *Sci. Rep.* **2015**, *5*. [CrossRef] [PubMed]

21. Guo, J.-P.; Zhang, X.-Y.; Wu, Y.-R.; Zhaxi, Y.; Che, H.-Z.; La, B.; Wang, W.; Li, X.-W. Spatio-temporal variation trends of satellite-based aerosol optical depth in China during 1980–2008. *Atmos. Environ.* **2011**, *45*, 6802–6811. [CrossRef]

22. Meng, F.; Cao, C.; Shao, X. Spatio-temporal variability of Suomi-NPP VIIRS-derived aerosol optical thickness over China in 2013. *Remote Sens. Environ.* **2015**, *163*, 61–69. [CrossRef]

23. Meng, F.; Xin, J.; Cao, C.; Shao, X.; Shan, B.; Xiao, Q. Seasonal variations in aerosol optical thickness over eastern China determined from VIIRS data and ground measurements. *Int. J. Remote Sens.* **2016**, *37*, 1868–1880. [CrossRef]

24. Xiao, Q.; Zhang, H.; Choi, M.; Li, S.; Kondragunta, S.; Kim, J.; Holben, B.; Levy, R.C.; Liu, Y. Evaluation of VIIRS, GOCI, and MODIS collection 6 AOD retrievals against ground sunphotometer observations over east Asia. *Atmos. Chem. Phys.* **2016**, *16*, 20709–20741. [CrossRef]

25. Wang, L.; Gong, W.; Xia, X.; Zhu, J.; Li, J.; Zhu, Z. Long-term observations of aerosol optical properties at wuhan, an urban site in central China. *Atmos. Environ.* **2015**, *101*, 94–102. [CrossRef]

26. Wang, W.; Gong, W.; Mao, F.; Pan, Z.; Liu, B. Measurement and study of lidar ratio by using a raman lidar in central China. *Int. J. Environ. Res. Public Health* **2016**, *13*, 508. [CrossRef] [PubMed]

27. Dubovik, O.; Smirnov, A.; Holben, B.N.; King, M.D.; Kaufman, Y.J.; Eck, T.F.; Slutsker, I. Accuracy assessments of aerosol optical properties retrieved from aerosol robotic network (AERONET) sun and sky radiance measurements. *J. Geophys. Res. Atmos.* **2000**, *105*, 9791–9806. [CrossRef]

28. Eck, T.F.; Holben, B.N.; Reid, J.S.; Dubovik, O.; Smirnov, A.; O'Neill, N.T.; Slutsker, I.; Kinne, S. Wavelength dependence of the optical depth of biomass burning, urban, and desert dust aerosols. *J. Geophys. Res. Atmos.* **1999**, *104*, 31333–31349. [CrossRef]

29. Tao, R.; Che, H.; Chen, Q.; Wang, Y.; Sun, J.; Zhang, X.; Lu, S.; Guo, J.; Wang, H.; Zhang, X. Development of an integrating sphere calibration method for cimel sunphotometers in China aerosol remote sensing network. *Particuology* **2014**, *13*, 88–99. [CrossRef]

30. Smirnov, A.; Holben, B.N.; Eck, T.F.; Dubovik, O.; Slutsker, I. Cloud-screening and quality control algorithms for the AERONET database. *Remote Sens. Environ.* **2000**, *73*, 337–349. [CrossRef]

31. Jackson, J.M.; Liu, H.; Laszlo, I.; Kondragunta, S.; Remer, L.A.; Huang, J.; Huang, H.-C. Suomi-NPP VIIRS aerosol algorithms and data products. *J. Geophys. Res. Atmos.* **2013**, *118*, 12673–12689. [CrossRef]

32. Bilal, M.; Nichol, J.E. Evaluation of MODIS aerosol retrieval algorithms over the Beijing-Tianjin-Hebei region during low to very high pollution events. *J. Geophys. Res. Atmos.* **2015**, *120*, 7941–7957. [CrossRef]

33. Remer, L.A.; Mattoo, S.; Levy, R.C.; Munchak, L. MODIS 3 km aerosol product: Algorithm and global perspective. *Atmos. Meas. Tech.* **2013**, *6*, 69–112. [CrossRef]

34. Chu, D.; Kaufman, Y.; Ichoku, C.; Remer, L.; Tanré, D.; Holben, B. Validation of MODIS aerosol optical depth retrieval over land. *Geophys. Res. Lett.* **2002**, *29*, 8007. [CrossRef]

35. He, Q.; Li, C.; Tang, X.; Li, H.; Geng, F.; Wu, Y. Validation of MODIS derived aerosol optical depth over the Yangtze river delta in China. *Remote Sens. Environ.* **2010**, *114*, 1649–1661. [CrossRef]

36. Tao, M.; Chen, L.; Wang, Z.; Tao, J.; Che, H.; Wang, X.; Wang, Y. Comparison and evaluation of the MODIS collection 6 aerosol data in China. *J. Geophys. Res. Atmos.* **2015**, *120*, 6992–7005. [CrossRef]

37. Remer, L.A.; Kleidman, R.G.; Levy, R.C.; Kaufman, Y.J.; Tanré, D.; Mattoo, S.; Martins, J.V.; Ichoku, C.; Koren, I.; Yu, H.; et al. Global aerosol climatology from the MODIS satellite sensors. *J. Geophys. Res. Atmos.* **2008**, *113*. [CrossRef]

38. Wang, W.; Gong, W.; Mao, F.; Zhang, J. Long-term measurement for low-tropospheric water vapor and aerosol by raman lidar in Wuhan. *Atmosphere* **2015**, *6*, 521–533. [CrossRef]

39. Li, Z.; Xia, X.; Cribb, M.; Mi, W.; Holben, B.; Wang, P.; Chen, H.; Tsay, S.-C.; Eck, T.F.; Zhao, F.; et al. Aerosol optical properties and their radiative effects in northern China. *J. Geophys. Res. Atmos.* **2007**, *112*, 321–341. [CrossRef]

40. Che, H.; Zhao, H.; Wu, Y.; Xia, X.; Zhu, J.; Wang, H.; Wang, Y.; Sun, J.; Yu, J.; Zhang, X.; et al. Analyses of aerosol optical properties and direct radiative forcing over urban and industrial regions in northeast China. *Meteorol. Atmos. Phys.* **2015**, *127*, 345–354. [CrossRef]

41. Xin, J.; Wang, L.; Wang, Y.; Li, Z.; Wang, P. Trends in aerosol optical properties over the Bohai rim in northeast China from 2004 to 2010. *Atmos. Environ.* **2011**, *45*, 6317–6325. [CrossRef]

42. Meng, F.; Cao, C.Y.; Shao, X.; Shi, Y.G. Spatial and temporal variation of visible infrared imaging radiometer suite (VIIRS)-derived aerosol optical thickness over Shandong, China. *Int. J. Remote Sens.* **2014**, *35*, 6023–6034. [CrossRef]

43. Xia, X.; Zong, X.; Sun, L. Exceptionally active agricultural fire season in mid-eastern China in June 2012 and its impact on the atmospheric environment. *J. Geophys. Res. Atmos.* **2013**, *118*, 9889–9900. [CrossRef]

44. Luo, Y.; Zheng, X.; Zhao, T.; Chen, J. A climatology of aerosol optical depth over China from recent 10 years of MODIS remote sensing data. *Int. J. Climatol.* **2014**, *34*, 863–870. [CrossRef]

remote sensing

MDPI

Article

Evaluation of Aerosol Optical Depth and Aerosol Models from VIIRS Retrieval Algorithms over North China Plain

Jun Zhu [1,2,3,4], Xiangao Xia [1,3,4,*], Jun Wang [2,5,*], Huizheng Che [6], Hongbin Chen [1,4], Jinqiang Zhang [1,3,4], Xiaoguang Xu [2,5], Robert C. Levy [7], Min Oo [8], Robert Holz [8] and Mohammed Ayoub [9]

[1] LAGEO, Institute of Atmospheric Physics, Chinese Academy of Sciences, Beijing 100029, China;
    junzhu@nuist.edu.cn (J.Z.); chb@mail.iap.ac.cn (H.C.); zjq@mail.iap.ac.cn (J.Z.)
[2] Department of Earth and Atmospheric Sciences, University of Nebraska–Lincoln, Lincoln, NE 68588, USA;
    xiaoguang-xu@uiowa.edu
[3] Collaborative Innovation Center on Forecast and Evaluation of Meteorological Disasters,
    Nanjing University of Information Science & Technology, Nanjing 210044, China
[4] College of Earth Sciences, University of Chinese Academy of Sciences, Beijing 100049, China
[5] Department of Chemical and Biochemical Engineering, Center for Global and Regional Environmental
    Studies, and Informatics Initiative, The University of Iowa, Iowa City, IA 52241, USA
[6] State Key Laboratory of Severe Weather (LASW) and Institute of Atmospheric Composition,
    Chinese Academy of Meteorological Sciences (CAMS), CMA, Beijing 100081, China; chehz@camscma.cn
[7] Climate and Radiation Laboratory, NASA GSFC, Greenbelt, MD 20771, USA; robert.c.levy@nasa.gov
[8] Space Science and Engineering Center, University of Wisconsin–Madison, Madison, WI 53706, USA;
    min.oo@ssec.wisc.edu (M.O.); reholz@ssec.wisc.edu (R.H.)
[9] Qatar Environment & Energy Research Institute, Qatar Foundation, Ar-Rayyan, Qatar; mayoub@qf.org.qa
*   Correspondence: xxa@mail.iap.ac.cn (X.X.); jun-wang-1@uiowa.edu (J.W.)

Academic Editors: Omar Torres, Yang Liu, Richard Müller and Prasad S. Thenkabail
Received: 13 February 2017; Accepted: 27 April 2017; Published: 2 May 2017

Abstract: The first Visible Infrared Imaging Radiometer Suite (VIIRS) was launched on Suomi National Polar-orbiting Partnership (S-NPP) satellite in late 2011. Similar to the Moderate resolution Imaging Spectroradiometer (MODIS), VIIRS observes top-of-atmosphere spectral reflectance and is potentially suitable for retrieval of the aerosol optical depth (AOD). The VIIRS Environmental Data Record data (VIIRS_EDR) is produced operationally by NOAA, and is based on the MODIS atmospheric correction algorithm. The "MODIS-like" VIIRS data (VIIRS_ML) are being produced experimentally at NASA, from a version of the "dark-target" algorithm that is applied to MODIS. In this study, the AOD and aerosol model types from these two VIIRS retrieval algorithms over the North China Plain (NCP) are evaluated using the ground-based CE318 Sunphotometer (CE318) measurements during 2 May 2012–31 March 2014 at three sites. These sites represent three different surface types: urban (Beijing), suburban (XiangHe) and rural (Xinglong). Firstly, we evaluate the retrieved spectral AOD. For the three sites, VIIRS_EDR AOD at 550 nm shows a positive mean bias (MB) of 0.04–0.06 and the correlation of 0.83–0.86, with the largest MB (0.10–0.15) observed in Beijing. In contrast, VIIRS_ML AOD at 550 nm has overall higher positive MB of 0.13–0.14 and a higher correlation (0.93–0.94) with CE318 AOD. Secondly, we evaluate the aerosol model types assumed by each algorithm, as well as the aerosol optical properties used in the AOD retrievals. The aerosol model used in VIIRS_EDR algorithm shows that dust and clean urban models were the dominant model types during the evaluation period. The overall accuracy rate of the aerosol model used in VIIRS_ML over NCP three sites (0.48) is higher than that of VIIRS_EDR (0.27). The differences in Single Scattering Albedo (SSA) at 670 nm between VIIRS_ML and CE318 are mostly less than 0.015, but high seasonal differences are found especially over the Xinglong site. The values of SSA from VIIRS_EDR are higher than that observed by CE318 over all sites and all assumed aerosol modes, with a positive bias of 0.02–0.04 for fine mode, 0.06–0.12 for coarse mode and 0.03–0.05 for bi-mode at 440 nm. The overestimation of SSA but positive AOD MB of VIIRS_EDR indicate that other factors (e.g., surface reflectance characterization or cloud contamination) are important sources of error in the VIIRS_EDR algorithm, and their effects on aerosol retrievals may override the effects from non-ideality in these aerosol models.

Keywords: aerosol optical depth; aerosol models; VIIRS; NCP region

---

1. Introduction

Atmospheric aerosols have important impacts on climate, air quality and human health [1–3]. Their properties are highly variable in both space and time. Space-based platforms provide a global view of the aerosol system, unmatched by any other measurement system in terms of the spatial coverage [4]. With the long-history of aerosol products derived from the Moderate resolution Imaging Spectroradiometer (MODIS) and the aging of the instrument, there is programmatic interest in continuing similar aerosol retrieval capabilities. Since the launch of the Visible Infrared Imaging Radiometer Suite (VIIRS) instrument on the Suomi National Polar-orbiting Partnership (S-NPP) satellite in late 2011, there has been great interest in retrieving aerosol properties from VIIRS [5–7].

Currently, there are multiple algorithms available for deriving aerosol optical depth (AOD) and other aerosol properties from VIIRS data [6,8]. Here, we consider two. The VIIRS Environmental Data Record (VIIRS_EDR) is being produced by the United States National Oceanic and Atmospheric Administration (NOAA) [6]. At the same time, the National Aeronautics and Space Administration (NASA) is considering long-term continuity for developing an aerosol climate data record. For this purpose, Levy et al. are experimenting with a "MODIS-like" dark-target algorithm for use on VIIRS data (VIIRS_ML) [8]. How does each of these algorithms perform for retrieving AOD and other aerosol properties over China, a region of extreme diversity of aerosol sources, compositions, and loadings?

Preliminary evaluation of VIIRS_EDR and VIIRS_ML derived AOD, has been performed separately using co-located sunphotometer data from the AErosol Robotic NETwork (AERONET) and other networks [9]. For the period of 1 May 2012–30 April 2013, Jackson et al. showed VIIRS_EDR underestimated the AOD over land with global mean bias of −0.02 [6]. However, the Suomi NPP VIIRS aerosol data product assessment report showed VIIRS_EDR overestimated the AOD in land of 0.06 to 0.11 for the period of 1 May 2012–14 October 2012 [10]. In addition, larger biases were found in western Asia and India [9]. Since experimental VIIRS_ML has only been available since 2015, the evaluation is limited. Levy et al. briefly validate the VIIRS_ML AOD at 550 nm by comparing it to AERONET observations from March 2013 to February 2014 and showed VIIRS_ML overestimates the AOD over land with global positive bias of 0.005 [8].

The above studies have limited scope in that they only provide estimates of the global expected error over land. They do not focus on evaluating products over regional scales, especially where AERONET data are sparse. Hence, the focus of this study is to evaluate VIIRS_EDR and VIIRS_ML data over the North China Plain (NCP: 114–120°E, 34.5–41°N), one of the most densely populated regions in China that has experienced enormous economic growth in past two or three decades [11–14]. Indeed, the NCP is one of the most severely polluted areas in the world with frequent heavy haze events in recent years [11,15,16]. Given that retrieved aerosol optical properties are often used as a proxy for assessing climate and air quality in the NCP region, a regional validation of VIIRS aerosol products has important consequences [12,17,18].

The organization of this paper is as follows. We introduce the VIIRS instrument and the two aerosol retrieval algorithms and corresponding data in Section 2. The validation data set and methods for inter-comparison are in Section 3. The AOD evaluation results are presented in Section 4. Section 5 states the aerosol model types and the optical properties comparison between the CE318 sunphotometer and VIIRS, including VIIRS_EDR and VIIRS_ML. The conclusions are presented in Section 6.

2. VIIRS Satellite Data

2.1. What Is VIIRS?

VIIRS is a cross-track scanning radiometer with 22 spectral bands covering the visible/infrared spectrum from 0.412 to 12.05 μm. The design and concept of VIIRS operations combine aspects from several legacy instruments, including the NOAA's Advanced Very High Resolution Radiometer (AVHRR), NASA's MODIS, Sea-viewing Wide Field-of-view Sensor (SeaWiFS), and the Department of Defense's Operational Linescan System (OLS) sensors [19]. It has a wider swath (~3000 km) than MODIS, which allows a global sample of the Earth everywhere every day. It flies in a Sun-synchronous near-circular ascending polar orbit 829 km above the Earth with the local equator-crossing time at 13:30.

VIIRS has three types of bands: imagery bands (I-bands), moderate resolution bands (M-bands), and the day-night band [6]. The M-bands (total 16 bands) have 0.742 km × 0.776 km nadir resolution and 1.60 km × 1.58 km at the edge of scan. Other bands are used to create the VIIRS Cloud Mask (VCM), which is used as input to aerosol algorithms, as well as in internal tests to characterize environmental conditions. Most of M-bands are used to derive the aerosol parameters. Specifically, M1 (0.412 μm), M2 (0.445 μm), M3 (0.488 μm), M5 (0.672 μm), and M11 (2.25 μm) bands are used over land; and M5, M6 (0.746 μm), M7 (0.865 μm), M8 (1.24 μm), M10 (1.61 μm), and M11 are used over ocean. A detailed description of the VIIRS bands is shown in [19].

2.2. Overview of the Two Retrieval Algorithms over Land

The VIIRS_EDR and VIIRS_ML algorithms are similar in many ways. Both algorithms start with the satellite measurements of spectral reflectance at the top-of-atmosphere (TOA), and are compared to a look-up table (LUT) to determine the most plausible solutions for aerosol and surface properties. The measured reflectance at the TOA is a summation of scattering events from the surface and the atmosphere. In both algorithms, the aerosol optical properties of the aerosol models are essential for radiative transfer computing to generate the atmospheric LUT that is needed for AOD retrievals.

However, there are also differences. The VIIRS_EDR algorithm has the heritage from the MODIS atmospheric correction algorithm for land surface reflectance, in which the expected surface reflectance ratio at different wavelengths are used as a constraint in retrieving AOD [20,21]. In contrast, VIIRS_ML has the heritage from the MODIS Collection-6 aerosol algorithm, in which surface reflectance ratios at different wavelengths are prescribed and the TOA reflectance is used as the most important constraints [8,22]. Aerosol model type assignment in the two algorithms is also different. In VIIRS_EDR algorithm, the aerosol model type is selected at each pixel for each inversion by using extra blue wavelengths to constrain the aerosol type, while in the VIIRS_ML algorithm, the aerosol model type is assigned to each region and each season prior to retrieval based on the past cluster analysis of AERONET inversions [22,23]. In our assessment of the VIIRS AOD in the NCP region, we will also use aerosol properties from CE318 inversions to evaluate the aerosol model types and associated optical properties in the VIIRS_EDR and VIIRS_ML algorithms, and thereby, analyze one likely source for AOD retrieval uncertainties.

Based on the climatology of AERONET inversion data, the VIIRS_EDR algorithm defines a set of five microphysical aerosol model types. These five models are denoted as dust (for example, observed at Cape Verde), high absorption smoke (African savanna, Zambia), low absorption smoke (Amazonian forest, Brazil), clean urban (Goddard Space Flight Center, Greenbelt, MD, USA), and polluted urban (Mexico City, Mexico) aerosols. All model types have size distributions defined by bimodal lognormal distributions of spherical particles [6]. As explained by Jackson et al. [6], the retrieval LUT is created by starting with a Mie scattering code, for which aerosol inputs include a real part and an imaginary part of refractive indices and size parameters of aerosol fine and coarse mode (i.e., volume mean radius, standard deviation and volume concentration). During the retrieval, the algorithm selects the aerosol models with the lowest residual which is computed based on deviations between the 412 nm, 445 nm, 488 nm, and 2250 nm surface reflectances predicted from the 672 nm surface reflectance and the computed surface reflectances using the retrieved AOD for that model type.

For the VIIRS_ML, aerosol model types are also derived from AERONET inversion climatology. However, the retrieval algorithm uses that information in a different way. Levy et al. clustered and classified AERONET retrieval products into statistics that represented the most likely aerosol conditions for a particular region and season [23]. These aerosol model types are separated into fine-mode dominated (fine model) and coarse-mode dominated (coarse model), and the fine model is further separated into being strongly absorbing, moderately absorbing and weakly absorbing aerosol models. In the classification, the moderately absorbing aerosol model is set as the default, overwritten only if clear dominance of one of the other two aerosol model types is observed. By clustering, it is shown that the single scattering albedo (SSA) values at 670 nm of three fine models is ~0.85 for strongly absorbing, ~0.90 for moderately absorbing and ~0.95 for weakly absorbing and ~0.95 for the coarse model. The global type classification was updated for Collection-6, by classifying the AERONET data through 2010 [22,23]. Note that the categories of aerosol model type used for VIIRS_ML are not exactly analogous to those used for VIIRS_EDR.

2.3. VIIRS_EDR Data

The VIIRS_EDR level 2 aerosol products are obtained from NOAA Comprehensive Large Array-data Stewardship System (CLASS) at http://www.nsof.class.noaa.gov. The VIIRS_EDR aerosol parameters are derived primarily from the M-bands of the radiometric channels covering the visible through the shortwave infrared spectral regions (412 nm to 2250 nm). As explained by Jackson et al. [6], the VIIRS_EDR AOD is generated from 8×8 pixel aggregations of the intermediate product (IP), where in turn the IP represents retrieved AOD for each and every native resolution (e.g., 0.75 km) pixel. The pixels with clouds, cloud shadows, snow, ice, subpixel water, bright land surface, fire, sunglint, suspended sediments or shallow water, and large solar zenith angle are screened out using the internal tests and the external VIIRS VCM before proceeding with the aerosol retrieval [6]. The VIIRS_EDR product represents the statistics of the 8×8 aggregation, which is a retrieve then average strategy. Consequently, the resolution of the VIIRS_EDR data is ~6×6 km^2 at nadir (~12.8×12.8 km^2 at the edge of scan). Data screening and aggregation methods can be seen in [6].

VIIRS_EDR AOD has been collected from 2 May 2012 to 31 March 2014 over the NCP. The data between 15 October 2012 and 27 November 2012 are rejected because of an inadvertent error introduced in the operational aerosol code during this period [6]. The AOD at 488 nm, 550 nm and 672 nm and AOD Quality Flags (QF1), as well as Land Model Aerosol Index flag (QF4), are used in this study. The values of QF1 refer to the estimated "quality" of the retrieval product, so that QF1 = 0, 1, 2, and 3, represent not value produced, low, medium and high quality, respectively. The values of QF4 refer to which aerosol model type was used in the AOD retrieval, where QF4 = 0, 1, 2, 3, and 4, refer to dust, high absorption smoke, low absorption smoke, clean urban, and polluted urban aerosol model, respectively.

2.4. VIIRS_ML Data

VIIRS_ML data are available from the NASA Atmosphere Science Investigator-led Processing System at the University of Wisconsin (A-SIPS; http://sips.ssec.wisc.edu/). Following the strategy of the Dark-Target retrieval, the VIIRS_ML follows an average, then retrieve once logic [8]. This means that the averaging is upon observations (spectral reflectance) within the box, and following the MODIS protocol, the aerosol retrieval is performed only once. Using 10×10 pixel aggregations, the VIIRS_ML aerosol product is reported at 7.5 km (at nadir) resolution based on M-band pixel resolution. The VIIRS_ML algorithm does the cloud masking by applying the internal spatial variability and reflectance threshold tests (e.g., 3 × 3 pixel spatial variability and visible/1024/1038 nm tests). The strategies for masking, selecting and aggregating pixels for VIIRS_ML are described in [8]. Similar to the VIIRS_EDR, the AOD at 550 nm, 488 nm and 672 nm over land and the Quality Flag (QF) of aerosol retrievals during 2 May 2012 to 31 March 2014 are used. In order to be consistent with the analysis of VIIRS_EDR, VIIRS_ML data during 15 October 2012 to 27 November 2012 were not used for evaluation. The values of this QF are similar to the VIIRS_EDR QF1: 0 = bad, 1 = marginal, 2= good, and 3 = very good.

As for the aerosol model type, Levy et al. note that the aerosol model in NCP region is assumed to be moderately absorbing fine model during Winter (DJF) and Spring (MAM), and weakly absorbing fine model during Summer (JJA) and Autumn (SON) [22]. Since the aerosol type may differ day-to-day, this assumption is meant to be climatologically representative and can lead to errors in instantaneous AOD retrieval. These same assumptions are used for the VIIRS_ML.

3. Ground-Truth Data and Methods for Satellite-Sunphotometer Comparison

3.1. Sunphotometer Data

The ground data used to evaluate the VIIRS aerosol products in this study consists of CE318 sunphotometer (CE318) observations and retrievals. The CE318 instrument performs direct sun extinction measurements at eight wavelengths ranging from 340 to 1020 nm and sky radiance measurements at four wavelengths, i.e., 440, 675, 870, and 1020 nm. The AOD data were calculated from direct sun observations with an accuracy of 0.01 to 0.02 [24,25]. Refractive index, volume mean radius, volume concentration and single scattering albedo (SSA) retrieved from the CE318 sky measurements characterize the aerosol type. The uncertainties of refractive index are 30–50% for the imaginary part and 0.04 for the real part when AOD at 440 nm (AOD_{440nm}) > 0.4 and solar zenith angle > 50°, and the uncertainties increase for lower AODs [26,27]. SSA uncertainty is estimated to be less than 0.03 for AOD_{440nm} > 0.4 and the uncertainty increases for lower AODs [26,27]. Note that inversion data (size/optics) are sparse compared to direct sun observations of spectral AOD.

CE318 sunphotometer data (including the corresponding inversion products) over three sites in NCP region during the period of 2 May 2012–31 March 2014 were used. The location and description of the three CE318 sunphotometer sites are provided in Table 1. These three sites can be considered as representative of urban (Beijing), suburban (XiangHe) and regional background (Xinglong) environments, respectively. Affected by Asian monsoons, the NCP region has a moderate continental climate with cold winters and hot summers. Heavy anthropogenic pollution from urbanization, industrial, and agricultural activities mixed with coarse dust particles (most occurring in spring) result in a rather complex nature of aerosol physical and optical properties in the NCP [12]. Notably, the regional background station Xinglong is located at a mountain with the elevation of 970 m which is higher than the other two sites. However, even at this station, urban/industrial and dust aerosol could occur through aerosol regional transportation [28] and secondary aerosol formation. Therefore, as will be shown in our analysis, the complex features of aerosol properties may help explain some of the uncertainties in satellite retrievals of AOD in this region.

Table 1. Site location and description of the CE318.

Station Name	Lon (°)	Lat (°)	Site Description
Beijing	116.381 E	39.977 N	Urban station, 92 m a.s.l., located in urban area of Beijing
XiangHe	116.962 E	39.754 N	Suburban station, 36 m a.s.l., 50 km to the east of Beijing
Xinglong	117.578 E	40.396 N	Regional back-ground station, 970 m a.s.l., on the top of a mountain, 100 km to the north of Beijing

Data at Beijing and XiangHe are downloaded from AERONET (http://aeronet.gsfc.nasa.gov/) and the data at Xinglong are obtained from China Aerosol Remote Sensing Network (CARSNET) [29]. AERONET level 1.5 inversion data (from sky-light measurements) are used since the level 2 inversion data are very less frequent and unsuitable for data statistics. At the same time, we used the conditions of $AOD_{440nm} > 0.4$ and solar zenith angle $> 50°$ to constrain the data quality according to [26,27]. The AOD in these two networks are consistent with one another as the correlation coefficients are larger than 0.999 and have a 99.9% significance level [29]. The CARSNET calibration and comparison with AERONET are described in detail in other references [29–31].

It is noted that Beijing and XiangHe are part of AERONET stations, and their aerosol data during 2005 were used in the cluster analysis for the VIIRS_ML aerosol model assignment, but the Xinglong site was not used [23]. Moreover, the retrieved aerosol properties in recent years may change from those used in 2005 and 2010 due to the rapid development in the past few years over the NCP region [11,12]. Therefore, using aerosol property data from more ground sites during recent years over this region can be used to help evaluate whether those past analyses from shorter periods (e.g., only one year) and fewer sites are still representative. Notably, none of these three sites were used in the analysis by Dubovik et al. [27], which means they do not characterize typical aerosol properties of smoke, dust, and urban particles that have been adopted in the VIIRS_EDR algorithm. Thus, these three sites are better suited to evaluate the aerosol model type used in both the VIIRS_EDR and VIIRS_ML algorithms over the NCP region.

3.2. Method for Data Matchup

The spatiotemporal collocation between satellite and CE318 measurements follows the method of the Multi-sensor Aerosol Products Sampling System (MAPSS), in which sunphotometer data with ±30 min of satellite overpass are compared with satellite data within 25 km radius of the sunphotometer [32]. Minimum requirements for a matchup are at least two observations from AERONET and 5 pixels from the satellite. The CE318 AOD at 550 nm and at VIIRS blue (488 nm) and red (672 nm) bands are interpolated from 440 nm, 675 nm, 870 nm and 1020 nm by using an established fitting method [33]. The results for comparison of AOD values between VIIRS (VIIRS_EDR and VIIRS_ML) and ground CE318 observations are presented with various statistical parameters, including the number of matchup data (N), the mean bias (MB), root mean squared error (RMSE), correlation coefficient (R), and the percentage of data within the expected error 0.05 + 0.15 AOD (%EE) which is used as the MODIS AOD expected uncertainty over land [34], the slope (Slope) and intercept at y-axis (y-int) of linear regression.

3.3. Methods for Aerosol Model Evaluation and Aerosol Properties Analysis

Due to constraints placed on the inversion of CE318 sky-radiance data ($AOD_{440nm} > 0.4$; solar zenith angle $> 50°$, etc.), statistics for aerosol optical properties are sparse. To collocate aerosol optical properties from sunphotometer with aerosol models assumed by either VIIRS retrieval algorithm, we require different averaging domains. Here, we use daily-averaged aerosol optical properties retrieved from CE318 sky radiance measurements. Since the aerosol model types used for satellite AOD retrievals may vary spatially, we select only the model type assumed at the pixel that includes the site.

In the extraction of AOD from the VIIRS_EDR, those with quality QF < 1 retrievals are rejected (these with QF < 1 are not products and are mostly with cloud contamination and sunglint). Seasonal and total frequencies of each aerosol model type occurrence in the three sites are calculated to show the typical aerosol model types used in the VIIRS_EDR land algorithm over NCP sites.

The aerosol model type evaluation of VIIRS (VIIRS_EDR and VIIRS_ML) is based on the SSA comparisons between VIIRS and CE318. The SSA values at four wavelengths (440 nm, 670 nm, 870 nm and 1020 nm) of the five aerosol model types used in VIIRS_EDR can be obtained from Dubovik et al. [27]. The values of SSA at 670 nm (SSA_{670nm}) are 0.98, 0.84, 0.93, 0.97 and 0.88 for dust, high absorption smoke, low absorption smoke, clean urban and polluted urban aerosol model, respectively. Thus, the SSA values at the NCP sites in the VIIRS_EDR retrival can be derived from the aerosol model type assumed in the VIIRS_EDR pixel that includes the site. The VIIRS_ML aerosol model type is assumed globally based on the cluster analysis of SSA_{670nm} derived from all AERONET inversions and it is fixed in each season for the NCP region (i.e., weakly absorbing fine model for Summer and Autumn: $SSA_{670nm} \sim 0.95$, moderately absorbing fine model for Spring and Winter: $SSA_{670nm} \sim 0.90$) [22,23], so it is unnecessary to extract the aerosol model from VIIRS_ML pixel-by-pixel. Thus, we do the seasonal comparison; that is, the seasonal mean SSA_{670nm} values of CE318 inversion are calculated and compared with the seasonal SSA_{670nm} values of the VIIRS_EDR and VIIRS_ML. Only the VIIRS SSAs with the $AOD_{550nm} > 0.25$ are used to meet the requirement of CE318 $AOD_{440nm} > 0.4$ [23].

The SSA is also used to classify the aerosol type of the CE318 inversion, which is to evaluate the aerosol model type for each retrieval from the VIIRS algorithms. We firstly collocate the daily matchup data between the VIIRS_EDR and CE318 and between the VIIRS_ML and CE318. To evaluate the aerosol model type of VIIRS_EDR, the CE318 inversions are classified to the five aerosol types as the VIIRS_EDR. The CE318 inversion with Angstrom exponent < 0.6 and AOD at 1020 nm > 0.3 (according to Dubovik et al. [27]) is classified as dust type. If not dust type, the CE318 inversion with $SSA_{670nm} < 0.86$ is the high absorption smoke, with $0.86 < SSA_{670nm} < 0.905$ is polluted urban, with $0.905 < SSA_{670nm} < 0.95$ is low absorption smoke, and with $SSA_{670nm} > 0.95$ is clean urban aerosol type. As for the evaluation of aerosol model type used in the VIIRS_ML, the CE318 inversions are classified to the four aerosol types as the VIIRS_ML. Use the same way to find out the CE318 data with coarse model (same as the dust). For the rest CE318 data, that with $SSA_{670nm} < 0.875$ is regarded as strong absorbing fine model, $0.875 < SSA_{670nm} < 0.925$ is the moderately absorbing fine model, and $SSA_{670nm} > 0.925$ is the weakly absorbing fine model. The threshold values of the classification are based on the SSA values of the aerosol models used in the VIIRS algorithms. This method is actually using the aerosol size and scattering properties to classify the aerosol type, which has been studied by Giles et al. [35]. After classifying the CE318 aerosol type, the comparisons of aerosol type between the VIIRS_EDR and CE318 and between the VIIRS_ML and CE318 are done to show the accuracy rate of aerosol model type used in the VIIRS_EDR and VIIRS_ML. If the VIIRS aerosol model type is same as that of CE318, the aerosol model type used in the VIIRS is deemed as accurate. The accuracy rate is defined as the ratio of the number of accurate to the number of all daily matchups. The accuracy rate reflects the applicability of aerosol model type used in the VIIRS algorithms.

We also conduct the SSA comparison of different modes (fine, coarse and bi-mode) between the VIIRS_EDR and CE318 retrieval. We compute the SSA for all aerosol modes at 440 nm in the VIIRS_EDR by inputting the aerosol parameters (refractive indices, size parameters and volume concentrations of each mode) into Mie scattering calculation [36]. The reason for using 440 nm is that the aerosol model properties in the VIIRS_EDR algorithm are mostly referred at 440 nm [6]. The input aerosol parameters of the VIIRS_EDR at each site are obtained by extracting the aerosol model type over the site pixel and calculated according to the Table 2 in reference [6]. Although SSA_{440nm} is available from CE318 inversion, for consistency we also use the Mie code to compute the SSA_{440nm} based on the aerosol optical properties inversed from CE318 sky radiances. We have compared the SSAs between the CE318 inversion and the Mie scattering calculation and the result shows that the bias of the two SSA values is very low (less than 0.01). That is because the CE318 inversion also uses Mie scattering calculation to obtain SSA. The resultant SSA_{440nm} of CE318 is compared with that of the VIIRS_EDR SSA_{440nm}. Since the CE318 sky-radiance inversion product is only reliable for $AOD_{440nm} > 0.4$, we also choose the aerosol properties of the VIIRS_EDR when $AOD_{550nm} > 0.25$ [23].

Table 2. Statistics of matchup of CE318 and VIIRS AOD at 550 nm in the NCP region during 2 May 2012–31 March 2014. N is the number of matchup data. MB is the mean bias. RMSE is root mean squared error. R is correlation coefficient. %EE stands the percentage of data within the expected error of 0.05 + 0.15 AOD. Slope and y-int are the slope and intercept at y-axis of the linear regression, respectively.

QF	N	MB	RMSE	R	%EE	Slope	y-int
VIIRS_EDR vs. CE318							
QF > 0	860	0.05	0.24	0.83	44.3	0.85	0.07
QF > 1	762	0.04	0.22	0.84	46.5	0.88	0.05
QF = 3	564	0.06	0.23	0.86	48.9	0.91	0.07
VIIRS_ML vs. CE318							
QF > 0	817	0.14	0.25	0.94	51.0	1.27	0.02
QF > 1	755	0.13	0.24	0.94	53.1	1.26	0.01
QF = 3	683	0.13	0.25	0.93	54.0	1.26	−0.00

4. Results of AOD Inter-Comparison

4.1. Evaluation of the VIIRS_EDR AOD at 550 nm

Table 2 reports the validation results of the two VIIRS algorithms compared to the collocated ground CE318 observations, for the period of 2 May 2012–31 March 2014 over the NCP region. There are 860, 762 and 564 instantaneous VIIRS_EDR–CE318 matchups of QF > 0, QF > 1 and QF = 3 at the NCP sites during the period, respectively. Starting with gross statistics, the slope and intercept of the best-fit equation between the VIIRS_EDR and CE318 AOD are 0.85–0.91 and 0.05–0.07, respectively, with R ranging from 0.83–0.86. The VIIRS_EDR data are well correlated with CE318 observations. However, the VIIRS_EDR AOD showed a positive MB of 0.04–0.06 and a rather large RMSE of 0.22–0.24. Only 44.3–48.9% of the compared AODs meet the expected error envelope of 0.05 + 0.15 AOD. Filtering by quality flag (QF > 1), the comparison improves for all statistics, however constraining to only high quality flags (QF = 3) does not improve the overall agreement any further. These issues need to be studied at each site.

The MB and RMSE in the NCP are both larger than the counterparts in the global assessment statistics [6,9]. Table 3 presents the evaluation results of each NCP site, separately. Clearly, all properties (MB, RMSE, R and %EE) demonstrate the worst performance over Beijing. The MB in XiangHe site is (−0.02)–0.00 with high R of 0.89–0.92, which is more comparable to the global agreement [9]. Beijing is an urban site, while the VIIRS_EDR uses the global surface reflectance ratios as the expected spectral surface reflectance relationship, which may cause the largest error at the Beijing site [6].

Table 3. Statistics of the matchup between CE318 and VIIRS_EDR AOD at 550 nm over each site during 2 May 2012–31 March 2014.

VIIRS_EDR QF	Site	N	MB	RMSE	R	%EE	Slope	y-int
	Beijing	336	0.10	0.29	0.76	36.9	0.83	0.17
QF > 0	XiangHe	323	−0.02	0.21	0.89	48.3	0.88	0.04
	Xinglong	201	0.05	0.17	0.81	50.2	0.86	0.08
	Beijing	291	0.10	0.27	0.79	38.8	0.90	0.14
QF > 1	XiangHe	289	−0.02	0.19	0.90	50.2	0.90	0.02
	Xinglong	182	0.04	0.15	0.82	52.7	0.81	0.08
	Beijing	204	0.15	0.30	0.80	36.8	0.82	0.22
QF = 3	XiangHe	220	0.00	0.18	0.92	53.6	0.89	0.05
	Xinglong	140	0.02	0.14	0.83	59.3	0.81	0.06

In the previous global evaluation for VIIRS_EDR products, it was recommended to use data with a higher QF [9]. In the NCP, comparisons at both XiangHe (suburban) and Xinglong (rural) support this recommendation (R and %EE increase and MB decreases with increasing QF). However, for the Beijing site, the matchup statistics are poor, and increasing QF does not help. Thus, for sites that are not optimal for aerosol retrieval in the first place (e.g., urban), QF may not be a useful diagnostic.

4.2. Evaluation of the VIIRS_ML AOD at 550 nm

Comparing the VIIRS_ML to CE318 (Table 2), there are 817, 755, and 683 matchups for QF > 0, QF > 1, and QF = 3, respectively. Since each algorithm has its own definition of QF, there are different relative contributions of each QF level [6,8]. Overall, the VIIRS_ML AODs show a high correlation with CE318 (R is 0.93–0.94) but overestimates over this region with high MBs (0.13–0.14) and large slope values for the equations of best fit (1.26–1.27). More than half of the VIIRS_ML data are within the expected error envelope (%EE > 50%). As compared to VIIRS_EDR, the VIIRS_ML has a higher bias, but has larger correlation with more data within the EE.

Following site-by-site comparison for the VIIRS_EDR, we evaluate the VIIRS_ML AOD at each site (Table 4). Like the VIIRS_EDR, the VIIRS_ML shows the largest bias over the Beijing urban site. Since the urban surface reflectance may be underestimated in the "dark-target" algorithm, this can lead to an overestimation of AOD [37]. The VIIRS_ML AOD over XiangHe performs the best with the highest values of R and %EE but the MB is not the lowest and it is higher than that between the VIIRS_EDR and CE318 (MB of (−0.02)–0.00). However, the R values between the VIIRS_ML and CE318 at all three sites are higher than those found for the VIIRS_EDR and CE318. The values of %EE of the VIIRS_ML over the XiangHe and Xinglong sites are higher but lower over the Beijing site compared to those between the VIIRS_EDR and CE318.

Table 4. Statistics of the matchup of the CE318 and VIIRS_ML AOD at 550 nm over each site during 2 May 2012–31 March 2014.

VIIRS_ML QF	Site	N	MB	RMSE	R	%EE	Slope	y-int
QF>0	Beijing	292	0.26	0.33	0.94	21.2	1.19	0.18
	XiangHe	340	0.09	0.20	0.97	70.6	1.26	−0.03
	Xinglong	185	0.03	0.13	0.89	62.2	1.07	0.02
QF>1	Beijing	263	0.26	0.33	0.91	22.8	1.14	0.20
	XiangHe	322	0.08	0.20	0.97	72.0	1.26	−0.04
	Xinglong	170	0.03	0.12	0.92	64.1	1.19	−0.01
QF=3	Beijing	227	0.28	0.36	0.89	19.8	1.08	0.25
	XiangHe	302	0.07	0.19	0.97	73.2	1.26	−0.04
	Xinglong	154	0.01	0.11	0.92	66.9	1.15	−0.02

The results of the quality flag analysis of the VIIRS_ML AOD are similar to those for the VIIRS_EDR. Using high quality data leads to the best performance at the XiangHe and Xinglong sites but it is not suitable at the Beijing site.

4.3. Evaluation of the VIIRS AOD at Red and Blue Bands

While the AOD is retrieved at 550 nm, neither algorithm uses reflectance at 550 to derive AOD. This is because the Earth's surface tends to be brighter in green wavelengths (e.g., vegetation), and not suitable for aerosol retrieval. The VIIRS_EDR algorithm is based on the calculation of surface reflectance at blue (488 nm) and red (672 nm) and three other bands (412 nm, 445 nm and 2250 nm) [6]. As for the VIIRS_ML algorithm, AOD at 550 nm is inversed by using 488 nm, 672 nm and 2257 nm measured TOA reflectance to find a the optimal solution [8]. Hence, the VIIRS AOD spectral dependence between blue and red bands are also evaluated with the CE318 data. According to quality flag analysis in Sections 4.1 and 4.2, we choose the matchup data of QF > 1 for both the VIIRS_EDR-CE318 and the

VIIRS_ML-CE318. Then, we selected the data of VIIRS_EDR, VIIRS_ML and CE318 with the dates in common. That was to select the matchup data of VIIRS_EDR, CE318 and VIIRS_ML.

Figure 1 shows the average AODs of the matchup data between VIIRS_EDR, CE318 and VIIRS_ML at three wavelengths over the three sites. Compared to CE318 measurements, the VIIRS_EDR AOD at 488 nm overestimates over Beijing and Xinglong but performs well for the XiangHe site. At 672 nm, the VIIRS_EDR AOD also overestimates over Beijing, but slightly undervalues AOD at XiangHe. However, the VIIRS_ML AOD overestimates at all three wavelengths and over all the three sites, especially over Beijing. The biases of each wavelength can reach 0.2. One of the possible reasons for the larger bias over Beijing may be that the VIIRS_ML algorithm is actually the Dark-Target aerosol retrieval algorithm and this algorithm may overestimate the AOD values over bright surfaces such as urban centers by 0.2 [22,37]. The large VIIRS_ML biases may also be related to the errors of aerosol model type (to be discussed in next section).

We also calculate the aerosol Angstrom Exponent (AE) between 488 nm and 672 nm (AE = $\log(AOD_{488nm}/AOD_{672nm})/\log(672nm/488nm)$) to describe the AOD spectral dependence. AE is often used as an indicator of aerosol size distribution which is related to aerosol type: AE ~ 0 corresponds to large particles; and AE ~ 2 corresponds to small particles. The average AE values in Figure 1 are calculated from each matchup data set. The AE values of the VIIRS_EDR and the VIIRS_ML show large differences when comparing to that from CE318. AE biases of the VIIRS_EDR are 0.06 in Beijing, 0.31 in XiangHe and 0.55 in Xinglong, while the biases for the VIIRS_ML are −0.42 over Beijing, −0.17 in XiangHe and 0.20 over Xinglong. The AE of the VIIRS_EDR is larger than that from CE318 over all the three sites, but the AE of the VIIRS_ML is often lower than CE318 except for over Xinglong site. These indicate that the aerosol size of aerosol model type used in the VIIRS_EDR algorithm is smaller while VIIRS_ML is larger except for Xinglong (the aerosol model type used in VIIRS AOD retrieval will be discussed in next section).

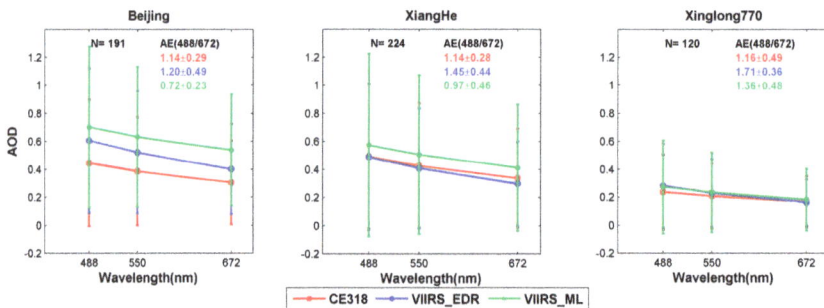

Figure 1. The average aerosol optical depth (AOD) of matchup data at blue (488 nm), red (672 nm) and 550 nm wavelengths between VIIRS_EDR, CE318 and VIIRS_ML. N is the number of matchup data. The average and standard deviation of Angstrom Exponent (AE) from 488 nm and 672 nm of each matchup data set is also shown in this figure: red for CE318, blue for VIIRS_EDR, and green for VIIRS_ML.

5. Results of Aerosol Model and the Optical Properties Inter-Comparison

The aerosol model types used in the VIIRS_EDR and the VIIRS_ML over the NCP are evaluated by comparing SSA values with those derived from the CE318 inversion. The comparisons of aerosol optical properties between the VIIRS_EDR and CE318 are also shown in this section to help explain the error from aerosol model used in VIIRS_EDR.

5.1. Evaluation of the VIIRS_EDR Aerosol Model Type

Figure 2 shows the frequencies of each aerosol model type used in the VIIRS_EDR (hereafter called M_VIIRS_EDR) AOD retrieval at the Beijing, XiangHe and Xinglong sites during the evaluation period. For M_VIIRS_EDR, the dust and clean urban aerosol models are the two dominant model types used in NCP region. In Beijing, the M_VIIRS_EDR shows that dust and clean urban models account for more than 80% of the aerosols during the evaluation period. The frequency of the polluted urban model is less than 1%. However, Beijing is a mega city with a population of approximately 21 million and five million vehicles are located in the heavy polluted NCP region. It is undisputed that polluted urban aerosol is the dominant aerosol [38]. Thus, the M_VIIRS_EDR is unsuitable at the Beijing site. Because XiangHe is located 50 km to the east of Beijing, the M_VIIRS_EDR for XiangHe shows similar results to Beijing. However, XiangHe also shows some differences from Beijing: dust and clean urban models decrease and other models increase. As for Xinglong station (regional back ground station), the M_VIIRS_EDR shows more polluted urban and low absorption smoke and less dust models than the Beijing and XiangHe stations. From Beijing to Xinglong, the M_VIIRS_EDR shows that the frequency of polluted urban aerosol models increase, which is inconsistent with the fact that pollution decreases from Beijing to Xinglong according to past study results in the NCP region [12,28,39].

To show the aerosol model type differences between the VIIRS retrievals and CE318 sunphotometer observations, the seasonal values of SSA_{670nm} of the CE318 inversion, VIIRS_EDR and VIIRS_ML in the NCP three sites are shown in Table 5. Comparing the VIIRS_EDR and CE318, the SSA_{670nm} values from the VIIRS_EDR are higher than those from the CE318 during almost all seasons and over all the three sites. This result reflects more frequent weakly absorbing aerosol model type used in VIIRS_EDR retrievals in the NCP sites, as Figure 2 shows more frequency of dust and clean urban aerosol models. The largest difference between the VIIRS_EDR and CE318 at Beijing and XiangHe sites is shown during winter with 0.07 at Beijing and 0.08 at XiangHe. However, the difference at Xinglong during winter is smallest. The largest difference at the Xinglong site occurred during the spring and summer and the difference value is 0.03, which is less than those at Beijing and XiangHe. These indicate that the M_VIIRS_EDR at Beijing and XiangHe have more errors than at Xinglong.

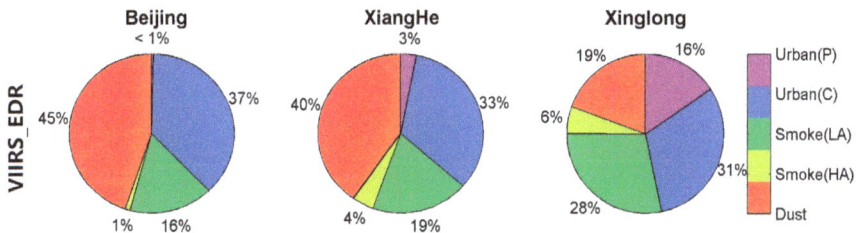

Figure 2. The frequencies of the aerosol model types used in the VIIRS_EDR at Beijing, XiangHe and Xinglong. Urban(P) is the polluted urban aerosol model, Urban(C) is the clean urban aerosol model, Smoke(LA) is the low absorption smoke aerosol model, Smoke(HA) is the high absorption smoke aerosol model, and the last aerosol model is for Dust.

To know how many aerosol types used in VIIRS_EDR are suitable, we calculated the accuracy rate of the M_VIIRS_EDR at the NCP three sites and the results are shown in Figure 3. The accuracy rates of the M_VIIRS_EDR over Beijing, XiangHe, and Xinglong are 0.24, 0.20, and 0.37, respectively (Figure 3a). The accuracy rate of the M_VIIRS_EDR over Xinglong is highest, which is consist with the lowest SSA difference between the VIIRS_EDR and CE318 at Xinglong. As for each model type, although the dust model is more used in the M_VIIRS_EDR (Figure 2), the accuracy rate of the dust model is very small at Beijing and XiangHe and even equal to zero at Xinglong site (Figure 3a,c).

The accuracy rate of low absorption smoke is relatively higher than other models of M_VIIRS_EDR. The accuracy rate of the polluted urban aerosol model is practically zero because the frequency of polluted urban model occurred in the M_VIIRS_EDR over Beijing and XiangHe stations is very low (Figure 2) and it is different from the aerosol type of the CE318 of the matchup. All these results indicate that the M_VIIRS_EDR selected more weakly absorbing aerosol models in the NCP sites.

It is worth to noting that with more frequencies of clean and dust aerosol models (higher SSA values) used in the VIIRS_EDR retrieval, the VIIRS_EDR AOD should have an underestimation if the surface reflectance characterization in VIIRS_EDR algorithm is perfect. The fact that the VIIRS_EDR AOD in Beijing has a high bias reflects that other factors (e.g., surface reflectance characterization or cloud contamination) are important error sources in the VIIRS_EDR algorithm, and their effects on aerosol retrievals override the effects from non-ideality in aerosol model types. This can be an interesting topic for future studies.

Table 5. The seasonal values of SSA at 670 nm from CE318 inversion, VIIRS_EDR and VIIRS_ML during 2 May 2012–31 March 2014.

Station	Sensor	SSA at 670 nm			
		Spring	Summer	Autumn	Winter
Beijing	CE318	0.94 ± 0.02	0.96 ± 0.03	0.95 ± 0.03	0.91 ± 0.04
	VIIRS_EDR	0.96 ± 0.02	0.96 ± 0.03	0.97 ± 0.02	0.98 ± 0.01
	VIIRS_ML	~0.9	~0.95	~0.95	~0.9
XiangHe	CE318	0.91 ± 0.04	0.95 ± 0.03	0.92 ± 0.04	0.89 ± 0.04
	VIIRS_EDR	0.96 ± 0.02	0.95 ± 0.03	0.96 ± 0.03	0.97 ± 0.01
	VIIRS_ML	~0.9	~0.95	~0.95	~0.9
Xinglong	CE318	0.93 ± 0.03	0.92 ± 0.03	0.92 ± 0.03	0.96 ± 0.03
	VIIRS_EDR	0.96 ± 0.02	0.94 ± 0.02	0.95 ± 0.03	0.97 ± 0.00
	VIIRS_ML	~0.9	~0.95	~0.95	~0.9

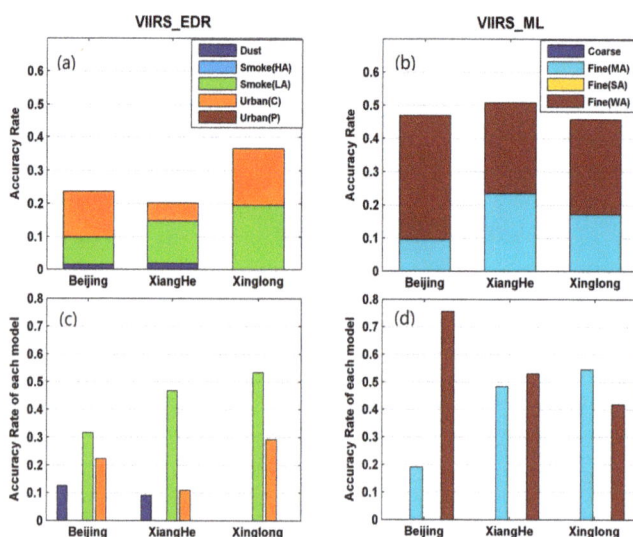

Figure 3. The accuracy rate of the aerosol model used in: the VIIRS_EDR (**a,c**); and the VIIRS_ML (**b,d**). The accuracy rate stands for the ratio of the number of accurate model type used in the VIIRS to all the matchups between the VIIRS and CE318. The accuracy rate of each model is the ratio of the number of accurate model to the matchups of its corresponded model type.

5.2. Evaluation of the VIIRS_ML Aerosol Model Type

Using the same way of evaluating the M_VIIRS_EDR, the aerosol model type assumed in the VIIRS_ML (M_VIIRS_ML) in the NCP sites is evaluated and shown in Table 5 and Figure 3. In Table 5, it can be found that the biases of SSA_{670nm} between the VIIRS_ML and CE318 are $\leq$0.04 in almost all sites and all seasons except for Xinglong during the winter. The average bias for all seasons is -0.015, 0.0075 and -0.0075 at Beijing, XiangHe and Xinglong, respectively. These biases are less than the differences between the VIIRS_EDR and CE318 at the corresponding site. That is likely because the M_VIIRS_ML is defined according to the AERONET sunphotometer inversion in local regions while M_VIIRS_EDR is from the predefined aerosol model types at five sites located in other places. Notably, the obvious undervaluation of the VIIRS_ML SSA over Beijing during almost all seasons may cause the overestimation of the VIIRS_ML AOD over Beijing (MB $\geq$ 0.26 shown in Table 2) versus the VIIRS_EDR.

However, there are some large differences of SSA_{670nm} between the VIIRS_ML and CE318 in some seasons. The moderately absorbing model in spring over Beijing used in the VIIRS_ML may be inappropriate because the CE318 inversion shows weakly absorbing type. The largest difference is found at Xinglong (regional background site). The absolute biases in all seasons over this site are $\geq$0.3 and the highest bias 0.6 is occurred in the winter. The CE318 shows weakly absorbing aerosols in the winter that is in contrast to the moderately absorbing aerosols used by the VIIRS_ML. This may indicate that the M_VIIRS_ML over the NCP may be unsuitable for regional background sites.

In Figure 3, it can be found that the accuracy rates of the M_VIIRS_ML over Beijing, XiangHe, and Xinglong are 0.47, 0.51, and 0.46, respectively (Figure 3b). The average accuracy rate of the M_VIIRS_ML over the NCP region (0.48) is higher than that of M_VIIRS_EDR (0.27). As for each aerosol model type (Figure 3d), the accuracy rate of the weakly absorbing fine model is higher than the moderately absorbing fine model in Beijing and XiangHe. However, in Xinglong, the accuracy rate of the weakly absorbing fine model is lower than the moderately absorbing fine model. The large difference of the accuracy rate of the two model types in Beijing reflects that the moderately absorbing fine model assumed in the spring and winter may require careful consideration because of the dust in the spring and more strongly absorbing aerosols in the winter cannot be neglected [39].

For different sites, the accuracy rates of the M_VIIRS_EDR over Beijing and XiangHe are higher than their corresponding values of the M_VIIRS_EDR. From Beijing to Xinglong, the accuracy rate of the M_VIIRS_EDR decreases, while M_VIIRS_ML varies little with lowest value over Xinglong site. This reflects that the M_VIIRS_EDR is unsuitable over the NCP urban and suburban sites, while the M_VIIRS_ML is suitable over urban and suburban sites. The highest accuracy rate of the M_VIIRS_EDR over the NCP sites is only 0.37. Thus, the aerosol model type selection in the VIIRS_EDR algorithm is inappropriate in the NCP region, which may cause an important error in the AOD inversion.

5.3. Inter-Comparison of Aerosol Properties between CE318 and the VIIRS_EDR

Since the M_VIIRS_EDR performs less well in the NCP region based on above analysis and M_VIIRS_ML is defined according to the CE318 inversion in local regions, we only compare the aerosol properties between CE318 and the VIIRS_EDR. Table 6 shows the averages of the SSA_{440nm} of fine, coarse and bi-modal aerosols derived from CE318 (AOD_{440nm} > 0.4) and the VIIRS_EDR (AOD_{550nm} > 0.25) at the three sites over the time period of evaluation (2 May 2012–31 March 2014). The values of SSA_{440nm} from the CE318 are less than that of the VIIRS_EDR for all modes and all sites, which indicates that VIIRS_EDR overestimates SSA_{440nm} values over all sites (consistent with SSA_{670nm} in Table 5). The difference of SSA_{440nm} in the coarse mode (i.e., 0.09 in Beijing, 0.12 in XiangHe, and 0.06 in Xinglong) is larger than SSA_{440nm} in fine mode (i.e., 0.02 in Beijing, 0.04 in XiangHe and 0.03 in Xinglong). This indicates that aerosol properties in the coarse mode in the VIIRS_EDR need to be revised for the NCP region. The biases of bi-mode SSA_{440nm} are 0.03 (Beijing), 0.05 (XiangHe) and 0.03 (Xinglong). The overestimation of SSA but largest positive AOD MB of the VIIRS_EDR over Beijing site indicate again that other positive bias factor (e.g. surface reflectance characterization) overpowers the negative bias due to SSA of aerosol model type over Beijing. The largest overestimation of SSA and negative MB of AOD (in Table 3) are occurred at XiangHe, which may indicate that errors from the aerosol model type are overpowering at the XiangHe site.

Table 6. The fine (f), coarse (c) and bi-mode (Bi) aerosol SSA at 440 nm for CE318 and the VIIRS_EDR during 2 May 2012–31 March 2014.

Station		SSA(f)	SSA(c)	SSA(Bi)
	CE318	0.95	0.72	0.93
Beijing	VIIRS_EDR	0.97	0.81	0.96
	Bias(VIIRS-CE318)	0.02	0.09	0.03
	CE318	0.93	0.66	0.91
XiangHe	VIIRS_EDR	0.97	0.78	0.96
	Bias	0.04	0.12	0.05
	CE318	0.95	0.72	0.93
Xinglong	VIIRS_EDR	0.97	0.78	0.96
	Bias	0.02	0.06	0.03

Since the SSA is calculated by inputting the aerosol parameters in Table 2 from Jackson et al. [6] to the Mie scattering code and large differences of SSA between the VIIRS_EDR and CE318 are found above, it is necessary to compare the aerosol properties between the VIIRS_EDR and CE318. The constraints of CE318 $AOD_{440nm} > 0.4$ and VIIRS_EDR $AOD_{550nm} > 0.25$ are also used and the daily aerosol optical properties retrieved from CE318 sky radiance measurements are averaged before the followed analysis. Table 7 shows the average aerosol physical properties from CE318 and the VIIRS_EDR. The refractive index is RI with real part is RI(r), while imaginary part is RI(i). Volume mean radius, standard deviation and volume concentration are r, σ and V, respectively. The fine and coarse mode aerosols are shown by f and c in bracket pairs. Figure 4 shows the seasonal comparison of normalized aerosol physical properties from CE318 and the VIIRS_EDR at the three sites. Each parameter is normalized between 0.1 and 1 to well show the difference of CE318 and the VIIRS_EDR. The length of each radius in the circle equals 1 and each radius direction stands for one aerosol parameter.

Table 7. The average aerosol physical properties (at 440 nm) of CE318 and VIIRS_EDR during 2 May 2012–31 March 2014.

Station	Sensor	RI(r)	RI(i)	r(f)	σ(f)	V(f)	r(c)	σ(c)	V(c)
Beijing	CE318	1.49	0.0089	0.20	0.53	0.14	2.70	0.60	0.18
	VIIRS_EDR	1.43	0.0043	0.19	0.44	0.13	3.15	0.69	0.19
XiangHe	CE318	1.49	0.013	0.19	0.53	0.13	2.81	0.62	0.16
	VIIRS_EDR	1.43	0.0053	0.19	0.41	0.15	3.50	0.74	0.10
Xinglong	CE318	1.47	0.0083	0.22	0.56	0.10	2.84	0.61	0.10
	VIIRS_EDR	1.43	0.0056	0.18	0.40	0.13	3.51	0.75	0.08

Distinctly differences of various parameters between CE318 and the VIIRS_EDR can be found. For the total averages in three sites (Table 7), CE318 shows high values of RI(r) 1.47–1.49 and RI(i) 0.0083–0.013, which indicates that aerosol in the NCP is more absorptive than that used in the VIIRS_EDR, which corresponds the overestimation of SSA for the VIIRS_EDR (Table 5). The r(f) values from CE318 (0.19–0.22) are slightly higher than the values of the VIIRS_EDR (0.18–0.19), while the r(c) values of CE318 (2.70–2.84) are significantly lower than the values of the VIIRS_EDR (3.15–3.51), which may cause more dust aerosol for the VIIRS_EDR (Figure 2) and also reflects the dust aerosol model in Cape Verde maybe different from the dust in Asia [40,41]. The σ(f) in CE318 is higher than the

VIIRS_EDR but lower for σ(c). The V(f) values of CE318 are lower than that of the VIIRS_EDR except for Beijing station. The V(c) of the VIIRS_EDR is comparable with that of CE318 except in XiangHe, which consists with the lowest accuracy rate of the M_VIIRS_EDR in XiangHe (Figure 3a).

In Figure 4, the shapes generated by eight parameters (1, RI(r); 2, RI(i); 3, r(f); 4, σ(f); 5, V(f); 6, r(c); 7, σ(c); and 8, V(c)) for CE318 and the VIIRS_EDR differ from each other and each of them (CE318 and VIIRS_EDR) shows a different seasonal variation. For Beijing, CE318 shows similar shapes in summer and autumn but significantly different in spring and winter; higher RI(r) in spring and winter and higher RI(i) in winter. While the VIIRS_EDR shows two pairs of similar shapes; spring–summer and autumn–winter. As for XiangHe station, CE318 shapes are similar to Beijing's but the VIIRS_EDR shows some difference from Beijing's; autumn is not similar to winter but similar to spring and summer. As for Xinglong, CE318 shapes show a distinct difference in winter compared to XiangHe's due to the lower RI(r), RI(i) and r(c) and higher r(f), while the VIIRS_EDR shapes are similar to XiangHe's in all seasons.

The significant differences of the aerosol microphysical properties between the VIIRS_EDR and CE318 over the NCP indicate that the aerosol microphysical properties in the VIIRS_EDR algorithm are not suitable over the NCP region. Furthermore, this also reflects that the five models based on the five AERONET stations in the reference [27] can not be applied globally.

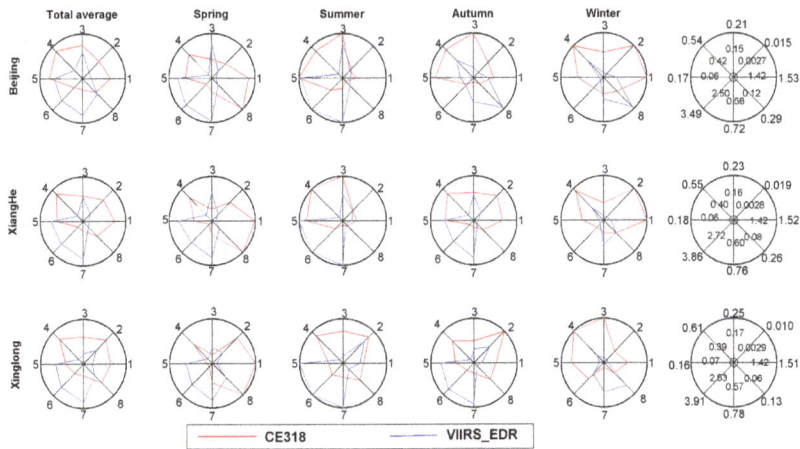

Figure 4. The comparisons of normalized aerosol physical properties from CE318 and VIIRS_EDR at the three sites. Each parameter at each site is normalized between 0.1 and 1 to well show the difference of CE318 and VIIRS_EDR. The length of each radius in the circle equals to 1 and each radius direction stands for one aerosol parameter: 1, real part of refractive indices; 2, imaginary part of refractive indices; 3, volume mean radius of fine mode; 4, standard deviation of fine mode; 5, volume concentration of fine mode; 6, volume mean radius of coarse mode; 7, standard deviation of coarse mode; and 8, volume concentration of coarse mode, respectively. The values in last column in each site stand for the values of the eight aerosol properties at the radiuses of 0.1 and 1 positions.

6. Conclusions

Using CE318 data at three sites in the NCP region to evaluate the VIIRS AOD, aerosol model types and aerosol optical properties used in the VIIRS_EDR and VIIRS_ML algorithms at three sites over the NCP, we conclude:

a.   The VIIRS_EDR AOD at 550 nm has a positive MB of 0.04–0.06 with R of 0.83–0.86. Among the three sites, the bias at Beijing is largest with MB of 0.10–0.15, RMSE of 0.27–0.30 and low %EE of 36.8–38.8%. The quality flags analysis shows that using the high quality products of

AOD at XiangHe and Xinglong are recommended but not at Beijing site. The VIIRS_ML AOD overestimates more over the NCP region with higher positive MB of 0.13–0.14 but shows higher correlation (0.93–0.94) with ground-based AOD. The results of evaluation of the VIIRS_ML for each site and quality flags analysis are found to be similar to that for the VIIRS_EDR.

b. The aerosol model types used in the VIIRS_EDR AOD retrieval in the three sites are mostly dust and clean urban aerosol models (the frequencies of these two models in three sites are all larger than 50%) with less frequency of polluted urban aerosol models used (less than 1%). The accuracy rates of the M_VIIRS_EDR over the Beijing, XiangHe and Xinglong sites (0.24, 0.20 and 0.37) are lower than that of the M_VIIRS_ML (0.47, 0.51 and 0.46) during the evaluation period.

c. The values of SSA_{440nm} from CE318 are less than the VIIRS_EDR for all modes and sites, with differences of 0.02–0.04 for fine mode, 0.06–0.12 for coarse mode and 0.03–0.05 for bi-modes. The overestimation of SSA but positive AOD mean bias of the VIIRS_EDR indicate that other factors (e.g., surface reflectance characterization or cloud contamination) are important error sources in the VIIRS_EDR algorithm, and their effects on aerosol retrievals override the effects from non-ideality in aerosol model types. The differences of SSA_{670nm} between VIIRS_ML and CE318 in the NCP are mostly less than 0.015 but high seasonal differences are also found. The undervaluation of SSA used in the VIIRS_ML algorithm over the NCP causes the overestimation of AOD, especially at Beijing site.

We recommend that the aerosol model types and the microphysical properties in the VIIRS_EDR algorithm in NCP region are not representative and need to be refined. The AOD bias in Beijing is largest but we do not find the lowest accuracy rate of the M_VIIRS_EDR. In addition, the higher values of the VIIRS_EDR SSA versus CE318, which should lead to lower AODs from the satellite inversion, are inconsistent with the positive MBs of AOD in the NCP region. All these points indicate that there are other error sources that need to be examined in the AOD retrieval for the VIIRS_EDR algorithm, especially for Beijing site. Future studies should investigate these potential sources of error including the surface reflectance and cloud contamination.

Acknowledgments: This research was supported by the national key research and development program of China (No.2016YFC0200403), the Strategic Priority Research Program (A) of the Chinese Academy of Sciences (XDA05100301), the National Science Foundation of China (41175031, 41590874 & 41375153), National Key R&D Program Pilot Projects of China (2016YFA0601901) and the CAMS Basis Research Project (2016Z001 & 2014R17). J. Wang's participation is supported by UNL's internal funding.

Author Contributions: Xiangao Xia, Jun Wang and Robert C. Levy conceived and designed the experiments and participated in the analysis of the data; Mohammed Ayoub participated in the design of the experiments; Huizheng Che, Hongbin Chen, Jingqiang Zhang, Min Oo and Robert Holz contributed the materials; Xiaoguang Xu contributed analysis tool; and Jun Zhu performed the experiments, analyzed the data and wrote the paper.

References

1. Charlson, R.J.; Schwartz, S.E.; Hales, J.M.; Cess, R.D.; Coakley, J.A.; Hansen, J.E.; Hofmann, D.J. Climate Forcing by Anthropogenic Aerosols. *Science* **1992**, *255*, 423–430. [CrossRef] [PubMed]

2. Kaufman, Y.J.; Tanre, D.; Boucher, O. A Satellite View of Aerosols in the Climate System. *Nature* **2002**, *419*, 215–223. [CrossRef] [PubMed]

3. Pope, C.A.; Burnett, R.T.; Thun, M.J.; Calle, E.E.; Krewski, D.; Ito, K.; Thurston, G.D. Lung Cancer, Cardiopulmonary Mortality, and Long-Term Exposure to Fine Particulate Air Pollution. *JAMA J. Am. Med. Assoc.* **2002**, *287*, 1132–1141. [CrossRef]

4. Diner, D.J.; Ackerman, T.P.; Anderson, T.L.; Bösenberg, J.; Braverman, A.J.; Charlson, R.J.; Collins, W.D.; Davies, R.; Holben, B.N.; Hostetler, C.A.; et al. Paragon: An Integrated Approach for Characterizing Aerosol Climate Impacts and Environmental Interactions. *Bull. Am. Meteorol. Soc.* **2004**, *85*, 1491–1501. [CrossRef]

5. Cao, C.Y.; Xiong, J.; Blonski, S.; Liu, Q.H.; Uprety, S.; Shao, X.; Bai, Y.; Weng, F.Z. Suomi Npp Viirs Sensor Data Record Verification, Validation, and Long-Term Performance Monitoring. *J. Geophys. Res. Atmos.* **2013**, *118*. [CrossRef]

6. Jackson, J.M.; Liu, H.Q.; Laszlo, I.; Kondragunta, S.; Remer, L.A.; Huang, J.F.; Huang, H.C. Suomi-Npp Viirs Aerosol Algorithms and Data Products. *J. Geophys. Res. Atmos.* **2013**, *118*, 12673–12689. [CrossRef]

7. Meng, F.; Cao, C.Y.; Shao, X.; Shi, Y.G. Spatial and Temporal Variation of Visible Infrared Imaging Radiometer Suite (Viirs)-Derived Aerosol Optical Thickness over Shandong, China. *Int. J. Remote Sens.* **2014**, *35*, 6023–6034. [CrossRef]

8. Levy, R.C.; Munchak, L.A.; Mattoo, S.; Patadia, F.; Remer, L.A.; Holz, R.E. Towards a Long-Term Global Aerosol Optical Depth Record: Applying a Consistent Aerosol Retrieval Algorithm to Modis and Viirs-Observed Reflectance. *Atmos. Meas. Tech.* **2015**, *8*, 4083–4110. [CrossRef]

9. Liu, H.Q.; Remer, L.A.; Huang, J.F.; Huang, H.C.; Kondragunta, S.; Laszlo, I.; Oo, M.; Jackson, J.M. Preliminary Evaluation of S-Npp Viirs Aerosol Optical Thickness. *J. Geophys. Res. Atmos.* **2014**, *119*, 3942–3962. [CrossRef]

10. Hsu, C.; Levy, R.; Sayer, A.; Lee, J.; Laszlo, I.; Bettenhausen, C.; Gautam, R.; Mattoo, S.; Munchak, L.; Liu, H. *Suomi NPP VIIRS Aerosol Data Product Assessment Report*; NASA: Washington, DC, USA, 2013; pp. 1–17.

11. Quan, J.; Tie, X.; Zhang, Q.; Liu, Q.; Li, X.; Gao, Y.; Zhao, D. Characteristics of Heavy Aerosol Pollution During the 2012–2013 Winter in Beijing, China. *Atmos. Environ.* **2014**, *88*, 83–89. [CrossRef]

12. Xia, X.; Chen, H.; Goloub, P.; Zong, X.; Zhang, W.; Wang, P. Climatological Aspects of Aerosol Optical Properties in North China Plain Based on Ground and Satellite Remote-Sensing Data. *J. Quant. Spectrosc. Radiat. Transf.* **2013**, *127*, 12–23. [CrossRef]

13. Luo, Y.; Lu, D.; Zhou, X.; Li, W.; He, Q. Characteristics of the Spatial Distribution and Yearly Variation of Aerosol Optical Depth over China in Last 30 Years. *J. Geophys. Res. Atmos.* **2001**, *106*, 14501–14513.

14. Li, Z.; Xia, X.; Cribb, M.; Mi, W.; Holben, B.; Wang, P.; Chen, H.; Tsay, S.-C.; Eck, T.F.; Zhao, F. Aerosol Optical Properties and Their Radiative Effects in Northern China. *J. Geophys. Res. Atmos.* **2007**, *112*, 321–341. [CrossRef]

15. Tao, M.; Chen, L.; Su, L.; Tao, J. Satellite Observation of Regional Haze Pollution over the North China Plain. *J. Geophys. Res. Atmos.* **2012**, *117*. [CrossRef]

16. Che, H.; Xia, X.; Zhu, J.; Li, Z.; Dubovik, O.; Holben, B.; Goloub, P.; Chen, H.; Estelles, V.; Cuevas-Agulló, E. Column Aerosol Optical Properties and Aerosol Radiative Forcing During a Serious Haze-Fog Month over North China Plain in 2013 Based on Ground-Based Sunphotometer Measurements. *Atmos. Chem. Phys.* **2014**, *14*, 2125–2138. [CrossRef]

17. Wang, Y.; Che, H.Z.; Ma, J.Z.; Wang, Q.; Shi, G.Y.; Chen, H.B.; Goloub, P.; Hao, X.J. Aerosol Radiative Forcing under Clear, Hazy, Foggy, and Dusty Weather Conditions over Beijing, China. *Geophys. Res. Lett.* **2009**, *36*. [CrossRef]

18. Ma, N.; Zhao, C.S.; Nowak, A.; Müller, T.; Pfeifer, S.; Cheng, Y.F.; Deng, Z.Z.; Liu, P.F.; Xu, W.Y.; Ran, L.; et al. Aerosol Optical Properties in the North China Plain During Hachi Campaign: An In-Situ Optical Closure Study. *Atmos. Chem. Phys.* **2011**, *11*, 5959–5973. [CrossRef]

19. Cao, C.; de Luccia, F.J.; Xiong, X.; Wolfe, R.; Weng, F. Early on-Orbit Performance of the Visible Infrared Imaging Radiometer Suite Onboard the Suomi National Polar-Orbiting Partnership (S-Npp) Satellite. *IEEE Trans. Geosci. Remote Sens.* **2014**, *52*, 1142–1156. [CrossRef]

20. Vermote, E.F.; El Saleous, N.; Justice, C.O.; Kaufman, Y.J.; Privette, J.L.; Remer, L.; Roger, J.C.; Tanre, D. Atmospheric Correction of Visible to Middle-Infrared Eos-Modis Data over Land Surfaces: Background, Operational Algorithm and Validation. *J. Geophys. Res. Atmos.* **1997**, *102*, 17131–17141. [CrossRef]

21. Vermote, E.F.; Kotchenova, S. Atmospheric Correction for the Monitoring of Land Surfaces. *J. Geophys. Res. Atmos.* **2008**, *113*. [CrossRef]

22. Levy, R.C.; Mattoo, S.; Munchak, L.A.; Remer, L.A.; Sayer, A.M.; Patadia, F.; Hsu, N.C. The Collection 6 Modis Aerosol Products over Land and Ocean. *Atmos. Meas. Tech.* **2013**, *6*, 2989–3034. [CrossRef]

23. Levy, R.C.; Remer, L.A.; Dubovik, O. Global Aerosol Optical Properties and Application to Moderate Resolution Imaging Spectroradiometer Aerosol Retrieval over Land. *J. Geophys. Res. Atmos.* **2007**, *112*. [CrossRef]

24. Holben, B.N.; Eck, T.F.; Slutsker, I.; Tanre, D.; Buis, J.P.; Setzer, A.; Vermote, E.; Reagan, J.A.; Kaufman, Y.J.; Nakajima, T.; et al. Aeronete a Federated Instrument Network and Data Archive for Aerosol Characterization. *Remote Sens. Environ.* **1998**, *66*, 1–16. [CrossRef]

25. Eck, T.F.; Holben, B.N.; Reid, J.S.; Dubovik, O.; Smirnov, A.; O'Neill, N.T.; Slutsker, I.; Kinne, S. Wavelength Dependence of the Optical Depth of Biomass Burning, Urban, and Desert Dust Aerosols. *J. Geophys. Res.* **1999**, *104*, 31333–31349. [CrossRef]

26. Dubovik, O.; Smirnov, A.; Holben, B.N.; King, M.D.; Kaufman, Y.J.; Eck, T.F.; Slutsker, I. Accuracy Assessments of Aerosol Optical Properties Retrieved from Aerosol Robotic Network (Aeronet) Sun and Sky Radiance Measurements. *J. Geophys. Res. Atmos.* **2000**, *105*, 9791–9806. [CrossRef]

27. Dubovik, O.; Holben, B.; Eck, T.F.; Smirnov, A.; Kaufman, Y.J.; King, M.D.; Tanre, D.; Slutsker, I. Variability of Absorption and Optical Properties of Key Aerosol Types Observed in Worldwide Locations. *J. Atmos. Sci.* **2002**, *59*, 590–608. [CrossRef]

28. Zhu, J.; Che, H.; Xia, X.; Chen, H.; Goloub, P.; Zhang, W. Column-Integrated Aerosol Optical and Physical Properties at a Regional Background Atmosphere in North China Plain. *Atmos. Environ.* **2014**, *84*, 54–64. [CrossRef]

29. Che, H.Z.; Zhang, X.Y.; Chen, H.B.; Damiri, B.; Goloub, P.; Li, Z.Q.; Zhang, X.C.; Wei, Y.; Zhou, H.G.; Dong, F.; et al. Instrument Calibration and Aerosol Optical Depth Validation of the China Aerosol Remote Sensing Network. *J. Geophys. Res. Atmos.* **2009**, *114*, 106510–106566. [CrossRef]

30. Tao, R.; Che, H.; Chen, Q.; Wang, Y.; Sun, J.; Zhang, X.; Lu, S.; Guo, J.; Wang, H.; Zhang, X. Development of an Integrating Sphere Calibration Method for Cimel Sunphotometers in China Aerosol Remote Sensing Network. *Particuology* **2014**, *13*, 88–99. [CrossRef]

31. Xie, Y.; Zhang, Y.; Xiong, X.X.; Qu, J.J.; Che, H.Z. Validation of Modis Aerosol Optical Depth Product over China Using Carsnet Measurements. *Atmos. Environ.* **2011**, *45*, 5970–5978. [CrossRef]

32. Petrenko, M.; Ichoku, C.; Leptoukh, G. Multi-Sensor Aerosol Products Sampling System (MAPSS). *Atmos. Meas. Tech.* **2012**, *5*, 913–926. [CrossRef]

33. Ångström, A. On the Atmospheric Transmission of Sun Radiation and on Dust in the Air. *Geogr. Ann.* **1929**, *11*, 156–166. [CrossRef]

34. Remer, L.A.; Kaufman, Y.J.; Tanre, D.; Mattoo, S.; Chu, D.A.; Martins, J.V.; Li, R.R.; Ichoku, C.; Levy, R.C.; Kleidman, R.G.; et al. The Modis Aerosol Algorithm, Products, and Validation. *J. Atmos. Sci.* **2005**, *62*, 947–973. [CrossRef]

35. Giles, D.M.; Holben, B.N.; Eck, T.F.; Sinyuk, A.; Smirnov, A.; Slutsker, I.; Dickerson, R.R.; Thompson, A.M.; Schafer, J.S. An Analysis of Aeronet Aerosol Absorption Properties and Classifications Representative of Aerosol Source Regions. *J. Geophys. Res.* **2012**, *117*, 127–135. [CrossRef]

36. Wang, J.; Xu, X.; Ding, S.; Zeng, J.; Spurr, R.; Liu, X.; Chance, K.; Mishchenko, M. A Numerical Testbed for Remote Sensing of Aerosols, and Its Demonstration for Evaluating Retrieval Synergy from a Geostationary Satellite Constellation of Geo-Cape and Goes-R. *J. Quant. Spectrosc. Radiat. Transf.* **2014**, *146*, 510–528. [CrossRef]

37. Oo, M.M.; Jerg, M.; Hernandez, E.; Picon, A.; Gross, B.M.; Moshary, F.; Ahmed, S.A. Improved Modis Aerosol Retrieval Using Modified Vis/Swir Surface Albedo Ratio over Urban Scenes. *IEEE Trans. Geosci. Remote Sens.* **2010**, *48*, 983–1000. [CrossRef]

38. Lee, J.; Kim, J.; Song, C.H.; Kim, S.B.; Chun, Y.; Sohn, B.J.; Holben, B.N. Characteristics of Aerosol Types from Aeronet Sunphotometer Measurements. *Atmos. Environ.* **2010**, *44*, 3110–3117. [CrossRef]

39. Sun, L.; Xia, X.-A.; Wang, P.-C.; Chen, H.-B.; Goloub, P.; Zhang, W.-X. Identification of Aerosol Types and Their Optical Properties in the North China Plain Based on Long-Term Aeronet Data. *Atmos. Ocean. Sci. Lett.* **2013**, *6*, 216–222.

40. Su, L.; Toon, O.B. Saharan and Asian Dust: Similarities and Differences Determined by Calipso, Aeronet, and a Coupled Climate-Aerosol Microphysical Model. *Atmos. Chem. Phys.* **2011**, *11*, 3263–3280. [CrossRef]

41. Formenti, P.; Schütz, L.; Balkanski, Y.; Desboeufs, K.; Ebert, M.; Kandler, K.; Petzold, A.; Scheuvens, D.; Weinbruch, S.; Zhang, D. Recent Progress in Understanding Physical and Chemical Properties of African and Asian Mineral Dust. *Atmos. Chem. Phys.* **2011**, *11*, 8231–8256. [CrossRef]

remote sensing

MDPI

Article

High Resolution Aerosol Optical Depth Retrieval Using Gaofen-1 WFV Camera Data

Kun Sun [1], Xiaoling Chen [1,2,*], Zhongmin Zhu [1,3,*] and Tianhao Zhang [1]

[1]   State Key Laboratory of Information Engineering in Surveying, Mapping and Remote Sensing,
      Wuhan University, Wuhan 430079, China; bigdianya@foxmail.com (K.S.); tianhaozhang@whu.edu.cn (T.Z.)
[2]   The Key Laboratory of Poyang Lake Wetland and Watershed Research, Ministry of Education,
      Jiangxi Normal University, Nanchang 330022, China
[3]   College of Information Science and Engineering, Wuchang Shouyi University, Wuhan 430064, China
*   Correspondence: Xiaoling_chen@whu.edu.cn (X.C.); zhongmin.zhu@whu.edu.cn (Z.Z.);
      Tel.: +86-139-0719-5381 (X.C.)

Academic Editors: Yang Liu, Jun Wang, Omar Torres, Richard Müller and Prasad S. Thenkabail
Received: 5 December 2016; Accepted: 15 January 2017; Published: 19 January 2017

Abstract: Aerosol Optical Depth (AOD) is crucial for urban air quality assessment. However, the frequently used moderate-resolution imaging spectroradiometer (MODIS) AOD product at 10 km resolution is too coarse to be applied in a regional-scale study. Gaofen-1 (GF-1) wide-field-of-view (WFV) camera data, with high spatial and temporal resolution, has great potential in estimation of AOD. Due to the lack of shortwave infrared (SWIR) band and complex surface reflectivity brought from high spatial resolution, it is difficult to retrieve AOD from GF-1 WFV data with traditional methods. In this paper, we propose an improved AOD retrieval algorithm for GF-1 WFV data. The retrieved AOD has a spatial resolution of 160 m and covers all land surface types. Significant improvements in the algorithm include: (1) adopting an improved clear sky composite method by using the MODIS AOD product to identify the clearest days and correct the background atmospheric effect; and (2) obtaining local aerosol models from long-term CIMEL sun-photometer measurements. Validation against MODIS AOD and ground measurements showed that the GF-1 WFV AOD has a good relationship with MODIS AOD (R^2 = 0.66; RMSE = 0.27) and ground measurements (R^2 = 0.80; RMSE = 0.25). Nevertheless, the proposed algorithm was found to overestimate AOD in some cases, which will need to be improved upon in future research.

Keywords: Gaofen-1; aerosol optical depth; deep blue; Wuhan; urban aerosol

---

1. Introduction

As an important component of atmosphere, aerosols play a vital role in climate change, earth radiation budget and air quality [1–3]. Atmospheric aerosol is a major source of uncertainty in the global climate system for its high spatial and temporal variability and short lifetime [4]. Satellite remote sensing provides a convenient way to estimate aerosol optical properties in space and time [5]. Numerous satellite data-based aerosol retrieval algorithms have been developed in recent years [6–9].

A key physical parameter which can be retrieved from satellite data is Aerosol Optical Depth (AOD), which is defined as the integrated light extinction over vertical path through the atmosphere. AOD has been widely applied in many aspects such as atmospheric correction of satellite images [6], air quality assessment [10] and haze pollution monitoring [11]. To estimate AOD from satellite data, the core problem is to separate the atmospheric and surface scattering contributions from the total signal observed by satellite. Generally, there exist two uncertain factors in the separating process: the determinations of aerosol model and surface reflectance. Aerosol model is usually derived from long-term AErosol RObotic NETwork (AERONET) or other regional ground measurements [12,13].

Many methods were successfully developed in the estimation of surface reflectance from satellite observation. One of the earliest and more classic techniques is called Dark Dense Vegetation (DDV) or Dark Target (DT), a method which utilizes the stable correlations between visible bands and the shortwave infrared (SWIR) band at 2.1 μm [14]. The method and its improved version have been successfully applied in the AOD retrieval of moderate-resolution imaging spectroradiometer (MODIS) [15,16], Landsat TM [17,18], and many other sensors [19–21]. The DT method shows a good performance over dark surfaces (e.g., dense vegetation), but tends to overestimate AOD over bright surfaces (especially urbanized areas) [22,23]. Another widely adopted technique is called the Deep Blue (DB) method, which is based on the assumption that the surface reflectance keeps unchanged or changes little during a specified period. The DB method is able to retrieve AOD over bright surfaces (e.g., desert, arid/semiarid and urban regions). The DB algorithms have been developed for many satellite sensors such as SeaWiFS [24], MODIS [25,26] and GOCI [27].

With the rapid growth of industrialization and urbanization in China during recent years, aerosol pollution problems such as haze and high particulate matter concentration have become increasingly severe for most regions of China [28,29]. Valid and high resolution AOD retrieval covering all land surface types (including urban areas) is very important for regional air quality monitoring in China. However, retrieving AOD over urban areas in China remains a challenging problem. The primary difficulty is the estimation of surface reflectance. The DT method is limited over urban areas; moreover, it is difficult to acquire completely "clear" images when using the DB method to establish surface reflectance database over urban areas with heavy aerosol loadings. Furthermore, the determination of the aerosol model is also difficult in many urban regions of China due to the lack of enough ground measurements and complex aerosol sources.

Most of the currently released AOD products (e.g., MODIS, SeaWiFS, MISR and MERIS) are usually at a spatial resolution of a few kilometers, which is too coarse for regional-scale applications. High spatial resolution sensors such as the Landsat series usually have a long re-visiting period (16 days), which makes it more difficult in terms of algorithm design. Gaofen-1 (GF-1) satellite, launched by the Chinese government in April 2013, has four Wide-Field-of-View (WFV) cameras onboard. Four WFV cameras provide multi-spectral images from visible to near-infrared (NIR) band, with a high spatial resolution of 16 m and a re-visiting period of 4 days. Although the absence of SWIR band makes it difficult to retrieve AOD using the DT method, the high temporal resolution makes it possible to use the DB method.

In this paper, we attempted to retrieve high resolution AOD (160 m × 160 m) from GF-1 WFV data by using an improved DB method. The algorithm was implemented over Wuhan, an urban area in central China. In the proposed algorithm, surface reflectance of GF-1 WFV was estimated by establishing a seasonal surface reflectance database with the support of the MODIS AOD product. An aerosol model was determined by statistical analysis of long-term ground measurements. To evaluate the retrieved results from the proposed algorithm, the spatial distribution of retrieved AOD was presented and analyzed in detail, and the retrieved results were compared with corresponding MODIS AOD and ground measurements respectively. Finally, the performance and limitations of the algorithm were also discussed.

2. Study Area and Datasets

2.1. Study Area

Wuhan (113°41′E–115°05′E, 29°58′N–31°22′N), provincial capital of Hubei province, is the largest city in central China (Figure 1). It is situated on the middle-lower Yangtze Plain and eastern Jianghan Plain, with two large rivers (Yangtze River and Han River) flowing through the main city. As a significant regional economic center in central China, Wuhan has a population of more than ten million, and covers an area of about 8594 km^2. According to the classification result from GF-1 WFV data in Wuhan, the built-up area and water occupy about 23% and 15% of the total area, respectively.

With the rapid growth of urbanization and industrialization, Wuhan has been suffering from severe air pollution in recent years, including high particulate matter concentrations and haze pollution [30,31]. The annual mean AOD at 500 nm over Wuhan is up to 1.0, and the region is mainly populated with fine-mode particles [32].

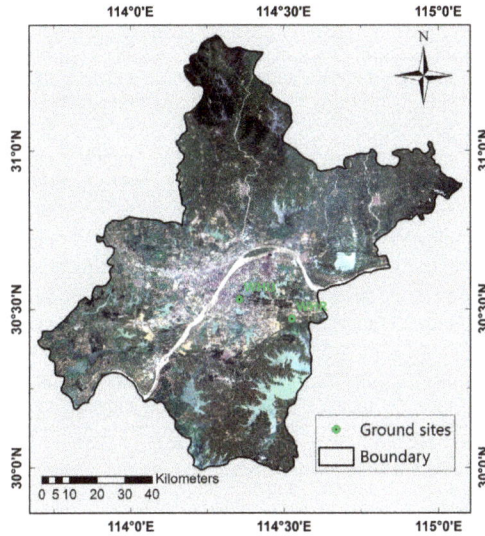

Figure 1. Geolocation of Wuhan city shown by Gaofen-1 Wide-Field-of-View (GF-1 WFV) image (RGB composited).

2.2. Datasets

2.2.1. Gaofen-1 Wide-Field-of-View (GF-1 WFV) Data

GF-1 WFV data are provided from four WFV cameras onboard the GF-1 satellite, with a spatial resolution of 16 m and a temporal resolution of 4 days. The detailed characteristics of GF-1 WFV instruments are summarized in Table 1. Here we collected all the available GF-1 WFV data over Wuhan from July 2013 to January 2016 (download from http://www.cresda.com/CN/). All data need to be well pre-processed before beginning the AOD retrieval process. The pre-processes include geometric correction, image cutting and mosaic, and radiometric calibration. The accuracy requirement of geometric correction is 1~2 pixels. The accuracy and stability of radiometric calibration is crucial in the AOD retrieval using satellite instruments, especially for GF-1 WFV cameras without on-board calibration. Feng et al. [33] performed a cross-calibration of GF-1 WFV cameras using Landsat 8 Operational Land Imager (OLI) images. Validations with satellite data and in situ measurements showed that the calibration uncertainty is ~8%. The calibration coefficients from the work by Feng et al. were adopted in this paper. In addition, according to the research by Li et al. [34], the signal-to-noise ratio (SNR) of GF-1 WFV cameras at four bands are 294, 125, 77 and 34 respectively, which are slightly lower than the SNR of corresponding MODIS bands. However, considering the difference in spatial resolution, the SNR of GF-1 WFV cameras could have an obvious improvement when resampled to a resolution of 160 m. Therefore, the radiometric performance of GF-1 WFV cameras is in a good condition and could meet the requirement in the AOD retrieval.

Table 1. Characteristics of WFV (Wide-Field-of-View) cameras aboard on Gaofen-1 satellite.

Band	Band Range (µm)	Spatial Resolution (m)	Re-Visiting Period (Days)	Swath (km)
Blue	045–0.52			
Green	0.52–0.59	16	4	800
Red	0.63–0.69			
NIR	0.77–0.89			

2.2.2. Moderate-Resolution Imaging Spectroradiometer (MODIS) Aerosol Optical Depth (AOD) Data

The AOD data, derived from MODIS sensors onboard Terra and Aqua, are provided by two independent algorithms (DT and DB) at a nominal spatial resolution of 10 km × 10 km (MOD04 for Terra and MYD04 for Aqua, available from https://ladsweb.nascom.nasa.gov/search/index.html). The recent version of the MODIS AOD product has been updated to Collection 6. As the DT retrieval cannot provide accurate results over urban areas, here we choose MODIS DB AOD in the research for its expansion of spatial coverage over urban areas and improved retrieval accuracy [26]. The spatial coverage of MODIS DB retrieval can reach up to 100% over the Wuhan region in cloudless conditions, which makes it suitable to be used to evaluate the average aerosol loading and its spatial variation. Corresponding to the overpassing time of GF-1 (10:30 local solar time), all MODIS DB AOD data from Terra (MOD04) over Wuhan during 2013–2016 were collected; and then the regional average AOD, its standard deviation and spatial coverage were calculated to assist the automatic selection of GF-1 WFV images in a clean and cloudless condition.

2.2.3. Ground Measurements

Ground measurements from a Cimel sun photometer CE-318 and handheld MICROTOPS-II sun photometer were included in the research. The information of the two instruments is summarized in Table 2, and the geolocation is shown in Figure 1. The sun photometer CE-318 was installed at the campus of Wuhan University, an urban site of Wuhan (30°32′N, 114°21′E, named as WHU here). The instrument provides a long-term observation of aerosols over Wuhan since 2007. The detailed analysis of observation results from the instrument has been described in Wang et al. [32]. The MICROTOPS-II sun photometer was used to measure AOD at a rural site of Wuhan (30°28′N, 114°32′E, named as WHR here) during December 2014 to June 2015. Multiyear observation results from WHU were used to analyze the aerosol model over Wuhan. Measurements from both WHU and WHR were used to evaluate the algorithm accuracy.

Table 2. The list of input variables used to calculate the look-up table.

Site	Lat/Lon	Terrain	Instrument	Observing Period
WHU	30°32′N, 114°21′E	Urban	CE-318	2008–2012; December 2014–June 2015
WHR	30°28′N, 114°32′E	Rural	MICROTOPS-II	December 2014–June 2015

3. AOD Retrieval Algorithm

The basic principle of AOD retrieval algorithm is based on radiative transfer theory. Assuming a lambertian surface under a plane-parallel atmosphere, the reflectance at the top of atmosphere (TOA) can be expressed by the following equation [35]:

$$\rho_\lambda^{TOA}(\mu_s, \mu_v, \phi) = T_g \left[\rho_\lambda^{atm}(\mu_s, \mu_v, \phi) + \frac{T(\mu_s)T(\mu_v)\rho_\lambda^s}{(1 - S_\lambda \rho_\lambda^s)} \right] \tag{1}$$

where $\mu_s = cos\theta_s$, $\mu_v = cos\theta_v$, θ_s is solar zenith angle, θ_v is view zenith angle, ϕ is relative azimuth angle, λ represents the corresponding sensor band, ρ_λ^{TOA} is the apparent reflectance, T_g is the gaseous transmission, ρ_λ^{atm} is the atmospheric path reflectance (Rayleigh and aerosol), S_λ is the spherical albedo of the atmosphere, $T(\mu_s)$ and $T(\mu_v)$ correspond to the downward and upward atmospheric transmission respectively, and ρ_λ^s is the Lambertian surface reflectance.

To solve the radiative transfer equation, a radiative transfer model Second Simulation of the Satellite Signal in the Solar Spectrum (6S) was adopted in the study [35]. Given specific band, view geometry, atmospheric model, AOD, aerosol type and surface reflectance, the TOA reflectance can be simulated by 6S according to the calculation of three key parameters: ρ, s and $T(T(\mu_s)T(\mu_v))$. For satellite remotely sensed data, if aerosol type and surface reflectance are determined, AOD can be estimated by comparing measured and simulated TOA reflectance.

Generally, in order to speed up the calculation process, the Look-Up Table (LUT) technique is adopted. The LUT for AOD retrieval is a multidimensional data table, which contains pre-calculated parameters, ρ, s and T, under specific view geometry, atmospheric model, AOD, aerosol type and surface reflectance. The detailed set of intervals and ranges for input variables in the LUT is listed in Table 3. Utilizing the pre-calculated LUT, the TOA reflectance for any given pixel can be simulated by linear/nonlinear interpolation between the neighboring bins for geometry, AOD and surface reflectance.

Table 3. The list of input variables used to calculate the look-up table.

Input Variables	No. of Entries	Entries
SZA	14	0, 6, 12, ... , 78
VZA	14	0, 6, 12, ... , 78
RAA	16	0, 12, 24, ... , 180
AOD	9	0, 0.25, 0.50, 0.75, 1.0, 1.5, 2.0, 3.0, 5.0
Atmospheric model	2	Mid-latitude summer/winter
Surface reflectance	4	0.0, 0.1, 0.2, 0.3

SZA: solar zenith angle; VZA: view zenith angle; RAA: relative azimuth angle.

Retrieving AOD from satellite data requires an accurate determination of surface reflectance and local aerosol type. As GF-1 WFV cameras do not have a SWIR band, it is difficult to estimate the surface reflectance from GF-1 WFV data using the DT method. However, a re-visiting period of 4 days makes it possible to estimate the surface reflectance from GF-1 WFV data using the DB method. By compositing the clearest GF-1 WFV images over Wuhan during a season, the seasonal surface reflectance of GF-1 WFV data can be determined. Based on the aerosol optical properties derived from long-term ground observations over Wuhan, aerosol types from each season and different AOD ranges are determined by statistical analysis.

A complete flowchart of the GF-1 WFV AOD retrieval algorithm is shown in Figure 2. The main processes include: pre-process of GF-1 WFV data; calculation of LUT containing local aerosol types and establishment of surface reflectance database. According to the LUT and surface reflectance database established for GF-1 WFV cameras, the TOA reflectance at any geometry and AOD can be simulated. Thus, AOD at 550 nm can be estimated by matching the measured and simulated TOA reflectance.

To demonstrate the availability of GF-1 WFV data in AOD retrieval, a sensitive analysis was performed by simulating the TOA reflectance in GF-1 WFV blue band under different AOD and surface conditions (as shown in Figure 3). In the simulation, the view geometry condition was set as: $\theta_s = 30°$, $\theta_v = 30°$, $\phi = 100°$; and the aerosol type was set as continental mode in the 6S. Obviously, the GF-1 WFV blue band is very sensitive to the AOD in the range of 0 to 3 in various surface conditions. When AOD is greater than 3, the TOA reflectance increases slowly with AOD in all surface conditions. It indicates that AOD could be accurately estimated from GF-1 WFV data in most cases when the surface reflectance and aerosol type are accurately estimated.

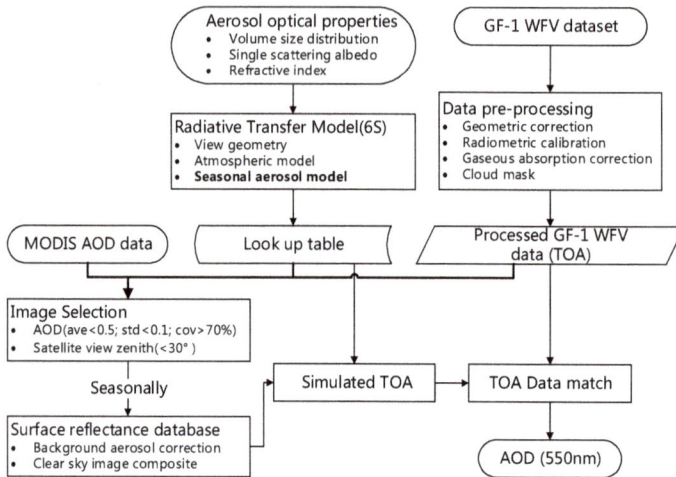

Figure 2. Flowchart of AOD retrieval algorithm for GF-1 WFV data. Ave, std and cov represent average value, standard deviation and spatial coverage of MODIS AOD, respectively.

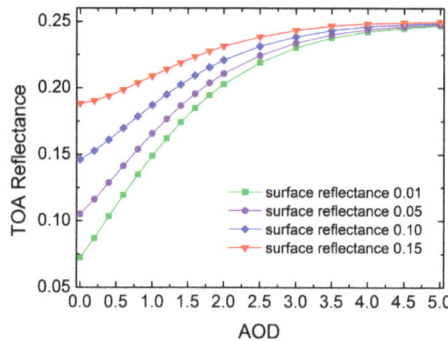

Figure 3. The change of simulated TOA reflectance in GF-1 WFV blue band with different AOD and surface reflectance.

Two important issues in the GF-1 WFV data process need to be noted here. One is cloud mask; the other is gaseous absorption correction (including water vapor and ozone). Since there is no thermal infrared band in GF-1 WFV data, a tailor-made cloud masking method was devised to take advantage of TOA reflectance and corresponding surface reflectance in the three visible bands. The method screens clouds according to the high reflectivity by cloud and high contrast between clouds and the underlying surface. The detailed criterion for cloud masking is listed in Equation (2).

$$if\left(\rho_{red}^{TOA} > 0.2\right) \ and \ \left(\rho_{red}^{TOA} - \rho_{red}^{Surf} > 0.1\right) then \ mask \tag{2}$$

where ρ_{red}^{TOA} and ρ_{red}^{Surf} are the TOA reflectance and surface reflectance of GF-1 WFV red band respectively. The thresholds determined for cloud mask are based on a trial-and-error approach. Figure 4 shows the cloud mask results for some selected GF-1 WFV images using the proposed approach. Generally, the method shows a good performance in the cloud screening of GF-1 WFV data.

Since there is spatio-temporal variation of water vapor and ozone over Wuhan, gaseous absorption correction using standard atmospheric profile from 6S may bring uncertain error. Here an accurate correction of gaseous absorption for water vapor and ozone was performed for all GF-1 WFV data. The water vapor data were collected from NCEP (National Center for Environmental Prediction) reanalysis daily average products (http://www.esrl.noaa.gov/psd/data/reanalysis/reanalysis.shtml); and the ozone data were extracted from OMI (Ozone Monitoring Instrument) $1° \times 1°$ grid ozone products (http://disc.sci.gsfc.nasa.gov/Aura/data-holdings/OMI).

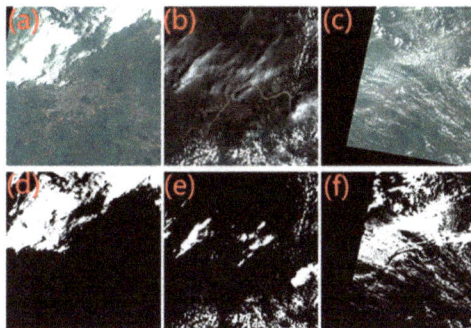

Figure 4. The origin maps (**a–c**) and corresponding cloud mask results (**d–f**) for three selected GF-1 WFV images. The white regions in (**d–f**) represent cloud covering areas.

After a description of basic principle of the AOD retrieval algorithm for GF-1 WFV data, two key parts of the algorithm are described in detail. The analysis of aerosol optical properties over Wuhan is described in Section 3.1. The methodology of surface reflectance determination for GF-1 WFV data is addressed in Section 3.2.

3.1. Aerosol Optical Properties over Wuhan

The determination of local aerosol types is crucial for AOD retrieval. It has been reported in many researches that inappropriate aerosol models have a significant impact on the accuracy of MODIS AOD product in China [36,37]. The aerosol optical properties over Wuhan present a complex variation pattern for various aerosol sources such as industries, traffic, biomass burning and dust [32]. According to the aerosol optical properties derived from long-term ground measurements at WHU, the average AOD (550 nm) during 2008–2012 is up to 0.7, and nearly 50% of AOD values are in the range of 0.5–1.0 (Figure 5a); the scattering map of Single Scattering Albedo (SSA) and Fine Mode Fraction of AOD indicates that this region is dominated by urban/industrial, biomass burning and mixed aerosol types (Figure 5b) [38,39].

To derive local aerosol types over Wuhan, a method following the research of Kim et al. [40] was adopted in this paper. The method discriminates various aerosol types by statistical analysis of aerosol optical properties in different seasons and AOD ranges. The average volume size distribution for four seasons (spring, summer, autumn and winter) in different AOD ranges is shown in Figure 6, and the corresponding refractive index and SSA data are listed in Table 4. Generally, the aerosol volume size distribution over Wuhan follows a bimodal lognormal size distribution with obvious seasonal variability in volume concentration and peak radius. Fine-mode particles are dominant in all seasons except spring, when frequent dust events cause an increase in coarse-mode particles. It is consistent with the analysis results using AERONET measurements over East Asia [40]. The refractive index and SSA at 670 nm in different seasons and AOD bins, as listed in Table 4, also present a seasonal variability. The average refractive index (real part) for spring (MAM: March, April and May), summer (JJA: June, July and August), autumn (SON: September, October and November), and winter (DJF: December, January and February) are 0.87, 0.86, 0.83 and 0.86 respectively; while the average SSA are 1.50, 1.40, 1.43 and 1.46 for each season in the same order as above.

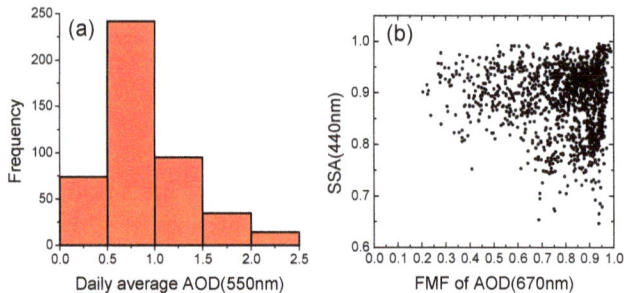

Figure 5. (**a**) Frequency distribution of ground-measured daily AOD at 550 nm over Wuhan from 2008 to 2013; (**b**) Scatter map of SSA at 440 nm and FMF of AOD at 670 nm (SSA: Single Scattering Albedo; FMF: Fine Mode Fraction).

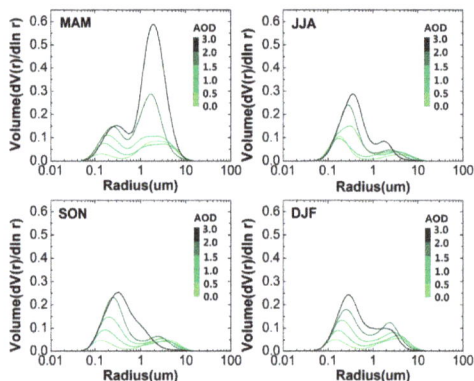

Figure 6. The average volume size distribution for each season under different AOD ranges. Spring (MAM: March, April and May); summer (JJA: June, July and August); autumn (SON: September, October and November); and winter (DJF: December, January and February).

Table 4. The refractive index and single scattering albedo at 670 nm for each season under different AOD ranges.

Parameters	Season	AOD Range				
		0–0.5	0.5–1.0	1.0–1.5	1.5–2.0	2.0–3.0
Refractive index (real) at 670 nm	Spring	1.54	1.45	1.48	1.48	1.51
	Summer	1.33	1.41	1.41	1.42	1.41
	Autumn	1.46	1.44	1.42	1.40	1.41
	Winter	1.46	1.46	1.46	1.45	1.44
Refractive index (imaginary) at 670 nm	Spring	0.0122	0.0428	0.0175	0.0094	0.0065
	Summer	0.0258	0.0538	0.0121	0.0112	0.0039
	Autumn	0.0425	0.0539	0.0247	0.0180	0.0220
	Winter	0.0268	0.0304	0.0245	0.0139	0.0141
Single scattering albedo at 670 nm	Spring	0.86	0.78	0.86	0.90	0.92
	Summer	0.78	0.72	0.91	0.93	0.97
	Autumn	0.81	0.75	0.85	0.89	0.86
	Winter	0.84	0.82	0.85	0.90	0.90

Spring (MAM: March, April and May); summer (JJA: June, July and August); autumn (SON: September, October and November); winter (DJF: December, January and February).

3.2. Surface Reflectance Determination

Assuming a constant surface reflectance over one region during a specific period, the regional surface reflectance can be estimated from time series of satellite images by using clear sky composite technique. The method has been widely applied in the AOD retrieval algorithm for many satellite sensors in global and regional scale. For example, the method was applied in MODIS DB algorithm to provide global land AOD product [41], and it was also used to retrieve regional high resolution AOD from MODIS and Landsat data [42,43]. Allowing for the high spatial and temporal resolution of GF-1 WFV data, it is feasible to retrieve high resolution AOD using clear sky composite method. The main difficulties lie in the complex surface reflectivity (especially over urban areas) brought from high spatial resolution, which may result in complicated variations in satellite signals caused by cloud and topographic shadow, surface feature change and Bidirectional Reflectance Distribution Function (BRDF) effect.

To estimate surface reflectance from GF-1 WFV data, we proposed a novel strategy in the selection and compositing of GF-1 WFV images. The compositing period was determined to be three months in consideration of observing frequency of GF-1 WFV cameras. To avoid directly composite all GF-1 WFV images, MODIS AOD product was introduced in the selection of clear sky images. We calculated the regional mean, minimum, maximum, standard deviation and spatial coverage value of MODIS AOD over Wuhan corresponding to the overpassing time of GF-1 WFV data. Three criteria were adopted to ensure the selection of clearest and cloudless GF-1 WFV images. They were: (1) mean value < 0.5; (2) standard deviation < 0.1; and (3) spatial coverage > 70%. Furthermore, in order to minimize the BRDF effect, only GF-1 WFV images with satellite view zenith less than 30° were selected. The eligible GF-1 WFV images with a total number of 31 are listed in Table 5. A small number of GF-1 WFV images that failed to meet the above criteria were also included to ensure enough eligible images during each season.

As the heavy loading of aerosols over Wuhan, the aerosol effect is non-negligible on clear days. After a first selection of clear sky GF-1 WFV images with the support of prior knowledge from MODIS AOD product, atmospheric correction was performed for all eligible images by using 6S with the corresponding averaged MODIS AOD and established seasonal aerosol model. Finally, the atmospheric corrected images were composited seasonally by using minimum reflectance technique. It should be noted here that the compositing process was not pixel-based but window-based in a 10 × 10 size (corresponding to a spatial resolution of 160 m × 160 m). In each processing window, the 30% brightest and 30% darkest pixels were excluded, and the average value of remaining 40% pixels was used to represent the reflectance of the entire window. The same calculation was performed in the AOD retrieval process. The window-based calculation method can effectively avoid the impact of retained clouds, topographic shadow and changing of surface features on the AOD retrieval.

Figure 7 shows the seasonal surface reflectance images (red, green, blue (RGB) composited) derived from GF-1 WFV data over Wuhan during 2014, and Figure 8 shows the seasonal surface reflectance variation of typical land cover types (urban, forest and farmland). It can be seen from the two figures that the surface reflectance over Wuhan presents an obvious seasonal variation pattern.

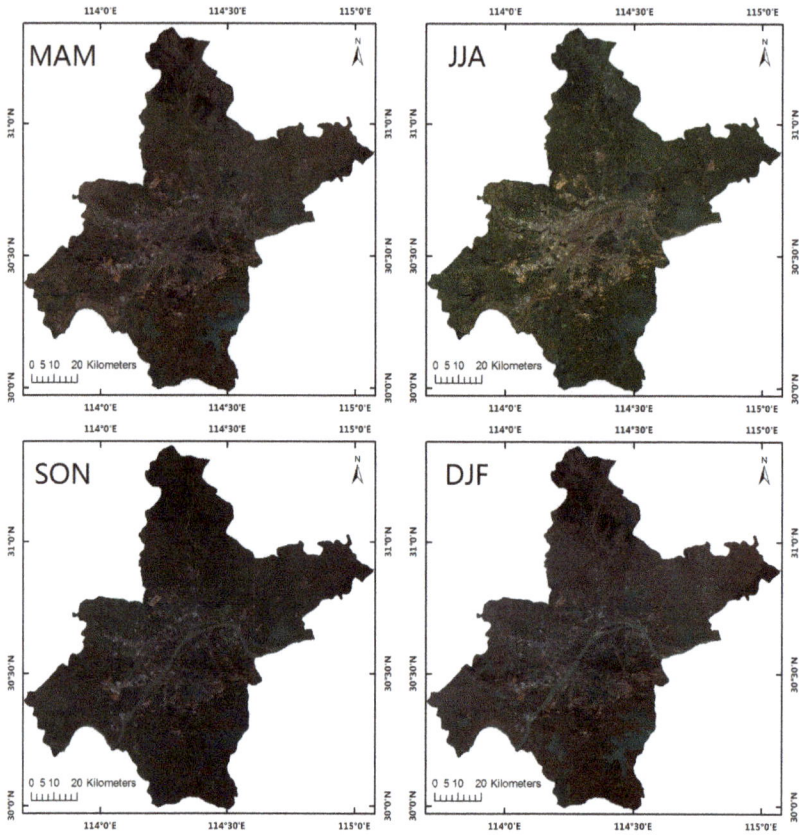

Figure 7. Seasonal surface reflectance images (red, green, blue (RGB) composited) derived from GF-1 WFV data over Wuhan during 2014. MAM (March, April and May), JJA (June, July and August), SON (September, October and November) and DJF (December, January and February) represent the season of Spring, Summer, Autumn and Winter respectively.

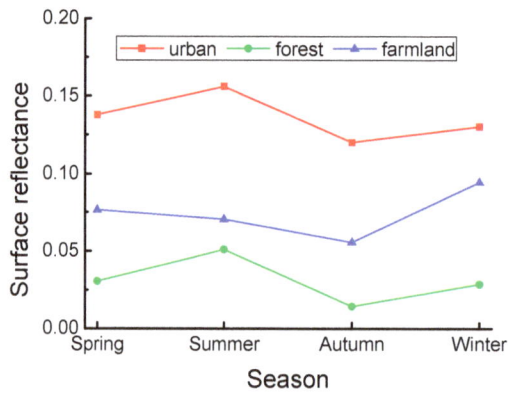

Figure 8. Seasonal surface reflectance variation of typical land cover types.

Table 5. List of GF-1 WFV candidate images for surface reflectance database.

Year	Season	GF-1 WFV Image Date	MODIS AOD					View Zenith(°)
			Mean	Min	Max	Std	Coverage	
2013	JJA	8 July 2013	0.27	0.11	0.70	0.11	99.5	31.0
		1 August 2013	0.12	0.05	0.48	0.09	84.7	38.1
		10 August 2013	0.09	0.05	0.59	0.07	100.0	7.5
	SON	19 September 2013	0.28	0.14	0.88	0.10	95.4	12.4
		1 October 2013	0.49	0.22	0.96	0.10	92.9	18.4
		6 October 2013	0.42	0.09	0.92	0.16	100.0	31.4
		9 October 2013	0.35	0.09	0.72	0.14	99.5	22.3
		14 October 2013	0.42	0.21	0.92	0.15	100.0	27.6
		7 November 2013	0.20	0.12	0.34	0.06	48.0	14.5
		19 November 2013	0.38	0.14	0.58	0.10	100.0	6.7
2014	DJF	23 January 2014	0.32	0.11	0.60	0.13	85.7	7.0
	MAM	17 March 2014	0.43	0.25	0.64	0.09	100.0	5.2
	JJA	22 July 2014	0.31	0.09	0.84	0.12	92.9	0.9
		26 July 2014	0.40	0.05	1.25	0.37	75.0	4.7
		30 July 2014	0.11	0.05	0.50	0.07	100.0	10.2
	SON	21 September 2014	0.08	0.05	0.57	0.09	71.4	24.8
		8 October 2014	0.38	0.26	0.61	0.07	100.0	6.0
		24 October 2014	0.36	0.17	0.58	0.09	100.0	14.7
		14 November 2014	0.27	0.18	0.39	0.05	100.0	11.1
2015	DJF	8 December 2014	0.23	0.05	0.50	0.09	99.5	20.6
		16 December 2014	0.14	0.05	0.22	0.03	100.0	29.8
		17 December 2014	0.06	0.04	0.22	0.03	100.0	22.1
		21 December 2014	0.22	0.03	0.46	0.12	99.5	17.3
	MAM	25 March 2015	0.19	0.05	0.62	0.11	100.0	1.0
		14 April 2015	0.25	0.13	0.44	0.06	100.0	25.0
	JJA	3 August 2015	0.08	0.05	0.25	0.04	100.0	10.0
		23 August 2015	0.17	0.05	0.65	0.15	99.5	33.4
	SON	20 October 2015	0.53	0.17	0.84	0.10	100.0	4.4
		1 November 2015	0.54	0.36	0.79	0.13	72.4	20.2
2016	DJF	16 December 2015	0.08	0.03	0.15	0.03	100.0	24.3
		17 December 2015	0.06	0.03	0.20	0.04	100.0	26.6

MAM: March, April and May; JJA: June, July and August; SON: September, October and November; DJF: December (last year), January (current year), and February (current year); Mean, min, max and std represent regional mean, minimum, maximum value and standard deviation of MODIS AOD respectively. Coverage represents the percentage of valid AOD over Wuhan; View zenith is the satellite view zenith angle of corresponding GF-1 WFV image at the center position of Wuhan.

4. Results and Discussion

4.1. Retrieved Results from the Proposed Algorithm

Utilizing the algorithm described above, we processed all the GF-1 WFV data over Wuhan from July 2013 to January 2016. Figures 9 and 10 show continuously retrieved results during haze periods in the summer and winter of 2014 respectively. It can be seen that the proposed algorithm could obtain valid AOD over all land surface types in cloudless condition (Figure 10g,h). Water bodies were masked in the algorithm for their rapid variation in surface reflectance. Because the looser thresholds were set in the cloud screening process, the algorithm could retrieve AOD effectively in hazy conditions (Figures 9e,f and 10i). Nevertheless, looser thresholds also lead to incomplete cloud screening in some cases, and the retained clouds could further result in some extremely high AOD values (Figures 9d and 10j). The retrieved AOD presents a uniform spatial distribution with relatively low value (<0.5) on cloudless and clear days (Figure 10g,h), whereas on polluted and hazy days the AOD rapidly increases, thus presenting an obviously non-uniform spatial distribution (Figures 9f and 10f). The AOD during heavy hazy days ranges from 2 to 4, indicating a severe air pollution (Figures 9e and 10i).

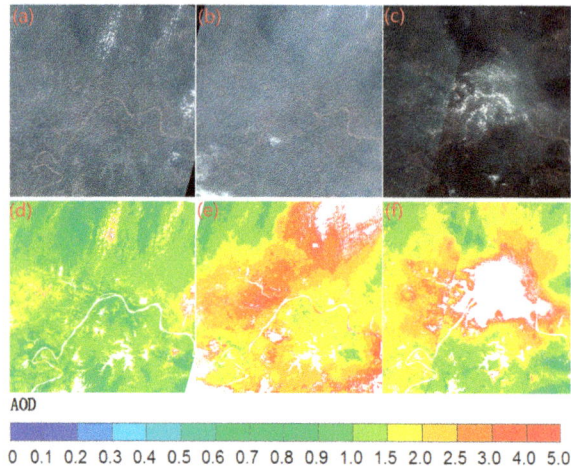

Figure 9. The AOD retrieval results from GF-1 WFV data during a haze period in summer. (**a–c**) are the RGB composited GF-1 WFV images on 3 June, 6 June and 11 June of 2014 respectively; (**d–f**) are the corresponding retrieved AOD at 550 nm. Water bodies were masked in the algorithm.

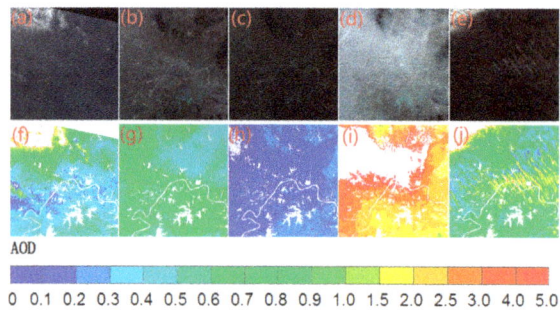

Figure 10. The AOD retrieval results from GF-1 WFV data during a haze period in winter. (**a–e**) are the RGB composited GF-1 WFV images on 15 January, 19 January, 23 January, 27 January and 31 January of 2014 respectively; (**f–j**) are the corresponding retrieved AOD at 550 nm.

Based on the complete AOD results in 2014 and 2015, we obtained the spatial distribution of yearly average AOD over Wuhan (Figure 11). There exists an obvious difference in the spatial distribution of yearly average AOD over urban and non-urban areas of Wuhan. The high AOD value zone is mainly distributed in the urban area of Wuhan (located in the center region), with a regional mean value of about 0.9, whereas most AOD values over non-urban areas of Wuhan range from 0.5 to 0.8. This situation is common in many cities of the world, where the anthropogenic aerosol emissions sourced from traffic, industry and cooking are dominant year round [44]. Annual difference also can be seen in the yearly average AOD map over Wuhan. The high AOD zone was distributed in the east of the urban area in Wuhan during the year 2014, whereas it was found to be partly transported to the western urban area of Wuhan during 2015. This may be caused by the change of aerosol source, wind direction and other environmental factors over this region.

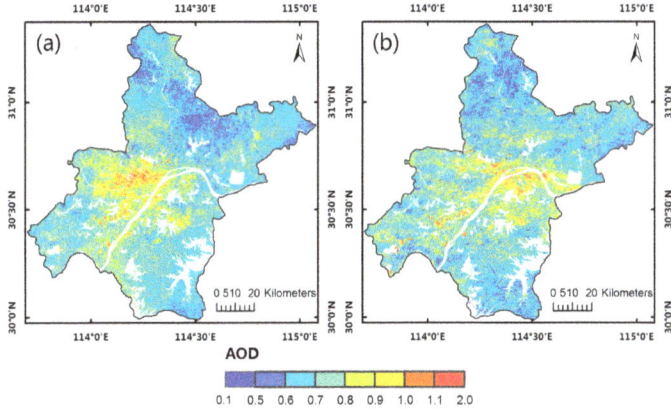

Figure 11. The spatial distribution yearly average AOD over Wuhan derived from GF-1 WFV data in 2014 (**a**) and 2015 (**b**).

4.2. Comparison of GF-1 WFV AOD with MODIS AOD

A spatial and temporal comparison between GF-1 WFV AOD and MODIS AOD is presented in this section. First, we perform an intercomparison of spatial distribution between GF-1 WFV AOD (160 m), MODIS DB AOD (10 km) and MODIS DT AOD (3 km). Figure 12 shows the spatial distribution of GF-1 WFV, MODIS DB and MODIS DT AOD on 8 July 2013 (cloudless, clear day) and 7 June 2014 (cloudless, hazy day). It can be seen that the spatial distribution patterns and values of GF-1 WFV, MODIS DB and MODIS DT AOD are roughly consistent with each other. The GF-1 WFV AOD tends to be a little higher than the MODIS DB and DT AOD, which may be caused by the difference in spatial resolution, overpassing time and retrieval algorithm. Similar to GF-1 WFV AOD, the MODIS DB AOD could also cover all land surface types in Wuhan; nevertheless, the GF-1 WFV AOD at a spatial resolution of 160 m provides more abundant details than MODIS DB AOD in the spatial distribution. The MODIS DT AOD has a finer spatial resolution compared to MODIS DB AOD, but it fails to cover bright surfaces (including urban areas) in Wuhan.

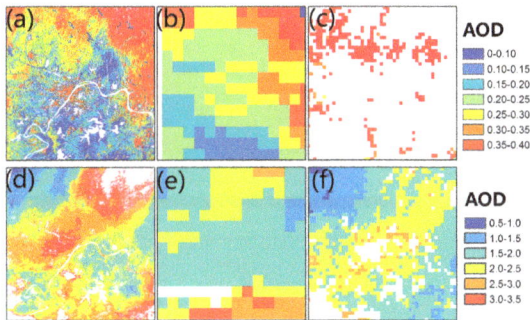

Figure 12. Comparison between GF-1 WFV AOD at 160 m, MODIS DB AOD at 10 km and MODIS DT AOD at 3 km. (**a–c**) are the GF-1 WFV AOD, MODIS DB AOD and MODIS DT AOD respectively on 8 July 2013; (**d–f**) are the GF-1 WFV AOD, MODIS DB AOD and MODIS DT AOD respectively on 7 June 2014.

To further validate the spatial and temporal consistency between GF-1 WFV AOD and MODIS AOD, we resampled all the GF-1 WFV AOD data to the same spatial resolution of MODIS DB AOD, and then compared the GF-1 WFV AOD with the spatio-temporally matched MODIS DB AOD. Figure 13 shows the relationship between GF-1 WFV AOD and MODIS DB AOD. They present a reasonable relationship, with a correlation coefficient (R^2) of 0.66 and root mean square error (RMSE) of 0.266. The slope (0.54) is less than 1, indicating that GF-1 WFV AOD is generally higher than MODIS DB AOD.

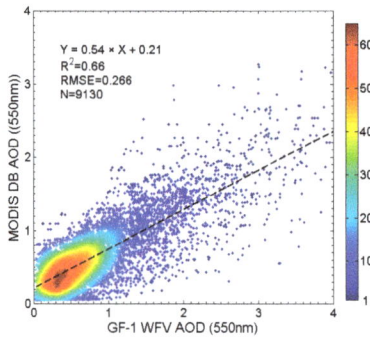

Figure 13. Relationship between GF-1 WFV AOD and MODIS DB AOD at 550 nm. RMSE and N represent Root Mean Square Error and total number of matched pixels respectively.

4.3. Comparison of GF-1 WFV AOD with Ground Measurements

Using the ground measurements at WHU and WHR (See Figure 1) from December 2014 to June 2015, we evaluated the accuracy of GF-1 WFV and MODIS AOD. The criterion of collocation between satellite-derived and ground-measured AOD is: (1) the time difference between satellite and ground measurements is less than 30 minutes; (2) the collocated satellite-derived AOD is the average value of pixels within a sampling widow (3 × 3 for MODIS DB AOD, 5 × 5 for GF-1 WFV AOD) centered on the ground site. The scatter plots between GF-1 WFV AOD, MODIS DB AOD and ground-measured AOD are shown in Figure 14. Because the satellite-derived AOD was reported to show different accuracies and biases in different regions with various surface types [45], we therefore performed independent statistics over ground sites WHU and WHR.

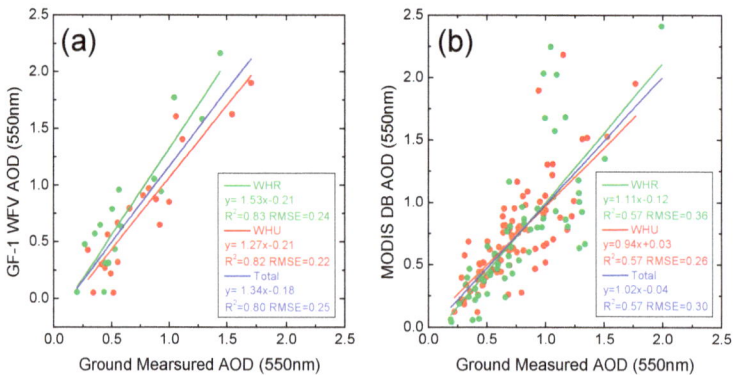

Figure 14. A comparison between GF-1 WFV AOD, MODIS DB AOD and ground-measured AOD at 550nm. (**a**) Scatter plot between GF-1 WFV AOD and ground-measured AOD; (**b**) Scatter plot between MODIS DB AOD and ground-measured AOD.

106

Compared with MODIS DB AOD ($R^2 = 0.57$; RMSE = 0.30), GF-1 WFV AOD shows a better relationship with ground measurements, with a higher correlation coefficient ($R^2 = 0.80$) and a smaller RMSE (RMSE = 0.25). However, the slope between GF-1 WFV AOD and ground measurements is 1.34, indicating that the proposed algorithm tends to overestimate AOD, whereas the slope between MODIS DB AOD and ground measurements is close to 1, indicating that the MODIS DB algorithm has smaller bias error. Generally, the relationship of satellite-derived and ground-measured AOD is approximately the same at urban (WHU) and rural (WHR) site for GF-1 WFV and MODIS data. Moreover, the slope at WHU is higher than the slope at WHR for both GF-1 WFV and MODIS AOD, thereby indicating that the proposed algorithm and MODIS DB algorithm have better accuracy over urban areas when compared to rural areas.

4.4. Performance and Limitations of the Proposed Algorithm

To our knowledge, retrieving high resolution AOD from satellite data is still uncommon at the present time. In related literature, Sun et al. [9,46] made an attempt to retrieve AOD from Chinese HJ-1 CCD and Landsat 8 OLI data with the support of MODIS surface reflectance products. The method may contain specific uncertainties in the transformation of different satellite data, and is limited in the spatial resolution of retrieved AOD (up to 500 m corresponding to the resolution of MODIS data). Luo et al. [43] proposed a mixed algorithm of the DT and DB methods to retrieve high resolution AOD from Landsat TM data over Beijing. The application potential of the method is limited, however, due to the long re-visiting period (16 days) of Landsat TM images.

By making full use of high spatial and temporal resolution of GF-1 WFV data, an operational AOD retrieval algorithm was proposed in this paper. The algorithm is capable of retrieving AOD over both bright and dark land surfaces in cloudless condition. The retrieved GF-1 WFV AOD has a spatial resolution of 160 m and temporal resolution of 4 days, which makes it possible to capture the detailed spatial variety and complete process of haze pollution over Wuhan region (as shown in Figures 9 and 10). The retrieved GF-1 WFV AOD achieve a high consistency with collocated MODIS AOD ($R^2 = 0.66$; RMSE = 0.27) and ground measurements ($R^2 = 0.80$; RMSE = 0.25). The validation results are on the same level as the performance of regional high resolution MODIS AOD retrieval over north and south China [42,47]. Nevertheless, the proposed algorithm tends to overestimate AOD in the comparison with MODIS AOD and ground measurements, which need to be further analyzed and improved upon in future research.

To analyze the error sources of the proposed algorithm, we visualized the relative errors ((GF-1 WFV AOD − Ground-measured AOD)/Ground-measured AOD) from view of different WFVs, seasons and scattering angles (Figure 15). First, GF-1 satellite has four WFV cameras all used in the AOD retrieval. Satellite data from four WFV cameras have different view geometries and sensor response characteristics, which may result in errors in the retrieval algorithm. The relative error distribution of GF-1 WFV AOD classified by WFVs is shown in Figure 15a. WFV1 has the most collocated AODs, 81% of which show a positive relative error. WFV4 has only 4 collocated AODs, all of them showing a negative relative error. It seems that the algorithm accuracy changes with different WFV sensors, though more data is needed to confirm this conclusion. Second, As GF-1 WFV AOD is retrieved by the seasonal surface reflectance database, there may exist seasonal difference in the algorithm accuracy. It can be seen from the limited collocated AODs that the algorithm tends to overestimate AOD in spring (MAM) and underestimate AOD in winter (DJF) (Figure 15b). The seasonal difference is mainly caused by impact factors such as background aerosol correction and minimum reflectance image composite in the establishment of surface reflectance database. Finally, the BRDF effect correction is significant for the MODIS DB algorithm [26,47], whereas it was also found that the effect of BRDF correction was not significant for MODIS data in a regional study [40]. As for the GF-1 WFV data, the BRDF effect seems to be severer for the high spatial resolution and wide swath. In the proposed algorithm, the BRDF effect was minimized by limiting satellite view zenith angle (<30°) and adopting window-based calculation strategy in the surface reflectance determination process. In order to evaluate the impact of BRDF effect on the algorithm accuracy, the relative error distribution of GF-1 WFV AOD along with corresponding scattering angles were analyzed in Figure 15c. Nearly all AODs with relatively low scattering angles (100°–120°) have a small positive relative error, whereas AODs with high scattering angles (120°–150°) show an irregular variation in the relative error. More specifically, the small number of high relative errors in absolute value (>80%) all occur in high scattering angles (>130°). This analysis demonstrates the impact of the BRDF effect on the accuracy of the proposed algorithm, which needs to be taken into account in future research.

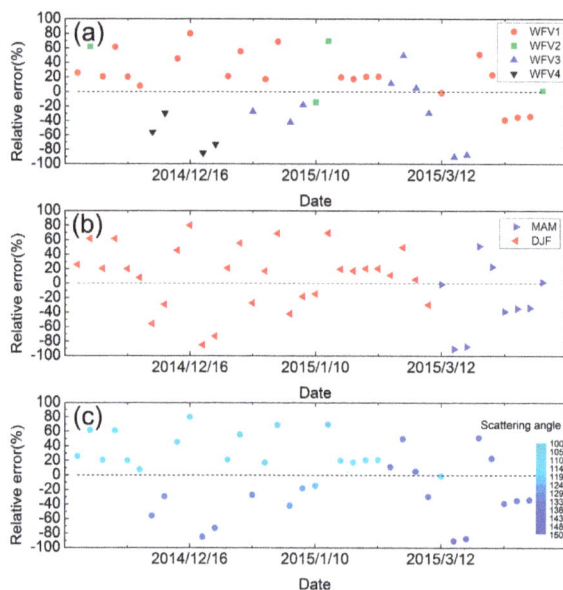

Figure 15. The relative error distribution of GF-1 WFV AOD. (**a**) Relative error distribution classified by WFV sensors; (**b**) relative error distribution classified by seasons; (**c**) relative error distribution classified by scattering angles (°).

5. Conclusions

To fill in the blank of high resolution Aerosol Optical Depth (AOD) for regional air quality studies, we proposed an improved approach to retrieve AOD from Gaofen-1 Wide-Field-of-View (GF-1 WFV) data. The proposed algorithm can work effectively over all land surface types in Wuhan area, with a spatial resolution of 160 m × 160 m and a temporal resolution of 4 days. The improved techniques used in the algorithm include: (1) Seasonal aerosol model over Wuhan was obtained and introduced in 6S model based on long-term ground measurements; (2) The moderate-resolution imaging spectroradiometer (MODIS) AOD product was introduced to support the image selection and background aerosol correction in the establishment of seasonal surface reflectance database for GF-1 WFV data.

Retrieved results show that the proposed algorithm works well during clear and hazy days, and provides more abundant details in spatial distribution than the MODIS AOD product. The GF-1 WFV AOD data were compared with MODIS and ground-measured AOD respectively for validation. Generally, GF-1 WFV AOD presents a good relationship with MODIS AOD ($R^2 = 0.66$; RMSE = 0.27) and ground measurements ($R^2 = 0.80$; RMSE = 0.25). Moreover, compared with MODIS AOD, GF-1 WFV AOD has a finer spatial resolution and covers more land surface types, rendering it more suitable for application in the studies of regional atmospheric environments such as air quality and haze monitoring.

There still exist some uncertainties to be further determined in the proposed algorithm. Although a high correlation is achieved in the validation against MODIS and ground-measured AOD, the proposed algorithm tends to overestimate AOD in many cases. Key factors including season, different WFV cameras and Bidirectional Reflectance Distribution Function (BRDF) effect were found to have potential impact on the accuracy of the algorithm. In order to improve the stability and accuracy of the algorithm, the impact of these factors need to be further analyzed after a collection of more satellite and ground measurements in the future.

Finally, it should be noted here that in recent years the Chinese government has launched an increasing number of satellites (e.g., HJ-1A/B, CBERS-04, GF-2, and GF-4) for ecological environment monitoring (for detail information, please see http://www.cresda.com/CN/index.shtml). These satellites have similar orbits, band sets, as well as spatial and temporal resolutions. The algorithm described here provides a good example for the application of these satellites to atmospheric environment monitoring.

Acknowledgments: This study was financially supported by the program of Key Laboratory for National Geographic Census and Monitoring, National Administration of Surveying, Mapping and Geo-information (No. 2014NGCM), the Public Interest Fund of Surveying & Mapping and Geoinformation Research (No. 201512026), the China Postdoctoral Science Foundation (Grant No. 2015M572198), the Hubei Provincial Natural Science Foundation of China(No. 2016CFB620), and the National Science Foundation of China (Grant No. 41331174). We thank China Centre for Resources Satellite Data and Application for providing GF-1 WFV data, and MODIS team for the AOD data used in the research.

Author Contributions: Kun Sun designed and performed the experiments, and prepared the manuscript draft. Xiaoling Chen helped to organize the research and outline the manuscript structure. Zhongmin Zhu contributed the data analysis and preparation of the manuscript. Tianhao Zhang helped to process the GF-1 data in the research.

Conflicts of Interest: The authors declare no conflict of interest.

References

1. Haywood, J.; Boucher, O. Estimates of the direct and indirect radiative forcing due to tropospheric aerosols: A review. *Rev. Geophys.* **2000**, *38*, 513–543. [CrossRef]

2. Kaufman, Y.J.; Tanre, D.; Boucher, O. A satellite view of aerosols in the climate system. *Nature* **2002**, *419*, 215–223. [CrossRef] [PubMed]

3. Sun, Y.; Wang, Z.; Wild, O.; Xu, W.; Chen, C.; Fu, P.; Du, W.; Zhou, L.; Zhang, Q.; Han, T.; et al. "APEC Blue": Secondary aerosol reductions from emission controls in Beijing. *Sci. Rep.* **2016**, *6*, 20668. [CrossRef] [PubMed]

4. Mishchenko, M.I.; Geogdzhayev, I.V.; Cairns, B.; Carlson, B.E.; Chowdhary, J.; Lacis, A.A.; Liu, L.; Rossow, W.B.; Travis, L.D. Past, present, and future of global aerosol climatologies derived from satellite observations: A perspective. *J. Quant. Spectrosc. Radiat. Transf.* **2007**, *106*, 325–347. [CrossRef]

5. Yang, J.; Gong, P.; Fu, R.; Zhang, M.; Chen, J.; Liang, S.; Xu, B.; Shi, J.; Dickinson, R. The role of satellite remote sensing in climate change studies. *Nat. Clim. Chang.* **2013**, *3*, 875–883, Erratum in **2014**, *4*, 74. [CrossRef]

6. Holben, B.; Vermote, E.; Kaufman, Y.J.; Tanré, D.; Kalb, V. Aerosol retrieval over land from AVHRR data-application for atmospheric correction. *IEEE Trans. Geosci. Remote Sens.* **1992**, *30*, 212–222. [CrossRef]

7. Remer, L.A.; Kaufman, Y.; Tanré, D.; Mattoo, S.; Chu, D.; Martins, J.V.; Li, R.-R.; Ichoku, C.; Levy, R.; Kleidman, R. The MODIS aerosol algorithm, products, and validation. *J. Atmos. Sci.* **2005**, *62*, 947–973. [CrossRef]

8. Lee, K.H.; Li, Z.; Kim, Y.J.; Kokhanovsky, A. Atmospheric aerosol monitoring from satellite observations: A history of three decades. In *Atmospheric and Biological Environmental Monitoring*; Springer: Heidelberg, Germany, 2009; pp. 13–38.

9. Sun, L.; Sun, C.K.; Liu, Q.H.; Zhong, B. Aerosol optical depth retrieval by HJ-1/CCD supported by MODIS surface reflectance data. *Sci. China Earth Sci.* **2010**, *53*, 74–80. [CrossRef]

10. Ma, Z.; Hu, X.; Sayer, A.M.; Levy, R.; Zhang, Q.; Xue, Y.; Tong, S.; Bi, J.; Huang, L.; Liu, Y. Satellite-based spatiotemporal trends in PM$_{2.5}$ concentrations: China, 2004–2013. *Environ. Health Perspect.* **2016**, *124*, 184–192. [CrossRef] [PubMed]
11. Tao, M.; Chen, L.; Su, L.; Tao, J. Satellite observation of regional haze pollution over the North China Plain. *J. Geophys. Res. Atmos.* **2012**, *117*. [CrossRef]
12. Omar, A.H.; Won, J.G.; Winker, D.M.; Yoon, S.C.; Dubovik, O.; McCormick, M.P. Development of global aerosol models using cluster analysis of Aerosol Robotic Network (AERONET) measurements. *J. Geophys. Res. Atmos.* **2005**, *110*. [CrossRef]
13. Li, C.; Lau, A.; Mao, J.; Chu, D. Retrieval, Validation, and Application of the 1-km Aerosol Optical Depth from MODIS Measurements over Hong Kong. *IEEE Trans. Geosci. Remote Sens.* **2005**, *43*, 2650–2658.
14. Kaufman, Y.J.; Wald, A.E.; Remer, L.A.; Gao, B.-C.; Li, R.-R.; Flynn, L. The MODIS 2.1-μm channel-correlation with visible reflectance for use in remote sensing of aerosol. *IEEE Trans. Geosci. Remote Sens.* **1997**, *35*, 1286–1298. [CrossRef]
15. Levy, R.C.; Remer, L.A.; Mattoo, S.; Vermote, E.F.; Kaufman, Y.J. Second-generation operational algorithm: Retrieval of aerosol properties over land from inversion of Moderate Resolution Imaging Spectroradiometer spectral reflectance. *J. Geophys. Res. Atmos.* **2007**, *112*. [CrossRef]
16. Kaufman, Y.J.; Tanré, D.; Remer, L.A.; Vermote, E.F.; Chu, A.; Holben, B.N. Operational remote sensing of tropospheric aerosol over land from EOS moderate resolution imaging spectroradiometer. *J. Geophys. Res. Atmos.* **1997**, *102*, 17051–17067. [CrossRef]
17. Liang, S.; Fallah-Adl, H.; Kalluri, S.; JáJá, J.; Kaufman, Y.J.; Townshend, J.R.G. An operational atmospheric correction algorithm for Landsat Thematic Mapper imagery over the land. *J. Geophys. Res. Atmos.* **1997**, *102*, 17173–17186. [CrossRef]
18. Gillingham, S.; Flood, N.; Gill, T.; Mitchell, R. Limitations of the dense dark vegetation method for aerosol retrieval under Australian conditions. *Remote Sens. Lett.* **2012**, *3*, 67–76. [CrossRef]
19. Santer, R.; Ramon, D.; Vidot, J.; Dilligeard, E. A surface reflectance model for aerosol remote sensing over land. *Int. J. Remote Sens.* **2007**, *28*, 737–760. [CrossRef]
20. Nichol, J.E.; Wong, M.S.; Chan, Y.Y. Fine resolution air quality monitoring from a small satellite: CHRIS/PROBA. *Sensors* **2008**, *8*, 7581–7595. [CrossRef] [PubMed]
21. Wang, Z.; Chen, L.; Gong, H.; Gao, H. Modified DDV method of aerosol optical depth inversion over land surfaces from CBERS02B. *J. Remote Sens.* **2009**, *13*, 1047–1059.
22. Munchak, L.; Levy, R.; Mattoo, S.; Remer, L.; Holben, B.; Schafer, J.; Hostetler, C.; Ferrare, R. MODIS 3 km aerosol product: Applications over land in an urban/suburban region. *Atmos. Meas. Tech.* **2013**, *6*, 1747–1759. [CrossRef]
23. Levy, R.C.; Remer, L.A.; Kleidman, R.G.; Mattoo, S.; Ichoku, C.; Kahn, R.; Eck, T.F. Global evaluation of the Collection 5 MODIS dark-target aerosol products over land. *Atmos. Chem. Phys.* **2010**, *10*, 10399–10420. [CrossRef]
24. Hsu, N.C.; Gautam, R.; Sayer, A.M.; Bettenhausen, C.; Li, C.; Jeong, M.J.; Tsay, S.C.; Holben, B.N. Global and regional trends of aerosol optical depth over land and ocean using SeaWiFS measurements from 1997 to 2010. *Atmos. Chem. Phys.* **2012**, *12*, 8037–8053. [CrossRef]
25. Hsu, N.C.; Tsay, S.-C.; King, M.D.; Herman, J.R. Deep blue retrievals of Asian aerosol properties during ACE-Asia. *IEEE Trans. Geosci. Remote Sens.* **2006**, *44*, 3180–3195. [CrossRef]
26. Hsu, N.; Jeong, M.J.; Bettenhausen, C.; Sayer, A.; Hansell, R.; Seftor, C.; Huang, J.; Tsay, S.C. Enhanced deep blue aerosol retrieval algorithm: The second generation. *J. Geophys. Res. Atmos.* **2013**, *118*, 9296–9315. [CrossRef]
27. Lee, J.; Kim, J.; Song, C.H.; Ryu, J.; Ahn, Y.; Song, C.K. Algorithm for retrieval of aerosol optical properties over the ocean from the Geostationary Ocean Color Imager. *Remote Sens. Environ.* **2010**, *114*, 1077–1088. [CrossRef]
28. Guo, J.P.; Zhang, X.Y.; Wu, Y.R.; Zhaxi, Y.; Che, H.Z.; La, B.; Wang, W.; Li, X.W. Spatio-temporal variation trends of satellite-based aerosol optical depth in China during 1980–2008. *Atmos. Environ.* **2011**, *45*, 6802–6811. [CrossRef]
29. Lu, Z.; Zhang, Q.; Streets, D.G. Sulfur dioxide and primary carbonaceous aerosol emissions in China and India, 1996–2010. *Atmos. Chem. Phys.* **2011**, *11*, 9839–9864. [CrossRef]

30. Cheng, H.; Gong, W.; Wang, Z.; Zhang, F.; Wang, X.; Lv, X.; Liu, J.; Fu, X.; Zhang, G. Ionic composition of submicron particles (PM1.0) during the long-lasting haze period in January 2013 in Wuhan, central China. *J. Environ. Sci.* **2014**, *26*, 810–817. [CrossRef]

31. Song, J.; Guang, W.; Li, L.; Xiang, R. Assessment of air quality status in Wuhan, China. *Atmosphere* **2016**, *7*, 56. [CrossRef]

32. Wang, L.; Gong, W.; Xia, X.; Zhu, J.; Li, J.; Zhu, Z. Long-term observations of aerosol optical properties at Wuhan, an urban site in Central China. *Atmos. Environ.* **2015**, *101*, 94–102. [CrossRef]

33. Feng, L.; Li, J.; Gong, W.; Zhao, X.Z.; Chen, X.; Pang, X. Radiometric cross-calibration of Gaofen-1 WFV cameras using Landsat-8 OLI images: A solution for large view angle associated problems. *Remote Sens. Environ.* **2016**, *174*, 56–68. [CrossRef]

34. Li, J.; Chen, X.L.; Tian, L.Q.; Huang, J.; Feng, L. Improved capabilities of the Chinese high-resolution remote sensing satellite GF-1 for monitoring suspended particulate matter (SPM) in inland waters: Radiometric and spatial considerations. *ISPRS J. Photogramm. Remote Sens.* **2015**, *106*, 145–156. [CrossRef]

35. Vermote, E.F.; Tanré, D.; Deuze, J.L.; Herman, M.; Morcette, J. Second simulation of the satellite signal in the solar spectrum, 6S: An overview. *IEEE Trans. Geosci. Remote Sens.* **1997**, *35*, 675–686. [CrossRef]

36. Li, Z.; Niu, F.; Lee, K.; Xin, J.; Hao, W.; Nordgren, B.; Wang, Y.; Wang, P. Validation and understanding of Moderate Resolution Imaging Spectroradiometer aerosol products (C5) using ground-based measurements from the handheld Sun photometer network in China. *J. Geophys. Res. Atmos.* **2007**, *112*. [CrossRef]

37. He, Q.; Li, C.; Tang, X.; Li, H.; Geng, F.; Wu, Y. Validation of MODIS derived aerosol optical depth over the Yangtze River Delta in China. *Remote Sens. Environ.* **2010**, *114*, 1649–1661. [CrossRef]

38. Giles, D.M.; Holben, B.N.; Eck, T.F.; Sinyuk, A.; Smirnov, A.; Slutsker, I.; Dickerson, R.R.; Thompson, A.M.; Schafer, J.S. An analysis of AERONET aerosol absorption properties and classifications representative of aerosol source regions. *J. Geophys. Res. Atmos.* **2012**, *117*. [CrossRef]

39. Russell, P.B.; Bergstrom, R.W.; Shinozuka, Y.; Clarke, A.D.; DeCarlo, P.F.; Jimenez, J.L.; Livingston, J.M.; Redemann, J.; Dubovik, O.; Strawa, A. Absorption angstrom exponent in AERONET and related data as an indicator of aerosol composition. *Atmos. Chem. Phys.* **2010**, *10*, 1155–1169. [CrossRef]

40. Kim, M.; Kim, J.; Wong, M.S.; Yoon, J.; Lee, J.; Wu, D.; Chan, P.W.; Nichol, J.; Chung, C.; Ou, M. Improvement of aerosol optical depth retrieval over Hong Kong from a geostationary meteorological satellite using critical reflectance with background optical depth correction. *Remote Sens. Environ.* **2014**, *142*, 176–187. [CrossRef]

41. Levy, R.; Mattoo, S.; Munchak, L.; Remer, L.; Sayer, A.; Patadia, F.; Hsu, N. The Collection 6 MODIS aerosol products over land and ocean. *Atmos. Meas. Tech.* **2013**, *6*, 2989–3034. [CrossRef]

42. Wong, M.S.; Nichol, J.; Lee, K.H. An operational MODIS aerosol retrieval algorithm at high spatial resolution, and its application over a complex urban region. *Atmos. Res.* **2011**, *99*, 579–589. [CrossRef]

43. Luo, N.; Wong, M.S.; Zhao, W.; Yan, X.; Xiao, F. Improved aerosol retrieval algorithm using Landsat images and its application for PM_{10} monitoring over urban areas. *Atmos. Res.* **2015**, *153*, 264–275. [CrossRef]

44. Gupta, P.; Khan, M.N.; da Silva, A.; Patadia, F. Modis aerosol optical depth observations over urban areas in Pakistan: Quantity and quality of the data for air quality monitoring. *Atmos. Pollut. Res.* **2013**, *4*, 43–52. [CrossRef]

45. Nichol, J.; Bilal, M. Validation of MODIS 3 km resolution aerosol optical depth retrievals over Asia. *Remote Sens.* **2016**, *8*, 328. [CrossRef]

46. Sun, L.; Wei, J.; Bilal, M.; Tian, X.; Jia, C.; Guo, Y.; Mi, X. Aerosol optical depth retrieval over bright areas using Landsat 8 OLI images. *Remote Sens.* **2016**, *8*, 23. [CrossRef]

47. Li, S.; Chen, L.; Xiong, X.; Tao, J.; Su, L.; Han, D.; Liu, Y. Retrieval of the haze optical thickness in North China Plain using MODIS data. *IEEE Trans. Geosci. Remote Sens.* **2013**, *51*, 2528–2540. [CrossRef]

remote sensing

MDPI

Article

Interference of Heavy Aerosol Loading on the VIIRS Aerosol Optical Depth (AOD) Retrieval Algorithm

Yang Wang [1,2], Liangfu Chen [1,*], Shenshen Li [1,*], Xinhui Wang [3], Chao Yu [1,4], Yidan Si [1] and Zili Zhang [5]

[1] State Key Laboratory of Remote Sensing Science, Institute of Remote Sensing and Digital Earth, Chinese Academy of Sciences, Beijing 100101, China; wangyang01@radi.ac.cn (Y.W.); yuchao@radi.ac.cn (C.Y.); siyidan2014@126.com (Y.S.)
[2] University of the Chinese Academy of Sciences, Beijing 100049, China
[3] Beijing Municipal Environmental Monitoring Center, Beijing 100101, China; saint.tail.always@163.com
[4] State Key Joint Laboratory of Environment Simulation and Pollution Control, School of Environment, Tsinghua University, Beijing 100101, China
[5] Zhejiang Environmental Monitoring Center, Zhejiang 310000, China; xjholieagle@163.com
* Correspondence: chenlf@radi.ac.cn (L.C.); lishenshen@126.com (S.L.); Tel.: +86-135-0124-7025 (L.C.)

Academic Editors: Yang Liu, Jun Wang, Omar Torres, Richard Müller and Prasad S. Thenkabail
Received: 13 February 2017; Accepted: 19 April 2017; Published: 23 April 2017

Abstract: Aerosol optical depth (AOD) has been widely used in climate research, atmospheric environmental observations, and other applications. However, high AOD retrieval remains challenging over heavily polluted regions, such as the North China Plain (NCP). The Visible Infrared Imaging Radiometer Suite (VIIRS), which was designed as a successor to the Moderate Resolution Imaging Spectroradiometer (MODIS), will undertake the aerosol observations mission in the coming years. Using the VIIRS AOD retrieval algorithm as an example, we analyzed the influence of heavy aerosol loading through the 6SV radiative transfer model (RTM) with a focus on three aspects: cloud masking, ephemeral water body tests, and data quality estimation. First, certain pixels were mistakenly screened out as clouds and ephemeral water bodies because of heavy aerosols, resulting in the loss of AOD retrievals. Second, the greenness of the surface could not be accurately identified by the top of atmosphere (TOA) index, and the quality of the aggregation data may be artificially high. Thus, the AOD retrieval algorithm did not perform satisfactorily, indicated by the low availability of data coverage (at least 37.97% of all data records were missing according to ground-based observations) and overestimation of the data quality (high-quality data increased from 63.42% to 80.97% according to radiative simulations). To resolve these problems, the implementation of a spatial variability cloud mask method and surficial index are suggested in order to improve the algorithm.

Keywords: AOD; VIIRS; heavy aerosol loading; retrieval algorithm; remote sensing

---

1. Introduction

Atmospheric aerosols are solid and liquid particles that are suspended in the air and are often related to dust, smoke, soot, and sea salt. Climate models indicate that aerosols can significantly impact the radiation budget of the Earth [1], cloud formation [2], and precipitation [3]. However, the uncertainty associated with the average climate impacts of aerosols remains large [4,5]. Furthermore, aerosols can also impact human health in heavily polluted regions [6–8]. The aerosol optical depth (AOD) is a basic optical property of aerosol research and has been broadly applied in climate research and atmospheric environmental observations.

Satellite remote sensing has the advantage of observing and quantifying aerosol systems at a global scale from space [9]. The Visible Infrared Imaging Radiometer Suite (VIIRS) aboard the Suomi National Polar-orbiting Partnership (Suomi-NPP) spacecraft was launched in October of 2011 [10]. This instrument was largely built on the success of the Moderate Resolution Imaging Spectroradiometer (MODIS), which has successfully retrieved AOD for more than 15 years [11]. The VIIRS was designed to have many similar features as its predecessors, and its aerosol algorithm was also based on the MODIS Dark-Target algorithm [12]. VIIRS will perform tasks for climate and air quality applications after MODIS completes its mission.

The accuracy and availability of AOD products under polluted atmospheric environments, such as heavy polluted areas in China, are limited. Increasing fossil fuel consumption and biomass burning in China [13] have caused severe air pollution events and have worsened the atmospheric environment in northern China [14,15]. Haze is an atmospheric phenomenon in which aerosol particles obscure the clarity of the sky and decrease the visibility below 10 km. Frequent haze events can be detected by ground-based observations [16,17] such as the AErosol Robotic NETwork (AERONET) [18], Chinese Sun Haze-meter Network (CSHNET) [19,20], and others. In situ observations have shown that the haze frequency and affected area have significantly increased over recent decades [21,22]. However, limited in situ observations and uneven distributions could introduce considerable uncertainty. Satellite observations can provide wide spatial coverage and long-term data records. From a temporal perspective, Zhang et al. [23] used the Absorbing Aerosol Index (AAI) to show that the haze over northern and eastern China follows an increasing trend that is similar to the pattern that is observed from MODIS AOD [24,25]. Climate Data Records (CDR) and regular air quality observations require an accurate, consistent, and wide-coverage AOD product [26,27]. However, certain retrieval vacancies exist over areas of heavy aerosol loading, so the VIIRS AOD products are not acceptable under hazy conditions. Previous research has attempted to improve the ability to retrieve hazy AODs with MODIS data [11,28]; however, the quality of these products remains insufficient under polluted conditions.

AOD products may be influenced by unsuitable hypotheses for cloud masks and pixel selection and poor data quality assurance. The success of aerosol retrieval depends on the ability to screen out unsuitable pixels. The most important step is accurate cloud masking. The standard MODIS cloud mask (MxD35) is considered too cloud conservative and not clear-sky sufficient for aerosol retrieval [29,30]. Therefore, after applying the MODIS Collection_4 algorithm, Martins et al. (2002) developed a new independent cloud mask that was mainly based on a spatial variability test. With VIIRS, the cloud mask in aerosol retrieval is based on a VIIRS cloud mask product (VCM), which is similar to MxD35 [12,31]. Although the VCM product performs well as evaluated by MODIS and CALIPSO data [32], this product still has flaws when used in AOD retrieval. Furthermore, the ephemeral water body test calculates the top of the atmosphere (TOA) normalized difference vegetation index (NDVI) and excludes pixels below a certain threshold [12]. However, heavy aerosol loading likely affects the calculation of this parameter. The data quality for the product's aggregation strategy depends on the pixel number and greenness. The greenness is defined by another TOA NDVI [33] that is minimally affected by the AOD. However, under hazy conditions, the hypothesis must be reconsidered. Therefore, under heavy aerosol loading conditions, AOD products can be affected by cloud masking, the ephemeral water body test, and quality assurance issues.

This article focuses on analyzing how these three factors influence the AOD retrieval algorithm under polluted atmospheric conditions and provides feasible advice. The data from the retrieval algorithm and analytical method are described in Section 2. Then, a radiative transfer simulation and certain examples are used to illustrate the results of a cloud mask, ephemeral water body test, and data aggregation in Section 3. Section 4 analyzes the causes of these impacts and provides a quantitative evaluation. Finally, Section 5 provides concluding remarks.

2. Data and Methods

2.1. North China Plain

The study area is mainly located on the North China Plain (NCP), which is the largest alluvial plain in China. The NCP is surrounded by the Yanshan and Taihang mountains to the north and west and vast deserts in north-western China and Mongolia. Because its climate is characterized by both humid winds from the Pacific and dry winds from the interior of the Asian continent, the composition of the particulate matter in the air is complex and includes dust, sea salt, and industrial matter. The NCP is the most polluted area in China and one of the most polluted areas in the world because of weather conditions, terrain influences, and pollutant emissions (Figure 1).

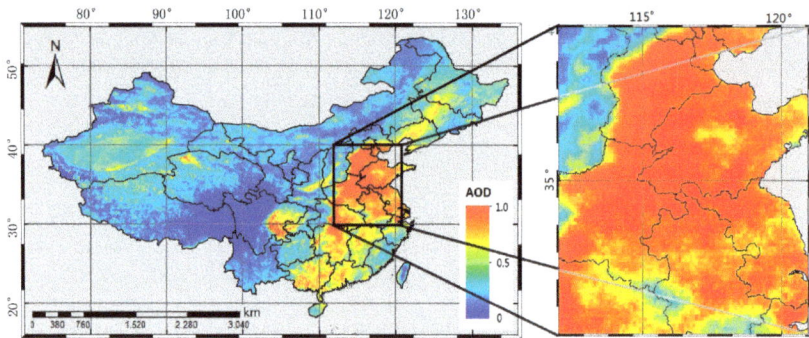

Figure 1. Annual average AOD distributions over the research area in 2015 (MODIS Collection 6 Deep Blue AOD at 550 nm). The right-hand figure shows the NCP, which is marked by a black square frame in the left-hand figure.

During periods of intense atmospheric pollutant emissions and calm and steady weather, particulate concentrations are extremely high, resulting in hazy days and inhibiting AOD satellite retrievals. Figure 2 shows the VIIRS AOD product on three polluted days over the NCP. Certain areas with heavy aerosol loading, such as the invalid values in the red ellipses in Figure 2, lack satellite-retrieved AOD information.

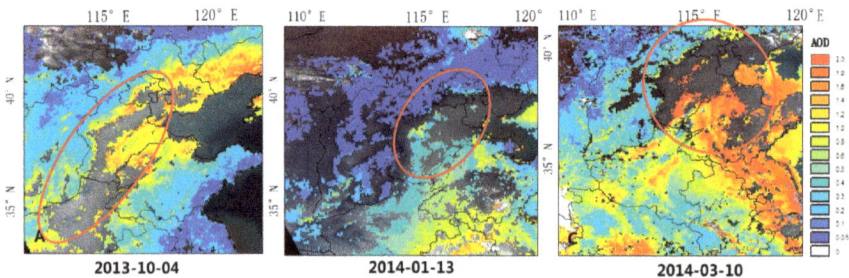

Figure 2. National Oceanic and Atmospheric Administration (NOAA)'s VIIRS AOD products (all data quality) over hazy areas. The AOD products were overlaid on the true color image, and no retrieval areas were set as transparent. Some AOD values are invalid, which are marked with red ellipses, because of heavy haze events.

2.2. Ground-Based Observations

AERONET uses sky scanning spectral radiometers to make ground-based observations of atmospheric aerosol optical properties and precipitable water [18,34]. The network provides a long-term and continuous dataset for satellite product validation. The level 2.0 AOD datasets have undergone cloud screening, calibration checks, and quality assurance. The level 2.0 dataset of Beijing_CAMS station from 2013 to 2016 was used in this study.

Because AERONET does not measure the 550 nm band, the AOD at 500 nm is used instead, denoted $AOD_{AERONET}$. We defined $AOD_{AERONET} > 0.6$ as polluted. The protocol requires at least six AERONET measurements within a 3-h period centered on the satellite overpass time. In total, 187 days were selected as polluted days. The VIIRS EDR data matchup requires retrievals within a 27.5-km-radius circle that is centered on the AERONET station [35]. If at least 20% of the pixels of all the potential retrievals are found to be clouds, the record is defined as being affected by cloud. The ephemeral water body and over-range tests are performed identically to the cloud test.

2.3. Satellite Data

The VIIRS instrument aboard the Suomi-NPP spacecraft was launched in October of 2011 and was designed to have similar capabilities as MODIS. Suomi-NPP orbits with a similar equator crossing time as Aqua. VIIRS data from 2013 to 2016 were used in this study, including sensor TOA reflectance (ρ_{TOA}), cloud mask data, geolocation data, and AOD data. The TOA measurement data were level 1b Sensor Data Records (SDR), including moderate-resolution bands (M-bands) with a spatial resolution of 750 m and imagery bands (I-bands) with a spatial resolution of 375 m. The cloud mask data were pixel-level Intermediate Product (IP) data. The AOD data were 6-km-resolution level 2 Environmental Data Records (EDR) aggregated as 8×8 IP retrievals. All of these data were downloaded from the National Oceanic and Atmospheric Administration (NOAA)'s Comprehensive Large Array-data Stewardship System (CLASS) website (http://www.class.ncdc.noaa.gov/saa/products/welcome).

MODIS is a key sensor aboard the Terra and Aqua satellites, which were launched in 2000 and 2002, respectively. MOD09 is the eight-day surface reflectance (ρ_S) product of MODIS/AQUA. The dataset was fused to the monthly average surface reflectance in January 2015 by using a minimum method [36]. These mature surficial reflectance data were used as input data in the radiative simulation. The dataset was downloaded from The Level-1 and Atmosphere Archive & Distribution System (LAADS) Distributed Active Archive Center (DAAC) managed by the National Aeronautics and Space Administration (NASA) (https://ladsweb.nascom.nasa.gov/search/).

2.4. Radiative Transfer Simulation

The Second Simulation of a Satellite Signal in the Solar Spectrum (6S) radiative transfer model (RTM) provides accurate simulations of satellite and plane observations [37]. The new vector version (6SV) of this code can work in both scalar and vector modes [38].

In this article, we used the 6SV RTM to simulate the radiative transfer procedure. The angle, aerosol type and target altitude were not important factors when analyzing the influences of high AOD values. Therefore, the satellite zenith angle, solar zenith angle, and relative azimuth angle were set to 30°, 30° and 60°, respectively. The aerosol type was assumed to be continental, and the target altitude was set to 0.

In Section 3.1, ρ_{TOA} was simulated under each AOD (in 550 nm) from 0 to 3 using different ρ_s values for VIIRS bands M1 and M3.

In Section 3.2, three typical land covers—soil, vegetation, and water—are selected. The surface reflectance information for several land cover types is listed in Table 1. We used 6SV to calculate ρ_{TOA} in the VIIRS bands I1 and I2 and then obtained the TOA NDVI.

Table 1. Surface reflectance information of five types of land cover.

ρ_s	Soil 1	Soil 2	Soil 3	Soil 4	Water	Vegetation
I1 (0.638 µm)	0.18	0.18	0.20	0.22	0.02	0.04
I2 (0.862 µm)	0.25	0.33	0.30	0.30	0.02	0.40

In Section 3.3, the MODIS surface reflectance product was used to simulate the TOA reflectance in the NCP area (113°E–116°E, 34°N–39°N) under three different atmospheric conditions. The air is assumed to be clean when AOD = 0.1, it is lightly polluted when AOD = 1, and it is heavily polluted when AOD = 2.

2.5. Method and Algorithm

2.5.1. Cloud Mask Algorithm

The VIIRS cloud mask depends on an external identification result, the VCM-IP product, which is not robust enough for aerosol retrievals. The VCM technique incorporates several cloud detection tests to determine whether a pixel is obstructed by a cloud, and the VIIRS pixels are assigned a label depending on the cloud confidence level, i.e., confidently cloudy, probably cloudy, probably clear, or confidently clear [39,40]. These tests include reflectance, brightness temperature (BT), brightness temperature difference (BTD), and spatial tests using M-band and I-band data.

The spatial variability in the reflectance at the TOA is suitable for a cloud mask that is devoted to the retrieval of aerosol data [41]. The spatial test uses the absolute standard deviation of every 3 × 3 pixel (3 × 3 STD) threshold to identify clouds. The 3 × 3 STD (σ) is calculated as follows:

$$\sigma = \sqrt{\frac{\sum_{i=1}^{9}(\rho_i - \bar{\rho})^2}{9}} \tag{1}$$

where ρ_i is the TOA reflectance of each pixel and $\bar{\rho}$ is the average TOA reflectance of all nine pixels.

In this study, we calculated the 3 × 3 STD in bands M1 (0.412 µm) and M3 (0.486 µm).

2.5.2. Ephemeral Water Body Test Method

The presence of surface water over land can affect retrieval algorithms; thus, an ephemeral water body detection test was applied to overcome this deficiency. This test is based on the TOA NDVI values calculated using bands I1 (0.638 µm) and I2 (0.862 µm) with the following equation [12]

$$NDVI = (\rho_{I2} - \rho_{I1})/(\rho_{I2} + \rho_{I1}) \tag{2}$$

where ρ_{I1} and ρ_{I2} are the TOA reflectances of bands I1 and I2, respectively. The TOA NDVI threshold is 0.1. If the TOA NDVI value of a pixel is less than the threshold, the pixel is identified as an ephemeral water body [12].

2.5.3. EDR Product Aggregation Strategy

The VIIRS AOD EDR product was constructed by aggregating 8 × 8 arrays of pixel levels retrieved from AOD IP. The overall quality (high, medium, and low) of the EDR depends on the pixel number, which is based on the IP quality within the EDR cell.

The quality of IP retrievals may be affected by the zenith and azimuth angles, clouds, pixel greenness, and other factors. Among these parameters, greenness was the only factor affected by aerosol loading. Thus, we only considered greenness and assumed that the other factors remained unchanged. The brightness index was used to identify when a pixel was dominated by a bright surface, less vegetated soil, or vegetation-dominated soil [12]. This index is also referred to as $NDVI_{SWIR}$ [33] and is calculated by using the equation

$$NDVI_{SWIR} = (\rho_{M8} - \rho_{M11})/(\rho_{M8} + \rho_{M11}) \tag{3}$$

where ρ_{M8} and ρ_{M11} are the TOA reflectance values of bands M8 and M11, respectively.

The VIIRS algorithm identifies a bright pixel when

$$NDVI_{SWIR} < 0.05 \text{ AND } \rho_{M11} > 0.3 \tag{4}$$

A vegetation-dominated pixel corresponds to $NDVI_{SWIR} > 0.2$, and the other pixels with intermediate values are defined as less vegetated. "Vegetation-dominated" is a necessary condition for "high" quality, and less vegetated conditions should be sufficient to assign the "degraded" quality flag [42].

The EDR product was labelled "high/medium/low" quality, depending on the number of AOD IP retrievals of different quality in the 8×8 EDR cell. The aggregation logic is displayed as a flowchart in Figure 3 according to the Algorithm Theoretical Basis Document (ATBD) [42].

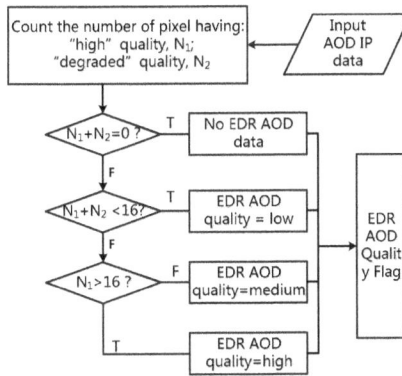

Figure 3. IP to EDR aggregation flow chart.

3. Results

3.1. Cloud Mask

The aerosol retrieval only works under cloud-free conditions. Thus, cloudy pixels must be identified and removed. The VIIRS AOD retrieval depends on the cloud mask when using the information from the VCM input.

We obtained cloud mask information in the form of "Confidently Cloudy" and "Probably Cloudy" from the AOD IP (IVAOT) quality flag (QF) data. To reveal inadequate cloud tests under heavy aerosol loading, we chose two locally hazy days—23 December 2013 and 18 March 2016—which are shown in Figure 4. On 23 December 2013 (Figure 4a), the NCP was covered by heavy haze but was almost cloud free. However, the VCM result indicates clouds in the area. A similar indication was found on 18 March 2016 (Figure 4b). These pixels would be excluded, which would result in no AOD retrieval. The VCM performs well in clear areas, although heavy aerosol loading may mislead the tests.

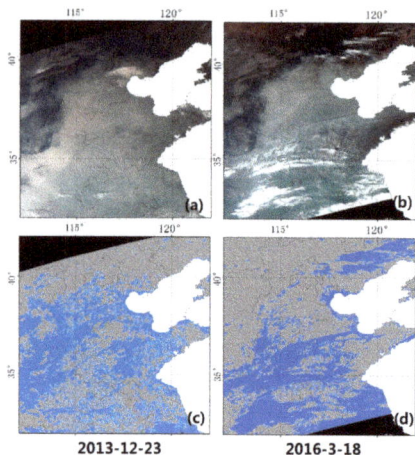

Figure 4. Two-day VIIRS true color images (**a,b**) and NOAA cloud mask result (**c,d**) over the NCP on 23 December 2013 and 18 March 2016. The cloud pixels are represented in blue in the cloud mask result.

In the visible channel, aerosols show a highly homogeneous spatial structure that can be easily separated from most clouds. Thus, the spatial variability test is efficient at masking clouds during aerosol retrieval [41]. We used a 6SV radiative transfer core to simulate ρ_{TOA} in the VIIRS bands M1 and M3. As shown in Figure 5, the difference in ρ_{TOA} values is smaller under high AOD conditions than under low AOD conditions. Therefore, the ρ_{TOA} values under heavy aerosol conditions are more homogeneous than those under clear sky conditions. The high degree of homogeneity under heavy aerosol conditions makes these pixels easier to distinguish from clouds.

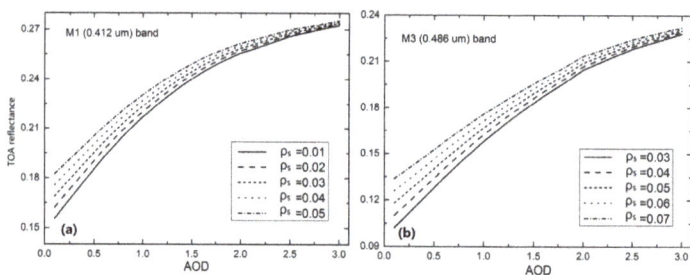

Figure 5. ρ_{TOA} simulation for the M1 (**a**) and M3 (**b**) bands under different AOD values (ranging from 0 to 3). The different lines represent several surface reflectance values.

We pre-selected more than 200,000 pixels from the RGB image in eastern Asia and classified them into three groups—clouds, haze, and clear sky—by visual interpretation. Figure 6 shows the statistical STD histograms of every 3×3 set of pixels in the VIIRS bands M1 and M3. The upper row is the frequency distribution diagram, and the bottom row is the cumulative frequency distribution diagram. In Figure 6a,b, the frequency peak of haze was located farther from that of clouds than that of clear sky. In Figure 6d, the differences among clouds, haze, and clear sky are more significant. Therefore, clouds and haze are easily separated in standard deviation histograms. Based on the histogram in Figure 6, the thresholds were defined as the separator between clouds and clear sky. We hope to reserve clear sky pixels as much as possible on the basis of the majority of cloud pixels being screened out. The red vertical lines are the suggested thresholds based on the 3×3 STD test for VIIRS bands M1 and

M3. The threshold in M1 is 0.005, and the threshold in M3 is 0.01. These thresholds generally do not exclude aerosol pixels (less than 2% of these samples) and only allow a small amount little cloud contamination (less than 5%).

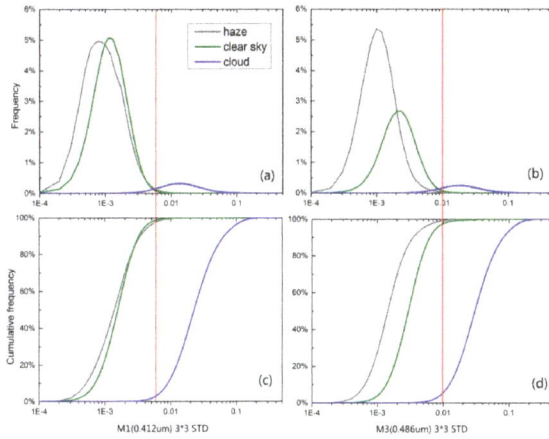

Figure 6. Histograms (**a,b**) and cumulative histograms (**c,d**) of the ρ_{TOA} STD in the VIIRS bands M1 (left column) and M3 (right column) for three types of pixels, including clouds (blue), haze (grey), and clear sky (green). The red lines are the suggested thresholds of the spatial variability test, which are 0.005 for M1 and 0.01 for M3.

3.2. Ephemeral Water Body Test

We extracted the results of the ephemeral water body test from the "IVAOT" QF data from 13 January 2014 and 10 March 2014, as shown in Figure 7. In these days of low precipitation, the NCP area could not contain as large a range of ephemeral water bodies as the algorithm test. These identification errors only occur in heavy aerosol loading areas.

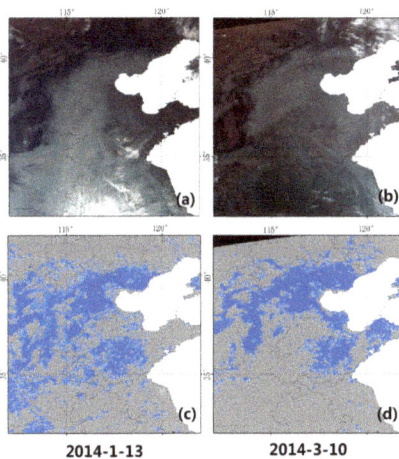

Figure 7. VIIRS true color image on 13 January 2014 (**a**) and 10 March 2014 (**b**) and the corresponding ephemeral water body test results (**c,d**) over the NCP. The ephemeral water body pixels are represented in blue.

Figure 8 shows the TOA NDVI simulation results for different land cover types. As seen in the vegetation line, the NDVI decreases with increasing AOD, and the NDVI value is always greater than 0. The TOA NDVI of water exhibits an almost constant negative pattern at any AOD range. Additionally, the TOA NDVI value of soil is obviously influenced by AOD: the values are greater than zero at low AOD values and close to or even less than zero at higher AOD values.

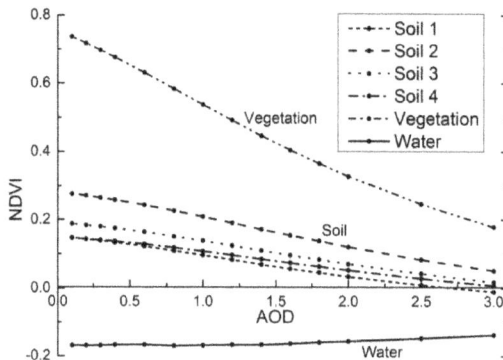

Figure 8. TOA NDVI simulation results for six types of land cover. The satellite zenith angle was 30°, the solar zenith angle was 30°, and the relative azimuth angle was 120°. In this simulation, the aerosol type was assumed to be continental, and the AOD ranged from 0 to 3.

The threshold of the VIIRS ephemeral water body test is 0.1. If the TOA NDVI value of a pixel is less than the threshold, then the pixel is identified as ephemeral water body. As the simulation in Figure 8 shows, certain soil pixels tended to be identified as ephemeral water bodies at high AOD levels.

3.3. Available Retrievals

Based on the NOAA AOD distribution image (Figure 2), the retrievals failed in certain areas with heavy aerosol conditions. The cause of this failure was demonstrated by radiative simulation. Here, we use AERONET ground-based observation data from 2013 to 2016 to quantify this deficiency. The high AOD values were not retrieved because of the mistaken identification as clouds or ephemeral water bodies. Moreover, this process might have occurred because the AOD retrieval range (0–2) was exceeded. Table 2 lists the number of retrieval results compared with the AERONET dataset. Clouds, ephemeral water bodies, and the retrieval ranges are denoted as a, b, and c, respectively.

Table 2. Number of VIIRS AOD EDR retrievals compared with the AERONET Beijing_CAMS station.

	Factor	Count	Total
Complete retrieval	-	-	67
Partial retrieval	a	19	
	b	14	
	c	2	49
	a, b	2	
	a, b, c	12	
No retrieval	a	16	
	b	22	
	a, b	29	71
	a, c	3	
	a, b, c	1	

Of the 187 AERONET observations, only 67 days have complete retrievals, and 71 days have no retrievals at all. The no-retrieval days comprised 37.97% of the dataset. Additionally, 49 days (26.20%) feature partial retrievals. The factors that were responsible for the lack of retrieval varied and included clouds, ephemeral water bodies, retrieval ranges, or combinations of multiple factors. In the partial retrieval group, the area with the 27.5-km-radius circle centered on the Beijing_CAMS station contains at least 20% pixels with retrieval data and at least 20% pixels with no retrieval data. We cannot confirm whether the clouds were real or whether pixels were mistakenly identified as clouds. However, some misidentification had to be included for clarity. In this example, the interference percentage of clouds and ephemeral water bodies was approximately equal. In summary, at least 37.97% of the high AOD data were not retrieved at the Beijing_CAMS station because of heavy aerosol loading.

3.4. Quality Assurance

The reason why NDVI$_{SWIR}$ was selected as a quality indicator was that the radiance in these bands is only slightly influenced by aerosol loading [33]. However, this hypothesis is inappropriate when the AOD is high.

Figure 9a shows the MODIS 1.23 μm surface reflectance of the NCP in January 2015. Then, we used the reflectance to calculate NDVI$_{SWIR}$ when AOD = 0.1, 1 and 2, which are denoted NDVI$_{SWIR}$_0.1, NDVI$_{SWIR}$_1 and NDVI$_{SWIR}$_2, respectively. Figure 9b is the difference between NDVI$_{SWIR}$ values under AOD = 0.1 and AOD = 1 (NDVI$_{SWIR}$_1-NDVI$_{SWIR}$_0.1). Figure 9c shows the difference between the NDVI$_{SWIR}$ values under AOD = 0.1 and AOD = 2 (NDVI$_{SWIR}$_2-NDVI$_{SWIR}$_0.1). As shown in Figure 9b,c, the NDVI$_{SWIR}$ values of aerosol loads of AOD = 1 and 2 were much higher than those of AOD = 0.1. These differences can also be observed in Figure 10, which is a histogram of NDVI$_{SWIR}$ values under different aerosol loading conditions. The frequency peak moves towards higher NDVI$_{SWIR}$ values as the AOD increases. The pixels with higher NDVI$_{SWIR}$ values probably tended to be identified as vegetation-dominated pixels. Although longer wavelengths are less influenced by aerosols, the influence cannot be neglected when the AOD is high. Therefore, the TOA NDVI$_{SWIR}$ is not suitable for land cover detection at high AOD levels.

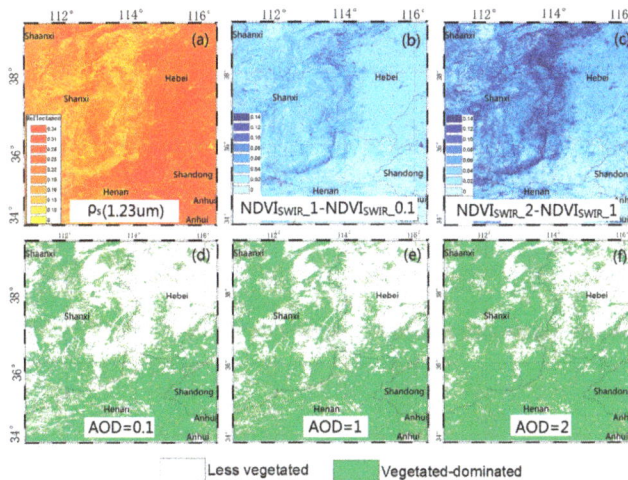

Figure 9. (a) MODIS surface reflectance at 1.23 μm over the NCP. The NDVI$_{SWIR}$ values were simulated by using the surface reflectance under aerosol conditions of AOD = 0.1, 1 and 2. (b) Difference in the NDVI$_{SWIR}$ simulation values between AOD = 1 and 0.1. (c) Difference in the NDVI$_{SWIR}$ simulation values between AOD = 2 and 0.1. The surface type identification results under different aerosol loads of (d) AOD = 0.1, (e) AOD = 1 and (f) AOD = 2 were also identified.

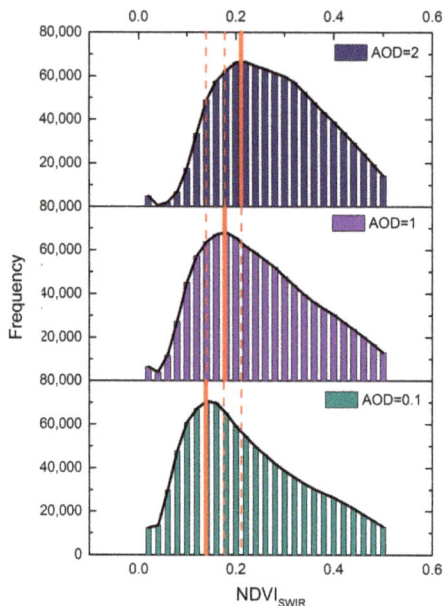

Figure 10. Histograms of NDVI$_{SWIR}$ when AOD = 0.1, 1, and 2. The red lines represent the NDVI$_{SWIR}$ frequency peak under the three different atmospheric conditions.

Every pixel was classified as "less vegetated" or "vegetated-dominated" according to the conditions in Section 2.5.3. Figure 9d–f show the classification results for AOD = 0.1, 1 and 2, respectively. The vegetation-dominated areas under AOD = 2 conditions are much larger than those under AOD = 0.1.

The EDR data QF depends on the quality and quantity of the IP data. Because this work was only concerned with surficial dominant types, we calculated the QF with 6-km-resolution EDR cells (including 8 × 8 pixels) using the aggregation strategy referred to in Section 2.5.3. Figure 11 shows the EDR QF when AOD = 0.1, 1, and 2. In Figure 11a, the data quality is better because the AOD is low. However, in Figure 11b,c, more pixels are identified as vegetation-dominated areas (see Figure 9e,f) when the AOD is higher, resulting in higher-quality EDR data. Certain EDR cells were identified as high quality in the central and western areas at AOD = 2, as shown in Figure 11c, whereas the cells were identified as medium quality at AOD = 0.1.

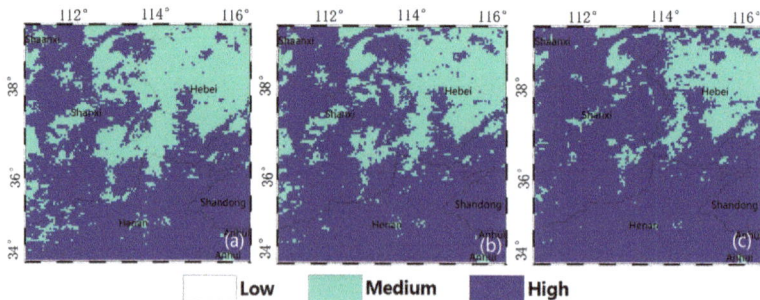

Figure 11. EDR data quality simulation results over the NCP under different aerosol loading: (**a**) AOD = 0.1, (**b**) AOD = 1, and (**c**) AOD = 2.

Table 3 lists the numbers and percentages of high- and medium-quality pixels under different aerosol loads, which correspond to the data in Figure 11. The high-quality EDR data percentage increased from 63.42% at AOD value of 0.1 to 80.91% at AOD value of 2.

Table 3. EDR data quality statistics under three aerosol loading conditions.

	AOD = 0.1		AOD = 1		AOD = 2	
	Number	Percentage	Number	Percentage	Number	Percentage
High	9909	63.42%	11,125	71.20%	12,651	80.97%
Medium	5716	36.58%	4500	28.80%	2974	19.03%

4. Analysis and Discussion

The AOD retrieval algorithm has three key basic scientific problems: cloud masks, aerosol models, and surface reflectance. Furthermore, the algorithm faces complex issues under high aerosol loading conditions. We analyzed and investigated these influences from the perspective of cloud masks, ephemeral water body tests, and data quality. These main effects influenced the retrieval availability and data quality.

4.1. Impact on Retrieval Availability

Accurate pixel selection is an important step before AOD inversions. Inappropriate pixels, such as clouds, bright surfaces, ice, snow, sun glint, among others, must be screened out. The cloud mask represents the most uncertain part of the AOD retrieval. Certain cloud mask methods [43–45] may have difficulty differentiating between clouds and high aerosol concentrations [41]. The VCM identification results also exhibit this limitation under high aerosol loading conditions, such as those shown in Figure 4. This type of error diminishes the retrieval area because of the masking of certain hazy pixels as clouds.

Identifying ephemeral water bodies is also an important component of pixel selection. However, when aerosol loading is high, the TOA radiative characteristics are obviously affected by the atmosphere and result in inaccurate identifications. As shown in Figure 8, the "soil 2" and "soil 3" lines decreased to values close to 0 when the AOD is high (AOD > 2.5). For less vegetated NCP areas, the NDVI of bare soil may decrease close to or less than zero under the influence of high AOD values. Therefore, these pixels would be screened out as ephemeral water body pixels. In certain sub-tropical and temperate climate regions, these mistaken identifications will occur under heavy aerosol loading when the vegetation coverage decreases in winter.

The influence of these two issues discussed above is focused on the available amount of AOD retrieval data. Cloud masks and ephemeral water body tests are influenced by heavy aerosol loading, resulting in the elimination of certain suitable pixels before retrieval. Therefore, AOD datasets tend to lack information in areas with high AOD values, resulting in underestimations in the spatial and temporal average AOD. This influence would lead to lower concentrations compared to ground-based observations and chemistry transport model (CTM) simulation results, which is very important for radiation and climate research as well as for air quality monitoring.

4.2. Impact on Data Quality

Using the Dark-Target retrieval algorithm, the AOD accuracy depends on surficial type and whether the type is consistent with the dark dense vegetation hypothesis. Therefore, the greenness parameter (TOA NDVI$_{SWIR}$) is an important indicator for determining the data quality. The NDVI$_{SWIR}$ is only slightly affected by the atmosphere at low AOD values, but the effects at high AOD values cannot be ignored. An investigation into the effects of different AOD values on data quality was performed for the NCP. The high-quality data percentage increased from 63.42% (AOD = 0.1) to 80.97% (AOD = 2). In this case, the data quality was different for each AOD assumption. However, in a realistic situation, the data quality should not change with increasing AOD. Furthermore, the data quality should be lower because of the lower accuracy of RTM simulations under high AOD conditions. Therefore, the quality degree is overestimated when the AOD is high. For scientific climate research that uses high-quality AOD data, overestimated data may introduce uncertainty errors that are related to high AOD values and may influence the uniformity of the data standard.

Aerosol models and radiative simulations are other influential factors. The aerosol model that is used in satellite observations is based on cluster analysis with global long-term AERONET data [46,47]. However, the aerosol components are complex and change when pollution episodes occur in extremely polluted countries or areas. Therefore, aerosol models are unlikely to be truly representative of the optical conditions viewed by satellites. Furthermore, compared to the Monte Carlo code, the 6SV RTM performs worse when AOD = 0.8 than when AOD = 0.2 [48]. Additionally, the error would be further magnified at higher AOD values because of multiple scattering.

4.3. Proposed Solution

Feasible methods are recommended for resolving these problems. On one hand, we should consider using a spatial variability test method that is effective for cloud tests with aerosol retrieval. According to the histogram analysis, the difference between clouds and haze is enhanced using 3 × 3 STDs. High aerosol concentration areas can be easily distinguished from clouds with this test. The suggested thresholds are 0.005 for the M1 band and 0.01 for the M3 band. On the other hand, the problems of ephemeral water bodies and data quality are both caused by atmospheric interference. Therefore, we recommend using the surficial parameters to overcome this issue. We could use a simplified retrieval algorithm to calculate the approximate values of AOD and use them as crude atmospheric corrections. Additionally, pre-calculated surface parameters could be used instead of TOA parameters to provide more accurate estimates of ephemeral water bodies and vegetation coverage and diminish the interference of high aerosol loading. Additionally, researchers should note that the quality of high AOD data is not credible when using satellite AOD products and should use these data carefully.

5. Conclusions

As a key optical and physical parameter of aerosols, AOD is critical for environment and climate research. However, under heavy aerosol loading conditions, cloud masks and ephemeral water body tests decrease the amount of available retrieval data, and the algorithm overestimates the data quality. Candidate pixels were identified as clouds or ephemeral water bodies because of heavy aerosol loading, indicating that certain crucial research areas that are associated with high AOD would not be retrieved. This retrieval coverage limitation can be attributed to incorrect cloud masks and water body tests. According to the statistical results from the AERONET Beijing_CAMS station, at least 37.97% of the high AOD data were not retrieved. Additionally, more high-quality (80.97% in radiative simulation) AOD data were retrieved under polluted atmospheric conditions because of improper TOA NDVI tests. These factors restrict the use of the AOD product. In environmental air quality assessments, a lack of high values may lead to underestimations in the spatial and temporal mean AOD values. This study used VIIRS AOD data as an example to qualitatively and semi-quantitatively analyze the limitations and deficiencies of AOD data under heavy aerosol loading. This work was necessary to improve algorithms and data applications.

Acknowledgments: This work was supported by the National Natural Science Foundation of China (Grant No. 91543128 and Grant No. 41301494) and the National Science and Technology Ministry (Grant No. 2014BAC21B03). The original VIIRS and MODIS data that were used in this paper are available for free through the links in Section 2. We are grateful for the AERONET data services.

Author Contributions: Y.W., L.C., and S.L. conceived and designed the experiments; Y.W. performed the experiments; Y.W. and X.W. analyzed the data; and Y.W., Y.S., and Z.Z. wrote the paper.

Conflicts of Interest: The authors declare no conflict of interest.

References

1. Ramanathan, V.; Crutzen, P.J.; Kiehl, J.T.; Rosenfeld, D. Atmosphere—Aerosols, climate, and the hydrological cycle. *Science* **2001**, *294*, 2119–2124. [CrossRef] [PubMed]
2. Rosenfeld, D.; Lohmann, U.; Raga, G.B.; O'Dowd, C.D.; Kulmala, M.; Fuzzi, S.; Reissell, A.; Andreae, M.O. Flood or drought: How do aerosols affect precipitation? *Science* **2008**, *321*, 1309–1313. [CrossRef] [PubMed]
3. Koren, I.; Feingold, G. Aerosol-cloud-precipitation system as a predator-prey problem. *Proc. Natl. Acad. Sci. USA* **2011**, *108*, 12227–12232. [CrossRef] [PubMed]
4. Bellouin, N.; Boucher, O.; Haywood, J.; Reddy, M.S. Global estimate of aerosol direct radiative forcing from satellite measurements. *Nature* **2005**, *438*, 1138–1141. [CrossRef] [PubMed]
5. Intergovernmental Panel on Climate Change. *Fifth Assessment Report: Climate Change 2013*; Cambridge University Press: Cambridge, NY, USA, 2013.
6. Pope, C.A.; Burnett, R.T.; Thun, M.J.; Calle, E.E.; Krewski, D.; Ito, K.; Thurston, G.D. Lung cancer, cardiopulmonary mortality, and long-term exposure to fine particulate air pollution. *Jama J. Am. Med. Soc.* **2002**, *287*, 1132–1141. [CrossRef]
7. Tie, X.X.; Wu, D.; Brasseur, G. Lung cancer mortality and exposure to atmospheric aerosol particles in Guangzhou, China. *Atmos. Environ.* **2009**, *43*, 2375–2377. [CrossRef]
8. Lim, S.S.; Vos, T.; Flaxman, A.D.; Danaei, G.; Shibuya, K.; Adair-Rohani, H.; Amann, M.; Anderson, H.R.; Andrews, K.G.; Aryee, M.; et al. A comparative risk assessment of burden of disease and injury attributable to 67 risk factors and risk factor clusters in 21 regions, 1990–2010: A systematic analysis for the global burden of disease study 2010. *Lancet* **2012**, *380*, 2224–2260. [CrossRef]
9. Mishchenko, M.I.; Geogdzhayev, I.V.; Cairns, B.; Carlson, B.E.; Chowdhary, J.; Lacis, A.A.; Liu, L.; Rossow, W.B.; Travis, L.D. Past, present, and future of global aerosol climatologies derived from satellite observations: A perspective. *J. Quant. Spectrosc. Radiat. Transf.* **2007**, *106*, 325–347. [CrossRef]
10. Cao, C.; De Luccia, F.J.; Xiong, X.; Wolfe, R.; Weng, F. Early on-orbit performance of the visible infrared imaging radiometer suite onboard the suomi national polar-orbiting partnership (S-NPP) satellite. *IEEE Trans. Geosci. Remote Sens.* **2014**, *52*, 1142–1156. [CrossRef]
11. Levy, R.C.; Mattoo, S.; Munchak, L.A.; Remer, L.A.; Sayer, A.M.; Patadia, F.; Hsu, N.C. The collection 6 MODIS aerosol products over land and ocean. *Atmos. Meas. Tech.* **2013**, *6*, 2989–3034. [CrossRef]
12. Jackson, J.M.; Liu, H.; Laszlo, I.; Kondragunta, S.; Remer, L.A.; Huang, J.; Huang, H.C. Suomi-NPP VIIRS aerosol algorithms and data products. *J. Geophys. Res. Atmos.* **2013**, *118*, 12673–12689. [CrossRef]
13. Wang, L.L.; Xin, J.Y.; Li, X.R.; Wang, Y.S. The variability of biomass burning and its influence on regional aerosol properties during the wheat harvest season in north China. *Atmos. Res.* **2015**, *157*, 153–163. [CrossRef]
14. Chen, H.P.; Wang, H.J. Haze days in north China and the associated atmospheric circulations based on daily visibility data from 1960 to 2012. *J. Geophys. Res. Atmos.* **2015**, *120*, 5895–5909. [CrossRef]
15. Tao, M.H.; Chen, L.F.; Su, L.; Tao, J.H. Satellite observation of regional haze pollution over the north China plain. *J. Geophys. Res. Atmos.* **2012**, *117*. [CrossRef]
16. Lee, K.H.; Li, Z.; Cribb, M.C.; Liu, J.; Wang, L.; Zheng, Y.; Xia, X.; Chen, H.; Li, B. Aerosol optical depth measurements in eastern China and a new calibration method. *J. Geophys. Res. Atmos.* **2010**. [CrossRef]
17. Zhao, X.J.; Zhao, P.S.; Xu, J.; Meng, W.; Pu, W.W.; Dong, F.; He, D.; Shi, Q.F. Analysis of a winter regional haze event and its formation mechanism in the north China plain. *Atmos. Chem. Phys.* **2013**, *13*, 5685–5696. [CrossRef]
18. Holben, B.N.; Eck, T.F.; Slutsker, I.; Tanre, D.; Buis, J.P.; Setzer, A.; Vermote, E.; Reagan, J.A.; Kaufman, Y.J.; Nakajima, T.; et al. Aeronet—A federated instrument network and data archive for aerosol characterization. *Remote Sens. Environ.* **1998**, *66*, 1–16. [CrossRef]
19. Xin, J.Y.; Wang, Y.S.; Pan, Y.P.; Ji, D.S.; Liu, Z.R.; Wen, T.X.; Wang, Y.H.; Li, X.R.; Sun, Y.; Sun, J.; et al. The campaign on atmospheric aerosol research network of china care-china. *Bull. Am. Meteorol. Soc.* **2015**, *96*, 1137–1155. [CrossRef]
20. Xin, J.Y.; Wang, Y.S.; Li, Z.Q.; Wang, P.C.; Hao, W.M.; Nordgren, B.L.; Wang, S.G.; Liu, G.R.; Wang, L.L.; Wen, T.X.; et al. Aerosol optical depth (aod) and angstrom exponent of aerosols observed by the chinese sun hazemeter network from august 2004 to september 2005. *J. Geophys. Res. Atmos.* **2007**, *112*, 13–16. [CrossRef]
21. Che, H.Z.; Zhang, X.Y.; Li, Y.; Zhou, Z.J.; Qu, J.J.; Hao, X.J. Haze trends over the capital cities of 31 provinces in china, 1981–2005. *Theor. Appl. Climatol.* **2009**, *97*, 235–242. [CrossRef]

22. Su, B.; Zhan, M.; Zhai, J.; Wang, Y.; Fischer, T. Spatio-temporal variation of haze days and atmospheric circulation pattern in china (1961–2013). *Quat. Int.* **2015**, *380*, 14–21. [CrossRef]

23. Zhang, X.; Wang, L.; Wang, W.; Cao, D.; Wang, X.; Ye, D. Long-term trend and spatiotemporal variations of haze over china by satellite observations from 1979 to 2013. *Atmos. Environ.* **2015**, *119*, 362–373. [CrossRef]

24. He, Q.S.; Li, C.C.; Geng, F.H.; Lei, Y.; Li, Y.H. Study on long-term aerosol distribution over the land of east china using modis data. *Aerosol Air Qual. Res.* **2012**, *12*, 304–319. [CrossRef]

25. Lin, J.T.; Li, J. Spatio-temporal variability of aerosols over east china inferred by merged visibility-geos-chem aerosol optical depth. *Atmos. Environ.* **2016**, *132*, 111–122. [CrossRef]

26. Li, Z.; Zhao, X.; Kahn, R.; Mishchenko, M.; Remer, L.; Lee, K.H.; Wang, M.; Laszlo, I.; Nakajima, T.; Maring, H. Uncertainties in satellite remote sensing of aerosols and impact on monitoring its long-term trend: A review and perspective. *Ann. Geophys.* **2009**, *27*, 2755–2770. [CrossRef]

27. Popp, T.; de Leeuw, G.; Bingen, C.; Bruhl, C.; Capelle, V.; Chedin, A.; Clarisse, L.; Dubovik, O.; Grainger, R.; Griesfeller, J.; et al. Development, production and evaluation of aerosol climate data records from european satellite observations (AEROSOL_CCI). *Remote Sens.* **2016**. [CrossRef]

28. Li, S.S.; Chen, L.F.; Xiong, X.Z.; Tao, J.H.; Su, L.; Han, D.; Liu, Y. Retrieval of the haze optical thickness in north china plain using modis data. *IEEE Trans. Geosci. Remote Sens.* **2013**, *51*, 2528–2540. [CrossRef]

29. Sayer, A.M.; Munchak, L.A.; Hsu, N.C.; Levy, R.C.; Bettenhausen, C.; Jeong, M.J. Modis collection 6 aerosol products: Comparison between aqua's e-deep blue, dark target, and "merged" data sets, and usage recommendations. *J. Geophys. Res. Atmos.* **2014**, *119*, 13965–13989. [CrossRef]

30. Remer, L.A.; Mattoo, S.; Levy, R.C.; Heidinger, A.; Pierce, R.B.; Chin, M. Retrieving aerosol in a cloudy environment: Aerosol product availability as a function of spatial resolution. *Atmos. Meas. Tech.* **2012**, *5*, 1823–1840. [CrossRef]

31. Frey, R.A.; Ackerman, S.A.; Liu, Y.H.; Strabala, K.I.; Zhang, H.; Key, J.R.; Wang, X.G. Cloud detection with modis. Part i: Improvements in the modis cloud mask for collection 5. *J. Atmos. Ocean. Technol.* **2008**, *25*, 1057–1072. [CrossRef]

32. Vermote, E.; Justice, C.; Csiszar, I. Early evaluation of the viirs calibration, cloud mask and surface reflectance earth data records. *Remote Sens. Environ.* **2014**, *148*, 134–145. [CrossRef]

33. Levy, R.C.; Remer, L.A.; Mattoo, S.; Vermote, E.F.; Kaufman, Y.J. Second-generation operational algorithm: Retrieval of aerosol properties over land from inversion of moderate resolution imaging spectroradiometer spectral reflectance. *J. Geophys. Res. Atmos.* **2007**, *112*. [CrossRef]

34. Dubovik, O.; Smirnov, A.; Holben, B.N.; King, M.D.; Kaufman, Y.J.; Eck, T.F.; Slutsker, I. Accuracy assessments of aerosol optical properties retrieved from aerosol robotic network (AERONET) sun and sky radiance measurements. *J. Geophys. Res. Atmos.* **2000**, *105*, 9791–9806. [CrossRef]

35. Ichoku, C.; Chu, D.A.; Mattoo, S.; Kaufman, Y.J.; Remer, L.A.; Tanre, D.; Slutsker, I.; Holben, B.N. A spatio-temporal approach for global validation and analysis of modis aerosol products. *Geophys. Res. Lett.* **2002**. [CrossRef]

36. Sun, L.; Wei, J.; Wang, J.; Mi, X.T.; Guo, Y.M.; Lv, Y.; Yang, Y.K.; Gan, P.; Zhou, X.Y.; Jia, C.; et al. A universal dynamic threshold cloud detection algorithm (udtcda) supported by a prior surface reflectance database. *J. Geophys. Res. Atmos.* **2016**, *121*, 7172–7196. [CrossRef]

37. Vermote, E.F.; Tanre, D.; Deuze, J.L.; Herman, M.; Morcrette, J.J. Second simulation of the satellite signal in the solar spectrum, 6s: An overview. *IEEE Trans. Geosci. Remote Sens.* **1997**, *35*, 675–686. [CrossRef]

38. Kotchenova, S.Y.; Vermote, E.F.; Matarrese, R.; Klemm, F.J., Jr. Validation of a vector version of the 6s radiative transfer code for atmospheric correction of satellite data. Part i: Path radiance. *Appl. Opt.* **2006**, *45*, 6762–6774. [CrossRef] [PubMed]

39. Hutchison, K.D.; Iisager, B.D.; Kopp, T.J.; Jackson, J.M. Distinguishing aerosols from clouds in global, multispectral satellite data with automated cloud classification algorithms. *J. Atmos. Ocean. Technol.* **2008**, *25*, 501–518. [CrossRef]

40. VCM ATBD, VIIRS Cloud Mask (VCM) algorithm theoretical basis document (Revision E): 474-00033. Released August 2014. Available online: https://www.star.nesdis.noaa.gov/jpss/documents/ATBD/D0001-M01-S01-011_JPSS_ATBD_VIIRS-Cloud-Mask_E.pdf (accessed on 22 April 2017).

41. Martins, J.V.; Tanre, D.; Remer, L.; Kaufman, Y.; Mattoo, S.; Levy, R. Modis cloud screening for remote sensing of aerosols over oceans using spatial variability. *Geophys. Res. Lett.* **2002**, *29*. [CrossRef]

42. Aerosol ATBD, VIIRS aerosol optical thickness and particle size parameter algorithm theoretical basis document (Revision B): 474-00049. Released May 2014. Available online: https://www.star.nesdis. noaa.gov/jpss/documents/ATBD/D0001-M01-S01-020_JPSS_ATBD_VIIRS-AOT-APSP_B.pdf (accessed on 22 April 2017).

43. Ackerman, S.A.; Strabala, K.I.; Menzel, W.P.; Frey, R.A.; Moeller, C.C.; Gumley, L.E. Discriminating clear sky from clouds with modis. *J. Geophys. Res. Atmos.* **1998**, *103*, 32141–32157. [CrossRef]

44. Platnick, S.; King, M.D.; Ackerman, S.A.; Menzel, W.P.; Baum, B.A.; Riedi, J.C.; Frey, R.A. The modis cloud products: Algorithms and examples from terra. *IEEE Trans. Geosci. Remote Sens.* **2003**, *41*, 459–473. [CrossRef]

45. King, M.D.; Menzel, W.P.; Kaufman, Y.J.; Tanre, D.; Gao, B.C.; Platnick, S.; Ackerman, S.A.; Remer, L.A.; Pincus, R.; Hubanks, P.A. Cloud and aerosol properties, precipitable water, and profiles of temperature and water vapor from modis. *IEEE Trans. Geosci. Remote Sens.* **2003**, *41*, 442–458. [CrossRef]

46. Dubovik, O.; Holben, B.; Eck, T.F.; Smirnov, A.; Kaufman, Y.J.; King, M.D.; Tanre, D.; Slutsker, I. Variability of absorption and optical properties of key aerosol types observed in worldwide locations. *J. Atmos. Sci.* **2002**, *59*, 590–608. [CrossRef]

47. Levy, R.C.; Remer, L.A.; Dubovik, O. Global aerosol optical properties and application to moderate resolution imaging spectroradiometer aerosol retrieval over land. *J. Geophys. Res. Atmos.* **2007**. [CrossRef]

48. Kotchenova, S.Y.; Vermote, E.F.; Levy, R.; Lyapustin, A. Radiative transfer codes for atmospheric correction and aerosol retrieval: Intercomparison study. *Appl. Opt.* **2008**, *47*, 2215–2226. [CrossRef] [PubMed]

remote sensing

MDPI

Article

Aerosol Retrieval Sensitivity and Error Analysis for the Cloud and Aerosol Polarimetric Imager on Board TanSat: The Effect of Multi-Angle Measurement

Xi Chen [1,2], Dongxu Yang [1,*], Zhaonan Cai [1], Yi Liu [1] and Robert J. D. Spurr [3]

[1] Key Laboratory of Middle Atmosphere and Global Environment Observation,
    Institute of Atmospheric Physics, Chinese Academy of Sciences, No. 40, Huayan Li, Chaoyang District,
    Beijing 100029, China; chenxilageo@mail.iap.ac.cn (X.C.); caizhaonan@mail.iap.ac.cn (Z.C.);
    liuyi@mail.iap.ac.cn (Y.L.)
[2] University of Chinese Academy of Sciences, No. 19A, Yuquan Lu, Shijing Shan District,
    Beijing 100049, China
[3] RT Solutions, Cambridge, MA 02138, USA; rtsolutions@verizon.net
* Correspondence: yangdx@mail.iap.ac.cn; Tel.: +86-186-1140-6910

Academic Editors: Jun Wang, Omar Torres, Yang Liu, Alexander A. Kokhanovsky,
Richard Müller and Prasad S. Thenkabail
Received: 22 November 2016; Accepted: 16 February 2017; Published: 22 February 2017

Abstract: Aerosol scattering is an important source of error in CO_2 retrievals from satellite. This paper presents an analysis of aerosol information content from the Cloud and Aerosol Polarimetric Imager (CAPI) onboard the Chinese Carbon Dioxide Observation Satellite (TanSat) to be launched in 2016. Based on optimal estimation theory, aerosol information content is quantified from radiance and polarization observed by CAPI in terms of the degrees of freedom for the signal (DFS). A linearized vector radiative transfer model is used with a linearized Mie code to simulate observation and sensitivity (or Jacobians) with respect to aerosol parameters. In satellite nadir mode, the DFS for aerosol optical depth is the largest, but for mode radius, it is only 0.55. Observation geometry is found to affect aerosol DFS based on the aerosol scattering phase function from the comparison between different viewing zenith angles or solar zenith angles. When TanSat is operated in target mode, we note that multi-angle retrieval represented by three along-track measurements provides additional 0.31 DFS on average, mainly from mode radius. When adding another two measurements, the a posteriori error decreases by another 2%–6%. The correlation coefficients between retrieved parameters show that aerosol is strongly correlated with surface reflectance, but multi-angle retrieval can weaken this correlation.

Keywords: aerosol; CAPI; DFS; retrieval error

1. Introduction

As one of the most important greenhouse gases, column-averaged CO_2 concentrations can be monitored by several space-based instruments, including the Scanning Imaging Absorption Spectrometer for Atmospheric Cartography (SCIAMACHY) [1,2] on the European Environmental Satellite (ENVISAT) and the Atmospheric Infrared Sounder (AIRS) [3,4] on the National Aeronautics and Space Administration (NASA) Aqua platform. However, to characterize CO_2 surface flux and the distribution of CO_2 sources and sinks, the uncertainty of column-averaged CO_2 dry air mole fractions (XCO$_2$) retrieval should be less than 1 ppm. This requirement can only be fulfilled by the dedicated CO_2 monitoring sensors, such as the Greenhouse gases Observing Satellite (GOSAT) launched in 2009 [5] and the Orbiting Carbon Observatory-2 (OCO-2) launched in July 2014 [6]. In this context, the Chinese Carbon Dioxide Observation Satellite (TanSat) will have been launched in the end of

2016 [7,8]. GOSAT, OCO-2 and TanSat measure in three near-infrared (NIR) bands around 0.76, 1.6 and 2.06 μm, which are more sensitive to CO_2 variations in the lower troposphere to meet the requirement of CO_2 surface flux retrieval [9].

The retrieval of column-averaged CO_2 dry air mole fractions (XCO_2) is biased due to uncertainties arising from atmospheric particle scattering [10], mainly caused by aerosols. As part of air pollution, aerosols are dramatically affected by human activities, especially for big cities developing quickly in China. Depending on the chemical components and particle size, aerosols with different absorbing and scattering properties could change the light path and have an impact on the radiation. While algorithm dependent, aerosols are shown to produce different patterns of bias in retrieved XCO_2 [11,12]. The instruments in NASA's Afternoon Constellation (A-Train) can provide near-simultaneous (same Equator-crossing time, i.e., 1:30 p.m.) observations of clouds and aerosols. As the newest member of the A-Train, OCO-2 retrievals will utilize the synergy with other missions in the A-Train, such as Cloud-Aerosol Lidar and Infrared Pathfinder Satellite Observations (CALIPSO) and MODIS (Moderate Resolution Imaging Spectroradiometer), to correct CO_2 retrieval bias [6]. Unlike OCO-2, however, GOSAT has two subunits: a Fourier-transform spectrometer (the main sensor) and the Cloud and Aerosol Imager (CAI). As an auxiliary sensor, CAI is essential for screening areas contaminated by clouds and correcting for the effects of scattering and absorption by aerosols, so reducing the retrieval errors of XCO_2 [13]. For TanSat, a similar concept for a synergistic observation instrument is the Cloud and Aerosol Polarimetric Imager (CAPI), which is designed to observe radiance in five bands from the ultraviolet to NIR (0.38, 0.67, 0.87, 1.375 and 1.64 μm), with additional measurements of the Stokes vector polarization quantities at 0.67 and 1.64 μm [14]. In addition to aerosol detection, CAPI also performs cloud screening, and the 1.375-μm channel is mainly used to detect cirrus. As with CAI, the additional aerosol information from simultaneous CAPI measurements is expected to deliver improvements in CO_2 retrieval accuracy. The wavelengths and signal-to-noise ratio (SNR) for each channel and some other important instrumental characteristics of CAPI are shown in Table 1.

Table 1. Instrument configuration for the Cloud and Aerosol Polarimetric Imager (CAPI).

Channels	Band Centre Wavelength (μm)	Band Range (μm)	Signal-to-Noise Ratio (SNR)	Radiance (W/m²/μm/sr)	Polarization Angle [1]
1	0.38	0.365–0.408	260	28.0	-
2	0.67	0.66–0.685	160	22	0°, 60°, 120°
3	0.87	0.862–0.877	400	25	-
4	1.375	1.36–1.39	180	6.0	-
5	1.64	1.628–1.654	110	7.3	0°, 60°, 120°

[1] Polarization angle represents the angle three polarizers placed in one axial direction.

The backscattered radiation varies substantially due to scattering by aerosols in the atmosphere. Aerosol optical depth (AOD), expressed in terms of integration of aerosol extinction coefficient over height, and the Ångström exponent, a dependency of aerosol optical depth on wavelength, are often used to describe aerosol optical properties depending on the chemical composition, microphysical parameters and the vertical distribution [15,16]. Several satellite instruments have been used to monitor aerosols from space by detecting multispectral reflected radiance. The MODIS, AVHRR (Advanced Very High Resolution Radiometer) and SCIAMACHY instruments can provide measurements from the visible to infrared in nadir viewing geometry [17–22]. Near-UV measurements from the TOMS (Total Ozone Mapping Spectrometer) and OMI (Ozone Monitoring Instrument) are suitable for aerosol detection over bright land surfaces [23,24]. AOD and the Ångström exponent can be derived from these observations [25–27], but the microphysical properties of aerosols, such as the refractive index and particle size distribution, cannot usually be determined exactly [21]. The Multi-angle Imaging SpectroRadiometer (MISR) measures radiance at various viewing angles along the track and combines these measurements in the retrieval to improve aerosol detection [28–30]. Furthermore, polarization has long been shown to be sensitive to the aerosol microphysical properties [31]. Therefore, simultaneous measurements of polarization and radiances, such as those obtained from the Polarization and Directionality of Earth Reflectances (POLDER) instrument [32,33] at 14 viewing angles, are shown to be valuable for characterizing aerosol microphysical properties [34].

The TOA radiances measured by satellites are affected by both aerosol backscattering and surface reflection. Therefore, the main challenge in improving satellite aerosol retrieval is to separate the contributions from aerosols and surface reflection [35]. Some algorithms use the relationship of surface reflectance between visible bands and the near-infrared band, like MODIS [18,19]. Others utilize the lower surface reflectance at shorter wavelengths, such as UV or the blue band, like the OMI aerosol algorithm [24,36,37]. A simultaneous retrieval approach for retrieving aerosol parameters along with XCO_2 has been proposed [38]. Simulation experiments applicable to measurements of GOSAT or OCO-2 have shown that residual aerosol-induced CO_2 errors can be reduced to some extent by this method [39].

In this study, we focus on analyzing the sensitivity to aerosol parameters and their retrieval errors from the a priori error and instrument noise using simulated CAPI observations. A numerical tool comprising a forward model, an instrument model and an error analysis model is established for CAPI simulation. This tool is similar in concept to the one developed for Geostationary (GEO) satellites [40] and used in simulation of CAPI [41]. The degrees of freedom for the signal (DFS) are used to evaluate the sensitivity of CAPI measurements to aerosol parameters in the state vector (AOD, refractive index and particle size distribution). Based on the optimal estimation inverse model, component retrieval errors are also calculated and analyzed. This theory has been used in estimation of information content from satellite measurements in some studies [38,42,43].

The wide field of view (FOV) for CAPI may cause some differences in aerosol retrieval for different viewing zenith angles (VZA); therefore, we check the variation in DFS and retrieval error with viewing angles. We focus on the comparison between two VZAs: $0°$ and $16°$ based on the FOV of CAPI. In addition, TanSat can be oriented to operate in target mode, to collect multi-angle observations at specific surface targets when the satellite moves overhead. The improvement in retrieval accuracy from the incorporation of these multi-angle measurements is also evaluated. We also conduct an analysis to compare the error patterns due to uncertainties from different sources for different types of aerosols.

The structure of the paper is as follows. Section 2 contains a description of the forward model. Section 3 summarizes the principles of our aerosol retrieval model and presents the retrieval sensitivity and error analysis methodology. Forward model simulation for CAPI is shown in Section 4. Section 5 presents a discussion of the information obtained from simulated measurements. The analysis of the a posteriori error and correlations between the retrieved elements are discussed in Section 6. The last section summarizes the paper.

2. Description of the Forward Model

A forward model is developed to simulate the TOA radiance and polarization measurements and to provide the necessary Jacobians with respect to the state vector. This forward model is a combination of a linearized aerosol Mie-scattering model [44] and a vector linearized discrete ordinate radiative transfer model (VLIDORT). The optical processes taken into consideration are Rayleigh scattering, gas absorption, particle scattering and surface reflection. We discuss the components of the forward model in the following subsections.

2.1. The Linearized Aerosol Scattering Model

A linearized aerosol model is an independent tool that can derive both the aerosol optical properties and their Jacobians with respect to aerosol microphysical properties [44]. These optical properties are aerosol extinction and scattering optical depth, the phase-function-normalized scattering matrices, as well as the corresponding coefficients expanded using generalized spherical functions.

The aerosol extinction or loading profile is specified at a reference wavelength. Instead of specifying aerosol loading at each level (which implies the retrieval of the entire aerosol profile), we use a parameterized aerosol profile described by only one or two parameters. In this study, we focus on the tropospheric aerosol, which is assumed to be distributed in the lower atmosphere (0–3 km). The total aerosol loading can be described either by the column number density or by the AOD. A distribution function $h(z_k)$ is used to parameterize the aerosol profile [39], so that AOD (or number density) in layer k at height z_k is expressed in terms of the total column AOD (or number density) and distribution function as follows:

$$\tau_{aer,k} = \tau_{aer} h(z_k) \Delta z_k, \tag{1}$$

where τ_{aer} is total AOD and Δz_k is the depth of layer k. The distribution function $h(z_k)$ can be selected as an exponential, linear or Gaussian function.

The extinction and scattering cross-section and normalized scattering matrix expansion coefficients are obtained from the linearized Mie code [44]. For simplicity, we use a mono-modal particle size distribution (PSD) like what the researchers have used in Frankenberg et al. [38]. Although aerosol particles in the actual atmosphere always exist as a mixture of several components [45], here we focus on the sensitivity of CAPI measurements to each aerosol component. Four typical types of aerosol (dust, soot, sea salt and sulfate) are used in our experiment. The Mie code will also calculate analytical partial derivatives of the optical properties with respect to the aerosol refractive index components and PSD parameters [44].

2.2. The Rayleigh Scattering and Gas Absorption Model

Atmospheric states are represented on a 25-level vertical grid, which includes the profiles of temperature and pressure and the volume mixing ratios of trace gases (O_2, H_2O and CO_2). For CO_2 absorption, we divided the CO_2 profile into two regimes. The lower regime is from the surface to 2 km, with the rest of the atmosphere comprising the upper regime. The concentration of CO_2 in the lower regime changes frequently due to various sources and sinks, while little change happens in the upper regime. Three ancillary parameters are also included to correct for the effect of using climatology data; these are (1) a single temperature shift S applied to all temperature levels uniformly, (2) the surface pressure; and (3) a scaling factor, F_{H2O}, for the total amount of water vapor.

Rayleigh scattering cross-sections and depolarization ratios are taken from Bodhaine et al. [46]. Jacobians of the Rayleigh optical depth are obtained by an appropriate differentiation.

For trace gas absorption, spectroscopic line parameters from the high-resolution transmission molecular absorption database (HITRAN) [47] are used as an input for the line-by-line (LBL) computation of absorption cross-sections. As with the Rayleigh optical inputs, the gas absorption optical properties are fully linearized.

2.3. The Surface Model

In this study, we take the surface reflection over land into consideration. Because CO_2 can only be retrieved at low AOD, in which circumstance single scattering dominates the radiative transfer process, and to keep consistent with CO_2 retrieval algorithm, we assume a Lambertian surface whose albedo is parameterized as follows. Based on the principle of the Medium Resolution Imaging Sensor (MERIS) [48] and LANDSAT TM data [49], wavelength-dependent surface reflectance data are constructed by assuming that any land surface is covered by both green vegetation and bare soil with a fraction. Thus, the spectral surface albedo $\alpha(\lambda)$ is taken to be a weighted linear mixture of the actual spectra of the vegetation albedo $\alpha_{veg}(\lambda)$ and bare soil albedo $\alpha_{soil}(\lambda)$:

$$\alpha(\lambda) = c\alpha_{veg}(\lambda) + (1-c)\alpha_{soil}(\lambda). \tag{2}$$

This is a simplified model that simulates most of the Earth's land surface, but ignores other less common land cover types. However, for MERIS, the fraction of soil and vegetation is derived from the normalized difference vegetation index (NDVI) in the satellite scene, while in our model, this fraction is treated as a state vector element.

The spectra of vegetation and bare soil are taken from the Advanced Spaceborne Thermal Emission and Reflection Radiometer (ASTER) database [50]. In our retrieval scheme, the surface spectral albedo is then characterized by the single parameter c in Equation (2) above. The analytical Jacobian with respect to this parameter is easy to obtain from the Lambertian albedo Jacobian.

2.4. The Radiative Transfer Model

The VLIDORT radiative transfer (RT) model is used to calculate radiances and their Jacobians with respect to atmosphere and surface parameters. VLIDORT is a linearized pseudo-spherical vector radiative transfer code based on the discrete ordinate method for the determination of single and multiple scattering radiation fields with solar-beam and/or thermal emission (Planck function) sources of radiation in a multilayer stratified atmosphere [51].

VLIDORT computes a four-element diffuse field of Stokes components {I, Q, U, V}, with I being the total intensity, Q and U describing linearly-polarized radiation and V characterizing circularly-polarized radiation. The magnitude of V is very small in the Earth's atmosphere, and therefore, we ignore this component in our calculations. For diffuse radiation, especially with a high aerosol concentration, multiple scattering has an obvious effect on the optical paths. Contributions from both attenuated solar beam single scattering and multiple scattering are included in VLIDORT [52]. For nadir-viewing at large solar zenith angles (SZA), the pseudo-spherical approximation, which treats solar beam attenuation for a curved atmosphere while all scattering events still take place in a plane-parallel medium, is deployed in VLIDORT instead of the pure plane-parallel assumption.

The optical properties, the layer total optical thickness, total single-scattering albedos and the 4×4 spherical-function expansion coefficient matrix characterizing the scattering phase function, and their corresponding derivatives with respect to retrieved parameters are required as inputs to VLIDORT. These optical and linearized inputs are from the aerosol model, Rayleigh scattering and gas absorption model introduced in previous sections, through suitable application of the chain rule.

3. Retrieval Method

3.1. Optimal Estimation Theory and Retrieval Error Analysis Method

The retrieval is done using optimal estimation theory. The DFS is used to represent the number of independently-retrievable quantities that can be derived from the inversion. Here, we summarize the main formulas used in the analysis.

The DFS is defined as the trace of the averaging kernel matrix, which describes the sensitivity of the retrieval to the true state vector:

$$A = \frac{\partial \hat{x}}{\partial x} = GK. \tag{3}$$

Here, K is a Jacobian matrix (derivatives of the simulated measurements with respect to elements of the state vector), and G is the contribution function matrix, defined as:

$$G = \left(K^T S_\varepsilon^{-1} K + S_a^{-1}\right)^{-1} K^T S_\varepsilon^{-1}, \tag{4}$$

in which S_ε is the observation error covariance matrix and S_a represents the a priori error covariance matrix. The averaging kernel matrix relates the retrieval state $\hat{x}$ to the a priori x_a and the true state vector x as follows:

$$\hat{x} = (I_n - A)x_a + Ax + G_y \varepsilon_y, \tag{5}$$

in which $\mathbf{I}_n$ is a unit matrix with dimension n (the rank of the state vector) and ε_y represents random error in the measurements.

In general, the averaging kernel quantifies the ability to infer the a posteriori state $\hat{x}$ for specified observation noise and a priori characterization. The closer $\mathbf{A}$ is to the unit matrix, the greater is the information that can be obtained from observation and the less it depends on the a priori characterization. We will discuss the DFS for a number of simulation scenarios in the following section.

From Equation (5), an expression for the retrieval error can be derived:

$$\hat{x} - x = (\mathbf{A} - \mathbf{I}_n)(x - x_a) + \mathbf{G_y}\mathbf{K_b}\left(\mathbf{b} - \hat{\mathbf{b}}\right) + \mathbf{G_y}\Delta f(x, \mathbf{b}, \mathbf{b}') + \mathbf{G_y}\varepsilon, \tag{6}$$

The retrieval error comprises four sources: the smoothing error, the model parameter error, the forward model error and the measurement noise. In this equation, the vector $\mathbf{b}$ is the ancillary model parameters that are not included in the state vector. The smoothing error is the component related to a priori uncertainties. Thus, the smoothing error covariance matrix $\mathbf{S_s}$ corresponding to the a priori error $\mathbf{S}_a$ is:

$$\mathbf{S}_s = (\mathbf{A} - \mathbf{I}_n)\mathbf{S}_a(\mathbf{A} - \mathbf{I}_n)^T. \tag{7}$$

Similarly, measurement noise $\mathbf{S}_m$ and the model parameter error covariance $\mathbf{S}_f$ can be expressed as Equations (8) and (9), respectively (assuming these are distributed normally):

$$\mathbf{S}_m = \mathbf{G}\mathbf{S}_\varepsilon\mathbf{G}^T. \tag{8}$$

$$\mathbf{S}_f = \mathbf{G}\mathbf{K_b}\mathbf{S_b}\mathbf{K_b}^T\mathbf{G}^T. \tag{9}$$

The error covariance matrix $\mathbf{S}_\varepsilon$ indicates instrument noise, which is usually random and uncorrelated between channels, and is therefore defined by the SNR in each channel of CAPI in our study. The evaluation of forward model error cannot be completed easily by some matrix calculation, so in this work, we do not take this error into consideration and only estimate the uncertainty introduced by ancillary parameters in the forward model. In addition, we ignore the uncertainties of other parameters, with auxiliary vector $\mathbf{b}$ only including the aerosol profile shape parameters.

In sum, the total posterior error of aerosol retrieval is usually expressed as:

$$\hat{\mathbf{S}} = \mathbf{S}_m + \mathbf{S}_s = \left(\mathbf{K}^T\mathbf{S}_\varepsilon^{-1}\mathbf{K} + \mathbf{S}_a^{-1}\right)^{-1}. \tag{10}$$

If the model parameter errors are included, $\mathbf{S}_f$ is added to this equation.

3.2. The State Vector and A Priori Uncertainty

In our analysis, the state vector x consists of aerosol optical parameters and surface parameters. The real and imaginary parts of the refractive index at the reference wavelength and the a priori of the mode radius and variance for lognormal PSD corresponding to the four mono-modal aerosol types are taken from the Optical Properties of Aerosols and Clouds (OPAC) database [45]. The land surface model parameter c (the fraction of vegetation cover) is also included in the state vector. Generally, the state vector includes six parameters: AOD, the real part and the imaginary part of the refractive index, PSD mode radius and variance and the fraction of vegetation surface albedo. The aerosol loading profile is assumed to take a Gaussian-shaped distribution, which is described by two parameters: the peak height and the half width of the profile. The AOD at the reference wavelength, aerosol profile parameters and surface parameters are consistent in simulations for each aerosol type. Table 2 lists the a priori and corresponding a priori uncertainties (square-root of the variance) for all state vector elements. To describe the fast and dramatic change of aerosol due to human activity, the a priori uncertainty of AOD is assumed 100%. The real part of refractive index of each type of aerosols has a small difference and changes little; thus, an assumption of 0.15 a priori uncertainty is available for each aerosol type. On the contrary, the uncertainty of the imaginary part of the refractive index is more difficult to estimate and is assumed large. The values are based on a previous study [38] and change with aerosol type. Similarly, the a priori uncertainties of the PSD mode radius and variance also follow the assumptions in [38]. Considering the relatively accurate estimation about the surface albedo, the a priori error of vegetation fraction is 0.2.

Table 2. State vector elements used in the simulation for each aerosol type.

State Vector Element	Aerosol Types	A Priori	A Priori Error (1σ)
AOD [1]	_[2]	0.3	0.3
Real refractive index [1]	Dust	1.53	0.15 [2]
	Soot	1.75	
	Sea salt	1.381	
	Sulfate	1.43	
Imaginary refractive index [1]	Dust	0.008	0.02
	Soot	0.44	1.0
	Sea salt	4.26×10^{-9}	1.0×10^{-8}
	Sulfate	1.0×10^{-8}	3.0×10^{-8}
PSD mode radius	Dust	0.39	0.4
	Soot	0.0118	0.01
	Sea salt	0.209	0.2
	Sulfate	0.0695	0.07
PSD variance	Dust	2.0	2.0 [2]
	Soot	2.0	
	Sea salt	2.03	
	Sulfate	2.03	
Vegetation fraction	_[2]	0.5	0.2

[1] Aerosol optical depth (AOD) and the refractive index are specified at a reference wavelength of 550 nm. [2] The a priori and a priori error of AOD and vegetation fraction, as well as the a priori error of real refractive index and particle size distribution (PSD) variance are the same for four aerosol types.

The measurement vector $\mathbf{y}_s$ for each viewing geometry is established using simulated (synthetic) CAPI measurements, including radiances $\mathbf{I}$ for the five CAPI bands and polarization quantities $\mathbf{Q}$ and $\mathbf{U}$ from Bands 2 and 5 (Table 1). Thus, the measurement vector for single-angle viewing is:

$$\mathbf{y}_s = \left[\mathbf{I}_{band1}, \mathbf{I}_{band2}, \mathbf{Q}_{band2}, \mathbf{U}_{band2}, \mathbf{I}_{band3}, \mathbf{I}_{band4}, \mathbf{I}_{band5}, \mathbf{Q}_{band5}, \mathbf{U}_{band5}\right]^T. \qquad (11)$$

To improve retrieval information using observations at multiple angles, we extend the measurement vector by concatenating individual measurements:

$$\mathbf{y} = \left[\mathbf{y}_{s1}{}^T, \mathbf{y}_{s2}{}^T, \mathbf{y}_{s3}{}^T, \cdots\right]^T \qquad (12)$$

where the subscripts s1, s2, s3, ... indicate observations at different viewing angles. Jacobians for the various bands and viewing geometries are concatenated in the same way.

In this study, we consider an ideal scenario and ignore calibration errors and systematic errors. The measurement noise is assumed to follow the SNR of each channel (Table 1). When considering the multi-angle observation, the instrument noise is accumulated in the same way as the measurement vector.

4. Simulated CAPI Measurements and Aerosol Sensitivity

CAPI measurements (radiances at the TOA) are simulated at multiple viewing angles by our forward model, and Jacobians with respect to the state vector are obtained. We select four viewing angles including the nadir (0°), the largest angle in the FOV (16°) and two larger angles (30°, 60°)

in the target mode. To compare the influence of surface albedo for the different bands and aerosols, the simulation is repeated over two different surfaces (large albedo of $\alpha(\lambda)$, called the bright surface, and small surface albedo of $0.5 \times \alpha(\lambda)$, called the dark surface). We also simulate observations in both winter and summer to investigate the impact of SZA on retrieval. Table 3 summarizes the simulation scenarios, with the corresponding observation geometries.

Table 3. Angles used in the simulation scenarios [1].

Scenario	Solar Zenith Angle (Degree)	Viewing Zenith Angle (Degree)	Scattering Angle (Degree)	Surface Albedo
Winter	65	0 (nadir), 16, 30, 60	115.0, 114.0, 111.5, 102.2	bright ($\alpha(\lambda)$), dark
Summer	20	0 (nadir), 16, 30, 60	160.0, 154.6, 144.5, 118.0	($\alpha(\lambda) \times 0.5$)

[1] Relative azimuth angles for the fore and aft satellite viewing are 90° and 270°, respectively.

The Jacobians of radiance and polarization, with respect to various aerosol properties, are calculated by linearization capabilities of the forward model. Figure 1 illustrates the Jacobians of radiance and polarization with respect to AOD. These Jacobians of radiance depend on the wavelengths and aerosol types and show little sensitivity to the surface albedo. Polarization at 670 nm is largely sensitive to the viewing angles and aerosol types. The Jacobian of radiance for soot is always negative due to its absorption of radiation, while the main scattering property of sea salt and sulfate causes their Jacobians to be positive regardless of the surface albedo. Moreover, the effect of surface albedo on the Jacobian of radiance is not consistent for each band. For example, the Jacobians for dust increase at 1640 nm, but decrease at 380 nm when the surface albedo is higher (Figure 1). This is the result of the balance of aerosol scattering and surface reflection in different bands.

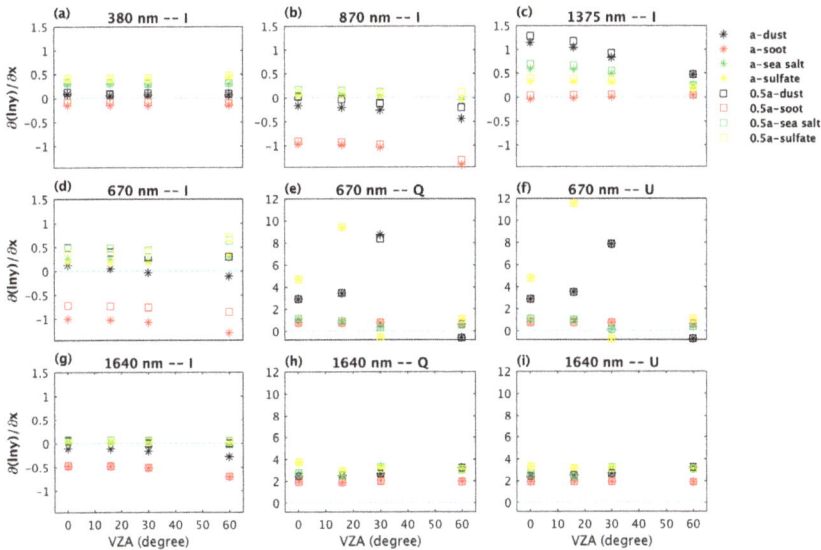

Figure 1. Jacobians of AOD in all channels of CAPI at four viewing angles (0°, 16°, 30°, 60°) and 20° solar zenith angle (SZA) with a bright and a dark surface for four types of aerosols. (**a–c**) represent Jacobians at 380 nm, 870 nm and 1375 nm, respectively. (**d–f**) and (**g–i**) are polarized Jacobians of Stokes vector I, Q and U at 670 nm and 1640 nm, respectively. The black, red, green and blue markers mean radiance or polarization for dust, soot, sea salt and sulfate, respectively. The asterisks represent the bright surface with albedo $\alpha(\lambda)$, and squares mean the dark surface with $0.5 \times \alpha(\lambda)$ as albedo. The thin blue dashed line represents zero Jacobian.

5. Impact of Observation Geometry and Multi-Angle Retrieval

In this section, we calculate the averaging kernels and discuss the diagonal elements of averaging kernel matrices for the four types of aerosol. Aerosol microphysical properties determine the different characteristics of aerosol scattering phase function as shown in Figure 2. The corresponding scattering phase functions for different simulation scenarios in Table 3 are also emphasized.

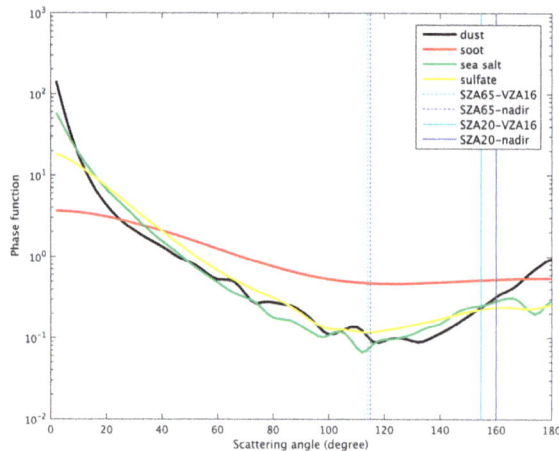

Figure 2. Scattering phase functions of four aerosol types used in this study. Black, red, green and yellow thick solid lines represent the scattering phase function of dust, soot, sea salt and sulfate. Scattering angles corresponding to different solar zenith angles and viewing zenith angles are indicated by cyan and blue thin solid or dashed lines, respectively. VZA, viewing zenith angle.

5.1. Influence of Observation Geometry

Figure 3 shows the DFS of each retrieved aerosol parameter for different observation geometries in both nadir viewing mode and target mode. It is clear that the DFS of AOD and PSD variance for all aerosol types are close to 1.0, while the DFS of the real and imaginary parts of the refractive indices and PSD mode radius are less and dependent on aerosol type. For example, among four aerosol types in the retrieval at nadir viewing with small SZA (20°), dust has the largest total information with 4.81 DFS, while the total DFS for soot is the smallest, only 3.05 (Figure 3). The difference of DFS between different aerosol types is determined by their scattering properties (Figure 2) and indicates the sensitivity of CAPI measurement to aerosol properties.

The ground pixels in the wide FOV of CAPI will be observed at different VZA (from 0°–16°) or different SZA. The viewing geometry can lead to some changes in aerosol information due to different aerosol scattering phase functions when scattering angle varies. Thus, we also compare the impact of different viewing geometries on the information obtained regarding aerosol parameters in Figure 3. For example, for dust at 20° SZA, DFS at the forward 16° VZA (blue bar) is lower than that at nadir, which is consistent with the lower scattering phase function at VZA of 16° (Figure 2). However, the DFS of soot for the four viewing scenarios in the nadir observation in our simulation are similar, due to the rather uniform distribution of its scattering phase function (Figure 2). Similarly, the comparison of the DFS at the same VZA, but different SZA (green and yellow bars in Figure 3) also proves the consistent relationship between phase function and aerosol DFS.

In addition to considering the vegetation fraction of surface albedo in the state vector, we also compare the DFS over two surface albedos, the bright and dark surface defined in Section 4, in Figure 3. Consistent with the sensitivity to AOD shown in Figure 1, the DFS for soot over the dark surface at 20° SZA is smaller than that over the bright surface due to less sensitivity (less absolute value of Jacobians at $0.5 \times \alpha(\lambda)$, such as the red markers in Figure 1). Similarly, the larger DFS for dust and sea salt at 20° SZA over the dark surface correspond to larger Jacobians (black and green markers in Figure 1). Therefore, although the wavelength dependence of surface albedo remains unchanged, the surface albedo value could still have an impact on aerosol information content.

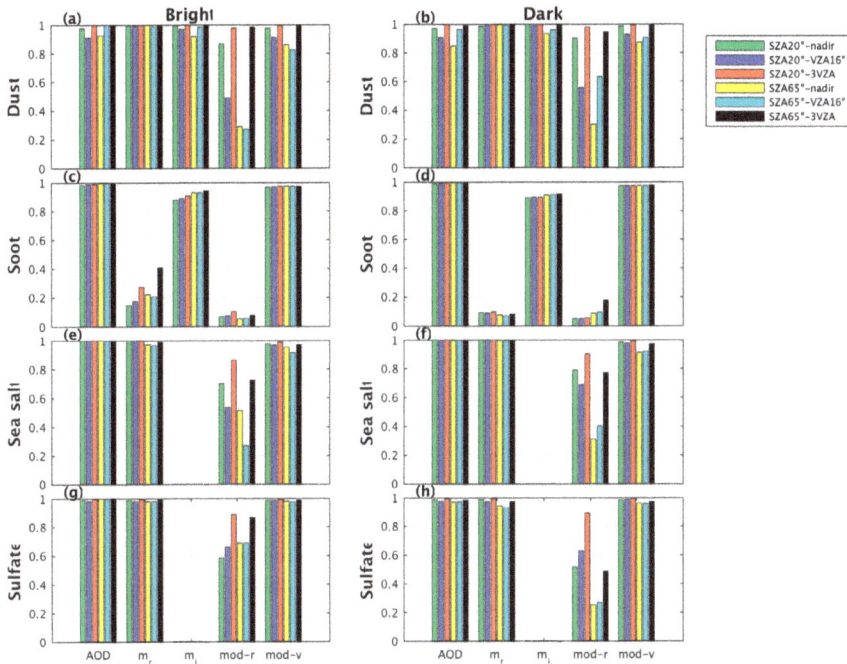

Figure 3. The degrees of freedom for the signal (DFS) of corresponding aerosol parameters at different viewing angles in high and low solar zenith angles for four types of aerosol over the bright and dark surface defined as in Section 4. The left column is for the bright surface, and the right column is for the dark surface. Four rows represent four aerosol types: dust (**a,b**), soot (**c,d**), sea salt (**e,f**) and sulfate (**g,h**). Different colored bars represent corresponding different observation geometries. The X-axis represents five retrieved aerosol parameters: AOD, real (m_r) and imaginary part of refractive index (m_i), mode radius (mod-r) and variance (mod-v) of PSD.

5.2. Improvement of Multi-Angle Measurement

In target mode, to validate satellite data with ground-based observation, TanSat can monitor the same location several times along an orbit. Therefore, several measurements of different viewing geometries for a specific ground pixel can be combined to improve the aerosol retrieval information and accuracy, similar to what has been achieved for POLDER [53]. For simplicity, three along-track measurements at the forward VZA 16°, backward VZA 16° and nadir viewing angles are combined for the multi-angle retrieval. From Figure 3, it is clear that the DFS in multi-angle mode increase compared to the single-view nadir mode result (such as the red and green bars). The additional information acquired in the multi-angle mode is most striking for the PSD mode radius, whose DFS is limited in nadir viewing and can be enhanced 0.22 on average. For the other parameters, we obtain a large amount of information at only-nadir viewing, so the improvement of multi-angle retrieval is not obvious. However, this improvement for soot is so little that only another 0.04 DFS is provided for the PSD mode radius. This situation can be explained by the weak scattering of soot and the small change of phase function in the chosen geometries. Generally, the DFS of the multi-angle retrieval of dust reaches 4.97, indicating we can retrieve almost all five aerosol parameters independently in multi-angle mode. For sea salt and sulfate, the total DFS are 3.85 and 3.87 in the multi-angle mode, respectively, so we can obtain a large amount of information for the four parameters other than the imaginary part of the refractive index. Unfortunately, for soot, the DFS is still only 3.25 in the multi-angle mode, and the ability to retrieve the real part of the refractive index and PSD radius is not improved. In summary, the DFS in multi-angle retrieval increases by 0.31, on average, compared to the nadir mode.

6. Error Analysis and Correlation Matrix for Aerosol Retrieval

In the previous section, we focused on the ability to retrieve aerosol properties from CAPI measurements. Another crucial issue to be considered is the source of error in this retrieval. In this section, we discuss the a posteriori error covariances, $\hat{S}$, and analyze its sources of a priori uncertainty and measurement noise, respectively, for all of the scenarios described in Section 5. Model parameter uncertainties only related to aerosol profile parameters (half-width, peak-height) are also taken into consideration. The impact of different observation geometries and satellite operational modes are presented. Finally, an error correlation matrix between all of the retrieved parameters is derived, to analyze the correlation between the surface and aerosol parameters.

6.1. Retrieval Error and Its Components

From Figures 4 and 5, it can be seen that the patterns of posterior errors and the corresponding smoothing errors are similar for all aerosol types. This implies that the a priori uncertainties rather than measurement noise are the dominated component of the posterior error. We also note that among all retrieved parameters, the imaginary part of the refractive index and the PSD mode radius have the largest posterior errors in nadir mode, i.e., 78% and 70% on average. Therefore, the retrieval accuracy of these two parameters cannot be guaranteed. Actually, for aerosols with little absorption, i.e., sea salt and sulfate, the DFS of the imaginary part of the refractive index are close to zero (Figure 3c,d), and the posterior errors are almost 100% (Figure 4l,p and Figure 5l,p); therefore, this parameter cannot be retrieved from measurements. In other words, its information is mainly from a priori, and the measurement noise has little effect on this parameter (Figure 4j,n and Figure 5j,n).

In addition to a posteriori error, measurement noise and smoothing errors, forward model parameter uncertainties from two aerosol profile parameters are also investigated in Figures 4 and 5. The apparent low values of these errors for sea salt and sulfate indicate that the inaccuracy in depicting the shape of the aerosol profile has little impact on the retrieval uncertainty for these two aerosols. The corresponding model parameter errors for dust and soot are larger, mainly because radiation in the UV band is sensitive to the height of aerosols with absorption.

Furthermore, we also note that in Figure 4, when the scattering phase function is larger at certain observation angles, lower retrieval errors are derived. However, purely noise-related errors are more random. From Figure 5, it is found that the use of three measurements in retrieval can effectively reduce posterior and smoothing errors, especially for those parameters with large errors. The posterior error of the PSD mode radius is reduced the most (23% on average), while the least improvement is achieved for the real part of the refractive index (less than 1%). Among all of the aerosols, the improvement of the multi-angle mode is most apparent for dust (76%), in relation to the large variability of its scattering phase function over the range of our sampled scattering angles. Similarly, the least improvement is achieved for soot (only 10%) due to both the small scattering property and the comparatively uniform phase function.

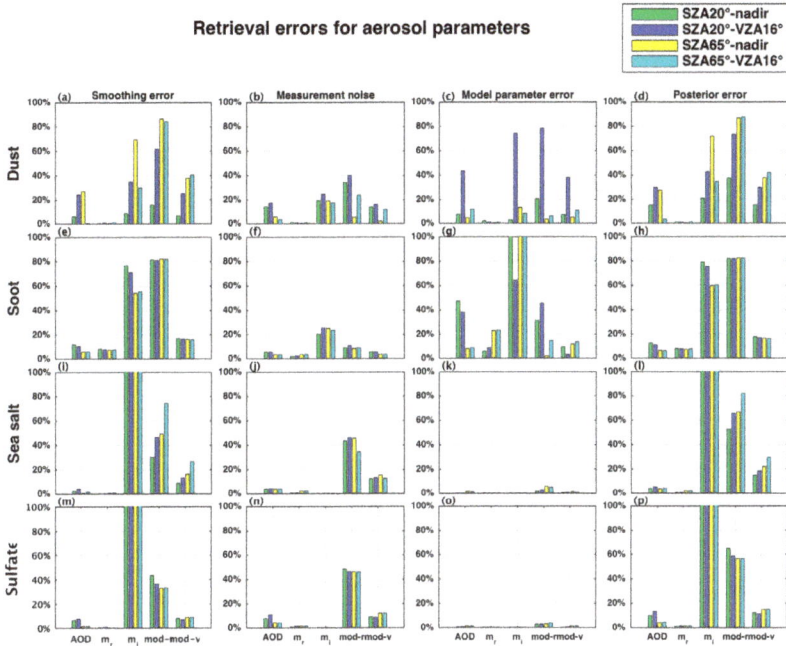

Figure 4. Different error estimates (1σ) of aerosol parameters at different observation geometries for all aerosols. Each row represents one type of aerosol: dust (**a–d**), soot (**e–h**), sea salt (**i–l**) and sulfate (**m–p**), respectively. The range of ordinate represents percentage.

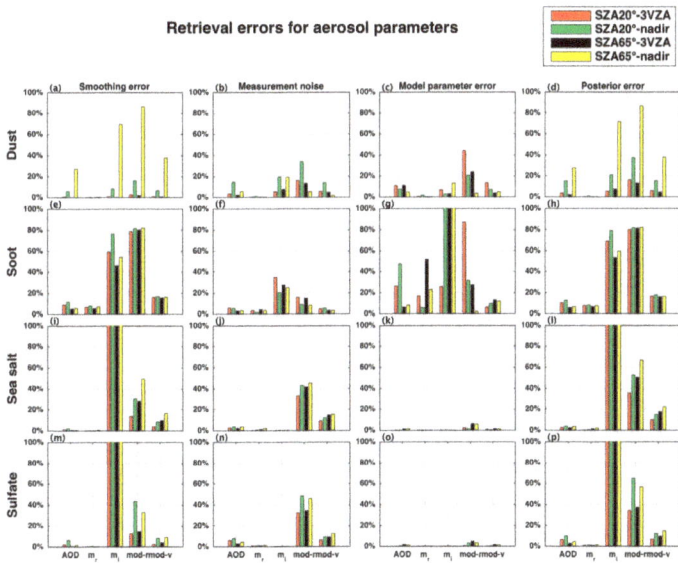

Figure 5. Different error estimates (1σ) of aerosol parameters at different solar zenith angles in nadir and multi-angle mode. Others are the same as Figure 4.

If another two measurements (forward VZA 30° and backward VZA 30°) are added in multi-angle retrieval, the retrieval errors will be reduced to a greater extent. Figure 6 shows that smoothing errors, and posterior errors are all improved again when five measurements are used in the retrieval. Measurement noise is also reduced, except for soot. Thus, adding more measurements in aerosol retrieval is not an efficient method to improve measurement noise for soot. For dust, which had the greatest improvement in multi-angle retrieval, the smoothing errors are less than 0.1% and almost can be ignored; meanwhile, the average posterior errors are as low as 3% when the number of measurement angles reaches five. The posterior errors for soot, sea salt and sulfate decrease by 4%, 6% and 2% on average, respectively. Finally, we find that a larger number of retrieval measurements can result in less smoothing and posterior errors, impacting on aerosols, especially dust.

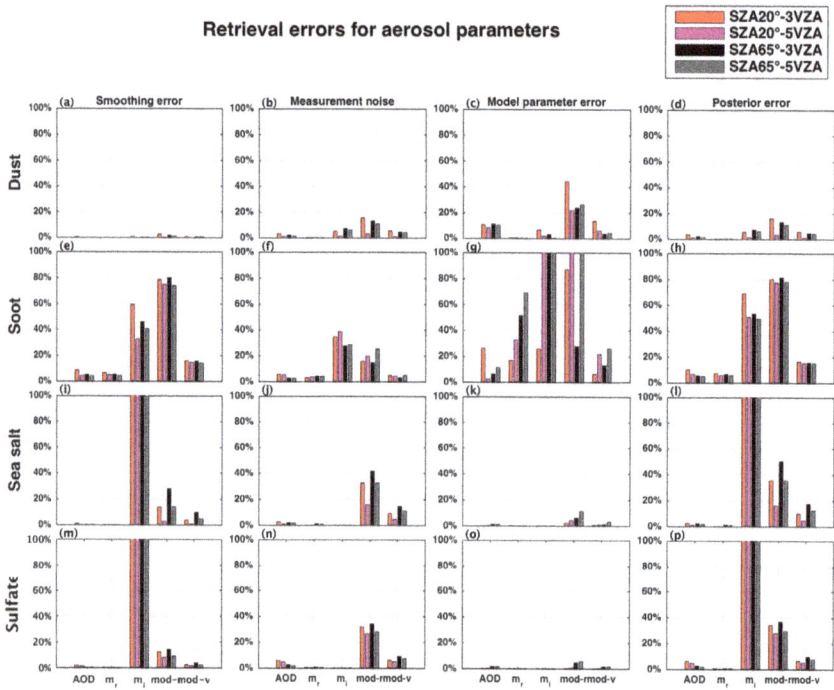

Figure 6. Four kinds of retrieval errors for multi-angle retrieval in three viewing angles and five viewing angles respectively. Others are the same as Figure 4.

6.2. Correlation of Surface and Aerosol

The error correlation matrices for elements of the state vector are shown in Figure 7. Our main objective is to study the correlation of surface and aerosol parameters (coefficient of vegetation fraction and AOD) and to analyze the strength of the correlation between surface reflection and aerosol scattering. The larger the correlation coefficient, the harder it is to separate the radiative effect from surface and aerosol. For all aerosols, the correlation coefficients between the vegetation fraction and AOD are not less than 0.6, indicating that surface reflection has some influence on aerosol retrieval. However, the correlation coefficients are positive for dust and soot, whereas negative for sea salt and sulfate. If the vegetation fraction is large, the surface albedo decreases, thereby resulting in a diminution of the radiance at TOA. When AOD increases, more scattering for sea salt and sulfate can lead to an enhancement of the radiance at the TOA. Thus, the vegetation fraction and AOD for non-absorbing sea salt and sulfate correlate negatively. When comparing nadir and multi-angle retrieval, the correlation of AOD and the vegetation fraction decreases to less than 0.4 for dust in multi-angle measurements. However, there is no obvious distinction for the other three types of aerosols. This proves that multi-angle retrieval can reduce the correlation between surface reflection and aerosol scattering for a strong scattering aerosol.

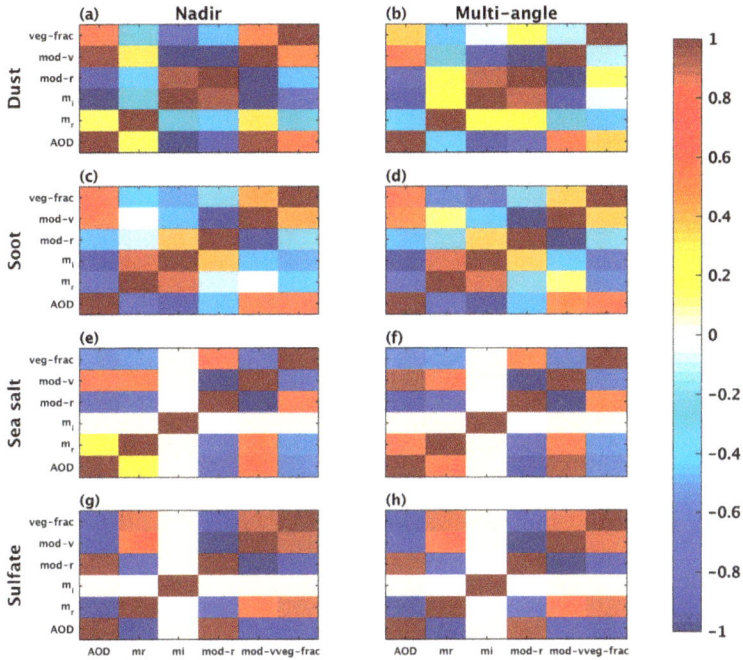

Figure 7. The correlation matrices of retrieval variables for four types of aerosols (four rows): dust (**a**,**b**), soot (**c**,**d**), sea salt (**e**,**f**) and sulfate (**g**,**h**). The left column is in nadir viewing mode, and the right column is in multi-angle mode.

7. Conclusions

In this study, we have investigated the potential to retrieve aerosol properties and the analysis of sources of retrieval errors, based on numerical experiments using simulated observations in five bands of CAPI on board TanSat. A forward model comprising a linearized Mie code, a trace gas absorption and Rayleigh scattering model, a surface parameterization and a linearized vector radiative transfer model are introduced. In the simulations, we assume a mono-modal aerosol, with a lognormal PSD and a Gaussian-model aerosol profile distribution. The retrieved parameters include five aerosol properties and one surface parameter. The Jacobians of simulated radiances and polarizations with respect to AOD at different observation geometries are discussed. The retrieval sensitivity specified by the DFS and the retrieval errors are analyzed and compared at different viewing geometries, as well as with multi-angle retrieval for four of the main components of the natural mixed aerosol.

In the nadir view mode of CAPI, we find that of all of the retrieval parameters, the lowest DFS is obtained for the PSD mode radius (0.55 on average), as well as large posterior error (70%), while the DFS for AOD is always close to 1.0. The range of the total DFS from 3.05–4.83 and the variability of retrieval errors among the different aerosol types confirm the sensitivity of CAPI observations to aerosol microphysical properties.

When retrieval is performed with measurements at three along-track viewing angles ($\pm16°$, nadir), there is additional information for all aerosol properties, with the average increase of the DFS being

0.31 and the posterior error reduced by a maximum of 23%. By adding another two measurements from other viewing geometries in the retrieval, the aerosol information obtained is improved, and the posterior error can decrease again, from 2%–6%, showing that it is worthwhile to fly CAPI in target mode instead of nadir-only mode. When CAPI works in nadir viewing, the different VZA in the wide FOV of the satellite has little impact on the aerosol information and retrieval error, related to the aerosol phase function. In addition to posterior error, we also analyze the smoothing errors, instrument noise and forward model parameter errors. The results show that the a priori uncertainty is the dominating source of posterior error. Moreover, retrieval errors for sea salt and sulfate are not sensitive to the inaccuracies of aerosol profile distribution, but they result in large errors for dust and soot. Additionally, not only the dependence of surface albedo on wavelength, but also the surface albedo value affects aerosol information content. The error correlation coefficients between the surface parameter and AOD for all aerosols indicate that their different optical properties result in a different strength of correlation between surface reflection and aerosol effects. Multi-angle retrieval can reduce this correlation too some extent for dust.

In recent CO_2 retrieval algorithms, an aerosol model has been added to produce a realistic atmosphere, and the performance of imperfect characterization of the aerosol is proven dominative in CO_2 retrieval uncertainties [11]. Thus, aerosol properties synchronously retrieved from CAPI measurements could help to reduce CO_2 retrieval errors for TanSat. It should be noted that the analysis of the retrieval of aerosol properties and error sources in this study is preliminary and accounts only for the impact of different observation geometries and the number of angles for multi-angle observation, as well as a number of pure aerosol types. The effect of aerosols on high precision greenhouse gas retrieval will be considered in future studies. More complex aerosol and surface models that reflect reality more closely will also be considered.

Acknowledgments: This study was supported by the National High-Tech Research and Development Program of China (2011AA12A104), the Chinese Academy of Sciences strategic priority program on space science (XDA04077300), the National Natural Science Foundation of China (41375035) and the External Cooperation Program of the Chinese Academy of Sciences (Grant No. GJHZ1507).

Author Contributions: Dongxu Yang, Zhaonan Cai and Yi Liu conceived of and designed the experiments. Xi Chen performed the experiments, analyzed the data and wrote the paper. Robert J. D. Spurr contributed analysis tools.

Conflicts of Interest: The authors declare no conflict of interest. The founding sponsors had no role in the design of the study; in the collection, analyses or interpretation of data; in the writing of the manuscript; nor in the decision to publish the results.

References

1. Bovensmann, H.; Burrows, J.P.; Buchwitz, M.; Frerick, J.; Noe'l, S.; Rozanov, V.V. SCIAMACHY mission objectives and measurement modes. *J. Atmos. Sci.* **1999**, *56*, 127–150. [CrossRef]
2. Buchwitz, M.; de Beek, R.; Burrows, J.P.; Bovensmann, H.; Warneke, T.; Notholt, J.; Meirink, J.F.; Goede, A.P.H.; Bergamaschi, P.; Kärner, S.; et al. Atmospheric methane and carbon dioxide from SCIAMACHY satellite data: Initial comparison with chemistry and transport models. *Atmos. Chem. Phys.* **2005**, *5*, 941–962. [CrossRef]
3. Chahine, M.T.; Pagano, T.S.; Aumann, H.H.; Atlas, R.E.A. AIRS: Improving weather forecasting and providing new data on greenhouse gases. *Bull. Am. Meteorol. Soc.* **2006**, *87*, 911–926. [CrossRef]
4. Crevoisier, C.; Heilliette, S.; Chédin, A.; Serrar, S.; Armante, R.; Scott, N.A. Midtropospheric CO_2 concentration retrieval from AIRS observations in the tropics. *Geophys. Res. Lett.* **2004**, *31*. [CrossRef]
5. Kuze, A.; Suto, H.; Nakajima, M.; Hamazaki, T. Thermal and near infrared sensor for carbon observation Fourier-Transform Spectrometer on the greenhouse gases observing satellite for greenhouse gases monitoring. *Appl. Opt.* **2009**, *48*, 6716–6733. [CrossRef] [PubMed]
6. Crisp, D.; Atlas, R.M.; Breon, F.-M.; Brown, L.R.; Burrows, J.P.; Ciais, P.; Connor, B.J.; Doney, S.C.; Fung, I.Y.; Jacob, D.J. The Orbiting Carbon Observatory (OCO) mission. *Adv. Space Res.* **2004**, *34*, 700–709. [CrossRef]
7. Chen, W.; Zhang, Y.; Yin, Z.; Zheng, Y.; Yan, C.; Yang, Z. In the tansat mission: Global CO_2 observation and monitoring. In Proceedings of the 63rd International Astronautical Congress, Naples, Italy, 1–5 October 2012.

8. Liu, Y.; Yang, D.X.; Cai, Z.N. A retrieval algorithm for TanSat XCO_2 observation: Retrieval experiments using GOSAT data. *Chin. Sci. Bull.* **2013**, *58*, 1520–1523. [CrossRef]

9. Yoshida, Y.; Ota, Y.; Eguchi, N.; Kikuchi, N.; Nobuta, K.; Tran, H.; Morino, I.; Yokota, T. Retrieval algorithm for CO_2 and CH_4 column abundances from short-wavelength infrared spectral observations by the greenhouse gases observing satellite. *Atmos. Meas. Tech.* **2011**, *4*, 717–734. [CrossRef]

10. Yang, D.; Liu, Y.; Cai, Z. Simulations of aerosol optical properties to top of atmospheric reflected sunlight in the near infrared CO_2 weak absorption band. *Atmos. Ocean. Sci. Lett.* **2013**, *6*, 60–64.

11. O'Dell, C.W.; Connor, B.; Bösch, H.; O'Brien, D.; Frankenberg, C.; Castano, R.; Christi, M.; Eldering, D.; Fisher, B.; Gunson, M.; et al. The ACOS CO_2 retrieval algorithm—Part 1: Description and validation against synthetic observations. *Atmos. Meas. Tech.* **2012**, *5*, 99–121. [CrossRef]

12. Houweling, S.; Hartmann, W.; Aben, I.; Schrijver, H.; Skidmore, J.; Roelofs, G.-J.; Breon, F.M. Evidence of systematic errors in SCIAMACHY-observed CO_2 due to aerosols. *Atmos. Chem. Phys.* **2005**, *5*, 3003–3013. [CrossRef]

13. Ishida, H.; Nakjima, T.Y.; Yokota, T.; Kikuchi, N.; Watanabe, H. Investigation of GOSAT TANSO-CAI cloud screening ability through an intersatellite comparison. *J. Appl. Meteorol. Clim.* **2011**, *50*, 1571–1586. [CrossRef]

14. Zhang, J.; Shao, J.; Yan, C. Cloud and Aerosol Polarimetric Imager. In Proceedings of the Conferences of the Photoelectronic Technology Committee of the Chinese Society of Astronautics: Optical Imaging, Remote Sensing, and Laser-Matter Interaction, Suzhou, China, 20–29 October 2013.

15. Herman, M.; Deuzé, J.L.; Devaux, C.; Goloub, P.; Bréon, F.M.; Tanré, D. Remote sensing of aerosols over land surfaces including polarization measurements and application to POLDER measurements. *J. Geophys. Res.* **1997**, *102*, 17039–17049. [CrossRef]

16. Maso, M.D.; Kulmala, M.; Riipinen, I.; Wagner, R.; Hussein, T.; Aalto, P.P.; Lehtinen, K.E.J. Formation and growth of fresh atmospheric aerosols: Eight years of aerosol size distribution data from SMEAR II, Hyytiälä, Finland. *Boreal. Environ. Res.* **2005**, *10*, 323–336.

17. Hasekamp, O.P.; Landgraf, J. Linearization of vector radiative transfer with respect to aerosol properties and its use in satellite remote sensing. *J. Geophys. Res.* **2005**, *110*. [CrossRef]

18. Kaufman, Y.J.; Tanré, D.; Remer, L.A.; Vermote, E.F.; Chu, A.; Holben, B.N. Operational remote sensing of tropospheric aerosol over land from EOS Moderate Resolution Imaging Spectroradiometer. *J. Geophys. Res.* **1997**, *102*, 17051–17067. [CrossRef]

19. Levy, R.C.; Remer, L.A.; Mattoo, S.; Vermote, E.F.; Kaufman, Y.J. Second-generation operational algorithm: Retrieval of aerosol properties over land from inversion of Moderate Resolution Imaging Spectroradiometer spectral reflectance. *J. Geophys. Res.* **2007**, *112*. [CrossRef]

20. Nagaraja Rao, C.R.; Stowe, L.L.; McClain, E.P. Remote sensing of aerosols over the oceans using AVHRR data theory, practice and applications. *Int. J. Remote Sens.* **1989**, *10*, 743–749. [CrossRef]

21. Mishchenko, M.I.; Geogdzhayev, I.V.; Cairns, B.; Rossow, W.B.; Lacis, A.A. Aerosol retrievals over the ocean by use of channels 1 and 2 AVHRR data: Sensitivity analysis and preliminary results. *Appl. Opt.* **1999**, *38*, 7325–7341. [CrossRef] [PubMed]

22. De Graaf, M.; Stammes, P.; Aben, E.A.A. Analysis of reflectance spectra of UV-absorbing aerosol scenes measured by SCIAMACHY. *J. Geophys. Res.* **2007**, *112*, 485–493. [CrossRef]

23. Torres, O.; Bhartia, P.K.; Herman, J.R.; Sinyuk, A.; Ginoux, P.; Holben, B. A long-term record of aerosol optical depth from TOMS observations and comparison to AERONET measurements. *J. Atmos. Sci.* **2002**, *59*, 398–413. [CrossRef]

24. Torres, O.; Tanskanen, A.; Veihelmann, B.; Ahn, C.; Braak, R.; Bhartia, P.K.; Veefkind, P.; Levelt, P. Aerosols and surface UV products from ozone monitoring instrument observations: An overview. *J. Geophys. Res.* **2007**, *112*. [CrossRef]

25. Chiapello, I.; Goloub, P.; Tanré, D.; Marchand, A.; Herman, J.; Torres, O. Aerosol detection by TOMS and POLDER over oceanic regions. *J. Geophys. Res.* **2000**, *105*, 7133–7142. [CrossRef]

26. Chu, D.A.; Kaufman, Y.J.; Ichoku, C.; Remer, L.A.; Tanre, D.; Holben, B.N. Validation of MODIS aerosol optical depth retrieval over land. *Geophys. Res. Lett.* **2002**, *29*. [CrossRef]

27. Abdou, W.A.; Diner, D.J.; Martonchik, J.V.; Bruegge, C.J.; Kahn, R.A.; Gaitley, B.J.; Crean, K.A. Comparison of coincident multiangle imaging spectroradiometer and Moderate Resolution Imaging Spectroradiometer aerosol optical depths over land and ocean scenes containing aerosol robotic network sites. *J. Geophys. Res.* **2005**, *110*. [CrossRef]

28. Diner, D.J.; Beckert, J.C.; Reilly, T.H.; Bruegge, C.J.; Conel, J.E.; Kahn, R.; Martonchik, J.V.; Ackerman, T.P.; Davies, R.; Gerstl, S.A. Multi-angle Imaging Spectroradiometer (MISR) instrument description and experiment overview. *IEEE Trans. Geosci. Remote Sens.* **1998**, *36*, 1072–1087. [CrossRef]

29. Kahn, R.A.; Gaitley, B.J.; Martonchik, J.V.; Diner, D.J.; Crean, K.A. Multiangle Imaging Spectroradiometer (MISR) global aerosol optical depth validation based on 2 years of coincident aerosol robotic network (AERONET) observations. *J. Geophys. Res.* **2005**, *110*, 1–16. [CrossRef]

30. Liu, Y.; Koutrakis, P.; Kahn, R. Estimating fine particulate matter component concentrations and size distributions using satellite-retrieved fractional aerosol optical depth: Part 1—Method development. *J. Air Waste Manag. Assoc.* **2007**, *57*, 1351–1359. [PubMed]

31. Hansen, J.E.; Travis, L.D. Light scattering in planetary atmospheres. *Space Sci. Rev.* **1974**, *16*, 527–610. [CrossRef]

32. Leroy, M.; Deuze, J.L.; Breon, F.M.; Hautecoeur, O.; Herman, M.; Buriez, J.C.; Tanre, D.; Bouffies, S.; Chazette, P.; Roujean, J.L. Retrieval of atmospheric properties and surface bidirectional reflectances over land from POLDER/ADEOS. *J. Geophys. Res.* **1997**, *102*, 17023–17037. [CrossRef]

33. Deuzé, J.L.; Bréon, F.M.; Devaux, C.; Goloub, P.; Herman, M.; Lafrance, B.; Maignan, F.; Marchand, A.; Nadal, F.; Perry, G.; et al. Remote sensing of aerosols over land surfaces from POLDER-ADEOS-1 polarized measurements. *J. Geophys. Res.* **2001**, *106*, 4913–4926. [CrossRef]

34. Hasekamp, O.P.; Landgraf, J. Retrieval of aerosol properties over land surfaces: Capabilities of multiple-viewing-angle intensity and polarization measurements. *Appl. Opt.* **2007**, *46*, 3332–3344. [CrossRef] [PubMed]

35. Sinyuk, A.; Dubovik, O.; Holben, B.; Eck, T.F.; Breon, F.-M.; Martonchik, J.; Kahn, R.; Diner, D.J.; Vermote, E.F.; Roger, J.-C.; et al. Simultaneous retrieval of aerosol and surface properties from a combination of AERONET and satellite data. *Remote. Sens. Environ.* **2007**, *107*, 90–108. [CrossRef]

36. Torres, O. Total Ozone Mapping Spectrometer measurements of aerosol absorption from space: Comparison to SAFARI 2000 ground-based observations. *J. Geophys. Res.* **2005**, *110*. [CrossRef]

37. Torres, O.; Ahn, C.; Chen, Z. Improvements to the OMI near-UV aerosol algorithm using A-Train CALIOP and AIRS observations. *Atmos. Meas. Tech.* **2013**, *6*, 3257–3270. [CrossRef]

38. Frankenberg, C.; Hasekamp, O.; O'Dell, C.; Sanghavi, S.; Butz, A.; Worden, J. Aerosol information content analysis of multi-angle high spectral resolution measurements and its benefit for high accuracy greenhouse gas retrievals. *Atmos. Meas. Tech.* **2012**, *5*, 1809–1821. [CrossRef]

39. Butz, A.; Hasekamp, O.P.; Frankenberg, C.; Aben, I. Retrievals of atmospheric CO_2 from simulated space-borne measurements of backscattered near-infrared sunlight: Accounting for aerosol effects. *Appl. Opt.* **2009**, *48*, 3322–3336. [CrossRef] [PubMed]

40. Wang, J.; Xu, X.; Ding, S.; Zeng, J.; Spurr, R.J.D.; Liu, X.; Chance, K.; Mishchenko, M.I. A numerical testbed for remote sensing of aerosols, and its demonstration for evaluating retrieval synergy from a geostationary satellite constellation of GEO-CAPE and GOES-R. *J. Quant. Spectrosc. Radiat.* **2014**, *146*, 510–528. [CrossRef]

41. Chen, X.; Wang, J.; Liu, Y.; Xu, X.; Cai, Z.; Yang, D.; Yan, C.-X. Angular dependence of aerosol information content in CAPI/TanSat observation over land: Effect of polarization and synergy with A-Train satellites. *Remote. Sens. Environ.* **2016**. under review.

42. Martynenko, D.; Holzer-Popp, T.; Elbern, H.; Schroedter-Homscheidt, M. Understanding the aerosol information content in multi-spectral reflectance measurements using a synergetic retrieval algorithm. *Atmos. Meas. Tech.* **2010**, *3*, 1589–1598. [CrossRef]

43. Geddes, A.; Bösch, H. Tropospheric aerosol profile information from high-resolution oxygen A-band measurements from space. *Atmos. Meas. Tech.* **2015**, *8*, 859–874. [CrossRef]

44. Spurr, R.J.D.; Wang, J.; Zeng, J.; Mishchenko, M.I. Linearized T-matrix and Mie scattering computations. *J. Quant. Spectrosc. Radiat.* **2012**, *113*, 425–439. [CrossRef]

45. Hess, M.; Koepke, P.; Schult, I. Optical Properties of Aerosols and Clouds: The software package OPAC. *Bull. Am. Meteorol. Soc.* **1998**, *79*, 831–844. [CrossRef]

46. Bodhaine, B.A.; Wood, N.B.; Dutton, E.G.; Slusser, J.R. On Rayleigh optical depth calculations. *J. Atmos. Ocean. Technol.* **1999**, *16*, 1854–1861. [CrossRef]

47. Rothman, L.S.; Gordon, I.E.; Barbe, A.; Benner, D.C.; Bernath, P.F.; Birk, M.; Boudon, V.; Brown, L.R.; Campargue, A.; Champion, J.P.; et al. The HITRAN 2008 molecular spectroscopic database. *J. Quant. Spectrosc. Radiat.* **2009**, *110*, 533–572. [CrossRef]

48. Von Hoyningen-Huene, W.; Freitag, M.; Burrows, J.B. Retrieval of aerosol optical thickness over land surfaces from top-of-atmosphere radiance. *J. Geophys. Res.* **2003**, *108*. [CrossRef]

49. Meer, F.V.D.; Jong, S.M.D. Improving the results of spectral unmixing of landsat thematic mapper imagery by enhancing the orthogonality of end-members. *Int. J. Remote. Sens.* **2000**, *21*, 2781–2797. [CrossRef]

50. Yamaguchi, Y.; Kahle, A.B.; Tsu, H.; Kawakami, T.; Pniel, M. Overview of advanced spaceborne thermal emission and reflection radiometer (ASTER). *Geosci. Remote Sens.* **1998**, *36*, 1062–1071. [CrossRef]

51. Spurr, R. Lidort and vlidort: Linearized pseudo-spherical scalar and vector discrete ordinate radiative transfer models for use in remote sensing retrieval problems. In *Light Scattering Reviews 3*; Kokhanovsky, D.A.A., Ed.; Springer: Berlin/Heidelberg, Germany, 2008; pp. 229–275.

52. Spurr, R.J.D. Vlidort: A linearized pseudo-spherical vector discrete ordinate radiative transfer code for forward model and retrieval studies in multilayer multiple scattering media. *J. Quant. Spectrosc. Radiat.* **2006**, *102*, 316–342. [CrossRef]

53. Mukai, S.; Sano, I. Retrieval algorithm for atmospheric aerosols based on multi-angle viewing of ADEOS/POLDER. *Earth Planets Space* **1999**, *51*, 1247–1254. [CrossRef]

remote sensing

MDPI

Article

Modelling Seasonal GWR of Daily PM$_{2.5}$ with Proper Auxiliary Variables for the Yangtze River Delta

Man Jiang [1], Weiwei Sun [1,2,*], Gang Yang [1] and Dianfa Zhang [1]

[1]   Department of Geography and Spatial Information Techniques, Ningbo University, 818 Fenghua Road, Ningbo 315211, China; jiangman126@126.com (M.J.); yanggang@nbu.edu.cn (G.Y.); zhangdianfa@nbu.edu.cn (D.Z.)
[2]   State Key Lab of Information Engineering on Survey, Mapping and Remote Sensing, Wuhan University, Wuhan 430079, China
*   Correspondence: sunweiwei@nbu.edu.cn; Tel.: +86-182-5879-6120

Academic Editors: Yang Liu, Jun Wang, Omar Torres, Richard Müller and Prasad S. Thenkabail
Received: 9 December 2016; Accepted: 1 April 2017; Published: 5 April 2017

Abstract: Over the past decades, regional haze episodes have frequently occurred in eastern China, especially in the Yangtze River Delta (YRD). Satellite derived Aerosol Optical Depth (AOD) has been used to retrieve the spatial coverage of PM$_{2.5}$ concentrations. To improve the retrieval accuracy of the daily AOD-PM$_{2.5}$ model, various auxiliary variables like meteorological or geographical factors have been adopted into the Geographically Weighted Regression (GWR) model. However, these variables are always arbitrarily selected without deep consideration of their potentially varying temporal or spatial contributions in the model performance. In this manuscript, we put forward an automatic procedure to select proper auxiliary variables from meteorological and geographical factors and obtain their optimal combinations to construct four seasonal GWR models. We employ two different schemes to comprehensively test the performance of our proposed GWR models: (1) comparison with other regular GWR models by varying the number of auxiliary variables; and (2) comparison with observed ground-level PM$_{2.5}$ concentrations. The result shows that our GWR models of "AOD + 3" with three common meteorological variables generally perform better than all the other GWR models involved. Our models also show powerful prediction capabilities in PM$_{2.5}$ concentrations with only slight overfitting. The determination coefficients R^2 of our seasonal models are 0.8259 in spring, 0.7818 in summer, 0.8407 in autumn, and 0.7689 in winter. Also, the seasonal models in summer and autumn behave better than those in spring and winter. The comparison between seasonal and yearly models further validates the specific seasonal pattern of auxiliary variables of the GWR model in the YRD. We also stress the importance of key variables and propose a selection process in the AOD-PM$_{2.5}$ model. Our work validates the significance of proper auxiliary variables in modelling the AOD-PM$_{2.5}$ relationships and provides a good alternative in retrieving daily PM$_{2.5}$ concentrations from remote sensing images in the YRD.

Keywords: seasonal GWR models; auxiliary variable selection; geographically weighted model; MODIS AOD; PM$_{2.5}$ concentrations; Yangtze River Delta

1. Introduction

Widespread air pollution has become a severe problem in China, with increasing population and pollution emissions. The Yangtze River Delta (YRD), as one of the most developed regions in eastern China, has been suffering deterioration of air quality and even more frequent haze episodes, severely threatening both life and health of its people. Particulate matter with an aerodynamic diameter less than 2.5 μm (PM$_{2.5}$) is one of most harmful components of pollution haze and it has severely toxic effects on climate, environment and human health [1,2]. Numerous epidemiological studies have validated the direct relation between high $PM_{2.5}$ concentrations and rising human health problems like asthma, tumors, and lung cancer [3–7]. Therefore, $PM_{2.5}$ concentration monitoring is a significant and pressing issue for both assessing human health exposure and making effective air pollution control measures in the YRD region.

Ground-based monitoring networks could provide accurate and real-time $PM_{2.5}$ concentrations. However, the discrete monitoring sites only measure $PM_{2.5}$ concentrations around a certain distance of the sites and cannot provide a spatial coverage of $PM_{2.5}$ concentrations. Moreover, major monitoring stations are scattered in urban environments and particularly in the metropolis, leaving most rural areas uncovered. Even though the number of monitoring stations in China has been clearly increasing in recent years, the sites are still insufficient to fill all space gaps of the YRD region [8,9].

In contrast, satellite remote sensing has distinct advantages in long-term monitoring and large-scale spatial coverage. Many satellite sensors like MODIS, MISR, and SeaWiFS collect the aerosol information in the atmosphere including aerosol scattering and absorption. They are widely used in estimating and monitoring $PM_{2.5}$ concentrations with aerosol optical depth (AOD). AOD measures the light extinction by aerosol scattering and absorption and it reflects the particle number and property of $PM_{2.5}$ in the total atmosphere. The satellite sensors then estimate the spatial coverage of daily coverage of daily $PM_{2.5}$ concentrations via the retrieval relations between ground-level $PM_{2.5}$ concentrations and satellite-based AOD [10–18].

The retrieval models of $PM_{2.5}$ concentrations coverage from satellite-based AOD can be divided into three main types: the scaling factor models [19,20], the physical analysis models [21,22], and the empirical statistical models [23]. Scaling factor models mainly originate from the chemical transport model (CTM), and they determine the scale factor between satellite-based AOD and ground-level $PM_{2.5}$ concentrations to estimate large-scale spatial distributions of satellite $PM_{2.5}$ concentrations. The models were designed for atmospheric regions without ground $PM_{2.5}$ monitoring data and the retrieval accuracy of $PM_{2.5}$ concentrations is relatively low [23]. Moreover, complicated parameters are mandatorily requiring to initialize and optimize the CTM. Different from scale factor models, physical analysis models analyze the AOD-$PM_{2.5}$ relationships and incorporate accountable physical parameters to construct quantitative functions of satellite $PM_{2.5}$ concentrations [24]. Unfortunately, it is a big challenge to collect these physical parameters in realistic applications. Furthermore, the physical mechanisms in reality are far more complicated than these ever-proposed formulas. Empirical statistical models bring about more accurate distribution retrievals of $PM_{2.5}$ concentrations when compared with the physical analysis models or scaling factor models [21]. Empirical statistical models [25] construct statistical regression functions between satellite-based AOD and in situ $PM_{2.5}$ concentration measurements, and they can be grouped into two classes including early-stage statistical models and advanced statistical models.

Early-stage statistical models are mainly referred to as simple or multiple linear regression models, whereas advanced statistical models develop features in delineating spatial and temporal variations in the relationships between AOD and $PM_{2.5}$ concentrations. Typical examples of advanced statistical models are the general additive line model (GAM) [26], the geographical weighted regression model (GWR) [27], the linear mixed effects (LME) [28], the geographically and temporarily weighted regression model (GTWR) [29], the two step models [10,14], and the three step models [30,31]. Amongst all of them, the GWR has a simple mathematical theory, low computational complexity, and relatively stable performance in considering unstable relationships between ground-level $PM_{2.5}$ concentrations and remote sensed AOD [8,9,15,16,26,32]. Moreover, the GWR shows good compatibility and it is always combined with other schemes to construct complicated statistical models such as the two steps models [14] and three step models [31]. The behaviors of GWR correlate closely with the above formulated models and therefore we focus our study on the GWR model and aim to promote its $PM_{2.5}$ retrieval performance in realistic applications of the YRD region.

The GWR model observes that the relation between $PM_{2.5}$ concentrations and AOD varies across different spatial locations, and additional factors in geography or meteorology are usually incorporated into the model to help explain the generation and dilution of $PM_{2.5}$ concentrations in the atmosphere [10]. Meteorological factors are mainly derived from physical models, and typical variables adopted in GWR include boundary layer height, relative humidity, temperature, and wind speed etc. Geographical factors, mainly referring to the land use the regression model, to explain spatial variations of air pollution in outdoor environments. Geographical variables mainly include demography, land use type, elevation, and vegetation coverage ratio etc. Table 1 lists representatives of auxiliary variables from meteorological or geographical factors in the GWR model for retrieving daily $PM_{2.5}$ concentrations. These auxiliary variables have been proven to enhance the stability of GWR in $PM_{2.5}$ concentration retrieval. However, unfortunately, two big problems still exist in how to properly select meteorological or geographical factors, and that severely hinders the performance improvement of the daily GWR model in realistic applications of the YRD region.

Table 1. Representatives of auxiliary variables in geographical weighted regression model (GWR) model for $PM_{2.5}$ concentrations.

Study Area	Meteorological Factors	Geographical Factors	References
China	relatively humidity, air temperature, wind speed, horizontal visibility	—	[16]
Global	GEOS–Chem chemical transport model (CTM)	urban land cover, elevation	[32]
China	boundary layer height, temperature, wind speed, relative humidity, air pressure	population density, monthly mean normalized difference vegetation index (NDVI)	[9]
Pearl River Delta region	temperature, wind speed, relative humidity	—	[15]
North American Regional	boundary layer height, relative humidity, air temperature, wind speed	percentage of forest cover	[27]

(1) The subjective scheme in selecting meteorological or geographical factors might result in unrepresentative or redundancy among different variables and that would reduce the retrieval performance of the daily GWR model. Different meteorological or geographical factors do have divergent contributions in the GWR model, but the contributions from these variables have never been carefully analyzed. Meanwhile, strong intra-correlations might exist among different meteorological or geographical factors. For example, air temperature has a negative correlation with air pressure, and the elevation correlates closely with demography on the same sites. However, current literatures have never carefully investigated the procedure in selecting proper meteorological or geographical variables for daily GWR modelling. Arbitrary selection of these factors would adversely degrade the accuracy of GWR in realistic applications.

(2) The subjective selection scheme has neglected metabolic contributions of these meteorological or geographical factors to the retrieval performance of the daily GWR model across four different seasons. Working mechanism and contributions of these meteorological or geographical variables vary across different seasons, causing the wide range of daily model performance [27]. For example, in eastern China, the wind dilutes air pollution in summer whereas it might show opposite influences in spatial distributions of $PM_{2.5}$ concentrations in winter. The reason is that the winter monsoon brings the articles from the north of China. Accordingly, it is of great necessity to consider the particularity of variable contributions of different factors in different seasons in order to guarantee good performance of the GWR model.

In our previous work of literature [33], we tested different potential influences from meteorological factors and geographical factors in retrieving $PM_{2.5}$ concentrations with the regular GWR model. In this manuscript, we design an automatic procedure to select proper variables from meteorological or geographical factors in the YRD region and construct specific GWR models for retrieving daily $PM_{2.5}$ concentrations in four different seasons. We validate our seasonal GWR models by comparing with regular GWR models with varying auxiliary variables and by comparing the predicted $PM_{2.5}$ concentrations with the observed. As far as we know, few relevant works have carefully explored the situation in current literatures especially for the YRD region. The recently proposed timely structure adaptive modeling (TSAM) tried to construct a daily AOD-PM$_{2.5}$ model by selecting daily varied auxiliary variables [34]. However, the prediction procedure of TSAM is too complicated to implement in the YRD region. Accordingly, the objects of this manuscript are to: (1) make clear different contributions of the main meteorological or geographical factors and their combinations to the GWR model of the YRD region; (2) propose an automatic procedure to find proper auxiliary variables for modelling GWR in different seasons; and (3) provide detailed equations of seasonal GWR models to benefit the retrieval of daily PM$_{2.5}$ concentrations in the YRD region.

2. Materials and Methods

2.1. Data

2.1.1. Ground-Level PM$_{2.5}$ Concentration Data

Our study region YRD includes Zhejiang, Jiangsu, Anhui provinces, and Shanghai city. We selected the year of 2013 as our study period because of the increasing public attention to haze episodes from 2013 [35] and the data accessibility of PM$_{2.5}$ monitoring sites in 2013. Figure 1 illustrates the 123 PM$_{2.5}$ monitoring sites of the YRD region in 2013. Ground-level hourly PM$_{2.5}$ concentration data was downloaded from China air quality real-time release system of the Chinese Ministry of Environmental Protection (available at http://106.37.208.233:20035). The PM$_{2.5}$ concentrations were measured by Tapered Element Oscillating Microbalances (TEOM) or beta attenuation method (BAM or β-gauge). The data has an uncertainty less than 0.75%, with its accuracy reaching up to ± 1.5 μg/m^3 for the hourly average, and hence it is accurate enough as ground truth for PM$_{2.5}$ concentration measures. From the consideration of simplicity and convenience, the PM$_{2.5}$ concentration data at Beijing time 11:00 AM from 1 January to 31 December 2013 was collected to match with the passing time of the MODIS Terra satellite (i.e., approximately 10:30 a.m. at local time).

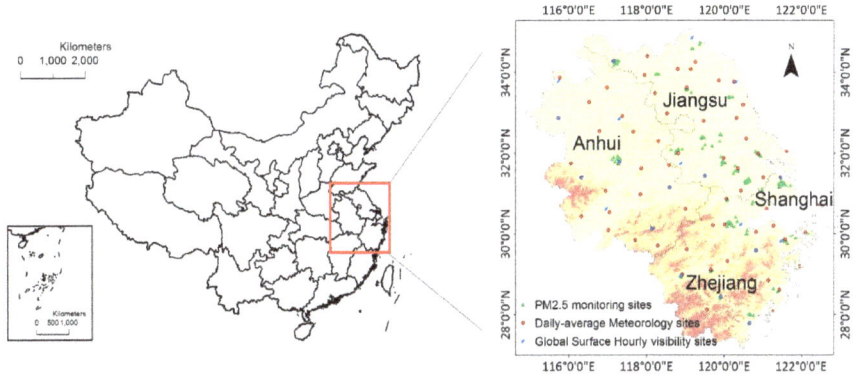

Figure 1. The ground PM$_{2.5}$ monitoring sites and meteorology stations in the Yangtze River Delta (YRD) region.

2.1.2. MODIS AOD

The MODIS sensors on the Terra and Aqua satellites provide global information of the Earth-atmosphere system in 36 spectral bands from visible to thermal infrared spectrum range (0.4–14 μm) with a swath width of 2330 km in 1–2 days. Compared with many other satellite-derived AOD products, the MODIS AOD has the greatest reputation because of its high temporal resolutions, relatively high spatial resolutions, good accuracy, and easy accessibility [36]. The latest MODIS AOD production version Collection 6 is constructed from MODIS imagery via both enhanced DB and DT algorithm and the AOD product is adaptable for both dark and bright surfaces. We in this study uses the Terra MODIS C6 DT 10-km AOD product and the data was download from Level-2 and Atmosphere Archive & Distribution System of NASA (available at http://ladsweb.nascom.nasa.gov/).

2.1.3. Meteorological Datasets

Referring to previous studies, we manually choose six factors as preliminarily auxiliary meteorological variables, including temperature (Temp), wind speed (WS), air pressure (Apre), vapor pressure (Vpre), relative humidity (RH), and surface horizontal visibility (VSB). The particle concentrations of $PM_{2.5}$ largely depend on the meteorological conditions. High surface temperature or high air pressure accelerates the atmospheric vertical motion to transport ground pollutants into higher places. Wind speed is an effective index of quantifying surface motions of air flow and affects the horizontal transport of ground pollutants. Relative humidity makes a correction of aerosol humidity in the atmosphere in order to better match with ground dry $PM_{2.5}$ concentrations. High relative humidity largely enhances the size and light extinction of particles, which comprise the sulfate, nitrate, and ammonium from coal and biomass burning, industrial, and vehicular sources. We also take vapor pressure as a meteorological variable because we regard it is a comprehensive variable and it correlates closely with the generation or aggregation of $PM_{2.5}$. The daily averaged data of above five variables were acquired from the China Daily Surface of Climatic Dataset in Chinese Meteorological Administration (available at http://data.cma.cn/). The YRD region has a total of 72 ground-level monitoring sites in Figure 1.

Although BLH has been a common used variable for AOD vertical correction in many previous studies, recent real experiments proved that BLH made unclear contributions in the GWR model for retrieving $PM_{2.5}$ concentrations in the YRD [28]. The biomass burning in the YRD greatly influences the aloft aerosol above the BLH, and the vertical correction of AOD might be underestimated by the BLH [37]. In contrast, the visibility directly reflects the relationship between AOD and ground-level extinction coefficient [16] and shows great significance in GWR modelling of $PM_{2.5}$ concentrations [8]. Therefore, we implemented VSB as a preliminary meteorological variable rather than BLH. The visibility data was acquired from 23 ground-level monitoring sites in the YRD and the dataset at 11:00 was collected from the National Climatic Data Center (NCDC) Global Surface Hourly database (available at http://gis.ncdc.noaa.gov/map/viewer/#app=clim&cfg=cdo&theme=hourly&layers=1&node=gis).

2.1.4. Geographical Datasets

We also manually selected three widely used geographical factors as preliminarily auxiliary geographical variables, geomorphy feature (GEOM), elevation, and vegetation coverage. We chose the geomorphy feature because it impacts the spread of air pollutants. The geomorphy feature dataset was obtained from the Institute of Geographical Sciences and Natural Resource Research, Chinese Academy of Sciences (available at http://www.resdc.cn/data.aspx?DATAID=124), with spatial resolutions of 10 km equal to the MODIS AOD product. The original geomorphy data was manually recategorized from 26 classes into four new classes in YRD, including plain, platform, hill, and mountain, in order to differentiate their influences in AOD-$PM_{2.5}$ relationships. The elevation is supposed to have negative effects on $PM_{2.5}$ distributions, because of the gravity sedimentation of air particles. The 90 m Digital Elevation Model (DEM) of SRTMDEM3 dataset was acquired from the Geospatial Data Cloud (available at http://www.gscloud.cn/). High vegetation coverage reduces the entry of aerosols into the atmosphere and absorbs particles in the atmosphere [38]. The 16-day synthesized normalized difference vegetation index (NDVI) product of MODIS (MOD13A2) was achieved from NASA (available at http://ladsweb.nascom.nasa.gov/) to represent the vegetation coverage with spatial resolutions of 1000 m. The time of MOD13_A2 product was carefully chosen to coincide with those of other daily datasets, including MODIS AOD, $PM_{2.5}$ concentrations, and the meteorological dataset prior to the day.

2.1.5. Data Pre-Processing and Integration

All the above datasets (Ground-level PM$_{2.5}$ concentrations, MODIS AOD, meteorological data, and geographical data) were transformed into the WGS84 geographic coordinate system. Meteorological data and ground-level PM$_{2.5}$ concentrations were acquired at different monitoring stations, and the average distance of two kinds of station was manually measured at 0.144 degrees. We argue that the meteorological condition varies insignificantly within the average distance and meteorological stations are evenly distributed in the study region. So it is reasonable to spatially join each PM$_{2.5}$ monitoring site with its nearest meteorological station. Accordingly, the ground-level PM$_{2.5}$ concentration measures and its meteorological data were then registered into the same monitoring site. Meanwhile, the YRD region was digitized into grid cells with a fixed grid size of 0.1 degrees. Using overlay analysis, the averages of MODIS AOD and geographical data (DEM, MODIS NDVI and geomorphy data) within each grid cell were assigned as corresponding values of its grid cell.

The MODIS AOD product has many missing pixels mainly due to cloud coverage, high surface reflectance above bright and urban areas, and model retrieval errors. That greatly reduces the usability of the AOD product in matching it with ground-level PM$_{2.5}$ concentration measures and also lowers the number of validated records in the GWR modelling. For the daily PM$_{2.5}$ concentrations retrieval, a proper number of daily validated records of ground-level PM$_{2.5}$ concentrations and AOD is a key point to ensure the robustness and accuracy of GWR modelling. A too small number of daily records could not reflect the realistic spatial coverage of PM$_{2.5}$ concentrations on the same day. According to practical experience from our preliminary trials, the daily threshold of 20 records was manually chosen to guarantee a sufficient number of validated daily records of AOD and ground-level PM$_{2.5}$ concentrations; the final records in our seasonal GWR modelling are 3482 scattering in 66 days of 2013.

2.2. Method

2.2.1. The Regular GWR Model

The GWR model is a spatial regression model that generates spatially continuous coefficients of all variables across the study area. It mainly contributes in analyzing the unstationary status of the spatially varied relationship between independent variables and dependent variables [39]. The GWR model assumes that the AOD-PM$_{2.5}$ relationship varies greatly with spatial locations in the study area, and it has been adapted to describe the unstable relations between PM$_{2.5}$ concentrations and AOD, as well as other geographical or meteorological factors. In this study, considering the nine auxiliary variables in Table 2 to constitute the preliminary variable set $AV = \{DNVI, Geom, Elev, Temp, RH, WS, Apre, Vpre, VSB\}$, the regular GWR model in retrieving daily PM$_{2.5}$ concentrations can be formulated as

$$PM_{2.5(i,j)} = \beta_{0(i,j)} + \beta_{AOD(i,j)}AOD_{(i,j)} + \sum_{k=1}^{c} \beta_{k(i,j)}AV^{k}_{SUB(i,j)} \qquad (1)$$

where $AOD_{(i,j)}$ and $PM_{2.5(i,j)}$ represent main variables of the daily GWR model at the position i on day j, with the coefficient of AOD as $\beta_{AOD(i,j)}$; $\beta_{0(i,j)}$ is a constant coefficient denoting the location-specific intercept at the position i on day j; AV^{k}_{SUB} denotes the k-th element of a subset selected from the set of auxiliary variables AV, with the subset size c no less than 9, and $\beta_{k(i,j)}$ is the location-specific slope or coefficient of its corresponding auxiliary variable AV^{k}_{SUB}.

Table 2. Main parameters of all involved variables in GWR modelling.

	Data	Variables (Abbreviation)	Unit	Time Frequency	Spatial Parameters
Main Variables	PM$_{2.5}$ concentration	PM$_{2.5}$	µg/m^3	Hourly	121 stations
	MODIS AOD	AOD	—	Daily	10 km
Preliminary Auxiliary Variables	Geographical data	NDVI	—	16 days	1 km
		Geomorphy (Geom)	—	—	10 km
		DEM(Elev)	m	—	90 m
	Meteorological data	Temperature (Temp)	°C		
		Relative humidity (RH)	%		
		Wind speed (WS)	m/s	Daily	72 stations
		Air pressure (Apre)	Pa		
		Vapor pressure (Vpre)	Pa		
		surface horizontal visibility (VSB)	km	Hourly	23 stations

2.2.2. Seasonal GWR Modelling with Proper Auxiliary Variables

Selecting proper auxiliary variables is of great significance to guarantee the performance of the GWR model as well as to maximize the contribution of each selected auxiliary variable. Previous works validated the seasonal variability of different auxiliary variables in affecting the AOD-PM$_{2.5}$ relations in the GWR model. Therefore, we would like to propose an automatic procedure to select the proper auxiliary variables and construct specific GWR models for retrieving daily PM$_{2.5}$ concentrations in the four different seasons. With main variables of daily PM$_{2.5}$ and daily AOD product, with preliminary auxiliary variables of NDVI, Geom, Elev, Temp, RH, WS, Apre, Vpre, and VSB, the main procedure of constructing seasonal GWR models includes the following steps shown in Figure 2.

1. The datasets of main variables and preliminary auxiliary variables are categorized into four different seasons, spring, summer, autumn, and winter.
2. Different regular GWR models are constructed with main variables and auxiliary variables in different seasons. For each model in each season, we take AOD and PM$_{2.5}$ as main variables and separately add each element of nine single auxiliary variables one at a time into the GWR model. The performance of obtained regular GWR models is quantified via the Determination Coefficient (R^2). By comparing with the simple seasonal GWR model without auxiliary variables, we rank the contributions of each auxiliary variable in the regular GWR modelling of daily PM$_{2.5}$ in descending order. Dominating auxiliary variables for GWR modelling in different seasons are then obtained.
3. Spearman correlation coefficient analysis is implemented into each pair of dominating auxiliary variables in different seasons. The operation is to reduce the collinearity and redundancy among dominating auxiliary variables. The spearman correlation coefficient is a nonparametric rank correlation coefficient, and it is a distribution-free version of the classical Pearson's product–moment correlation coefficient [40]. A higher coefficient means stronger relationships among different auxiliary variables and the coefficient at 0.3 is regarded as the threshold of weak correlations in our study. Once two dominating auxiliary variables have the spearman correlation coefficient over 0.3, and only one of them is chosen for further GWR modelling. The pruned auxiliary variables are obtained after the Spearman correlation coefficient analysis.
4. Factor analysis is carried out to verify the representativeness of pruned auxiliary variables. The idea of factor analysis is to group the variables having high correlations or close connections into the same class, where each class represents a basic structure called the common factor. The main common factors are able to reflect the major information of the original variables. In this study, the average of four season accumulated variance of the first four common factors is 70.97%. Moreover, the factor rotation in factor analysis provides actual physical meaning to explain working mechanisms of each pruned auxiliary variables. In the manuscript, we do not use uniform seasonal load matrix to construct new daily common variables and replace original variables because of the big probability of exaggerated errors.

5. The proper auxiliary variables are achieved for four different seasonal GWR models. The seasonal GWR models for daily PM$_{2.5}$ are finally obtained in the YRD region.

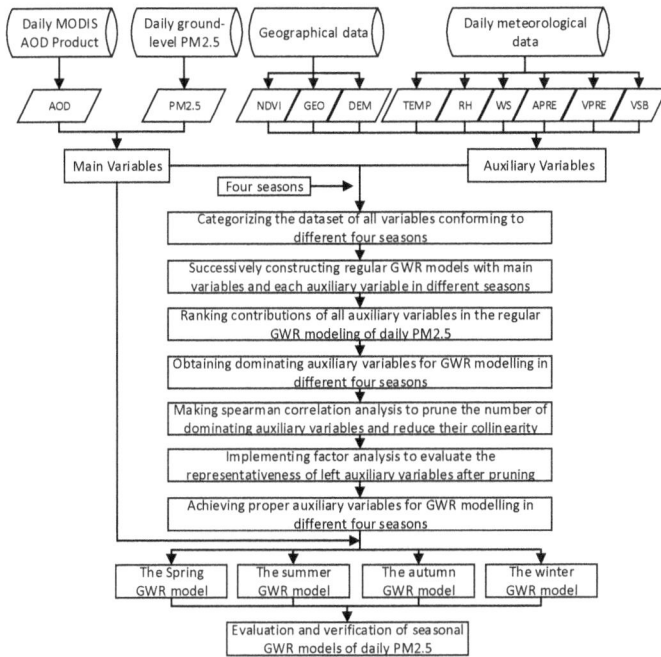

Figure 2. The procedure of constructing seasonal Geographically Weighted Regression (GWR) models with proper auxiliary variables.

2.2.3. Model Evaluation and Verification

In the manuscript, we employ two different schemes to comprehensively testify the performance of our seasonal GWR models: (1) comparison with other regular GWR models by varying the number of auxiliary variables; and (2) comparison with observed ground-level PM$_{2.5}$ concentrations.

We change the number of auxiliary variables to construct different GWR models and compare their performance with our obtained seasonal GWR models in the same season. To assess the performance of model fitting and evaluation, four popular measures are adopted, including R^2, corrected Akaike Information Criterion (AICc), Root Mean Squared Prediction Error (RMSE), and Mean Absolute Percentage Error (MAPE). R^2 is a common indicator for model fitting, and AICc is used to make comparisons between GWR models with different auxiliary variables. RMSE and MAPE describe the residuals between predicted PM$_{2.5}$ concentrations and the observed. MAPE weights the residual in terms of measured PM$_{2.5}$ concentrations, and smaller MAPE indicates higher PM$_{2.5}$ concentrations in the same residual condition. Compared with MAPE, RMSE is more sensitive to higher residual because it places more punishment on a higher residual than a lower one.

On the other hand, we compare the predicted daily PM$_{2.5}$ concentrations with the observed PM$_{2.5}$ measures in the testing sites to further verify the performance of our seasonal GWR models. From the above mentioned, some grids have no daily AOD from MODIS remote sensing to correspond with their PM$_{2.5}$ measures on the same ground-level monitoring sites because of cloud coverage and other reasons. The PM$_{2.5}$ concentrations in these grids are chosen to constitute the testing samples and their corresponding AOD measures are filled with the average AOD values of their nearest neighbors using buffer analysis via 0.3 degrees. The reason for filling absent AOD with buffer analysis is that many experiments from previous works validated that the AOD would not diverge greatly within a small distance. Moreover, with the proper auxiliary variable combination, the 10-fold cross validation is implemented to testify the predication performance of our seasonal GWR models.

3. Results

3.1. Descriptive Statistic of Datasets

Table 3 lists the statistic information of two major variables $PM_{2.5}$ and AOD for model fitting and evaluation in the experiments. The average $PM_{2.5}$ is highest in winter (97.76 µg/m³), then followed by spring (68.02 µg/m³), autumn (57.95 µg/m³), and summer (39.50 µg/m³). The averages of AOD reach the highest value in spring (0.82), then followed by summer, winter, and autumn. We argue that the inconsistent seasonal patterns of $PM_{2.5}$ and AOD are caused by different seasonal impacts of meteorological or geographical factors, rendering that the $PM_{2.5}$ concentrations and AOD are scattered to different extents even in opposite directions. Compared with model fitting in four seasons, the averages of $PM_{2.5}$ concentrations implemented in model evaluation are slightly higher. The reason for this is because some grids with higher $PM_{2.5}$ concentrations but without AOD caused by cloud coverage are grouped as testing samples for model evaluation.

Table 3. Description statistics of $PM_{2.5}$ and Aerosol Optical Depth (AOD) for GWR model fitting and evaluation.

	Variable	Model Fitting (N = 3482, day = 66)				Model Evaluation (N = 715, day = 66)			
Whole Year		Mean	Min	Max	SD	Mean	Min	Max	SD
	$PM_{2.5}$ (µg/m³)	61.75	3	400	40.43	67	21	267	32.
	AOD (Unit less)	0.69	0.03	3.51	0.41	0.62	0.04	2.98	0.35
	Variable	Model Fitting (N = 1237, day = 21)				Day-Site Evaluation (N = 198, day = 21)			
Spring		Mean	Min	Max	SD	Mean	Min	Max	SD
	$PM_{2.5}$ (µg/m³)	68.02	3	279	38.43	68.72	12	257	35.89
	AOD (Unit less)	0.82	0.08	3.51	0.41	0.69	0.11	3.21	0.39
	Variable	Model Fitting (N = 809, day = 16)				Day-Site Evaluation (N = 182, day = 16)			
Summer		Mean	Min	Max	SD	Mean	Min	Max	SD
	$PM_{2.5}$ (µg/m³)	39.50	3	400	23.25	41.50	14	400	26.25
	AOD (Unit less)	0.67	0.04	2.33	0.34	0.67	0.06	2.33	0.35
	Variable	Model Fitting (N = 1014, day = 18)				Day-Site Evaluation (N = 181, day = 18)			
Autumn		Mean	Min	Max	SD	Mean	Min	Max	SD
	$PM_{2.5}$ (µg/m³)	57.95	5	205	35.62	62.5	9	235	35.62
	AOD (Unit less)	0.58	0.035	2.50	0.42	0.59	0.04	2.70	0.51
	Variable	Model Fitting (N = 422, day = 11)				Day-Site Evaluation (N = 154, day = 11)			
Winter		Mean	Min	Max	SD	Mean	Min	Max	SD
	$PM_{2.5}$ (µg/m³)	96.76	5	284	50.10	102.3	25	267	45
	AOD (Unit less)	0.62	0.037	2.94	0.41	0.59	0.04	3.02	0.41

3.2. Proper Auxiliary Variables Analysis

We first respectively add each of the nine auxiliary variables into the simple GWR model. Figure 3 illustrates the determined coefficients R^2 of each auxiliary variable in its seasonal models. In four seasons, VSB contributes greatly to improving the GWR performance, whereas Georm and NDVI have little contributions to the GWR model in fitting the AOD-$PM_{2.5}$ relationships. The contributions from Vpre, Temp, WS, Elev, and Apre in GWR modelling diverge greatly across different seasons. In spring, the mean and median of WS, Vpre, VSB, and Temp variables show better contributions than other variables. The wider variations of R^2 from NDVI and Apre indicate more variations of their contributions in spring GWR modelling. In summer, Vpre, and VSB show remarkable performance, followed by Elev and Apre, while the WS variable shows less impact compared with its contribution in spring GWR modelling. In autumn, WS, Temp, Vpre, and VSB have remarkable contributions compared to other variables, regardless that Apre is less stable than the four variables above. In winter, all the variables have relatively unstable contributions in GWR modelling with wider ranges of R^2, and the reason for this might be the relatively smaller fitting samples and the fluctuating meteorological conditions in winter. From the above contributions of all the single auxiliary variables, we preliminarily selected the Elev, WS, Apre, Vpre, VSB, and Temp for further analysis.

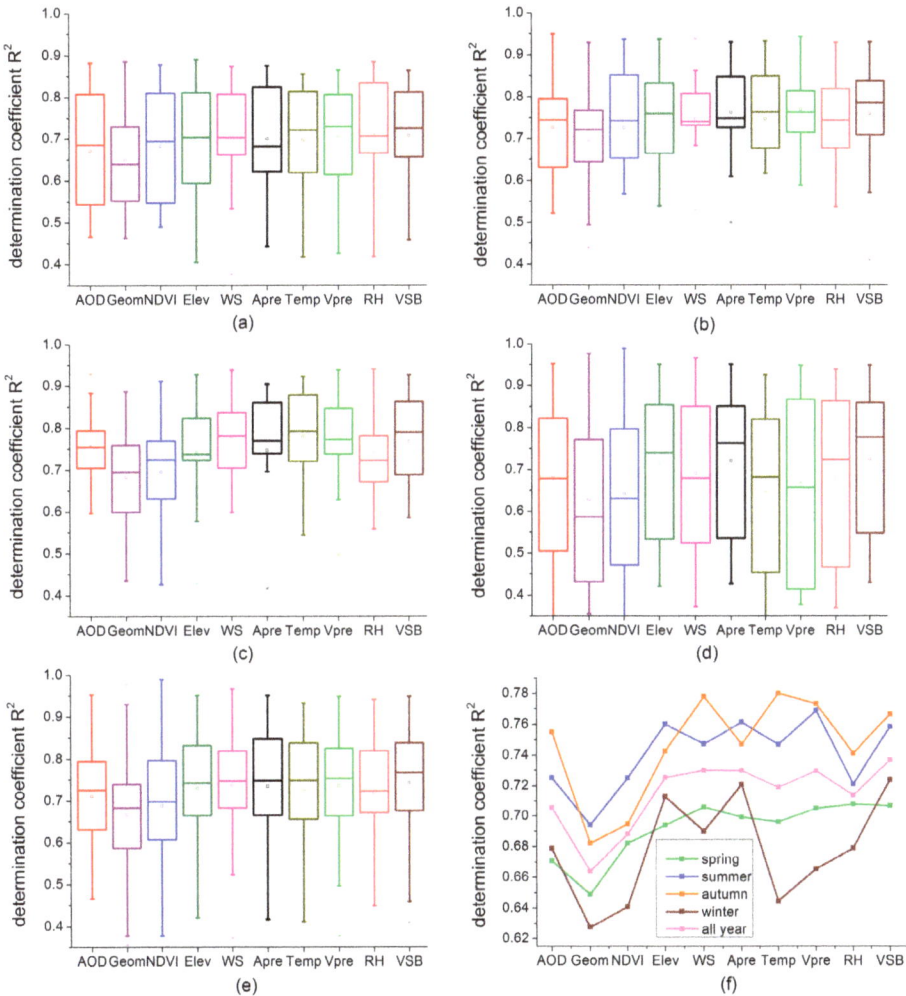

Figure 3. Contributions from nine auxiliary variables in GWR modelling of four seasons and the whole year. (**a**) spring, (**b**) summer, (**c**) autumn, (**d**) winter, (**e**) whole year, (**f**) curve line of contributions from all auxiliary variables in four seasons and the whole year. The box gives the 25%–75% percentile and the line in the box denotes the median. The whisker is the maximum and minimum of R^2, the points outside the box are outliers, inside the box an average of R^2.

With Spearman correlation coefficient analysis, the number of auxiliary variables is pruned to reduce the collinearity and redundancy among different variables. For the spring model, the Temp is removed because of its serious collinearly with Vpre. The Apre variable is discarded in the summer model because of its clear redundancy with Elev. The Vpre variable is dropped in the autumn model because of its collinearity with VSB. The variables in the winter model do not show clear collinearity. Three variables are left for four seasonal models and the yearly model respectively, WS, Vpre, and VSB for spring, Elev, Vpre, VSB for summer, WS, Temp, VSB for autumn, Apre, RH, VSB for winter and WS, Vpre and VSB for the whole year.

We implement factor analysis to verify the representatives of pruned auxiliary variables. We use the factor analysis method to extract the common factor and calculate the factor load matrix. The result shows the first four factors separately reach their 70.52% accumulated variance in spring, 71.31% in summer, 74.64% in autumn, and 67.38% in winter. The first common factor includes AOD and VSB, sometimes with RH in. The second common factor consistently includes Vpre and Temp. The third common factor usually includes WS and Elev. The last common factor includes NDVI and Geom. Apre is possible to be involved in both the first and third common factors. We respectively name these four common factor components comprehensive index, vertical diffusion effect, horizontal diffusion effect, and geographic effect according to their physical mechanism with the ground-level $PM_{2.5}$ concentrations. In this case, we find the pruned auxiliary variables exactly represent three common meteorology factors, and therefore factor analysis provides theoretical supports for our selected auxiliary variables. However, the geographic effect does not involve variables in our model, we will explain the reason latter.

From the above, we finally obtain three proper auxiliary variables for four seasonal GWR models and the year GWR model (listed in the supplementary file). The proper auxiliary variables are WS, Vpre, and VSB for spring GWR model, Elev, Vpre, VSB for summer model, WS, Temp, VSB for autumn model and Apre, RH, VSB for winter model. Integrating the proper auxiliary variables with the main variable, four seasonal GWR models and the year model for further comparison are finally obtained.

3.3. Evaluation and Verification of Seasonal GWR Models

In order to validate our four seasonal models, we implement two groups of experiments to validate our four seasonal models. The first experiment in comparison with other GWR models by changing different auxiliary variable combination is to verify the performance of our proper auxiliary variable combination in four seasonal models. The second experiment is to evaluate the behaviors of four seasonal GWR models in predicting daily $PM_{2.5}$ concentrations.

3.3.1. Comparison of Regular GWR Models with Varied Auxiliary Variables

In order to testify the performance of three variable combinations, we respectively change the number of auxiliary variables from 0–4 to construct eight comparison GWR models in each season. Table 4 lists auxiliary variables of 36 GWR models of four seasons in the comparison, and the nine models can be grouped into five types: "AOD + 0", "AOD + 1", "AOD + 2", "AOD + 3" (our model with three proper auxiliary variables), and "AOD + 4". The fourth variable in the comparison compared against our seasonal models is manually selected according to its contribution of a single variable in regular GWR modelling.

Table 4. The list of all GWR models with different variable combinations.

Model Groups	Models	Spring	Summer	Autumn	Winter	Year
AOD + 0	1			AOD		
AOD + 1	2	AOD, WS	AOD, Elev	AOD, WS	AOD, Apre	AOD, WS
	3	AOD, Vpre	AOD, Vpre	AOD, Temp	AOD, RH	AOD, Vpre
	4	AOD, VSB	AOD, VSB	AOD, VSB	AOD, VSB	AOD, VSB
AOD + 2	5	AOD, WS, Vpre	AOD, Elev, Vpre	AOD, WS, Temp	AOD, Apre RH	AOD, WS, Vpre
	6	AOD, WS, VSB	AOD, Elev, VSB	AOD, WS, VSB	AOD, Apre, VSB	AOD, WS, VSB
	7	AOD, Vpre, VSB	AOD, Vpre, VSB	AOD, Temp, VSB	AOD, RH, VSB	AOD, Vpre, VSB
AOD + 3 (Ours)	8	AOD, WS, Vpre, VSB	AOD, Elev, Vpre, VSB	AOD, WS, Temp, VSB	AOD, Apre, RH, VSB	AOD, WS, Vpre, VSB
AOD + 4	9	AOD, WS, Vpre, VSB, Elev	AOD, Elev Temp, Vpre, VSB,	AOD, WS, Temp, Vpre, VSB	AOD, WS, Apre, RH, VSB	AOD, WS , Temp, Vpre, VSB

Figure 4 demonstrates all the model fitting and evaluation results from all the models in four seasons. In spring, the R^2 of model fitting and model evaluation rises when the auxiliary variables gradually increase from 1–3, consistent with the decreasing RMSE and MAPE from 14.0 µg/m^3 to 12.8 µg/m^3 and from 27% to 22% in model fitting respectively. The optimal R^2, MAPE, and AICc reach the optimal result in model 5, although the RMSE has the minimum value in model 8 of "AOD + 3". With the variable number varying from 1–4, the overfitting degree of all models decreases from "AOD + 0", achieves the bottom at "AOD + 3" and then increases from "AOD + 3" to "AOD + 4". Specifically, our spring GWR model of "AOD + 3" has the slightest overfitting degree whereas its "AOD + 2" model encounters the most serious overfitting.

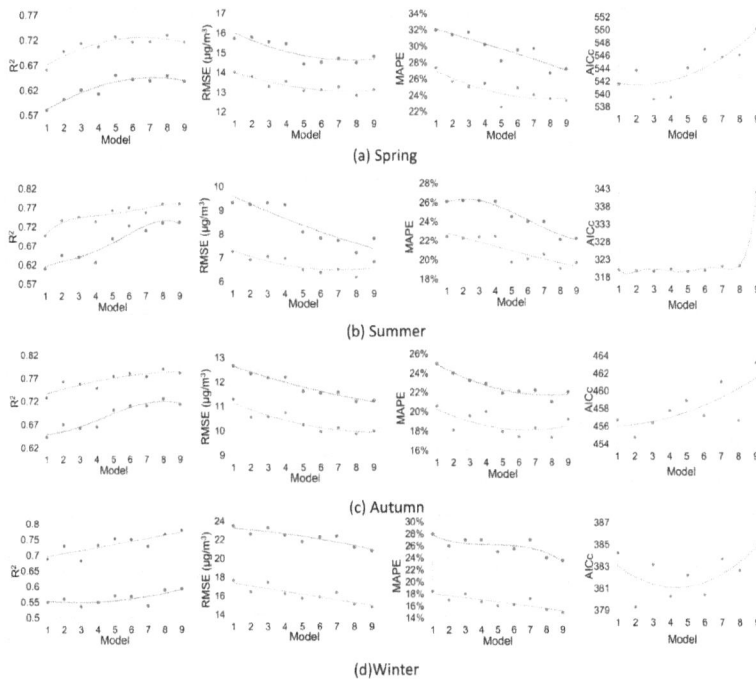

Figure 4. (**a–d**) The comparison of all seasonal GWR models with different auxiliary variables. Blue scattering points are quantitative measures (R^2, RMSE, MAPE and AICc) of model fitting and red points are those of model evaluation. The dash lines are fitted curves of all scattered points in different seasons.

In summer, an overall gradual improvement of model performance exists from "AOD + 0" to "AOD + 3", with a slight descending in "AOD + 4". For the GWR model fitting process, R^2 increases from 0.69 to 0.78, RMSE decreases from 7.25 µg/m^3 to 6.18 µg/m^3, and MAPE decreases from 22.5% to 19.0%. The model performance of GWR slightly descends from "AOD + 3" to "AOD + 4", and the model 8 of "AOD + 3" behaves best of all the models in summer. From model 1 to model 8, AICc values vary within a smaller range from 320.1 to 321.7, and the overall tendency of AICc in model evaluation is consistent with that of model fitting.

Similar to that of summer, the R^2, RMSE, and MAPE in autumn models show gradual improvement in performance from model 1 to model 8, with R^2 increasing from 0.72 to 0.75, RMSE decreasing from 11.35 µg/m^3 to 10.17 µg/m^3 and MAPE decreasing from 22.5% to 19.0%. Model 8 of "AOD + 3" has the highest R^2, lowest RMSE and MAPE and relatively less overfitting among all the models and it performs best of all.

In winter, the variation of R^2, RMSE, and MAPE diverges more than those of the other three seasons. The explanation for this is partly because of fewer samples in both model fitting and evaluation. In terms of R^2, Figure 4 shows a more apparent increasing tendency and does not have descending tendency at "AOD + 4" model. From model 1 to model 8, R^2 increases from 0.68 to 0.77, RMSE decreases from 17.67 µg/m^3 to 15.11 µg/m^3, MAPE decreases from 18.5% to 15.3%. More variation and severer overfitting also occur in the winter models, with the decreasing R^2 averaged at 0.173 from model fitting to model evaluation.

From the above, the comparison with changing numbers of auxiliary variables explains the best performance of our "AOD + 3" model, and verifies the effectiveness of our selected proper auxiliary variables in modelling seasonal GWR models.

Moreover, we listed key coefficients of auxiliary variables involved in nine GWR models from 1–9 on 16 September 2013 to further explain the special performance of our "AOD + 3" GWR model. We selected the day because of its high AOD coverage rate. Table 5 shows the coefficients of all involved auxiliary variables in the nine models. The results show that the auxiliary variables have clear effects in model fitting and evaluation. From model 1 to model 9, the changing combinations of the auxiliary variables explain their divergent contributions in improving the performance of the GWR model. Among all the nine models, the "AOD + 3" of model 8 performs best, with lowest overfitting in model fitting and evaluation. Figure 5 illustrates spatial distributions of PM$_{2.5}$ retrieved from these nine models on 16 September 2013.

Table 5. Parameter estimations and Mean Absolute Percentage Errors (MAPEs) for GWR models 1–9 on 16 September 2013.

Model	Parameter Estimate					MAPE (%)	
	β_{AOD}	β_{WS}	β_{Temp}	β_{VSB}	β_{Vpre}	Fitting	Evaluation
1	6.87	—	—	—	—	22.17	35.91
2	7.56	1.28	—	—	—	21.89	35.29
3	6.67	—	4.31	—	—	21.85	36.34
4	0.81	—	—	−6.65	—	22.64	33.18
5	6.38	0.55	2.24		—	22.24	34.34
6	2.77	1.33	—	−9.49	—	20.92	33.05
7	3.22	—	6.38	−5.82	—	20.99	34.04
8	3.09	0.61	3.89	−5.00	—	20.81	32.25
9	3.84	0.81	5.98	−4.81	1.45	20.92	32.79

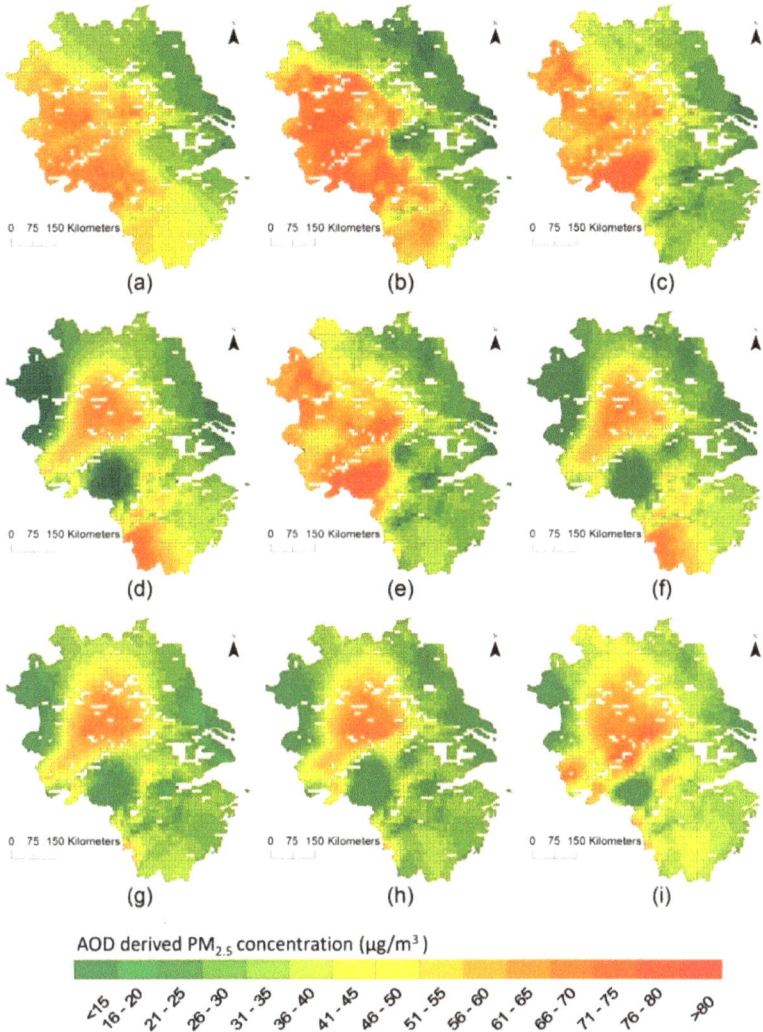

Figure 5. Spatial distribution maps of retrieved PM$_{2.5}$ concentrations of all 9 models on 16 September 2013. The radial basis interpolation (RBF) method was utilized to interpolate meteorological variables and grid them into cells with spatial resolution of 0.1 degree. (**a–i**) correspond to retrieved maps of PM$_{2.5}$ concentrations from models 1–9 in Table 4.

3.3.2. Comparison with the Observed PM$_{2.5}$ Concentrations

In this section, we validate the behaviors of four seasonal "AOD + 3" GWR models in predicting daily PM$_{2.5}$ concentrations, using 10-fold cross validation scheme.

Figure 6 depicts the regression results with zero intercept between our predicted PM$_{2.5}$ against observed PM$_{2.5}$ measures. All the four models behave as slightly overfitting when compared against their cross-validation results. The result also shows that our seasonal GWR models in summer and autumn perform better than those in spring and winter. In spring, the slope and R^2 for model fitting are 0.9618 and 0.8328. In summer, the slope and R^2 for model fitting are 0.9702 and 0.8503. In autumn, the slope and R^2 for model fitting are 0.9758 and 0.9156. In winter, the slope and R^2 for model fitting are 0.9706 and 0.8577. The slopes of four seasons are all less than 1, indicating that the estimated model generally underestimates the actual observed data. All the four models behave as slightly overfitting from model fitting to evaluation, with the slightest increase of RMSE in the summer and the most severe increase in the winter. The main reason for the worse performance in winter is partly because of the relatively small sample size. We argue that the frequent eruption of heavy pollution episodes in the winter of 2013 [35] also aggravated the possibility of unexpected situations in PM$_{2.5}$ estimation, causing higher RMSE values in the winter GWR model.

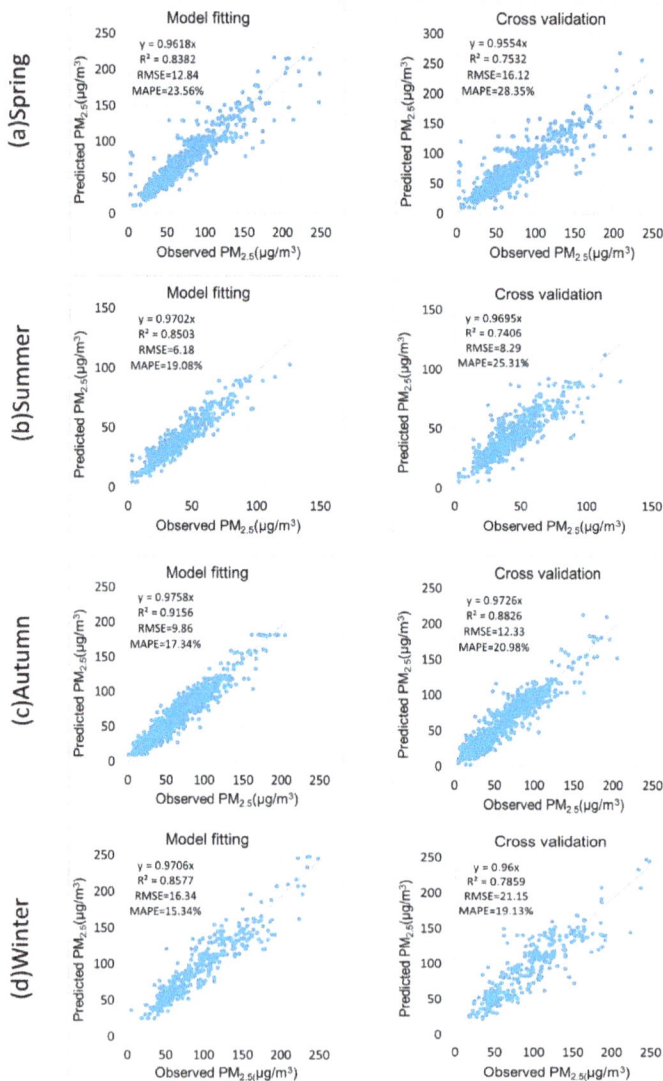

Figure 6. (**a–d**) Comparison between observed PM$_{2.5}$ and predicted PM$_{2.5}$ concentrations in the four seasonal models. The dashed lines are regression lines.

Moreover, we compare the autumn GWR model with the yearly model to further validate the specific seasonal variables pattern of our GWR models. The yearly model has the same auxiliary variables with our autumn model, and the reason for choosing the autumn model is because of its relatively larger sample size for GWR modelling. The result shows the autumn GWR model performs better than the yearly model, with 1.2 $\mu g/m^3$ lower RMSE and 2.79% lower MAPE in model fitting (Figure S1 in the supplementary file). The cases are similar in model evaluation. In addition, the overfitting is more severe in the yearly model, with a bigger gap between fitting and evaluation. The above also indicates that it is of great necessity to consider the seasonal auxiliary variables and establish seasonal models rather than the yearly model to guarantee the performance of the GWR model.

4. Discussion

Satellite images provide a reliable method to retrieve spatial coverages of $PM_{2.5}$ concentrations. One key problem is how to construct the GWR relations between satellite based AOD and ground-level $PM_{2.5}$ measures, with the help of auxiliary meteorological or geographical factors. The auxiliary variables arbitrarily adapted in the GWR model might have unclear contributions to the $PM_{2.5}$ concentration retrievals. Meanwhile, the contribution from single auxiliary variables or their combinations to the GWR modelling varies across different seasons. Therefore, we proposed an automatic procedure to select proper meteorological or geographical factors for GWR modelling in different seasons, rendering their proper auxiliary variable combinations to construct optimal daily GWR models in the YRD region. We made careful comparisons between our seasonal GWR models and regular GWR models with different auxiliary variables, between the predicted $PM_{2.5}$ concentrations from our models and the observed ones, in order to validate our model performance. The results explain that our seasonal models have better performance for $PM_{2.5}$ retrievals than the comparison models.

Theoretically speaking, the relationship between daily AOD and $PM_{2.5}$ is widely affected by few auxiliary variables. Also the proper choice of auxiliary variables could then upgrade the model accuracy as well as reduce the model redundancy. The effects from nine preliminary auxiliary variables which we adopted in $PM_{2.5}$ concentrations, can be grouped into four different aspects—comprehensive effect, vertical diffusion effect, horizontal diffusion effect, and geographical effect. The contribution of each selected auxiliary variable in our models is not constant, and it varies greatly among different seasons like that of TSAM [34]. VSB is a comprehensive indicator of air quality showing significant influences in all four seasons. Vpre is the partial pressure of water vapor in the whole atmosphere column, and our experimental results indicate that it is a good auxiliary variable for GWR modelling in spring, summer, and autumn when compared with the widely adopted RH in the current literature. The reason for this we guessed is because Vpre reflects and corrects the humidity of the whole aerosol column. Apre shows greater contribution in the winter GWR model than in those of other seasons. We guess the reason for this is that the high air pressure occurring in winter always comes along with cold air and sand-dust from the north or northwest of China, and that greatly increases the $PM_{2.5}$ concentrations in the YRD. Two geographical variables, Geom and NDVI, were not significant in the GWR modelling and accordingly were excluded from the auxiliary variable combinations. The relatively simple geomorphy features of the YRD, with 49% of plain and 7.9% of mountain, render that the Geom shows insignificant contributions in the GWR modelling. Although NDVI is relatively weaker with respect to other contemporary meteorology variables, it shows significant contributions in the spring and winter models. That indicates seasonal contribution patterns of NDVI in the $PM_{2.5}$ retrieval model. In general, our experimental results illustrate that the daily AOD-$PM_{2.5}$ relationships closely correlate with meteorological factors rather than geographical factors. The explanation for this is that the shifting meteorology conditions highly affect the generation or diffusion of daily $PM_{2.5}$ concentrations in the short term whereas geographical factors are crucial to long term prediction [30].

The combination of proper auxiliary variables in our seasonal models is carefully investigated in our study by comparing with regular GWR models via changing the number of auxiliary variables. The comparison between our seasonal models and the yearly model illustrates the importance of proper auxiliary variables. Generally, our seasonal models have the optimal performance with three auxiliary variables, and less or more variables than three would bring about instability or redundancy of the GWR models. Adding extra auxiliary variables shows little improvement in the GWR model but even reduces the model performance and aggravates the overfitting, due to the increasing risk of intra-correlations among variables [41]. We further analyzed the auxiliary variables in four seasonal model "AOD + 3" and guessed that they coincided with three meteorological effects in $PM_{2.5}$ concentrations. For the spring, WS determines the horizontal diffusion effect of $PM_{2.5}$, Vpre correlates closely with the vertical diffusion of $PM_{2.5}$, and VSB represents the comprehensive effect of $PM_{2.5}$ concentrations. For the summer season, Elev replaces WS in spring and has a close relation with the horizontal diffusion effect. For the autumn model, Temp correlates closely with the air vertical movement. For winter, Apre shows more correlation with the air vertical movement. The seasonal average $PM_{2.5}$ prediction of model evaluation R^2 is 0.79, and the result of our seasonal models is comparable to those of the previous works by Fang [34], but with less variables and computation cost. We also applied the specific seasonal GWR models to 2014, and the achieved result is satisfying (Figure S2 in the supplementary file).

Unfortunately, our study simultaneously has the following shortcomings needing improvement in the future. First, the selection of the seasonal variable largely depends on our numerous empirical statistical works and the analysis of common factors. Also, theoretical explanations of specific seasonal variables cannot be fully figured out and need to be further investigated. Second, only natural factors were involved in our study and which might result in neglecting some potential effects from anthropic factors especially social geographical factors. Actually, anthropic factors, especially social geographical variables have great potential in improving the $PM_{2.5}$ retrieval model [42]. The supplement work of more anthropic variables into our seasonal GWR models will be completed in a further study. Finally, the filling problem of absent pixels in the AOD product requires careful investigation in order to increase the number of validated records for more robust GWR modelling and to promote the utility of proposed seasonal GWR model in all daily conditions of the YRD region.

5. Conclusions

Auxiliary variables like meteorological or geographical factors show great potential in improving the GWR model accuracy for retrieving $PM_{2.5}$ concentrations from satellite AOD. However, the selection of proper variables and their different contributions in the four seasons have never been carefully investigated, hampering further applications of the GWR model in the YRD region. In this study, we put forward an automatic procedure of seasonal proper variable selection considering the contribution of each single variable and its inner structure in the GWR modelling. Moreover, we investigated the seasonal pattern of auxiliary variables, and constructed four seasonal GWR models with properly selected auxiliary variables. Our seasonal GWR models with proper variable combinations were tested with two groups of experiments. The seasonal GWR models was compared with regular GWR models by changing auxiliary variables and the predicted $PM_{2.5}$ concentrations from the four seasonal models were compared against the observed measures and that of the yearly GWR model. Our selected proper auxiliary variables in four models of "AOD + 3" reduce the redundancy of regular GWR models and simultaneously help to obtain better model accuracy than with other regular GWR models. The prediction performance of the four seasonal models behaves better than the yearly model and the predicted result coincides well with the observed ones, having high R^2 in model evaluation, averaged as 0.79 in 10-fold cross validation. Therefore, seasonal varied contribution of variables should be taken into consideration in $PM_{2.5}$ concentration retrieval and our seasonal GWR models could be a good alternative in modelling daily $PM_{2.5}$ concentrations from remote sensing in the YRD region.

Remote Sens. 2017, 9, 346

Supplementary Materials: The following are available online at www.mdpi.com/2072-4292/9/4/346/s1.

Acknowledgments: This research was funded by National Natural Science Foundation (41401389, 41671342), by the Chinese Postdoctoral Science Foundation (2015M570668, 2016T90732), by the Public Projects of Zhejiang Province (2016C33021), by the Zhejiang University Student Science and Technology Innovation and Xin-Miao Talented Plan Program (2016R405088), and by the K. C. Wong Magna Fund in Ningbo University. The authors thank Qingyue Open Environmental Data Center (https://data.epmap.org) for their generous support in hourly ground level PM$_{2.5}$ data processing. The authors also would like to thank the editor and referees for their suggestions that improved the manuscript.

Author Contributions: All the coauthors made significant contributions to the manuscript. Man Jiang carried on most experiments and wrote the manuscript. Weiwei Sun presented the key idea of seasonal GWR model and designed the experiments. Gang Yang compared the seasonal GWR model with the yearly GWR model and helped to revise the manuscript. Dianfa Zhang provided the background knowledge of atmospheric remote sensing and gave important suggestions for improving the manuscript.

Conflicts of Interest: The authors declare no conflict of interest.

References

1. Kaufman, Y.J.; Tanré, D.; Boucher, O. A satellite view of aerosols in the climate system. *Nature* **2002**, *419*, 215–223. [CrossRef] [PubMed]
2. An, X.; Hou, Q.; Li, N.; Zhai, S. Assessment of human exposure level to PM10 in China. *Atmos. Environ.* **2013**, *70*, 376–386. [CrossRef]
3. Dockery, D.W.; Pope, C.A.; Xu, X.; Spengler, J.D.; Ware, J.H.; Fay, M.E.; Ferris, B.G., Jr.; Speizer, F.E. An association between air pollution and mortality in six U.S. Cities. *N. Engl. J. Med.* **1993**, *329*, 1753–1759. [CrossRef] [PubMed]
4. Wang, Z.; Yang, L.; Mu, H.; Pan, X.; Jing, S.; Feng, C.; He, K.; Koutrakis, P.; Christiani, D.C. Acute health impacts of airborne particles estimated from satellite remote sensing. *Environ. Int.* **2013**, *51*, 150–159. [CrossRef] [PubMed]
5. Weichenthal, S.; Villeneuve, P.J.; Burnett, R.T.; van Donkelaar, A.; Martin, R.V.; Jones, R.R.; Dellavalle, C.T.; Sandler, D.P.; Ward, M.H.; Hoppin, J.A. Long-term exposure to fine particulate matter: Association with nonaccidental and cardiovascular mortality in the agricultural health study cohort. *Environ. Health Perspect.* **2014**, *122*, 609–615. [CrossRef] [PubMed]
6. Hu, Z. Spatial analysis of MODIS aerosol optical depth, PM$_{2.5}$, and chronic coronary heart disease. *Int. J. Health Geogr.* **2009**, *8*, 1–10. [CrossRef] [PubMed]
7. Shi, L.; Zanobetti, A.; Kloog, I.; Coull, B.A.; Koutrakis, P.; Melly, S.J.; Schwartz, J.D. Low-concentration PM$_{2.5}$ and mortality: Estimating acute and chronic effects in a population-based study. *Environ. Health Perspect.* **2016**, *124*, 46–52. [CrossRef] [PubMed]
8. Wei, Y.; Zang, Z.; Zhang, L.; Yi, L.; Wang, W. Estimating national-scale ground-level PM25 concentration in china using geographically weighted regression based on MODIS and MISR AOD. *Environ. Sci. Pollut. Res.* **2016**, 1–12. [CrossRef] [PubMed]
9. Ma, Z.; Hu, X.; Huang, L.; Bi, J.; Liu, Y. Estimating ground-level PM$_{2.5}$ in China using satellite remote sensing. *Environ. Sci. Technol.* **2014**, *48*, 7436–7444. [CrossRef] [PubMed]
10. Liu, Y.; Paciorek, C.J.; Koutrakis, P. Estimating regional spatial and temporal variability of PM$_{2.5}$ concentrations using satellite data, meteorology, and land use information. *Environ. Health Perspect.* **2009**, *117*, 886–892. [CrossRef] [PubMed]
11. Yao, L.; Lu, N. Spatiotemporal distribution and short-term trends of particulate matter concentration over China, 2006–2010. *Environ. Sci. Pollut. Res.* **2014**, *21*, 9665–9675. [CrossRef] [PubMed]
12. Wang, P.; Cao, J.-J.; Shen, Z.-X.; Han, Y.-M.; Lee, S.-C.; Huang, Y.; Zhu, C.-S.; Wang, Q.-Y.; Xu, H.-M.; Huang, R.-J. Spatial and seasonal variations of PM$_{2.5}$ mass and species during 2010 in Xi'an, China. *Sci. Total Environ.* **2015**, *508*, 477–487. [CrossRef] [PubMed]
13. Strawa, A.W.; Chatfield, R.B.; Legg, M.J.; Scarnato, B.V.; Esswein, R. In improving retrievals of regional PM$_{2.5}$ concentrations from MODIS and OMI multi-satellite observations. In Proceedings of the American Geophysical Union 2013 Fall Meeting, San Francisco, CA, USA, 9–13 December 2013.

14. Hu, X.; Waller, L.A.; Lyapustin, A.; Wang, Y.; Al-Hamdan, M.Z.; Crosson, W.L.; Estes, M.G.; Estes, S.M.; Quattrochi, D.A.; Puttaswamy, S.J. Estimating ground-level PM$_{2.5}$ concentrations in the southeastern united states using MAIAC AOD retrievals and a two-stage model. *Remote Sens. Environ.* **2014**, *140*, 220–232. [CrossRef]

15. Song, W.; Jia, H.; Huang, J.; Zhang, Y. A satellite-based geographically weighted regression model for regional PM 2.5 estimation over the Pearl River Delta region in China. *Remote Sens. Environ.* **2014**, *154*, 1–7. [CrossRef]

16. You, W.; Zang, Z.; Zhang, L.; Li, Y.; Pan, X.; Wang, W. National-scale estimates of ground-level PM2.5 concentration in China using geographically weighted regression based on 3 km resolution MODIS AOD. *Remote Sens.* **2016**, *8*, 184. [CrossRef]

17. Guo, H.; Cheng, T.; Gu, X.; Chen, H.; Wang, Y.; Zheng, F.; Xiang, K. Comparison of four ground-level PM$_{2.5}$ estimation models using parasol aerosol optical depth data from China. *Int. J. Environ. Res. Public Health* **2016**, *13*, 180. [CrossRef] [PubMed]

18. Ma, Z.; Hu, X.; Sayer, A.M.; Levy, R.; Zhang, Q.; Xue, Y.; Tong, S.; Bi, J.; Huang, L.; Liu, Y. Satellite-based spatiotemporal trends in PM$_{2.5}$ concentrations: China, 2004–2013. *Environ. Health Perspect.* **2016**, *124*, 184–192. [CrossRef] [PubMed]

19. Liu, Y.; Park, R.J.; Jacob, D.J.; Li, Q.; Kilaru, V.; Sarnat, J.A. Mapping annual mean ground-level PM$_{2.5}$ concentrations using multiangle imaging spectroradiometer aerosol optical thickness over the contiguous united states. *J. Geophys. Res. Atmos.* **2004**, *109*, 2285–2311.

20. Boys, B.L.; Martin, R.V.; Van, D.A.; Macdonell, R.J.; Hsu, N.C.; Cooper, M.J.; Yantosca, R.M.; Lu, Z.; Streets, D.G.; Zhang, Q. Fifteen-year global time series of satellite-derived fine particulate matter. *Environ. Sci. Technol.* **2014**, *48*, 11109–11118. [CrossRef] [PubMed]

21. Lin, C.; Li, Y.; Yuan, Z.; Lau, A.K.; Li, C.; Fung, J.C. Using satellite remote sensing data to estimate the high-resolution distribution of ground-level PM 2.5. *Remote Sens. Environ.* **2015**, *156*, 117–128. [CrossRef]

22. Zhang, Y.; Li, Z. Remote sensing of atmospheric fine particulate matter (PM$_{2.5}$) mass concentration near the ground from satellite observation. *Remote Sens. Environ.* **2015**, *160*, 252–262. [CrossRef]

23. Ma, Z. Study on Spatiotemporal Distribution of PM$_{2.5}$ in China Using Satellite Remote Sensing. Ph.D. Thesis, Nanjing University, Nanjing, China, 2015.

24. Li, Z.; Zhang, Y.; Shao, J.; Li, B.; Hong, J.; Liu, D.; Li, D.; Wei, P.; Li, W.; Li, L. Remote sensing of atmospheric particulate mass of dry PM$_{2.5}$ near the ground: Method validation using ground-based measurements. *Remote Sens. Environ.* **2016**, *173*, 59–68. [CrossRef]

25. Tian, J.; Chen, D. A semi-empirical model for predicting hourly ground-level fine particulate matter (PM 2.5) concentration in southern Ontario from satellite remote sensing and ground-based meteorological measurements. *Remote Sens. Environ.* **2010**, *114*, 221–229. [CrossRef]

26. You, W.; Zang, Z.; Zhang, L.; Li, Z.; Chen, D.; Zhang, G. Estimating ground-level PM10 concentration in northwestern China using geographically weighted regression based on satellite AOD combined with Calipso and MODIS fire count. *Remote Sens. Environ.* **2015**, *168*, 276–285. [CrossRef]

27. HuHu, X.; Waller, L.A.; Al-Hamdan, M.Z.; Crosson, W.L.; Estes, M.G.; Estes, S.M.; Quattrochi, D.A.; Sarnat, J.A.; Yang, L. Estimating ground-level PM(2.5) concentrations in the southeastern U.S. Using geographically weighted regression. *Environ. Res.* **2012**, *121*, 1–10. [CrossRef] [PubMed]

28. Ma, Z.; Liu, Y.; Zhao, Q.; Liu, M.; Zhou, Y.; Bi, J. Satellite-derived high resolution PM concentrations in Yangtze river delta region of China using improved linear mixed effects model. *Atmos. Environ.* **2016**, *133*, 156–164. [CrossRef]

29. Bai, Y.; Wu, L.; Qin, K.; Zhang, Y.; Shen, Y.; Zhou, Y. A geographically and temporally weighted regression model for ground-level PM$_{2.5}$ estimation from satellite-derived 500 m resolution AOD. *Remote Sens.* **2016**, *8*, 262. [CrossRef]

30. Kloog, I.; Koutrakis, P.; Coull, B.A.; Lee, H.J.; Schwartz, J. Assessing temporally and spatially resolved PM 2.5 exposures for epidemiological studies using satellite aerosol optical depth measurements. *Atmos. Environ.* **2011**, *45*, 6267–6275. [CrossRef]

31. Hu, X. Estimation of PM$_{2.5}$ concentrations in the conterminous U.S. using MODIS data and a three-stage model. In Proceedings of the American Geophysical Union 2015 Fall Meeting, San Francisco, CA, USA, 14–18 December 2015.

32. Van Donkelaar, A.; Martin, R.V.; Spurr, R.J.; Burnett, R.T. High-resolution satellite-derived $PM_{2.5}$ from optimal estimation and geographically weighted regression over North America. *Environ. Sci. Technol.* **2015**, *49*, 10482–10491. [CrossRef] [PubMed]

33. Jiang, M.; Sun, W. Investigating meteorological and geographical effect in remote sensing retrieval of $PM_{2.5}$ concentration in Yangtze River Delta. In Proceedings of the 2016 IEEE International Geoscience and Remote Sensing Symposium, Beijing, China, 10–18 July 2016; pp. 4108–4111.

34. Fang, X.; Zou, B.; Liu, X.; Sternberg, T.; Zhai, L. Satellite-based ground PM 2.5 estimation using timely structure adaptive modeling. *Remote Sens. Environ.* **2016**, *186*, 152–163. [CrossRef]

35. Mu, Q.; Zhang, S. Assessment of the Trend of Heavy $PM_{2.5}$ Pollution Days and Economic Loss of Health Effects during 2001–2013. *Acta Sci. Nat. Univ. Pekin.* **2015**, *51*, 694–706.

36. Remer, L.A.; Kaufman, Y.J.; Tanré, D.; Mattoo, S.; Chu, D.A.; Martins, J.V.; Li, R.R.; Ichoku, C.; Levy, R.C.; Kleidman, R.G. The MODIS aerosol algorithm, products, and validation. *J. Atmos. Sci.* **2005**, *62*, 947–973. [CrossRef]

37. Cheng, Z.; Wang, S.X.; Fu, X.; Watson, J.G.; Jiang, J.K.; Fu, Q.Y.; Chen, C.H.; Xu, B.Y.; Yu, J.S.; Chow, J.C. Impact of biomass burning on haze pollution in the Yangtze River Delta, China: A case study of summer in 2011. *Atmos. Chem. Phys.* **2014**, *14*, 4573–4585. [CrossRef]

38. Xu, J.H.; Jiang, H. Esitmation of $PM_{2.5}$ concentration over the Yangtze Delta using remote sensing: analysis of spatial and temporal variations. *Environ. Sci.* **2015**, *36*, 3119–3127. (In Chinese).

39. Fotheringham, A.S.; Charlton, M.E.; Brunsdon, C. Geographically weighted regression: A natural evolution of the expansion method for spatial data analysis. *Environ. Plan. A* **1998**, *30*, 1905–1927. [CrossRef]

40. Hauke, J.; Kossowski, T. Comparison of values of Pearson's and Spearman's correlation coefficients on the same sets of data. *Quaest. Geogr.* **2015**, *30*, 87–93. [CrossRef]

41. Tu, Y.K.; Kellett, M.; Clerehugh, V.; Gilthorpe, M.S. Problems of correlations between explanatory variables in multiple regression analyses in the dental literature. *Br. Dent. J.* **2005**, *199*, 457–461. [CrossRef] [PubMed]

42. Lin, G.; Fu, J.; Jiang, D.; Hu, W.; Dong, D.; Huang, Y.; Zhao, M. Spatio-temporal variation of $PM_{2.5}$ concentrations and their relationship with geographic and socioeconomic factors in China. *Int. J. Environ. Res. Public Health* **2013**, *11*, 173–186. [CrossRef] [PubMed]

remote sensing

MDPI

Article

Deriving Hourly PM2.5 Concentrations from Himawari-8 AODs over Beijing–Tianjin–Hebei in China

Wei Wang [1], Feiyue Mao [1,2,3,*], Lin Du [4,*], Zengxin Pan [1], Wei Gong [1,3] and Shenghui Fang [2]

[1] State Key Laboratory of Information Engineering in Surveying, Mapping and Remote Sensing (LIESMARS), Wuhan University, Wuhan 430079, China; wangweicn@whu.edu.cn (W.W.); pzx@whu.edu.cn (Z.P.); weigong@whu.edu.cn (W.G.)

[2] School of Remote Sensing and Information Engineering, Wuhan University, Wuhan 430079, China; shfang@whu.edu.cn

[3] Collaborative Innovation Center for Geospatial Technology, Wuhan 430079, China

[4] Faculty of Information Engineering, China University of Geosciences, Wuhan 430074, China

* Correspondence: maofeiyue@whu.edu.cn (F.M.); dulin@cug.edu.cn (L.D.)

Received: 10 June 2017; Accepted: 16 August 2017; Published: 19 August 2017

Abstract: Monitoring fine particulate matter with diameters of less than 2.5 μm (PM2.5) is a critical endeavor in the Beijing–Tianjin–Hebei (BTH) region, which is one of the most polluted areas in China. Polar orbit satellites are limited by observation frequency, which is insufficient for understanding PM2.5 evolution. As a geostationary satellite, Himawari-8 can obtain hourly optical depths (AODs) and overcome the estimated PM2.5 concentrations with low time resolution. In this study, the evaluation of Himawari-8 AODs by comparing with Aerosol Robotic Network (AERONET) measurements showed Himawari-8 retrievals (Level 3) with a mild underestimate of about −0.06 and approximately 57% of AODs falling within the expected error established by the Moderate-resolution Imaging Spectroradiometer (MODIS) (±(0.05 + 0.15AOD)). Furthermore, the improved linear mixed-effect model was proposed to derive the surface hourly PM2.5 from Himawari-8 AODs from July 2015 to March 2017. The estimated hourly PM2.5 concentrations agreed well with the surface PM2.5 measurements with high R^2 (0.86) and low RMSE (24.5 μg/m^3). The average estimated PM2.5 in the BTH region during the study time range was about 55 μg/m^3. The estimated hourly PM2.5 concentrations ranged extensively from 35.2 ± 26.9 μg/m^3 (1600 local time) to 65.5 ± 54.6 μg/m^3 (1100 local time) at different hours.

Keywords: air pollution; geostationary satellite; Himawari-8; hourly AOD; hourly PM2.5

1. Introduction

Ambient fine particulate matter with aerodynamic diameters less than 2.5 μm (PM2.5) are associated with adverse human health effects; thus, they are regarded worldwide as a public health threat [1,2]. Given the finer size of PM2.5 compared with PM10 (aerodynamic diameters less than 10 μm), PM2.5 can be breathed deeply into the lungs and seriously damage human organs [3]. The PM2.5 concentrations in the Beijing–Tianjin–Hebei (BTH) region, which is one of the most populated and polluted regions in North China, have increased significantly in the past few decades due to rapid economic growth and industrialization, further resulting in severe events of atmospheric pollution [1,4]. However, data on PM2.5 concentrations are often sparse because monitoring activities are often conducted in urban areas due to difficulties and high costs of technical application; thus, these data hardly reflect the real effects of local meteorology, topography, and the location of emission sources [5]. Satellite measurements can offer information on aerosol optical depths (AODs) with large-scale spatial coverage and different temporal–spatial resolutions. A promising correlation exists between AOD and atmospheric particles because AOD represents the quantity of light removed from a beam via aerosol particle scattering or absorption along the optical path [1,6–8]. Thus, satellite measurements have been widely employed as a proxy to infer surface PM2.5 concentrations [2,5,9].

Previous studies proposed establishing empirical models to correlate ground-level PM2.5 and satellite-derived AOD (e.g., linear, nonlinear, and logarithmic models) [10–12]. In addition to the AOD, predictors, such as aerosol types, meteorological factors, and land use information, have been incorporated into models to improve model performance [13–15]. Advanced statistical methods, such as generalized linear regression models [12], mixed effects models [16], generalized additive models [9], geographically-weighted regression [2], and semi-empirical models [7], have been employed to represent the relationships between the ground-level PM2.5 concentration and various predictors. Xin et al. [8] demonstrated the linear relationship of daily PM2.5 with the Moderate-resolution Imaging Spectroradiometer (MODIS) AODs (R^2 = 0.57) in North China from 2009 to 2011. Ma et al. [1] explored the relationship between the mass concentration of surface PM2.5 and MODIS AODs in the BTH region, and they suggested that the relation strongly depends on the season. Xie et al. [17] developed a mixed-effect model to derive daily estimations of surface PM2.5 using a 3 km MODIS AOD in Beijing, and the model performed well in cross-validations (CVs) with R^2 of 0.75−0.79. A similar study developed linear mixed-effect (LME) models to integrate MODIS AODs, meteorological parameters, and satellite-derived tropospheric NO_2 column density to estimate daily PM2.5 concentrations over the BTH region, in which model accuracy was calculated at R^2 = 0.77 with a mean error of 22.4% [18]. Other statistical models, including multiple linear regression [19], non-linear models [19], generalized additive models [9], and geographically-weighted regression [2,20] were developed to estimate the spatial distributions of PM2.5 and reduce the estimated errors. However, when PM2.5 was estimated and these models were applied in the BTH region, two issues were noted. First, the AODs obtained from polar orbit satellites were limited by observation frequency (e.g., MODIS conducted only twice a day) [21,22], so they were insufficient for understanding PM2.5 evolution. Second, it is still necessary to continue exploring more suitable models that can reflect the relationship between AOD and PM2.5.

A geostationary satellite can overcome the estimated PM2.5 with low time resolution [23]. Himawari-8, which is operated by the Japan Meteorological Agency and was launched on 7 October 2014 (operated on 7 July 2015), is a new geostationary meteorological satellite sensor that can characterize aerosols [24]. Himawari-8 can provide AODs with 10 min intervals and 5 km coverage over about one-third of the Earth (i.e., the Western Pacific Ocean, East and Southeast Asia, and Oceania) [25,26]. However, the accuracy evaluation of Himawari-8 aerosol production is limited, and bias and error characterization is a critical step in satellite aerosol production [21,27]. Therefore, we evaluated Himawari-8 retrievals by comparing them with the Aerosol Robotic Network (AERONET) sites before the AODs were applied in estimating PM2.5. In this study, a primary estimation of hourly PM2.5 based on the Himawari-8 hourly AODs over the BTH region in China was executed from July 2015 to March 2017. An improved LME model was proposed to estimate PM2.5 concentrations in the BTH region, and the model performance was assessed by a 10-fold CV method. The spatial distributions of hourly PM2.5 concentrations were derived from the improved LME model.

2. Study Area and Datasets

2.1. Study Area

The BTH region, also known as the Jing–Jin–Ji region, is the capital region of China. As the core area of the Bohai Economic Rim, the BTH region consists of two municipalities (Beijing and Tianjin) and 11 prefecture-level cities in Hebei Province. As shown in Figure 1, the BTH region with an area of 217,127 km^2 is located in northeastern mainland China between the longitudes of 113° to 120°E and latitudes of 36° to 43°N. With a temperate continental monsoon climate, the BTH region has humid and hot summers and dry and cold winters. In 2014, the annual average temperature in the BTH region was from 3.8 °C to 15.5 °C, whereas the annual average precipitation was around 400 mm. The dense population, industrialization, congested local traffic, and coal consumption of the BTH region all contributed to its status as the most concentrated PM2.5 region in China [1,18].

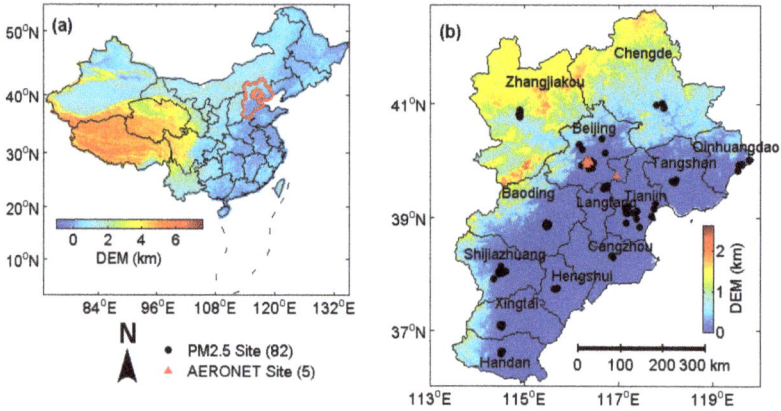

Figure 1. Elevation map of (**a**) China and (**b**) Beijing–Tianjin–Hebei region. (**b**) Spatial distributions of fine particulate matter (PM) and AERONET sites in Beijing–Tianjin–Hebei region.

2.2. Datasets

The datasets used in this study included Himawari-8 Level 3 hourly AOD data and hourly observation data of surface PM2.5 concentration in the BTH region (Figure 1). Datasets covering more than a year (from July 2015 to March 2017) were used. The center of Himawari-8 is 140.7°E over equator, and the observation area is located from 80°E to 160°W and from 60°N to 60°S [28]. Himawari-8 can provide AODs at 500 nm and Ångström exponents with 10 min intervals and 5 km coverage over about one-third of the Earth (i.e., Western Pacific Ocean, East and Southeast Asia, and Oceania) [25,26]. The AODs were subjected to quality assurance with four confidence levels, namely, "very good," "good," "marginal," and "no confidence" (or "no retrieval"). In this study, we only evaluated aerosol retrievals with the highest confidence level ("very good"). Himawari-8 hourly AODs with high quality were evaluated by comparing with AERONET measurements at level 1.5 because the accuracy reports of Himawari-8 retrievals were scarce. AERONET AODs could be used as a basis of comparison for satellite validation because their accuracy was less than 0.02 [29].

Hourly surface PM2.5 mass concentrations were obtained from the official website of the China Environmental Monitoring Center, which has been described in detail in a previous work [30]. Automated monitoring systems were installed in each site and used to measure the ambient concentration of SO_2, NO_2, O_3, CO, and PM2.5 and PM10 according to China Environmental Protection Standards. Meteorological data were obtained from reanalysis datasets (i.e., ERA-Interim) of the European Centre for Medium-Range Weather Forecasts (ECMWF). The ECMWF uses data assimilation systems and forecasting models to reanalyze observation datasets [31]. The ERA-Interim, one of the reanalysis datasets of ECMWF, offers a global atmosphere reanalysis since 1979. Meteorological data from the ERA-Interim include surface relative humidity (RH, %), and boundary layer height (BLH, m). The surface type is approximated by the Normalized Difference Vegetation Index (NDVI), which is obtained from MODIS 16-day NDVI production "CMG 0.05 Deg 16 days NDVI" in "MOD13C1/MYD13C1." An NDVI larger than 0.4 usually indicates vegetated areas, whereas a smaller value refers to soil-dominated surface in generally [27]. Additionally, the DEM covering the BTH region with a resolution of 90 m produced by the National Aeronautics and Space Administration (NASA) was downloaded from the Consortium for Spatial Information (http://srtm.csi.cgiar.org/index.asp). Detailed information of the datasets applied in this study is shown in Table 1.

Table 1. Summary of datasets applied in this study.

Dataset	Variable	Unit	Temporal Resolution	Spatial Resolution	Source
PM2.5	PM2.5	μg/m^3	1 h	Site	CEMC
AOD	Ground AOD	Unitless	~15 min	Site	AERONET
	Satellite AOD	Unitless	1 h	0.18	Himawari-8
Meteorological Factors	RH	%	6 h	0.125°	ECMWF
	BLH	m	3 h	0.125°	
Land	NDVI	Unitless	16 days	0.05°	MODIS
	DEM	m	not available	90 m	NASA

3. Method

3.1. Evaluation Method of the Himawari-8 AOD

Evaluation methods were applied as follows: (1) accuracy, which refers to the average difference between two datasets; (2) precision, which is the standard deviation of the difference; (3) uncertainty, which refers to root mean square deviation; (4) correlation coefficient (R), which refers to the correlation and dependence of the statistical relationships between two datasets; and (5) percentage of Himawari-8 AODs falling within the expected error (EE) range (±(0.05 + 0.15 AOD) over land), as established by MODIS (i.e., from the continuous validation of the MODIS aerosol team). The MODIS EE is a linear envelope line below and above the 1:1 line on a scatterplot, which can encompass at least 67% (about one standard deviation) of the collocations [27,32]. The MODIS uncertainty applied in this study can assess whether the high-quality Himawari-8 AOD can achieve the accuracy of MODIS. The spatiotemporal collocations between the Himawari-8 retrievals and AERONET AODs were consistent with those of other studies [27,33,34]. We averaged all of the Himawari-8 retrievals within the 20 km radius of an AERONET site to represent the satellite aerosol value. To obtain a representative Himawari-8 AOD around an AERONET site, the requirements are as follows: approximately 20% of the total Himawari-8 AODs within the 20 km radius circle centered on an AERONET site and at least two observations obtained from the AERONET within 30 min centered on the Himawari-8 measuring time. The threshold value of 20% can be found in the evaluation study of VIIRS (Visible Infrared Imaging Radiometer Suite) retrievals [35].

3.2. PM2.5 Estimated Model

The LME model with day-specific random effects for AOD was developed in [16], which can account for daily variations in the PM2.5-AOD relationship. The day-specific LME model has been widely applied in many studies because of its high accuracy [5,18,36]. The LME model is an extension of linear regression models for data that are collected and summarized in groups. The model describes the relationship between a response variable and independent variables, with coefficients that vary with respect to one or more grouping variables. The model consists of two parts: fixed effects and random effects [18]. Fixed-effects terms are the conventional linear regression part, and random effects are associated with individual experimental units drawn at random from a group (category). Random effects have prior normal distributions with mean 0 and constant variance, whereas fixed effects do not. The LME model can represent the covariance structure related to the grouping of data by associating the common random effects to observations that have the same level of a grouping variable [37].

Given that time-varying parameters, such as RH, PM2.5 vertical, and diurnal concentration profiles, and PM2.5 optical properties influence the PM2.5-AOD relationship, the statistical model allows for time variability in this relationship. If the spatial variability of these time-varying parameters is negligible, namely, the PM2.5-AOD relationship varies minimally spatially on a given time over the spatial scale, a quantitative relationship between PM2.5 concentrations and AOD values in their corresponding grid cells can be determined on a time basis [16]. Basic LME models were applied in previous studies [17,36]. We used the fitlme function of Matlab R2016b (MathWorks company), and the model structure is expressed by Model 1:

$$
\begin{aligned}
PM2.5_{n,m} &= [\beta_0 + b_{0,n,m}^{hour}] + [\beta_1 + b_{1,n,m}^{hour}] \times AOD_{n,m} + \beta_2 \times RH_{n,m} + \beta_3 \times BLH_{n,m} \\
&+ \beta_4 \times DEM_{n,m} + \beta_5 \times NDVI_{n,m} + \varepsilon_{n,m}; \\
&(b_{0,n,m}^{hour}, b_{1,n,m}^{hour}) \sim N[(0,0,\Sigma)], \varepsilon_{n,m} \sim N(0,\sigma^2);
\end{aligned} \tag{1}
$$

where n represents the monitoring grid index and m represents the hour (e.g., PM2.5$_{n,m}$ represents the hourly average ground-level PM2.5 measurements at time m at monitoring grid n); β_0 and $b_{0,n,m}$ are the fixed and random intercepts, respectively; β_1 and $b_{1,n,m}^{hour}$ are the fixed and hour-specific random slopes for AOD predictor, respectively; and β_2–β_5 are the fixed slopes for other predictors. Fixed effects correspond to the average effects of predictors on PM2.5 concentrations for the entire period. Random terms reflect the hour-to-hour variations in the AOD–PM2.5 relationship influenced by meteorology and satellite retrieval conditions. In addition, $\varepsilon_{n,m} \sim N(0,\sigma^2)$ represents the observation error, and Σ represents the variance–covariance matrix of the random effects.

The assumption of PM2.5-(AOD, predictors) relationships vary minimally spatially on a given day over a specific region, and neglect spatial non-stationarity in regional scales is the premise for estimating PM2.5 by Equation (1) [16,18]. Therefore, one of the limitations of the aforementioned model is that it does not consider spatial variabilities in large-region regressions, which is important for estimating geographical elements in large regions. Different cities are affected by various pollution sources, meteorological conditions, population densities, number of vehicles, and so on. All these factors influence the large-region regressions of LME models. Given that our study area was relatively large and our study period was relatively long, the relationship between PM2.5 and AOD was expected to vary in both space and time. To address both the spatial and temporal heterogeneity of the PM2.5–AOD relationship, we developed an improved LME model to fit the random (including hour- and location-specific) intercepts for the whole model and the random slopes for the AODs. We consider that the hour and location have corporate effect on the large-region regression for AOD-PM2.5 relation, which can be expressed as follows (Model 2):

$$
\begin{aligned}
PM2.5_{n,m} &= [\beta_0 + b_{0,n,m}^{hour*location}] + [\beta_1 + b_{1,n,m}^{hour*location}] \times AOD_{n,m} + \beta_2 \times RH_{n,m} \\
&+ \beta_3 \times BLH_{n,m} + \beta_4 \times DEM_{n,m} + \beta_5 \times NDVI_{n,m}; \\
&(b_{0,n,m}^{hour*location}, b_{1,n,m}^{hour*location}) \sim N[(0,0,\Sigma)], \varepsilon_{n,m} \sim N(0,\sigma^2);
\end{aligned} \tag{2}
$$

The term *hour* ∗ *location* used in the model is only the value of A times B, which represents the group-level parameters for calculating the random effects ($b_{0,n,m}^{hour*location}$ and $b_{1,n,m}^{hour*location}$). This term can be written as "*hour : location*" in Matlab. To represent a location (longitude and latitude) as a single value, we defined a complex number including location information as:

$$
Location = longitude + latitude * i \tag{3}
$$

where the real and imaginary parts of the expression correspond to longitude and latitude, respectively.

To obtain the PM2.5 estimation at a large region, $b_{0,n,m}^{hour*location}$ and $b_{1,n,m}^{hour*location}$ can be derived from the nearest location and corresponding hour training in the model. "Predict" function in Matlab can estimate predicted responses from the trained LME model at the values in the new datasets. However, the "predict" function cannot search for the nearest location for estimating PM2.5 in a new location. Therefore, we rewritten the "predict" function with the ability of search for the "nearest" PM2.5 site trained in the model. If we cannot find an appropriate random effect from a group-level in the trained model for estimating PM2.5 at a time and location, the PM2.5 in this case will be removed. As a practical technique that extends the ordinary LME model, Model 2 can examine the spatial variation at a regional scale.

In this study, the 10-fold CV was selected to compare and verify the performance of the LME and improved LME models. All of the samples were split into ten folds; that is, each fold was set approximately 10% of the total sample number. For each fold, one part was used for validation, whereas the remaining nine parts were used for training. This process was repeated for every fold. The predicted PM2.5 concentrations from all 10-fold processes were compared with the measured PM2.5 concentrations. Model performance was assessed by a determination coefficient (R^2), root mean square error (RMSE), and mean absolute error (MAE).

4. Results

4.1. Evaluation of Himawari-8 AOD

The results of the matchup comparison between Himawari-8 and AERONET are shown in Figure 2a–e, and the corresponding statistics are listed in Table 2. About 1000 instantaneous high-quality matchups of Himawari-8 and AERONET were determined for Beijing_CAMS, Beijing, Beijing_RADI, and Xianghe during the study period. The comparison of the Himawari-8 AODs against the AERONET observations showed the performances of Himawari-8 retrievals at the five sites, all of which exhibited high correlations (R^2: 0.74–0.81), low uncertainty (0.18–0.22), and a large percentage (54–59%) of retrievals falling within the EE. Himawari-8 also showed a slight underestimation with accuracy of about −0.06. The linear regression (yellow line in Figure 2) between AERONET and Himawari-8 retrievals demonstrates the slope from 0.58 to 0.65 and the positive intercept from 0.06 to 0.08. Overall, the performances of the current Himawari-8 AOD retrievals at the five sites were almost consistent with the AERONET AODs.

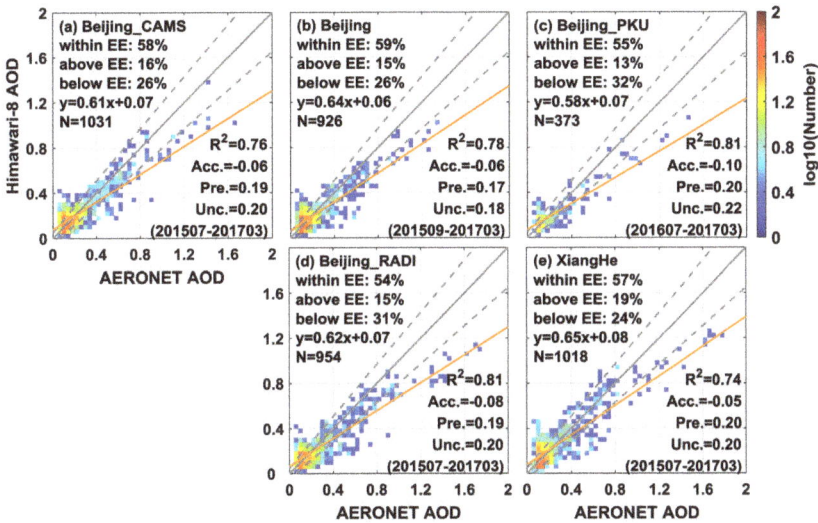

Figure 2. Collocations scatterplots of Himawari-8 and AERONET AODs at five sites of (a) Beijing_CAMS; (b) Beijing; (c) Beijing_PKU; (d) Beijing_RADI; and (e) Xianghe. The study period is from July 2015 to March 2017. The width of each pixel is 0.04 AOD, and the number of collocations falling within/above/below EE are represented in each figure. The yellow line is the regression line, the gray solid line is the 1:1 line, and the gray dashed lines are the expected errors (EE) envelopes.

Table 2. Comparative statistics of collocated Himawari-8 and AERONET AODs.

Site	N	Accuracy	Precision	Uncertainty	R^2	% Above/Within/Below EE
Beijing_CAMS	1031	−0.06	0.19	0.20	0.76	16/58/26
Beijing	926	−0.06	0.17	0.18	0.78	15/59/26
Beijing_PKU	373	−0.10	0.20	0.22	0.81	13/55/32
Beijing_RADI	954	−0.08	0.19	0.20	0.81	15/54/31
XiangHe	1018	−0.05	0.20	0.20	0.74	19/57/24

The time series of the hourly Himawari-8 AODs, AERONET measurements, and AOD bias with standard deviations (shadows) during the assessment period at the five AERONET sites are shown in Figure 3. The AODs of Himawari-8 and AERONET appeared to be coincident with each other. However, an underestimation of the Himawari-8 AOD was observed from 0900 to 1100 local time (LT) in the five AERONET sites.

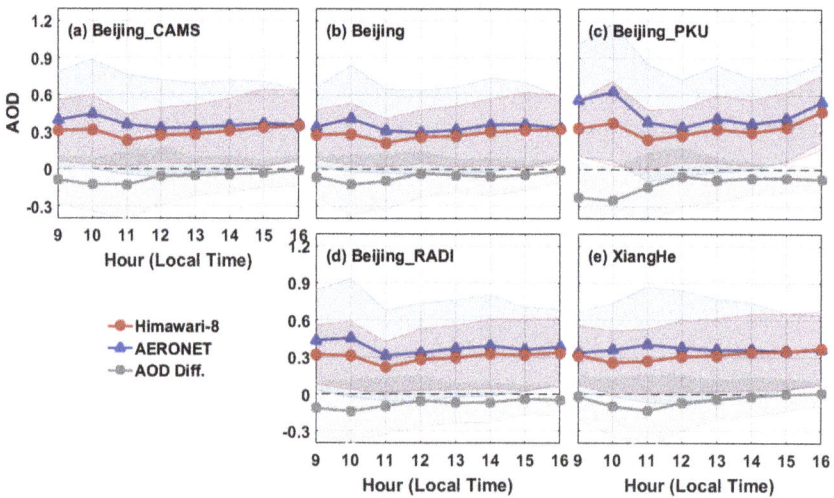

Figure 3. Time series of hourly AODs of Himawari-8 and AERONET, and hourly AOD difference between Himawari-8 and AERONET from the collocated matchups and standard deviations (shadows) over (**a**) Beijing_CAMS; (**b**) Beijing; (**c**) Beijing_PKU; (**d**) Beijing_RADI; and (**e**) Xianghe. The study period is from July 2015 to March 2017.

Figure 4 presents the spatial and hourly Himawari-8 AOD dataset at daytime from July 2015 to March 2017. The spatial distributions of the averaged AOD indicated large values in the BTH central and south regions, whereas small values emerged over the northwest region. The average AOD obtained from Himawari-8 was 0.32 ± 0.27. The average maximum AOD for the daytime was 0.38 ± 0.31 at 1500 LT. The mean AOD at 1100 LT was minimum with a mean AOD of 0.30 ± 0.26. Variations in mean AODs at the BTH region could be partially attributed to the underestimated Himawari-8 AODs from 0900 to 1200 LT (Figure 3).

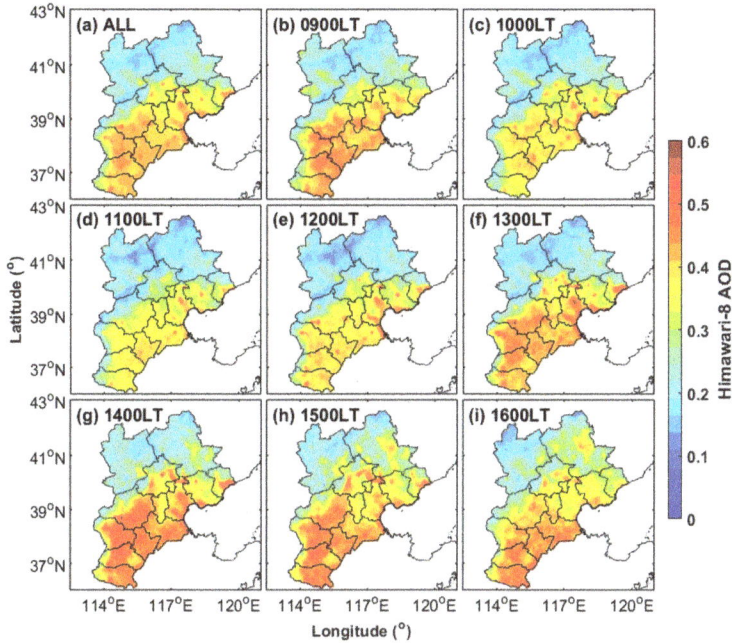

Figure 4. Spatial distributions of the averaged AOD derived from the Himawari-8 for all available data (**a**) and different hours (0900–1600 local time) (**b–i**). The study period is from July 2015 to March 2017.

4.2. Verification of Estimated PM2.5

To train the model, ground PM2.5 and these predictors require time–space consistency. Therefore, surface PM2.5 measurements should match the Himawari-8 AODs in space and time. We averaged all Himawari-8 retrievals within the 30 min and 5 km radius of a PM2.5 monitoring site to represent the satellite AOD value. To evaluate how much the AOD, meteorological, and land parameters used in the final model could improve the model performance, we fitted different models with various predictors as shown in Table 3. The Akaike information criterion (AIC) provides the relative quality of statistical models for a given dataset. The finalized LME model is generally determined based on the model performance denoted by fitting the R^2 (highest) and AIC (lowest) values [20]. The performances of models were assessed by coefficient of determination (R^2), MAEs, and RMSEs between the measured and estimated PM2.5 concentrations. The MAE was defined as (sum of absolute errors)/(the number of observations). The RMSE was defined as the square root of the mean of the squared errors.

The LME model with day-specific random effects is widely used in PM2.5 estimation [5,17]. Tests 1 and 2 using AOD as the only independent variable (the AOD-only model) showed that the LME model with hour-specific random effects exhibited better performance than that with day-specific effects. Tests 3–6 reveal that these predictors slightly improved for the model. We also fitted a model only using meteorological and land data without AOD (the non-AOD model) to determine how AOD could benefit the model performance (Tests 7 and 10). We found that AOD had an obvious positive effect on Models 1 and 2. If the intercept without random effects (Test 8), the performance is worse than Test 9 (Model 1). A comparison of Models 1 and 2 demonstrated that the improved model exhibited outstanding performance.

Table 3. Result comparison of model fitting with different predictors.

ID	Predictor(s)	Group Variable	AIC	R^2	RMSE	MAE
Test 1	AOD	Day	859,622.35	0.64	39.3	25.0
Test 2	AOD	Hour	845,625.88	0.74	33.1	20.4
Test 3	AOD; NDVI	Hour	845,589.45	0.74	33.1	20.4
Test 4	AOD; RH	Hour	845,569.14	0.74	33.1	20.4
Test 5	AOD; BLH	Hour	845,488.73	0.74	33.1	20.5
Test 6	AOD; DEM	Hour	844,323.77	0.74	32.9	20.4
Test 7	Model 1 without AOD	Hour	876,681.07	0.59	41.4	26.7
Test 8	Model 1 without $b_{0,n,m}^{hour*location}$	Hour	852,313.20	0.70	35.8	22.6
Test 9	Model 1	Hour	844,079.65	0.74	32.8	20.4
Test 10	Model 2 without AOD	Hour; location	815,754.76	0.83	26.6	15.8
Test 11	Model 2	Hour; location	809,533.48	0.93	17.1	10.1

Table 4 displays the analysis of variance (F-test and *p*-value) explained by each of the individual terms in Models 1 and 2. All *p*-values (<0.01) indicate significant effects of these predictors for the corresponding model. Beta is the coefficient of the fixed term for the two models.

Table 4. Analysis of variance (F-test and *p*-value) explained by each of the individual terms in different models. Beta is the coefficient of the fixed term for the two models.

Term	Model 1			Model 2		
	F-Test	*p*-Value	Beta	F-Test	*p*-Value	Beta
Intercept	767.6983	<0.01	39.43	2.0496×10^3	<0.01	54.44
AOD	1.2631×10^3	<0.01	96.38	4.9156×10^3	<0.01	104.34
RH	100.7208	<0.01	−0.15	38.2626	<0.01	0.11
DEM	1.2633×10^3	<0.01	−0.03	326.3786	<0.01	−0.02
BLH	216.8013	<0.01	−0.01	1.7214×10^3	<0.01	−0.02
NDVI	2.5560	0.11	−3.06	339.0167	<0.01	−28.91

Table 5 displays the estimate for the standard deviation of normal distribution for the random-effects term for intercept, AOD, and error grouped by hour for each day. Their confidence interval is small, which indicates that the random effects for intercept, AOD, and error grouped by hour for each day and $b_{1,m}^{hour*location}$ is significant.

Table 5. Standard deviations of normal distribution for the random-effects terms in the two models.

	Model 1			Model 2		
	$b_{0,m}^{hour}$	$b_{1,m}^{hour}$	$''_{n,m}$	$b_{0,m}^{hour*location}$	$b_{1,m}^{hour*location}$	$''_{n,m}$
Estimate	37.03	124.64	33.76	41.02	112.90	19.84
Lower	35.68	120.39	33.59	39.97	109.81	19.72
Upper	38.42	129.04	33.92	42.09	116.08	19.95

Previous studies on the PM2.5–AOD statistical model mainly used the site-based CV [16] and sample-based 10-fold CV methods [2]. For the site-based CV, one of the PM2.5 monitoring sites was used for validation, and the rest of the sites were used for model fitting; this process was conducted for each round of validation. In this section, the two CVs were selected to verify the performance of the proposed model. A comparison of the performance of the LME and improved LME models is presented in Table 6.

The predictive performances of the LME models during the study period were low with R^2 of 0.73 and 0.73, RMSE of 34.4 μg/m^3 and 34.5 μg/m^3, and MAE 21.7 μg/m^3 and 21.7 μg/m^3 for site-based CV and 10-fold CV, respectively. The superiority of the improved model to the LME model in estimating PM2.5 concentrations was confirmed by the site-based CVs and the 10-fold CVs, as evidenced by the RMSE and MAE values (Table 6). Site-based CVs could verify the performance of the improved LME model on location without any PM2.5 measurements. The 10-fold CV for the improved LME model presented higher R^2 (0.86 versus 0.73), lower RMSE (24.5 $\mu g/m^3$ versus 34.5 $\mu g/m^3$), and lower MAE (14.2 $\mu g/m^3$ versus 21.7 $\mu g/m^3$) compared with the ordinary LME model. Zheng et al. [18] used LME models to estimate daily PM2.5 concentrations in the BTH region with a R^2 of 0.77 in the 10-fold CV (predictor: MODIS AODs, meteorological factors, and tropospheric NO_2). Therefore, to ensure a relatively good estimation of PM2.5 concentrations, the improved LME model (Model 2) was applied in our study analysis.

Table 6. Result comparison of model fitting and cross-validation for Model 1 and 2.

Model	N	Site Cross-Validation			10-Fold Cross-Validation		
		R^2	RMSE	MAE	R^2	RMSE	MAE
Model 1	83,989	0.73	34.4	21.7	0.73	34.5	21.7
Model 2	83,989	0.87	24.1	14.0	0.86	24.5	14.2

Figure 5 presents the scatterplot of the comparison between measured and estimated PM2.5 concentrations in the BTH region at different hours. In these scatterplots, colors indicate the number of data points for a corresponding pixel. The high CV R^2 values of the improved LME model (i.e., 0.86 for all data and 0.81–0.90 for different hours) prove the acceptable performance of the model in the BTH region; that is, the model yielded reasonable predictions. However, the different CV R^2 values at different hours (e.g., maximum of 0.90 at 1500 LT; minimum of 0.81 at 0900 LT) imply that the performance of the improved LME model was higher at noon and in the afternoon than for the other hours during daytime. Figure 3 presents the underestimations of the Himawari-8 AODs from 0900 to 1100 LT at the five AERONET sites. Discrepancies in hourly results such as for the CV RMSE at 1600 LT (13.4 $\mu g/m^3$), which was smaller than for the other hours (above 20 $\mu g/m^3$), could be attributed to the relatively small number of matchups. Furthermore, the MAE in the BTH region was 14.2 $\mu g/m^3$ for all matchups, and the values ranged from 9.0 $\mu g/m^3$ to 17.4 $\mu g/m^3$ for the different hours. PM2.5 difference was equal to the estimated value minus the measured PM2.5 (Table 7).

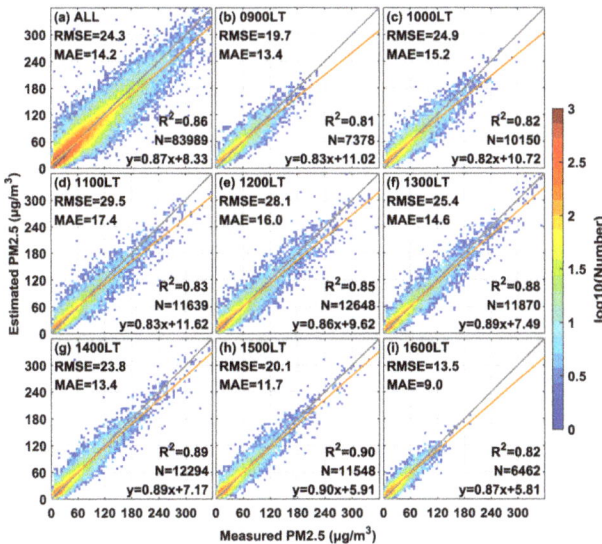

Figure 5. 10-fold cross-validation of estimated PM2.5 concentrations by comparing measured PM2.5 from all available data (**a**) and in different hours (0900–1600 local time) (**b**–**i**). The number of samplings (N), correlation coefficients (R), and linear regressions are included in the plot.

Table 7. Averages of estimated and measured PM2.5 at different hours.

Local Time	N	R^2	Estimated PM2.5	Measured PM2.5	PM2.5 Bias
ALL	83,989	0.86	61.6 ± 61.4	61.5 ± 65.8	0.1 ± 24.3
0900	7378	0.81	58.6 ± 41.5	57.6 ± 45.2	1.1 ± 19.7
1000	10,150	0.82	63.5 ± 53.2	64.5 ± 58.9	−1.0 ± 24.9
1100	11,639	0.83	71.0 ± 65.2	71.7 ± 71.8	−0.8 ± 29.5
1200	12,648	0.85	66.9 ± 67.1	67.0 ± 72.3	−0.1 ± 28.1
1300	11,870	0.88	66.7 ± 69.6	66.6 ± 73.4	0.1 ± 25.4
1400	12,294	0.89	62.9 ± 67.5	62.3 ± 71.0	0.6 ± 23.8
1500	11,548	0.90	54.8 ± 60.6	54.1 ± 63.8	0.6 ± 20.1
1600	6462	0.82	35.3 ± 30.3	34.0 ± 31.7	1.3 ± 13.4

Figure 6 shows the spatial distributions of the estimated PM2.5 errors in individual PM sites in the BTH region, which could be used to evaluate the accuracy of the improved LME model for each PM2.5 monitoring site. The red (blue) solid circles indicate that the estimated PM2.5 was overestimated (underestimated). Figure 6a shows the mean bias from all available data. Accordingly, the PM2.5 concentrations were overestimated in one of the sites in Qinghuangdao (~20 µg/m^3) and Shijiazhuang (~15 µg/m^3). Coasts with complex surfaces and aerosol types may reduce the performance of the Himawari-8 aerosol retrievals. In Qinghuangdao, the sites near the coast may result in overestimation at all hours. By contrast, sites with underestimations were observed in some parts of Tianjing. As shown in Figure 6, the rest of the sites displayed light-colored solid circles, which indicated unclear estimated biases. Despite the general consistency between estimated and measured PM2.5 concentrations in Figure 6a, site discrepancies were obvious in different hours, as shown in Figure 6b–i. In the ante meridiem (0900–1100 LT), most of the mild positive biases were observed in Beijing, Shijiazhuang, and Xingtai, whereas negative biases existed in Tianjing and Handan. The Himawari-8 AODs, which were highly accurate at noontime, might have contributed to the slight biases at 1200–1500 LT.

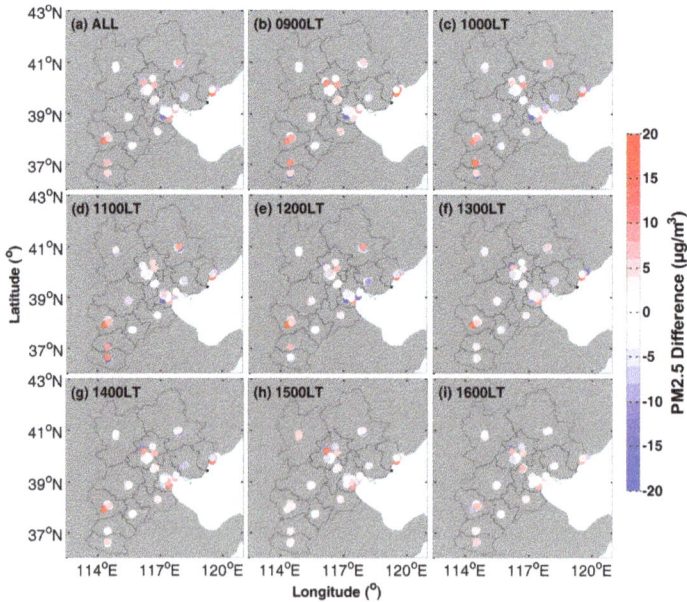

Figure 6. Differences in estimated and measured PM2.5 for individual PM monitoring sites: (**a**) all available data; (**b–i**) different hours (0900–1600 local time).

4.3. Spatial Distribution of PM2.5

The hourly spatial distributions of PM2.5 in the BTH region (Figure 7) were spatially heterogeneous, which implies the applicability of the improved LME model. Many fine-scale variations in AOD of Figure 4 show up as variations in the PM2.5 estimates in Figure 7. The consistency of spatial distribution between AOD and PM2.5 indicate the geographic correlations of AOD and PM2.5.

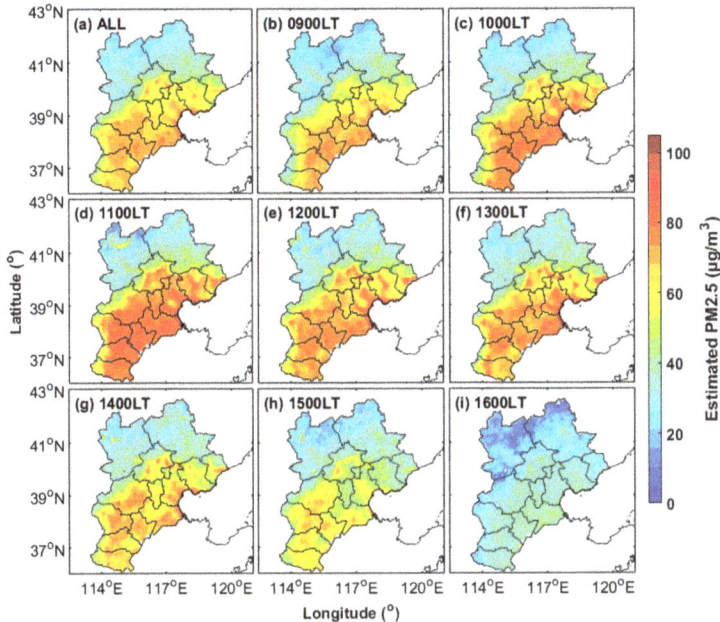

Figure 7. Spatial distribution of averaged PM2.5 estimations obtained from improved LME model for (**a**) all available data and (**b–i**) different hours (0900–1600 local time). The study period is from July 2015 to March 2017.

Urban areas with high PM2.5 concentrations, such as Beijing, Shijiazhuang, Xingtai, and Handan, were effectively captured by the improved LME model. The average PM2.5 in the BTH region was $58.2 \pm 52.7 \ \mu g/m^3$ (Table 7), which exceeded the World Health Organization Air Quality Interim Target-1 standard of $35 \ \mu g/m^3$. The average PM2.5 in southern BTH was larger than $50 \ \mu g/m^3$, which was considerably higher than those in the northern regions. Severely-polluted areas were located in Cangzhou and Hengshui, as evidenced by the large mean PM2.5 concentrations of 66.9 + 58.7 and 67.8 + 56.7 $\mu g/m^3$, respectively. Industrial production and high vehicle population contributed to high anthropogenic emissions, further resulting in a relatively high PM2.5 in southern BTH. Low PM2.5 in northern BTH (i.e., Zhangjiakou and Chengde) were observed at less than $35 \ \mu g/m^3$ on average; these areas have hilly topography and low human activities, resulting in low anthropogenic emissions. Our results were similar to that of a previous study [8], which reported an annual mean PM2.5 concentration of 45–55 $\mu g/m^3$ in the southern anthropogenic area. The satellite-derived population-weighted average of PM2.5 in Beijing was $51.2 \ \mu g/m^3$ during the study period (March 2013 to April 2014) [17]. A one-year study on the PM2.5 estimations in the BTH region using a generalized additive model presented an annual mean value of $69.4 \ \mu g/m^3$ with values ranging from $13.3 \ \mu g/m^3$ to $133.7 \ \mu g/m^3$ [9].

The spatial distributions of the hourly PM2.5 estimations differed across hours (Figure 7b–i); that is, 1100 LT was most polluted with a large mean PM2.5 of $65.5 \pm 54.6 \ \mu g/m^3$ (Table 8), as opposed to the minimum average of PM2.5 at 1600 LT. The hourly variations were consistent with the measured mean

PM2.5 values (Table 7), but these mean values were smaller because the estimated PM2.5 concentrations were averaged using datasets from urban and suburban regions. The variations in hourly Himawari-8 AODs also disagreed with the estimated PM2.5; for example, the maximum PM2.5 (65.5 ± 54.6 µg/m^3) corresponded to the minimal AOD (0.30 ± 0.26) at 1100 LT. Several factors might have influenced the variations, such as meteorological factors, which could synergistically affect PM2.5 concentrations.

Table 8. Average Himawari-8 AOD and estimated PM2.5 at different hours in the BTH region.

Local Time	Himawari-8 AOD	Estimated PM2.5 (µg/m^3)
ALL	0.32 ± 0.27	58.2 ± 52.7
0900	0.33 ± 0.25	52.6 ± 38.0
1000	0.31 ± 0.26	59.7 ± 47.6
1100	0.30 ± 0.26	65.5 ± 54.6
1200	0.32 ± 0.27	61.3 ± 56.8
1300	0.34 ± 0.28	62.2 ± 59.9
1400	0.37 ± 0.30	59.5 ± 58.1
1500	0.38 ± 0.31	52.3 ± 50.9
1600	0.37 ± 0.29	35.2 ± 26.9
MAM	0.30 ± 0.26	46.0 ± 38.4
JJA	0.38 ± 0.32	42.5 ± 29.8
SON	0.36 ± 0.31	57.8 ± 50.7
DJF	0.26 ± 0.23	71.5 ± 70.1

The spatial and seasonal images of averaged PM2.5 estimations obtained from the improved LME model are shown in Figure 8. The different seasons in Figure 8 are denoted as MAM (March, April, and May), JJA (June, July, and August), SON (September, October, and November), and DJF (December, January, and February). The void regions in northern regions in winter were due to the limitation of Himawari-8 in retrieving high-quality AOD under scarce vegetation in winter. Strong seasonality of PM2.5 concentrations was also found from the averaged PM2.5 concentration. Winter was the most polluted season with a mean estimated PM2.5 of 71.5 ± 70.1 µg/m^3, whereas summer was the cleanest season with a mean predicted concentration of 42.5 ± 29.8 µg/m^3. The mean predicted PM2.5 concentration was 46.0 ± 38.4 µg/m^3 in spring and 57.8 ± 50.7 µg/m^3 in autumn.

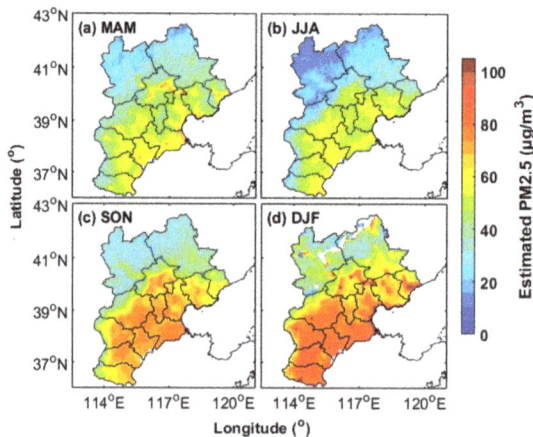

Figure 8. Spatial distribution of seasonally-averaged PM2.5 estimations obtained from the improved LME model.

5. Discussion

Given the sparse distribution of stationary PM2.5 sites, satellite data with wide spatial coverage are growing as one of the most important supplementary tools to estimate PM2.5 concentrations in a wide geographical space. The relationship between surface PM2.5 and column-integrated AOD is associated with vertical and size distribution of aerosols [38,39]. The particulate matter vertical distribution has been considered from a physics perspective, which could improve the correlation between PM2.5 and AOD [40,41]. Humidity correction for PM2.5 estimation is necessary because "dry" PM2.5 measurements after particles are heated to 50 °C may undervalue the aerosol particle mass (aerosol hygroscopicity results in AOD being affected by humidity) [42].

In this study, related factors, such as AOD, meteorological parameters, and land type, are individually integrated into the typical LME model to test their benefits on model performance. BLH and RH can be regarded as the correction factors of height and humidity, respectively [39]. Moreover, this study developed an improved LME model for satellite-based PM2.5 estimation. The improved LME model considered the spatial and temporal heterogeneity of the PM2.5–AOD relationship. As expected, the improved LME model clearly showed better performance in estimating PM2.5 concentrations from Himawari-8 AOD compared with the typical LME model in the BTH region. This result confirmed the necessity of the LME model in simultaneously considering spatial–temporal heterogeneity for PM2.5-(AOD, predictors) relationships.

The differences in performance of the improved model at 0900 and 1100 LT might be due to the underestimation of the Himawari-8 AODs at the specified times. PM2.5 underestimations predominantly occurred when the measured ground-level PM2.5 concentrations were high (i.e., greater than 80 µg/m³). Meanwhile, overestimated PM2.5 concentrations existed with slightly polluted levels, which was similar to those in the same region according to another study [18]. This result could be attributed to the nonlinear relationship between PM2.5 concentrations and AODs at different aerosol loadings [9]. Moreover, the predicted PM2.5 concentrations using the average AOD for the PM2.5 monitoring sites within the 5 km radius may not fully represent the site measurements.

For the hourly spatial distributions of PM2.5 in the BTH in Figure 7i, estimated PM2.5 at LT 16:00 showed a significant decrease. The Himawari-8 retrievals with high quality at 1600LT in winter were less than that in summer because the sun angle was probably too low for Himawari-8 retrievals at 1600LT in winter. The PM2.5 concentration in summer was lower than that in winter (Figure 8). Therefore, the limited collocations between ground PM2.5 and satellite-based AODs may result in the decrease at LT 16:00.

6. Conclusions

The spatial distributions of hourly PM2.5 concentrations are significant and necessary in understanding PM2.5 evolution. In this study, the primary estimation of hourly PM2.5 concentrations at daytime in the BTH region was executed with a proposed improved LME model using ground-based PM2.5 observations and collocated Himawari-8 (Level 3) with hourly AODs from July 2015 to March 2017.

(1) The Himawari-8 AOD with a "very good" confidence level was evaluated by comparing its values with the AERONET observations for the given study period. The Himawari-8 AODs at the five AERONET sites presented mild underestimations of about −0.06 with 57% of the AODs falling within the EE made from MODIS [±(0.05 + 0.15 AOD)].

(2) An improved LME model was developed for hourly PM2.5 estimation based on the relation between PM2.5 and AOD. The estimated PM2.5 concentrations agreed well with the surface PM2.5 measurements, as evidenced by the high R^2 (0.86) and low RMSE (24.5 µg/m³) based on 10-fold cross-validation.

(3) The average PM2.5 estimations of the improved LME model in the southern BTH were higher than those in the northern regions. The average PM2.5 concentration in the BTH region

was $58.2 \pm 52.7\ \mu g/m^3$. The estimated hourly PM2.5 concentrations ranged extensively from $35.2 \pm 26.9\ \mu g/m^3$ (1600 LT) to $65.5 \pm 54.6\ \mu g/m^3$ (1100 LT).

Future studies can focus on the improvement of the LME model to depict the spatial distributions of PM2.5 concentrations using fine spatial resolutions. The accurate derivation of surface PM2.5 concentrations from satellite retrievals largely depends on the quality of satellite aerosol products. Thus, efforts to improve the Himawari-8 AODs for hourly PM2.5 estimation or observation should be considered.

Acknowledgments: This study is supported by The National High Technology Research and Development Program of China (i.e., 863 Program, 2013AA102401), the National Key Research and Development Program of China (2016YFC0200900, 2017YFC0212600), the National Natural Science Foundation of China (41627804, 41701381), and the Program for Innovative Research Team in University of the Ministry of Education of China (IRT1278). The authors thank the AERONET principal investigators and site managers who provided the data used in the validation analysis (Huizheng Che from Chinese Academy of Meteorological Sciences, Jing Li from Peking University, Zhengqiang Li, Hong-Bin Chen, and Pucai Wang from the Chinese Academy of Sciences, etc.). The authors would also like to thank the China China Environmental Monitoring Center (CEMC: http: //106.37.208.233:20035) for providing the PM data. The authors would like to thank the National Aeronautics and Space Administration for the MODIS and DEM data.

Author Contributions: This work was conducted in collaboration with all authors. Feiyue Mao, Lin Du, Zengxin Pan, Wei Gong, Shenghui Fang and Wei Wang defined the research theme. This manuscript was written by Wei Wang. Feiyue Mao and Lin Du supervised and designed the research work. Feiyue Mao, Lin Du and Wei Wang improved the model for estimating PM2.5. Feiyue Mao and Zengxin Pan checked the experimental results. Wei Gong and Shenghui Fang provided the dataset applied in this study.

Conflicts of Interest: The authors declare no conflict of interest.

References

1. Ma, X.; Wang, J.; Yu, F.; Jia, H.; Hu, Y. Can MODIS AOD be employed to derive PM2.5 in Beijing-Tianjin-Hebei over China? *Atmos. Res.* **2016**, *181*, 250–256. [CrossRef]
2. Zhang, T.; Gong, W.; Wang, W.; Ji, Y.; Zhu, Z.; Huang, Y. Ground level PM2.5 estimates over China using satellite-based geographically weighted regression (GWR) models are improved by including NO$_2$ and enhanced vegetation index (EVI). *Int. J. Environ. Res. Public Health* **2016**, *13*, 1215. [CrossRef] [PubMed]
3. Pope, C.A. Epidemiology of fine particulate air pollution and human health: biologic mechanisms and who's at risk? *Environ. Health Perspect.* **2000**, *108*, 713–723. [CrossRef] [PubMed]
4. Zhang, Y.-L.; Cao, F. Fine particulate matter (PM2.5) in China at a city level. *Sci. Rep.* **2015**, *5*. [CrossRef] [PubMed]
5. Lee, H.J.; Chatfield, R.B.; Strawa, A.W. Enhancing the applicability of satellite remote sensing for PM2.5 estimation using MODIS deep blue AOD and land use regression in California, United States. *Environ. Sci. Technol.* **2016**, *50*, 6546–6555. [CrossRef] [PubMed]
6. Wang, W.; Gong, W.; Mao, F.; Zhang, J. Long-term measurement for low-tropospheric water vapor and aerosol by Raman lidar in Wuhan. *Atmosphere* **2015**, *6*, 521–533. [CrossRef]
7. Koelemeijer, R.B.A.; Homan, C.D.; Matthijsen, J. Comparison of spatial and temporal variations of aerosol optical thickness and particulate matter over Europe. *Atmos. Environ.* **2006**, *40*, 5304–5315. [CrossRef]
8. Xin, J.; Zhang, Q.; Wang, L.; Gong, C.; Wang, Y.; Liu, Z.; Gao, W. The empirical relationship between the PM2.5 concentration and aerosol optical depth over the background of north China from 2009 to 2011. *Atmos. Res.* **2014**, *138*, 179–188. [CrossRef]
9. Zou, B.; Chen, J.; Zhai, L.; Fang, X.; Zheng, Z. Satellite based mapping of ground PM2.5 concentration using generalized additive modeling. *Remote Sens.* **2017**, *9*, 1. [CrossRef]
10. Chu, D.A.; Kaufman, Y.J.; Zibordi, G.; Chern, J.D.; Mao, J.; Li, C.; Holben, B.N. Global monitoring of air pollution over land from the earth observing system-terra moderate resolution imaging spectroradiometer (MODIS). *J. Geophys. Res. Atmos.* **2003**, *108*. [CrossRef]
11. Gupta, P.; Christopher, S.A. Particulate matter air quality assessment using integrated surface, satellite, and meteorological products: Multiple regression approach. *J. Geophys. Res. Atmos.* **2009**, *114*. [CrossRef]
12. Liu, Y.; Sarnat, J.A.; Kilaru, V.; Jacob, D.J.; Koutrakis, P. Estimating ground-level PM2.5 in the Eastern United States using satellite remote sensing. *Environ. Sci. Technol.* **2005**, *39*, 3269–3278. [CrossRef] [PubMed]

13. Chen, Q.-X.; Yuan, Y.; Huang, X.; Jiang, Y.-Q.; Tan, H.-P. Estimation of surface-level PM2.5 concentration using aerosol optical thickness through aerosol type analysis method. *Atmos. Environ.* **2017**, *159*, 26–33. [CrossRef]

14. Guo, J.; Xia, F.; Zhang, Y.; Liu, H.; Li, J.; Lou, M.; He, J.; Yan, Y.; Wang, F.; Min, M.; et al. Impact of diurnal variability and meteorological factors on the PM2.5-AOD relationship: Implications for PM2.5 remote sensing. *Environ. Pollut.* **2017**, *221*, 94–104. [CrossRef] [PubMed]

15. Stafoggia, M.; Schwartz, J.; Badaloni, C.; Bellander, T.; Alessandrini, E.; Cattani, G.; de' Donato, F.; Gaeta, A.; Leone, G.; Lyapustin, A.; et al. Estimation of daily PM10 concentrations in Italy (2006–2012) using finely resolved satellite data, land use variables and meteorology. *Environ. Int.* **2017**, *99*, 234–244. [CrossRef] [PubMed]

16. Lee, H.J.; Liu, Y.; Coull, B.A.; Schwartz, J.; Koutrakis, P. A novel calibration approach of MODIS AOD data to predict PM2.5 concentrations. *Atmos. Chem. Phys. Discuss.* **2011**. [CrossRef]

17. Xie, Y.; Wang, Y.; Zhang, K.; Dong, W.; Lv, B.; Bai, Y. Daily estimation of ground-level PM2.5 concentrations over Beijing using 3 km resolution MODIS AOD. *Environ. Sci. Technol.* **2015**, *49*, 12280–12288. [CrossRef] [PubMed]

18. Zheng, Y.; Zhang, Q.; Liu, Y.; Geng, G.; He, K. Estimating ground-level PM2.5 concentrations over three megalopolises in China using satellite-derived aerosol optical depth measurements. *Atmos. Environ.* **2016**, *124 Pt B*, 232–242. [CrossRef]

19. Li, C.; Hsu, N.C.; Tsay, S.-C. A study on the potential applications of satellite data in air quality monitoring and forecasting. *Atmos. Environ.* **2011**, *45*, 3663–3675. [CrossRef]

20. Zou, B.; Pu, Q.; Bilal, M.; Weng, Q.; Zhai, L.; Nichol, J.E. High-resolution satellite mapping of fine particulates based on geographically weighted regression. *IEEE Geosci. Remote Sens. Lett.* **2016**, *13*, 495–499. [CrossRef]

21. Remer, L.A.; Kaufman, Y.; Tanré, D.; Mattoo, S.; Chu, D.; Martins, J.; Li, R.-R.; Ichoku, C.; Levy, R.; Kleidman, R. The MODIS aerosol algorithm, products, and validation. *J. Atmos. Sci.* **2005**, *62*, 947–973. [CrossRef]

22. Wang, W.; Mao, F.; Pan, Z.; Du, L.; Gong, W. Validation of VIIRS AOD through a comparison with a sun photometer and MODIS AODS over Wuhan. *Remote Sens.* **2017**, *9*, 403. [CrossRef]

23. Emili, E.; Popp, C.; Petitta, M.; Riffler, M.; Wunderle, S.; Zebisch, M. PM10 remote sensing from geostationary SEVIRI and polar-orbiting MODIS sensors over the complex terrain of the European Alpine region. *Remote Sens. Environ.* **2010**, *114*, 2485–2499. [CrossRef]

24. Bessho, K.; Date, K.; Hayashi, M.; Ikeda, A.; Imai, T.; Inoue, H.; Kumagai, Y.; Miyakawa, T.; Murata, H.; Ohno, T.; et al. An introduction to Himawari-8/9—Japan's new-generation geostationary meteorological satellites. *J. Meteorol. Soc. Jpn. Ser. II* **2016**, *94*, 151–183. [CrossRef]

25. Yumimoto, K.; Nagao, T.M.; Kikuchi, M.; Sekiyama, T.T.; Murakami, H.; Tanaka, T.Y.; Ogi, A.; Irie, H.; Khatri, P.; Okumura, H.; et al. Aerosol data assimilation using data from Himawari-8, a next-generation geostationary meteorological satellite. *Geophys. Res. Lett.* **2016**, *43*, 5886–5894. [CrossRef]

26. Shang, H.; Chen, L.; Letu, H.; Zhao, M.; Li, S.; Bao, S. Development of a daytime cloud and haze detection algorithm for Himawari-8 satellite measurements over central and eastern China. *J. Geophys. Res. Atmos.* **2017**. [CrossRef]

27. Liu, H.; Remer, L.A.; Huang, J.; Huang, H.-C.; Kondragunta, S.; Laszlo, I.; Oo, M.; Jackson, J.M. Preliminary evaluation of S-NPP VIIRS aerosol optical thickness. *J. Geophys. Res. Atmos.* **2014**, *119*, 3942–3962. [CrossRef]

28. Kurihara, Y.; Murakami, H.; Kachi, M. Sea surface temperature from the new Japanese geostationary meteorological Himawari-8 satellite. *Geophys. Res. Lett.* **2016**, *43*, 1234–1240. [CrossRef]

29. Dubovik, O.; Smirnov, A.; Holben, B.N.; King, M.D.; Kaufman, Y.J.; Eck, T.F.; Slutsker, I. Accuracy assessments of aerosol optical properties retrieved from aerosol robotic network (AERONET) sun and sky radiance measurements. *J. Geophys. Res. Atmos.* **2000**, *105*, 9791–9806. [CrossRef]

30. Xu, J.W.; Martin, R.V.; van Donkelaar, A.; Kim, J.; Choi, M.; Zhang, Q.; Geng, G.; Liu, Y.; Ma, Z.; Huang, L.; et al. Estimating ground-level PM$_{2.5}$ in Eastern China using aerosol optical depth determined from the goci satellite instrument. *Atmos. Chem. Phys.* **2015**, *15*, 13133–13144. [CrossRef]

31. Molteni, F.; Buizza, R.; Palmer, T.N.; Petroliagis, T. The ECMWF ensemble prediction system: Methodology and validation. *Q. J. R. Meteorol. Soc.* **1996**, *122*, 73–119. [CrossRef]

32. Levy, R.C.; Mattoo, S.; Munchak, L.A.; Remer, L.A.; Sayer, A.M.; Patadia, F.; Hsu, N.C. The collection 6 MODIS aerosol products over land and ocean. *Atmos. Meas. Tech.* **2013**, *6*, 2989–3034. [CrossRef]

33. Xiao, Q.; Zhang, H.; Choi, M.; Li, S.; Kondragunta, S.; Kim, J.; Holben, B.; Levy, R.C.; Liu, Y. Evaluation of VIIRS, GOCI, and MODIS collection 6 aod retrievals against ground sunphotometer observations over east Asia. *Atmos. Chem. Phys.* **2016**, *16*, 20709–20741. [CrossRef]

34. Nichol, J.; Bilal, M. Validation of MODIS 3 km resolution aerosol optical depth retrievals over Asia. *Remote Sens.* **2016**, *8*, 328. [CrossRef]

35. Huang, J.; Kondragunta, S.; Laszlo, I.; Liu, H.; Remer, L.A.; Zhang, H.; Superczynski, S.; Ciren, P.; Holben, B.N.; Petrenko, M. Validation and expected error estimation of Suomi-NPP VIIRS aerosol optical thickness and angström exponent with aeronet. *J. Geophys. Res. Atmos.* **2016**, *121*, 7139–7160. [CrossRef]

36. Ma, Z.; Liu, Y.; Zhao, Q.; Liu, M.; Zhou, Y.; Bi, J. Satellite-derived high resolution PM2.5 concentrations in Yangtze River delta region of China using improved linear mixed effects model. *Atmos. Environ.* **2016**, *133*, 156–164. [CrossRef]

37. Pinheiro, J.C.; Bates, D.M. *Mixed-Effects Models in S and S-PLUS*; Springer: Berlin/Heidelberg, Germany, 2004.

38. Li, S.; Joseph, E.; Min, Q. Remote sensing of ground-level PM2.5 combining AOD and backscattering profile. *Remote Sens. Environ.* **2016**, *183*, 120–128. [CrossRef]

39. Zhang, Y.; Li, Z. Remote sensing of atmospheric fine particulate matter (PM2.5) mass concentration near the ground from satellite observation. *Remote Sens. Environ.* **2015**, *160*, 252–262. [CrossRef]

40. Chu, D.A.; Ferrare, R.; Szykman, J.; Lewis, J.; Scarino, A.; Hains, J.; Burton, S.; Chen, G.; Tsai, T.; Hostetler, C.; et al. Regional characteristics of the relationship between columnar aod and surface PM2.5: Application of lidar aerosol extinction profiles over Baltimore–Washington corridor during discover-AQ. *Atmos. Environ.* **2015**, *101*, 338–349. [CrossRef]

41. Barnaba, F.; Putaud, J.P.; Gruening, C.; dell'Acqua, A.; Dos Santos, S. Annual cycle in co-located in situ, total-column, and height-resolved aerosol observations in the Po Valley (Italy): Implications for ground-level particulate matter mass concentration estimation from remote sensing. *J. Geophys. Res. Atmos.* **2010**, *115*. [CrossRef]

42. Song, W.; Jia, H.; Huang, J.; Zhang, Y. A satellite-based geographically weighted regression model for regional PM2.5 estimation over the Pearl river delta region in China. *Remote Sens. Environ.* **2014**, *154*, 1–7. [CrossRef]

remote sensing

MDPI

Article

Study of PBLH and Its Correlation with Particulate Matter from One-Year Observation over Nanjing, Southeast China

Yawei Qu [1], Yong Han [1,2,*], Yonghua Wu [3], Peng Gao [1] and Tijian Wang [1]

[1] School of Atmospheric Sciences, Nanjing University, Nanjing 210023, China; yawei_qu531@163.com (Y.Q.); 13813373839@163.com (P.G.); tjwang@nju.edu.cn (T.W.)
[2] School of Atmospheric Sciences, Sun Yat-sen University, Guangzhou 510275, China
[3] NOAA-CREST, City College of the City University of New York, New York, NY 10031, USA; yhwu@ccny.cuny.edu
* Correspondence: HanYong@nju.edu.cn; Tel.: +86-25-89681159

Received: 29 April 2017; Accepted: 23 June 2017; Published: 28 June 2017

Abstract: The Planetary Boundary Layer Height (PBLH) plays an important role in the formation and development of air pollution events. Particulate Matter is one of major pollutants in China. Here, we present the characteristics of PBLH through three-methods of Lidar data inversion and show the correlation between the PBLH and the $PM_{2.5}$ ($PM_{2.5}$ with the diameter <2.5 μm) in the period of December 2015 through November 2016, over Nanjing, in southeast China. We applied gradient method (GRA), standard deviation method (STD) and wavelet covariance transform method (WCT) to calculate the PBLH. The results show that WCT is the most stable method which is less sensitive to the signal noise. We find that the PBLH shows typical seasonal variation trend with maximum in summer and minimum in winter, respectively. The yearly averaged PBLH in the diurnal cycle show the minimum of 570 m at 08:00 and the maximum of 1089 m at 15:00 Beijing time. Furthermore, we investigate the relationship of the PBLH and $PM_{2.5}$ concentration under different particulate pollution conditions. The correlation coefficient is about −0.70, which is negative correlation. The average PBLH are 718 m and 1210 m when the $PM_{2.5} > 75$ μg/m^3 and the $PM_{2.5} < 35$ μg/m^3 in daytime, respectively. The low PBLH often occurs with condition of the low wind speed and high relative humidity, which will lead to high $PM_{2.5}$ concentration and the low visibility. On the other hand, the stability of PBL is enhanced by high PM concentration and low visibility.

Keywords: planetary boundary layer; $PM_{2.5}$; air pollution; Lidar

1. Introduction

In recent years, with the acceleration of urbanization and industrialization, air pollution is becoming more and more serious in China [1,2]. Fine Particulate Matter (PM) has become one of major pollutants because they can be inhaled into human body by respiration, resulting in various respiratory and cardiovascular disease [2]. Meanwhile, they can directly scatter and absorb solar radiance and indirectly modify cloud properties [3–6], thus, play an important role in Earth's energy budget, climate change and atmospheric environment. The PM concentration level also affects the stability of planetary boundary layer (PBL) [7,8]. Different numerical models and measurements have been applied to investigate the radiative forcing of nitrate [9], sulfate [10,11], and carbonaceous aerosols [12,13], as well as their mixtures [14–17] over East Asia. These studies demonstrated that aerosol particles can reduce the solar radiation reaching ground and augment the planetary albedo. The negative radiative forcing and cooling effects of aerosols in lower PBL and ground can suppress the development of PBL.

On the other hand, the concentration of aerosols is strongly affected by the meteorological conditions [18–20] and the PBLH plays an important role [21]. PBL is a strongly turbulent layer between earth's surface and free troposphere. The PBLH can weaken the exchange between boundary layer and free troposphere, because weak turbulence will occur in the bulk of the atmosphere due to the stable stratification between different layers [22]. Air pollutants released from non-buoyant ground sources, including aerosols, dust and other gaseous pollutants, are restricted within the boundary layer [23]. Therefore, the dispersion and transport of lower tropospheric particles mainly depend on the PBLH [24,25].

Consequently, the determination of PBLH is important to evaluate air-pollution events. The PBLH can be calculated from remote sensing observations methods, including satellite [26,27], wind profiler [28], ceilometer [29–31], sodar [32,33] and ground-based Lidar [34–37]. Lidar can provide continuous measurements with highly temporal-spatial resolution, and the continuous automatic inversion of PBLH from Lidar data is more feasible. Several methods have been employed to calculate the PBL height by using Lidar data, such as the gradient method [38–41], standard deviation method [42,43], wavelet analyses [44–46], and idealized profiles method [47,48]. Due to the big variation of aerosol concentration in boundary layer and free troposphere, the fundamental principle of these methods is to extract the height where the largest Lidar signal variance (i.e., strongest decrease of the backscatter signal) appears. However, each method has its own limitations (e.g., susceptible to noise and stratified aerosol structures). To our best knowledge, the study of multi-methods estimate of PBLH and the correlation between PBLH and $PM_{2.5}$ in Yangtze River Delta (YRD) is insufficient. In particular, the research on the $PM_{2.5}$–PBLH interaction helps better understand air pollution process and mechanism; this becomes very important for the severe haze episodes in the urban cities of China [8,20,23].

Here, we present a study of the PBLH variations and the correlation between PBLH and $PM_{2.5}$ by inversing PBLH through three different ways. The Lidar data were collected during December 2015–November 2016 in Nanjing, one of the megacities in YRD, China. Section 2 introduces the observation settings and the inversion methods, including the gradient method, standard deviation method and wavelet covariance transform method. In Section 3, we compare the PBLH calculated by different methods, show the characteristics of seasonal and diurnal PBLH variations, and further discuss the relationship between PBLH and $PM_{2.5}$ through statistics on one-year data and a case study. Finally, the conclusion and perspective are given in Section 4.

2. Materials and Methods

2.1. Observation

The LIDAR backscatter signal profile and the inversion for PBLH were carried out during December 2015–November 2016 in Nanjing, west part of Yangtze River Delta, China. A Raman Lidar system (LR112-D400) manufactured by Raymetrics of Greece was used at Atmospheric Parameters Vertical Detection Site (APVDS) in Nanjing University Xianlin Campus (32.12°N, 118.95°E). The Lidar system is based on a pulsed Nd:YAG laser, which transmits short pulses at 355 nm with a 10 Hz repetition rate and the maximum output energy of 85 mJ. The optical receiver is a Cassegrain telescope with 400 mm diameter and a field of view of 1.75 mrad. Four receiving channels are used to collect elastic scattering and polarization signals (355 parallel and 355 perpendicular channels) and Nitrogen (N_2) Raman scattering signals at 387 nm and water vapor Raman-scattering at 408 nm, respectively. The maximum detection height and minimum vertical resolution are 18 km and 7.5 m, respectively. The Lidar overlap area is around 255 m. The Lidar system worked in the rainless daytime during the one-year period. Due to the limitation of weather conditions and lack of operators, 63 days of effective samples were collected. The observational data covers 10 months of four seasons in Nanjing, including winter (December 2015–January 2016–February 2016), spring (March 2016–April 2016–May 2016), summer (June 2016–August 2016) and autumn (September 2016–November 2016), respectively. Lidar profiles obtained in this study are averaged over 4 min, which matches the typical time scale of atmospheric turbulence within the boundary layer [22].

The PM$_{2.5}$ concentration and visibility were measured at Xianlin Ambient Air Quality Monitoring Site (XAAQMS), which is located on the Xianlin Campus of Nanjing Normal University (32.11°N, 118.92°E) and only 4 km away from the APVDS. A continuous ambient particulate monitor (Thermo TEOM-1405) was used for the PM$_{2.5}$ measurement and the hourly-average data were collected. Meteorological parameters including temperature, relative humidity and wind speed over the same period were provided by the National Meteorological Station of Nanjing (NMSN ID: 58238, 32.00°N, 118.80°E). The location of APVDS, XAAQMS, and NMSN are shown in the map of Nanjing in Figure 1.

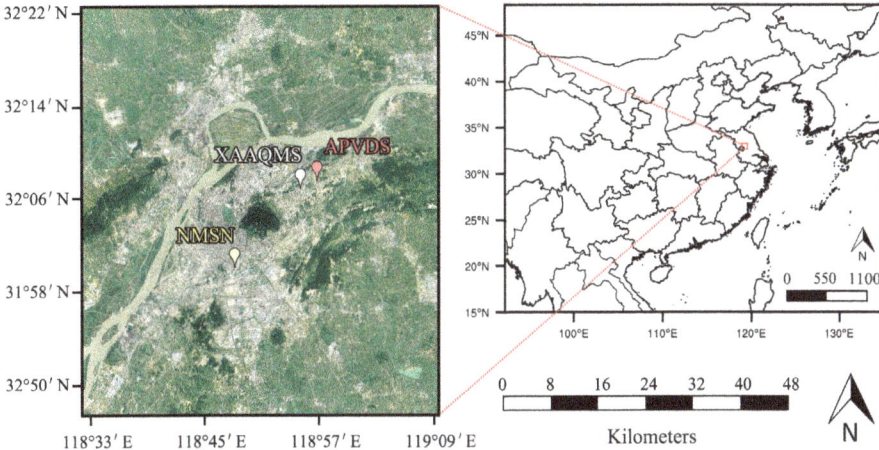

Figure 1. Map of Nanjing, China and the location of observation sites including APVDS (32.12°N, 118.95°E), XAAQMS (32.11°N, 118.92°E), and NMSN (32.00°N, 118.80°E).

2.2. Inversion of Backscatter Coefficient by Lidar

The Lidar equation [35] can be expressed as:

$$P(R) = P_0 \frac{c\tau}{2} An \frac{O(R)}{R^2} \beta(R) exp\left[-2\int_0^R \alpha(r)dr\right], \tag{1}$$

where *P(R)* is the power received from a distance *R* and P_0 is the average power of a single laser pulse. τ is the temporal pulse length, *A* is the area of primary receiver optics, *n* is the overall system efficiency and *O(R)* is the overlap function. $\beta(R)$ and $\alpha(r)$ represent backscatter coefficient and extinction coefficient, respectively.

The Lidar equation in this paper is solved by using the Klett–Fernard method [49]. Backscatter coefficient can be calculated through following equation:

$$\beta(R) = \frac{RCS(R)\cdot exp[2(L-L_{mol})\int_R^{R_{ref}}\beta_{mol}(r)dr]}{\frac{RCS(R_{ref})}{C\beta_{mol}(R_{ref})} + 2L\int_R^{R_{ref}} RCS(r\prime)\cdot exp[2(L-L_{mol})\int_R^{R_{ref}}\beta_{mol}(r\prime\prime)dr\prime\prime]dr\prime}, \tag{2}$$

where $C = \frac{\beta_{mol}(R_{ref})+\beta_{aer}(R_{ref})}{\beta_{mol}(r)}$, and β_{mol} and β_{aer} are backscatter coefficient of air molecules and aerosol, respectively. Reference point R_{ref} represents the clean atmosphere where $\beta_{aer}(R_{ref}) = 0$ and $C = 1$. The Range Corrected Signal (RCS) is defined as $RCS(R) = P(R)R^2$. $L = \frac{\alpha_{aer}(R)}{\beta_{aer}(R)}$ is the aerosol Lidar ratio and $L_{mol} = \frac{\alpha_{mol}(R)}{\beta_{mol}(R)}$.

2.3. Inversion of Planetary Boundary Layer Height

Aerosol is generally more abundant within the boundary layer than the upper atmosphere, thus for Lidar systems, the backscattered Lidar signals (e.g., RCS) within the PBL are much higher than that in the free troposphere. Based on this fact, several methods have been employed to determine the PBL height from Lidar data. In this study, we choose the GRA, WCT and STD method to estimate the PBLH where RCS abruptly decreases. The inversion methods are shown as follows:

The GRA method defines the position of the largest negative signal derivative ($D(z)$) as the instantaneous top of PBL or PBLH [37–41]. The $D(z)$ can be expressed as follows:

$$D(z) = \frac{dRCS}{dz}, \tag{3}$$

In the STD method, the PBLH is defined as the height of the maximum of Lidar signal variance [42,43]. The variance peaks of standard deviation (σ) are calculated from the variation in height of RCS, as follow:

$$\sigma = [\frac{1}{N} \sum_{i=1,N} (RCS_i - \overline{RCS})^2]^{\frac{1}{2}}, \tag{4}$$

In WCT method [37,44–46], the conversion covariance function $W_f(a, b)$ is defined as:

$$W_f(a, b) = \frac{1}{a} \int_{z_b}^{z_t} RCS(z)h(\frac{z - b}{a})dz, \tag{5}$$

where z is the height, z_t and z_b are the upper and lower limits of RCS profiles, a is the spatial dilation of the function, and b is the translation of the Haar function, i.e., the central position of the Haar function. The Haar function (h) is defined as follows:

$$h\left(\frac{z - b}{a}\right) = \begin{cases} +1, b - \frac{a}{2} \leq z < b \\ -1, b \leq z < b + \frac{a}{2} \\ 0, elsewhere \end{cases}. \tag{6}$$

In this paper, z_b is set as 255 m where the Lidar starts to collect full backscatter signals due to the limitation of the geometric overlap function. z_t is set as 2500 m to save the computing time and cloud contamination; this will not cut off the true maximum PBLH because we first visually see aerosol distribution gradient from the Lidar images. As shown in supplementary Figure S1, both the PBLH results from lidar and radiosonde agree well. The spatial extent (a) of the function is 150 m. The WCT method evaluates the similarity between RCS and Haar function. The abrupt change in RCS will occur at the height where $W_f(a, b)$ reaches the maxima, and the PBLH can be determined accordingly.

3. Results and Discussion

3.1. Comparison between Three PBLH Calculation Methods

We applied three methods above to calculate the PBLH. Figure 2 presents the results of Lidar RCS profiles and PBLHs, as well as their daily variation on 17 January 2016. At first, Figure 2a illustrates three PBLHs at 11:50 when cloud covers the Lidar detective region. The PBLH inversion is sensitive to the backscatter signal of boundary layer clouds [40]. The GRA method determines the PBLH at 918 m at the lower layer of cloud. The other two methods locate the PBLH at the upper layer of cloud with the value of 1049 m and 1076 m for STD and WCT method, respectively. Figure 2b,c compares the variation of PBLH calculation within 8 min (17:32–17:40). PBLHs derived from the GRA and STD method change abruptly from 1215 m to 1025 m in such a short period, while the RCS profile and the PBLH derived from WCT method are almost unchanged. The nearly 200 m difference of PBLHs from GRA and STD methods may be contributing to the signal noise. The GRA and STD methods are more mutable and more sensitive to noise when comparing with WCT method, which can also be observed in the diurnal variation in Figure 2d. Therefore, the WCT is the most stable method in PBLH determination. However, these methods are all able to show the variation of PBLH, which can be expressed as increasing in the morning and noon reaching, the maximum in the afternoon and decreasing after sunset. The average value of PBLH on 17 January 2016 is 1403 ± 156 m, and the PBLH from the above three methods are 1559 m, 1373 m and 1278 m for GRA, STD and WCT, respectively.

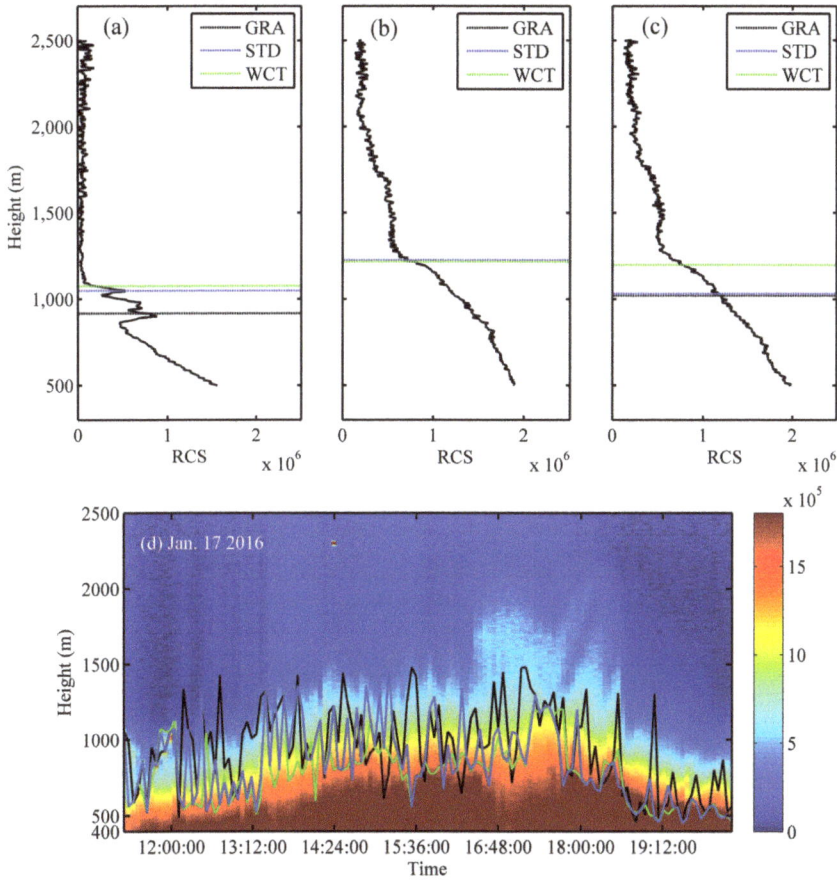

Figure 2. The RCS profile and PBLH at: 11:50 (**a**); 17:32 (**b**); and 17:40 (**c**) on 17 January 2016 and the diurnal variation of RCS and PBLH calculated by three methods (**d**) on 17 January 2016. The black, blue and green lines represent the PBLH calculated by GRA, STD and WCT, respectively.

3.2. PBL Statistical Characteristics

The PBLH generally shows seasonal and diurnal changes because of the variation in solar radiation, wind speed, atmospheric stability, etc. The following paragraphs will identify and discuss the statistical characteristics of PBLH over Nanjing.

The box plot in Figure 3 shows and compares statistical characteristics of PBLH seasonal variation, which is calculated by the GRA, STD and WCT methods and the average values of three methods. The figure reveals an annual variability of PBLH between 300 and 2433 m. Annual average boundary layer height is about 992 m, which is in good agreement with the experimental data in China [36,50] and is lower than the observational results from European countries [35,51]. The PBLH reaches lowest in winter with 822 m on average, and has its highest value 1351 m in summer. The PBLH values in spring and autumn are similar with the value of 1051 m and 1096 m, respectively. Schneider et al. [51] and Matthias et al. [52] also found the annual cycle with a maximum in summer and a minimum in winter. The possible explanation can be given as the higher solar radiation and heat flux in summer lead to stronger surface heating and then stronger turbulence and convection [51].

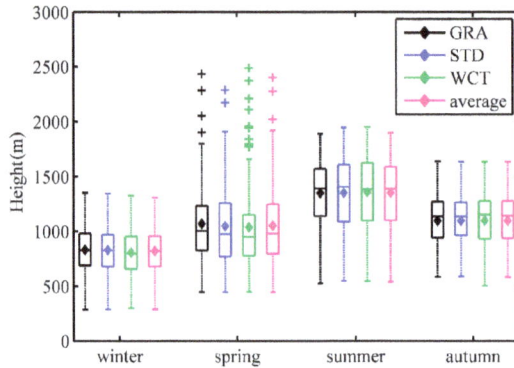

Figure 3. The seasonal variation of PBLH from the GRA, STD, and WCT methods, and their average of three above methods during one-year observation over Nanjing, China. The bottom and top of the box are the first and third quartiles, and the band inside the box is the median and the diamond is the average. The whisker is the lowest (highest) datum within 1.5 interquartile ranges (IQR) of the lower (upper) quartile, and data not included between the whiskers are plotted as an outlier with a plus.

The whiskers and outliers in Figure 3 show the variability in the PBLH in different seasons. It is clear that the PBLHs in winter are the most stable and have the minimum PBLH standard deviation (209 m). Though the maximum seasonal average PBLH is found in summer, the greatest variability of PBLH is found in spring with the standard deviation of 380 m. Spring is the only season with some outliers of PBLH and the maximum PBLH reaches 2433 m. Kamp et al. [53] found that the mean diurnal trend of PBLH in spring did not differ greatly from summer on clear days, while with the boundary-layer clouds the PBLH can be higher in spring than the one in summer. Thus, the variability in spring may be due to the existence of boundary-layer clouds. Considering the three different methods in seasonal PBLH inversion, the GRA method overestimates in winter and spring, and underestimates in summer and autumn, while WCT method shows an opposite trend with GRA in Figure 3. All three methods can reveal the characteristics of PBLH in different seasons.

Figure 4 depicts the hourly average PBLH and three-method average value of the PBL height and the related standard deviation during the daytimes (08:00–20:00). The diurnal cycle shows similar pattern in different seasons, which is generally minimum in the morning (08:00) and maximum in the afternoon. For annual average diurnal variation, the PBLH is 570 m at 08:00 and rises to a peak of 1089 m at 15:00. From 16:00 to 20:00, the annual average PBLH remains relatively stable and shows only a little lower after sunset, and finally decrease to 998 m at 20:00. The pattern in winter and spring are most similar to the annual cycle. The PBLH is kept at high level at 14:00 and 18:00 in summer, which leads to a two-peak pattern in this season. Strawbridge et al. [54] also observed the PBL peak at around 18:00–20:00 by using a Rapid Acquisition Scanning Aerosol Lidar (RASCAL) in August, 2001 in the Lower Fraser Valley (LFV) of British Columbia. The PBLH shows the greatest diurnal variations in summer and lowest in winter, which coincides with the results from Figure 3. The maximum among the year occurs at 15:00 in summer at 1554 m and the minimum occurs at 08:00 in winter at 552 m, respectively.

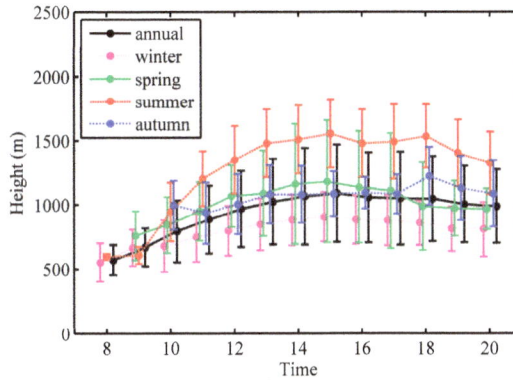

Figure 4. Diurnal variation of PBLH (three-method average) in different season observation in Nanjing, the error bar represents the standard deviation of each hour.

The error bar in Figure 4 represents the standard deviation of each hour. Therefore, larger error bars indicate more variability in PBLH, which is related to the more active convection within boundary layer. The standard deviation shows the greatest change in spring. The least stable PBLH can be found from 14:00 to 17:00 in spring, and the standard deviation is larger than 430 m. However, in the morning and evening of spring, the standard deviation is as low as 150 m. The diurnal variation of standard deviation resembles average PBLH; the higher the PBLH is, the larger the standard deviation is. This is possibly due to the strong turbulence and the weak stability of PBL [22]. The standard deviation is smaller than 210 m in winter's diurnal cycle, as the PBLH among four seasons, and winter PBL is the steadiest over the observation period.

3.3. PBLH Variation Properties under Different Particulate Pollution Conditions

In order to study the PBLH variation under different particulate pollution conditions, we classify the $PM_{2.5}$ concentration according to the new NAAQS of China. The air pollution level or category is often classified according to the $PM_{2.5}$ concentration. The NAAQS in China sets $PM_{2.5}$ concentration limits for 24-hour average with 35 $\mu g/m^3$ for Grade I and 75 $\mu g/m^3$ for Grade II [55]. Thus, in this study, three particulate pollution conditions are classified by the $PM_{2.5}$ concentrations levels as follows: (1) *good* condition, with $PM_{2.5}$ concentration less than 35 $\mu g/m^3$; (2) *slightly polluted*, with $PM_{2.5}$ concentration betwee 35 $\mu g/m^3$ and 75 $\mu g/m^3$; and (3) *polluted*, with $PM_{2.5}$ concentration exceeding 75 $\mu g/m^3$. We assume that the human-made emissions very little in Nanjing under the same season or month. We also exclude the data when the significant variations of weather or climate occur.

Figure 5 compares the PBLH calculated by the GRA, STD and WCT methods and the average of 3 methods under *the good*, *slightly polluted*, and *polluted* conditions. PBLH is relatively lower in *polluted* condition than that in *good* condition, and the average PBLH is 718 m and 1210 m, respectively. In *slightly polluted* days, the height of PBL is moderate with the value of 1027 m. Very high daytime average PBLH values can appear under *good* condition, however, the lowest value occurs under *slightly polluted* condition, though exceeds the 1.5 IQR. Moreover, the very low values of daytime average PBLH within the 1.5 interquartile ranges (IQR) are mostly limited to *the polluted* condition. The value of PBLH shows greater variability under the conditions with higher PBLH, while the value of PBLH is less variable when the PBLH is low. The standard deviation of PBLH under *the good* condition is 334 m, which is almost 3 times of the PBLH standard deviation under *the polluted* condition (106 m). Deng et al. [46] performed PBLH detection during a severe haze process in November 2009 in Guangzhou, China, and found that PBLH exceeded 1 km during the cleaning process and only 500 m during the severe haze, which agree well with our results. The explanation can be associated with the enhanced stability of urban boundary layer when the particulate matter concentration is high. Particles reduce the incoming solar radiation and lower the surface heating, leading to lower turbulent mixing and lower PBLH. PBLH will determines the level up to which the surface emissions are distributed, thus the shallow PBL further facilitate the particulate matter accumulation [23].

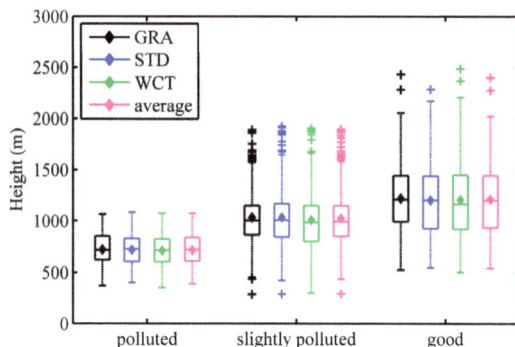

Figure 5. Average PBL height calculated by GRA, STD, and WCT methods under *good*, *slightly polluted* and *polluted* particulate pollution conditions, respectively.

Figure 6 displays the diurnal cycles of $PM_{2.5}$ concentration and PBLH from 08:00 to 20:00 under different particulate pollution conditions and their correlation. $PM_{2.5}$ concentrations are the lowest in *the good* condition, and the diurnal variation is not obvious and most of the hourly average $PM_{2.5}$ is around 23 $\mu g/m^3$. In *slightly polluted* and *polluted* conditions, the $PM_{2.5}$ level is high in both morning and evening. For *slightly polluted*, the maximum concentration of $PM_{2.5}$ is 61.43 $\mu g/m^3$, which appears at 09:00, while the greatest value of $PM_{2.5}$ in *polluted* condition occurs at 20:00 with 119.23 $\mu g/m^3$.

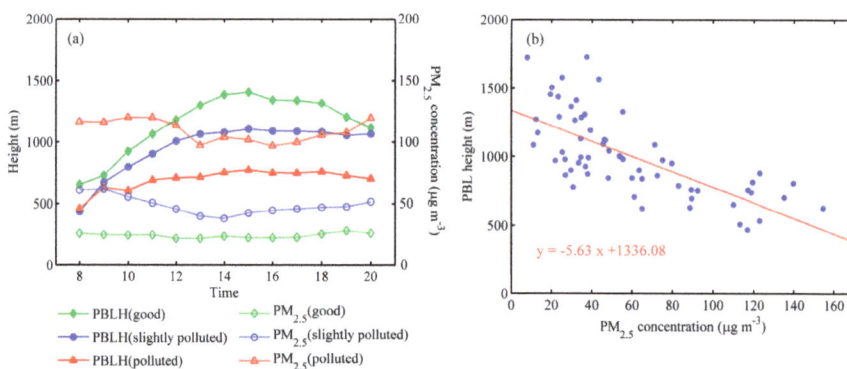

Figure 6. Diurnal variation of PBLH (three-method average) and $PM_{2.5}$ concentration under different particulate pollution conditions (**a**); and the correlation between daily average PBLH and $PM_{2.5}$ concentration (**b**). The correlation coefficient is −0.70, and the number of points is 63. The correlation is significant at the 0.01 level.

Generally, PBLH shows the opposite trend with the $PM_{2.5}$. The three conditions have comparable PBLH during the period 08:00–10:00 as well as at 20:00 of Beijing time. However, the PBLH varies a lot among different conditions in the afternoon. In *the polluted* condition, the PBLH rises from 459 m to 688 m in the first four hours, and then remains steady at around 730 m from 12:00. The PBLHs under *the good* condition exhibits the most apparent diurnal variation. It is located at 651 m at 08:00, and then significant increase to 1405 m at 15:00. It is remains steady from 16:00 to 18:00 and then drops to 1114 m at 20:00. The general increasing trend from 08:00 to 09:00 under *the slightly polluted* condition is similar to the one under *the polluted* condition, and the PBLH remains at this level untill 20:00.

Figure 6a shows the diurnal variations of PBLH and $PM_{2.5}$ under the different pollution conditions, which indicate a negative correlation between the PBLH and $PM_{2.5}$ concentrations. Figure 6b further compares their relationship with the daily averaged PBL height and $PM_{2.5}$ concentration. The correlation coefficient is up to -0.70 and significant at the 0.01 level, which means a strong anti-correlation between the PBLH and $PM_{2.5}$. This anti-correlation can be associated with two interaction ways. On the one hand, particulate matter can change the extinction capacity of atmosphere. The increasing concentration of atmospheric particulate matter (especially fine particles) weakens the solar radiation that can reach ground. Therefore, the turbulent kinetic energy in the air close to the ground decreases and the mixing of air is not strong enough to form a higher boundary layer [56]. One the other hand, when the PBLH is relatively low, the decreasing turbulence intensity in the PBL is not conducive to the diffusion of pollutants. Thus, the $PM_{2.5}$ concentration will accumulate within the PBL. In addition, we note that the effects from the cloud, seasons and extreme weather processes can also affect the PBLH [36,53,57]. At the same time, we analyzed a case in summer, the negative correlation between the PBLH and $PM_{2.5}$ concentration are shown in supplementary Figures S2 and S3. In order to rule out the impact of these factors, we performed a case study in four consecutive days without cloud cover in Section 3.4.

3.4. Case Study

We further explore the relationship between the PBLH and $PM_{2.5}$ concentration in a selected period, which includes the development and dismiss of a particle pollution case for a continuous 4-days long. Figure 7 displays the hourly averaged PBLH, $PM_{2.5}$ concentration, wind speed, visibility, temperature and relative humidity from 14 to 17 December 2015. $PM_{2.5}$ concentration drops dramatically from 200 $\mu g/m^3$ to 30 $\mu g/m^3$ in these days. In the meantime, PBLH correlates negatively with $PM_{2.5}$ and increases from around 620 m on 14 December to 1020 m on 17 December. The transition of PBLH and boundary layer structure between 15 and 16 December can also be observed from Figure 8. On 14 and 15 December, the PBLH is relatively low with the daily average value of 627 m and 699 m, respectively. Accompanied with the low PBLH, the average wind speed is only 2.5 m/s. The shallow boundary layer and weak wind impose restrictions on the diffusion of pollutants as well as water vapor. With a high relative humidity as 71.79 % on average, aerosols are more likely to accumulate through hygroscopic formation and increase $PM_{2.5}$ concentration [58]. Thus, more particle formation and less air diffusion lead to high $PM_{2.5}$ level, which are 137.58 $\mu g/m^3$ and 155.91 $\mu g/m^3$ on 14 and 15 December, respectively. On the first two polluted days, the visibility is as low as 6 km. The strong atmospheric extinction ability due to high level $PM_{2.5}$ results in the low visibility [57]. Therefore, the radiation will be impaired through the high particulate matter loading and further against the development of boundary layer. On 16 and 17 December, the development of PBL encourages the dispersion of particulate matter. High PBLH can be observed as 1029 m on 17 December, whereas $PM_{2.5}$ decreases to 37.93 $\mu g/m^3$ on the last two days accompanied with higher wind speed and lower relative humidity. With dry clean air on 16 and 17 December, the visibility grows up to 40 km, which represents an almost six-fold increase of visibility. Growing visibility indicates the weaken extinction effect of particulate matter, and will in turn facilitate the PBL development.

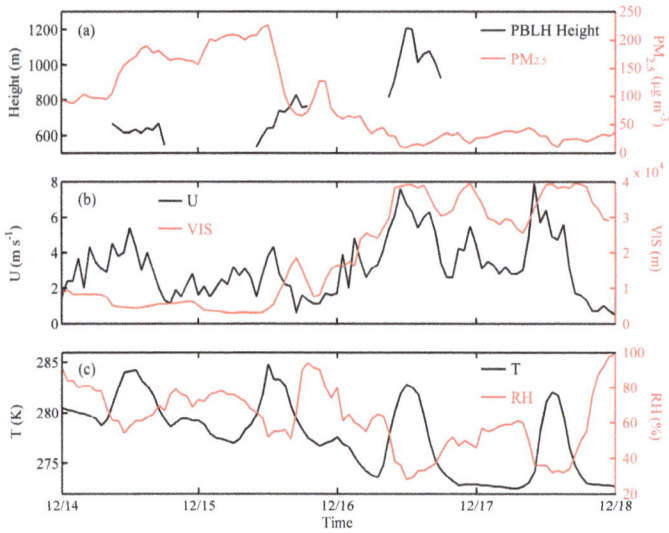

Figure 7. Time series of: (**a**) PBLH and PM$_{2.5}$ concentration; (**b**) surface wind speed (U) and visibility (VIS); and (**c**) temperature (T) and relative humidity (RH).

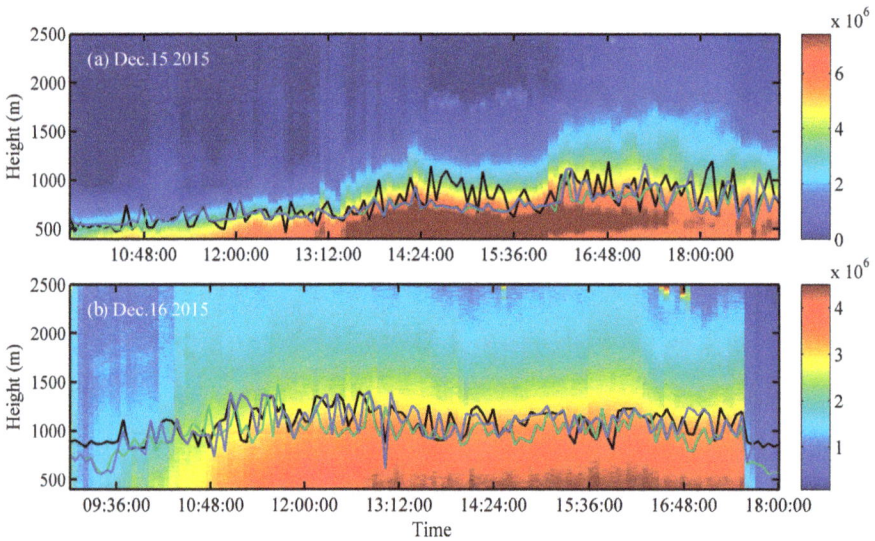

Figure 8. The diurnal variation of RCS and PBLH (calculated by three methods: black, blue and green lines represent GRA, STD and WCT method, respectively) on: 15 December 2015 (**a**); and 16 December 2015 (**b**).

4. Conclusions

In this paper, the seasonal and diurnal variations of daytime PBLH in Nanjing have been estimated from the one-year Lidar data by three different inversion methods, gradient method, standard deviation method and wavelet covariance transform method, and the correlation properties between PBLH and PM$_{2.5}$ were analyzed through both annual statistic and case study.

Remote Sens. **2017**, *9*, 668

Generally, the three methods show consistent variation of PBLH. The PBLH estimate can be affected by the backscatter signal of boundary layer clouds, and the GRA and STD methods are all more sensitive to the signal noise than the WCT method.

Annual average PBLH in the daytime during December 2015–November 2016 in Nanjing is 992 m. The daytime PBLH shows typical seasonal trend, highest in summer and lowest in winter, and the values of which are 1351 m and 822 m, respectively. PBLH shows the maximum variability in spring, with the standard deviation of 380 m. The diurnal cycle shows similar pattern in different seasons, and for annual average diurnal variation, the minimum PBLH is 580 m, which appears at 08:00, and the maximum is 1089 m at 15:00, respectively.

The PBLH is relatively lower when the ground $PM_{2.5}$ concentration is higher. The average daytime PBLH is 718 m and 1210 m in *the polluted* condition ($PM_{2.5} > 75$ μg/m^3) and *good* condition ($PM_{2.5} < 35$ μg/m^3), respectively. The diurnal variation of PBLH is the opposite to that of $PM_{2.5}$ concentration. Daily averaged PBLH and $PM_{2.5}$ concentration are anti-correlated with a correlation coefficient of -0.70.

In the case study, $PM_{2.5}$ concentration drops from 200 μg/m^3 to 30 μg/m^3 during 14–17 December 2015 while the PBLH increases from around 620 m to 1020 m. The polluted case accompanied with the low PBLH, while the clean case shows the opposite trend (high PBLH, high visibility, high wind speed, and low relative humidity).

This study revealed the variation characteristics of PBLH and its correlation between particulate matter concentrations on the ground, based on the one-year data over Nanjing of East China. We should note that the anthropogenic emissions in Nanjing are assumed to vary little by season. The significant or large-scale weather and/or climatology processes (e.g., cold front, monsoon, El Nino, La Nina, etc.) could be important issues in further study of the PBLH variation, and requires a longer period of observations.

Supplementary Materials: The following are available online at www.mdpi.com/2072-4292/9/7/668/s1, Figure S1: Comparison of PBL-height between the (a) Lidar and (b) radiosonde measurement on 15 August 2016, Figure S2: Time series of (a) PBL height and $PM_{2.5}$ concentration, (b) surface wind speed (U) and visibility (VIS), and (c) temperature (T) and relative humidity (RH), respectively, from 14 August to 17 August 2016, Figure S3: The diurnal variation of RCS and PBLH (calculated by three methods: black, blue and green lines represent GRA, STD and WCT method, respectively) in 15 August 2016 (a) and 16 August 2016 (b).

Acknowledgments: This work was jointly supported by the National Science and Technology Major Project (grant 2016YFC0203303), the National Science Foundation of Jiangsu Province (grant BE2015151), the National Science Foundation of China (grants 41075012, 40805006, 91544230, 41675030, and 2014CB441203). Y. Wu is supported by the NOAA-CREST grant #NA11SEC481008 and NYSERDA grant #100415. We thank for DuanYang Liu from Jiangsu Meteorological Observatory for providing the meteorological data and Samuel Lightstone from the City College of New York for revising the English-writing and comments. We gratefully acknowledged the constructive comments from three anonymous reviewers that greatly improve the manuscript.

Author Contributions: Yong Han conceived the study, supervised the data analysis, and edited the manuscript. Yawei Qu performed the data analysis and prepared the manuscript. Peng Gao carried out the Lidar observation experiment to obtain the observation data. Tijian Wang provided $PM_{2.5}$ data. Yonghua Wu provided the advices and discussions on the methodology and results.

Conflicts of Interest: The authors declare no conflict of interest.

References

1. Chen, P.; Wang, T.; Lu, X.; Yu, Y.; Kasoar, M.; Xie, M.; Zhuang, B. Source apportionment of size-fractionated particles during the 2013 Asian Youth Games and the 2014 Youth Olympic Games in Nanjing, China. *Sci. Total Environ.* **2017**, *579*, 860–870. [CrossRef] [PubMed]
2. You, W.; Zang, Z.L.; Zhang, L.F.; Li, Y.; Pan, X.B.; Wang, W.Q. National-scale estimates of ground-level $PM_{2.5}$ concentration in China using geographically weighted regression based on 3 km resolution MODIS aod. *Remote Sens.* **2016**, *8*, 13. [CrossRef]

3. Artaxo, P.; Bretherton, C.; Feingold, G.; Forster, P.; Kerminen, V.M.; Kondo, Y.; Liao, H.; Lohmann, U.; Rasch, P.; Satheesh, S.K.; et al. Cloud and Aerosols. In *Climate Change 2013: The Physical Science Basis: Working Group I Contribution to the Fifth Assessment Report of the Intergovernmental Panel on Climate Change*; Cambridge University Press: Cambridge, UK; New York, NY, USA, 2014; pp. 614–623.

4. Kan, H.D.; London, S.J.; Chen, G.H.; Zhang, Y.H.; Song, G.X.; Zhao, N.Q.; Jiang, L.L.; Chen, B.H. Season, sex, age, and education as modifiers of the effects of outdoor air pollution on daily mortality in Shanghai, China: The public health and air pollution in Asia (PAPA) study. *Environ. Health Perspect.* **2008**, *116*, 1183–1188. [CrossRef] [PubMed]

5. Deng, J.J.; Wang, T.J.; Liu, L.; Jiang, F. Modeling heterogeneous chemical processes on aerosol surface. *Particuology* **2010**, *8*, 308–318. [CrossRef]

6. Park, S.S.; Jung, Y.; Lee, Y.G. Spectral dependence on the correction factor of erythemal UV for cloud, aerosol, total ozone, and surface properties: A modeling study. *Adv. Atmos. Sci.* **2016**, *33*, 865–874. [CrossRef]

7. Atwater, M.A. The radiation budget for polluted layers of the urban environment. *J. Appl. Meteorol.* **1971**, *10*, 205–214. [CrossRef]

8. Gao, Y.; Zhang, M.; Liu, Z.; Wang, L.; Wang, P.; Xia, X.; Tao, M.; Zhu, L. Modeling the feedback between aerosol and meteorological variables in the atmospheric boundary layer during a severe fog-haze event over the North China Plain. *Atmos. Chem. Phys.* **2015**, *15*, 4279–4295. [CrossRef]

9. Wang, T.J.; Li, S.; Shen, Y.; Deng, J.J.; Xie, M. Investigations on direct and indirect effect of nitrate on temperature and precipitation in China using a regional climate chemistry modeling system. *J. Geophys. Res. Atmos.* **2010**, *115*, 13. [CrossRef]

10. Attwood, A.R.; Washenfelder, R.A.; Brock, C.A.; Hu, W.; Baumann, K.; Campuzano-Jost, P.; Day, D.A.; Edgerton, E.S.; Murphy, D.M.; Palm, B.B.; et al. Trends in sulfate and organic aerosol mass in the Southeast U.S.: Impact on aerosol optical depth and radiative forcing. *Geophys. Res. Lett.* **2014**, *41*, 7701–7709. [CrossRef]

11. Wang, J.; Park, S.; Zeng, J.; Ge, C.; Yang, K.; Carn, S.; Krotkov, N.; Omar, A.H. Modeling of 2008 kasatochi volcanic sulfate direct radiative forcing: Assimilation of omi SO_2 plume height data and comparison with MODIS and CALIOP observations. *Atmos. Chem. Phys.* **2013**, *13*, 1895–1912. [CrossRef]

12. Zhuang, B.L.; Jiang, F.; Wang, T.J.; Li, S.; Zhu, B. Investigation on the direct radiative effect of fossil fuel black-carbon aerosol over China. *Theor. Appl. Climatol.* **2011**, *104*, 301–312. [CrossRef]

13. Wang, Q.Y.; Huang, R.J.; Zhao, Z.Z.; Cao, J.J.; Ni, H.Y.; Tie, X.X.; Zhao, S.Y.; Su, X.L.; Han, Y.M.; Shen, Z.X.; et al. Physicochemical characteristics of black carbon aerosol and its radiative impact in a polluted urban area of China. *J. Geophys. Res. Atmos.* **2016**, *121*, 12505–12519. [CrossRef]

14. Li, S.; Wang, T.J.; Solmon, F.; Zhuang, B.L.; Wu, H.; Xie, M.; Han, Y.; Wang, X.M. Impact of aerosols on regional climate in southern and northern China during strong/weak East Asian summer monsoon years. *J. Geophys. Res. Atmos.* **2016**, *121*, 4069–4081. [CrossRef]

15. Xia, X.; Che, H.; Zhu, J.; Chen, H.; Cong, Z.; Deng, X.; Fan, X.; Fu, Y.; Goloub, P.; Jiang, H.; et al. Ground-based remote sensing of aerosol climatology in China: Aerosol optical properties, direct radiative effect and its parameterization. *Atmos. Environ.* **2016**, *124*, 243–251. [CrossRef]

16. Zhuang, B.L.; Wang, T.J.; Li, S.; Liu, J.; Talbot, R.; Mao, H.T.; Yang, X.Q.; Fu, C.B.; Yin, C.Q.; Zhu, J.L.; et al. Optical properties and radiative forcing of urban aerosols in Nanjing, China. *Atmos. Environ.* **2014**, *83*, 43–52. [CrossRef]

17. Zhuang, B.L.; Wang, T.J.; Liu, J.; Ma, Y.; Yin, C.Q.; Li, S.; Xie, M.; Han, Y.; Zhu, J.L.; Yang, X.Q.; et al. Absorption coefficient of urban aerosol in Nanjing, west Yangtze River delta, China. *Atmos. Chem. Phys.* **2015**, *15*, 13633–13646. [CrossRef]

18. Ma, J.Z.; Xu, X.B.; Zhao, C.S.; Yan, P. A review of atmospheric chemistry research in China: Photochemical smog, haze pollution, and gas-aerosol interactions. *Adv. Atmos. Sci.* **2012**, *29*, 1006–1026. [CrossRef]

19. Huang, R.J.; Zhang, Y.L.; Bozzetti, C.; Ho, K.F.; Cao, J.J.; Han, Y.M.; Daellenbach, K.R.; Slowik, J.G.; Platt, S.M.; Canonaco, F.; et al. High secondary aerosol contribution to particulate pollution during haze events in China. *Nature* **2014**, *514*, 218–222. [CrossRef] [PubMed]

20. Tao, M.H.; Chen, L.F.; Xiong, X.Z.; Zhang, M.G.; Ma, P.F.; Tao, J.H.; Wang, Z.F. Formation process of the widespread extreme haze pollution over Northern China in january 2013: Implications for regional air quality and climate. *Atmos. Environ.* **2014**, *98*, 417–425. [CrossRef]

21. Boynard, A.; Clerbaux, C.; Clarisse, L.; Safieddine, S.; Pommier, M.; Van Damme, M.; Bauduin, S.; Oudot, C.; Hadji-Lazaro, J.; Hurtmans, D.; et al. First simultaneous space measurements of atmospheric pollutants in the boundary layer from IASI: A case study in the North China Plain. *Geophys. Res. Lett.* **2014**, *41*, 645–651. [CrossRef]

22. Stull, R.B. *An Introduction to Boundary Layer Meteorology (Vol. 13)*; Kluwer Academic Publishers: Dordrecht, The Netherlands; Boston, MA, USA; London, UK, 1988.

23. Petaja, T.; Jarvi, L.; Kerminen, V.M.; Ding, A.J.; Sun, J.N.; Nie, W.; Kujansuu, J.; Virkkula, A.; Yang, X.Q.; Fu, C.B.; et al. Enhanced air pollution via aerosol-boundary layer feedback in China. *Sci. Rep.* **2016**, *6*, 6. [CrossRef] [PubMed]

24. Zilitinkevich, S.S.; Tyuryakov, S.A.; Troitskaya, Y.I.; Mareev, E.A. Theoretical models of the height of the atmospheric boundary layer and turbulent entrainment at its upper boundary. *Izv. Atmos. Ocean. Phys.* **2012**, *48*, 133–142. [CrossRef]

25. Tyagi, S.; Tiwari, S.; Mishra, A.; Singh, S.; Hopke, P.K.; Singh, S.; Attri, S.D. Characteristics of absorbing aerosols during winter foggy period over the national capital region of Delhi: Impact of planetary boundary layer dynamics and solar radiation flux. *Atmos. Res.* **2017**, *188*, 1–10. [CrossRef]

26. Leventidou, E.; Zanis, P.; Balis, D.; Giannakaki, E.; Pytharoulis, I.; Amiridis, V. Factors affecting the comparisons of planetary boundary layer height retrievals from CALIPSO, ECMWF and radiosondes over Thessaloniki, Greece. *Atmos. Environ.* **2013**, *74*, 360–366. [CrossRef]

27. Liu, J.J.; Huang, J.P.; Chen, B.; Zhou, T.; Yan, H.R.; Jin, H.C.; Huang, Z.W.; Zhang, B.D. Comparisons of PBL heights derived from CALIPSO and ECMWF reanalysis data over China. *J. Quant. Spectrosc. Radiat. Transf.* **2015**, *153*, 102–112. [CrossRef]

28. Bianco, L.; Wilczak, J.M. Convective boundary layer depth: Improved measurement by doppler radar wind profiler using fuzzy logic methods. *J. Atmos. Ocean. Technol.* **2002**, *19*, 1745–1758. [CrossRef]

29. Munkel, C.; Schafer, K.; Emeis, S. Adding confidence levels and error bars to mixing layer heights detected by ceilometer. In *Remote Sensing of Clouds and the Atmosphere XVI*; Kassianov, E.I., Comeron, A., Picard, R.H., Schafer, K., Eds.; Spie-Int Soc Optical Engineering: Bellingham, WA, USA, 2011; Volume 8177.

30. Lotteraner, C.; Piringer, M. Mixing-height time series from operational ceilometer aerosol-layer heights. *Bound. Layer Meteorol.* **2016**, *161*, 265–287. [CrossRef]

31. Uzan, L.; Egert, S.; Alpert, P. Ceilometer evaluation of the eastern mediterranean summer boundary layer height—First study of two Israeli sites. *Atmos. Meas. Tech.* **2016**, *9*, 4387–4398. [CrossRef]

32. Casasanta, G.; Pietroni, I.; Petenko, I.; Argentini, S. Observed and modelled convective mixing-layer height at Dome C, Antarctica. *Bound. Layer Meteorol.* **2014**, *151*, 597–608. [CrossRef]

33. Petenko, I.; Argentini, S.; Casasanta, G.; Kallistratova, M.; Sozzi, R.; Viola, A. Wavelike structures in the turbulent layer during the morning development of convection at Dome C, Antarctica. *Bound. Layer Meteorol.* **2016**, *161*, 289–307. [CrossRef]

34. De Tomasi, F.; Perrone, M.R. PBL and dust layer seasonal evolution by Lidar and radiosounding measurements over a peninsular site. *Atmos. Res.* **2006**, *80*, 86–103. [CrossRef]

35. Pal, S. Monitoring depth of shallow atmospheric boundary layer to complement Lidar measurements affected by partial overlap. *Remote Sens.* **2014**, *6*, 8468–8493. [CrossRef]

36. Deng, T.; Deng, X.; Li, F.; Wang, S.; Wang, G. Study on aerosol optical properties and radiative effect in cloudy weather in the Guangzhou region. *Sci. Total Environ.* **2016**, *568*, 147–154. [CrossRef] [PubMed]

37. Wang, W.; Gong, W.; Mao, F.Y.; Pan, Z.X. An improved iterative fitting method to estimate nocturnal residual layer height. *Atmosphere* **2016**, *7*, 106. [CrossRef]

38. Hoff, R.M.; GuiseBagley, L.; Staebler, R.M.; Wiebe, H.A.; Brook, J.; Georgi, B.; Dusterdiek, T. Lidar, nephelometer, and in situ aerosol experiments in Southern Ontario. *J. Geophys. Res. Atmos.* **1996**, *101*, 19199–19209. [CrossRef]

39. Flamant, C.; Pelon, J.; Flamant, P.H.; Durand, P. Lidar determination of the entrainment zone thickness at the top of the unstable marine atmospheric boundary layer. *Bound. Layer Meteorol.* **1997**, *83*, 247–284. [CrossRef]

40. Hennemuth, B.; Lammert, A. Determination of the atmospheric boundary layer height from radiosonde and Lidar backscatter. *Bound. Layer Meteorol.* **2006**, *120*, 181–200. [CrossRef]

41. Comeron, A.; Sicard, M.; Rocadenbosch, F. Wavelet correlation transform method and gradient method to determine aerosol layering from Lidar returns: Some comments. *J. Atmos. Ocean. Technol.* **2013**, *30*, 1189–1193. [CrossRef]

42. Hooper, W.P.; Eloranta, E.W. Lidar measurements of wind in the planetary boundary-layer—The method, accuracy and results from joint measurements with radiosonde and kytoon. *J. Clim. Appl. Meteorol.* **1986**, *25*, 990–1001. [CrossRef]

43. Menut, L.; Flamant, C.; Pelon, J.; Flamant, P.H. Urban boundary-layer height determination from Lidar measurements over the Paris area. *Appl. Opt.* **1999**, *38*, 945–954. [CrossRef] [PubMed]

44. Cohn, S.A.; Angevine, W.M. Boundary layer height and entrainment zone thickness measured by Lidars and wind-profiling radars. *J. Appl. Meteorol.* **2000**, *39*, 1233–1247. [CrossRef]

45. Granados-Munoz, M.J.; Navas-Guzman, F.; Bravo-Aranda, J.A.; Guerrero-Rascado, J.L.; Lyamani, H.; Fernandez-Galvez, J.; Alados-Arboledas, L. Automatic determination of the planetary boundary layer height using Lidar: One-year analysis over Southeastern Spain. *J. Geophys. Res. Atmos.* **2012**, *117*, 10. [CrossRef]

46. Deng, T.; Wu, D.; Deng, X.; Tan, H.; Li, F.; Liao, B. A vertical sounding of severe haze process in Guangzhou area. *Sci. China Earth Sci.* **2014**, *57*, 2650–2656. [CrossRef]

47. Steyn, D.G.; Boldi, M.; Hoff, R.M. The detection of mixed layer depth and entrainment zone thickness from Lidar backscatter profiles. *J. Atmos. Ocean. Technol.* **1999**, *16*, 953–959. [CrossRef]

48. Hägeli, P.; Steyn, D.G.; Strawbridge, K.B. Spatial and temporal variability of mixed-layer depth and entrainment zone thickness. *Bound. Layer Meteorol.* **2000**, *97*, 47–71. [CrossRef]

49. Klett, J.D. Lidar inversion with variable backscatter extinction ratios. *Appl. Opt.* **1985**, *24*, 1638–1643. [CrossRef] [PubMed]

50. Du, C.L.; Liu, S.Y.; Yu, X.; Li, X.M.; Chen, C.; Peng, Y.; Dong, Y.; Dong, Z.P.; Wang, F.Q. Urban boundary layer height characteristics and relationship with particulate matter mass concentrations in Xi'an, central China. *Aerosol Air Qual. Res.* **2013**, *13*, 1598–1607. [CrossRef]

51. Schneider, J.; Eixmann, R. Three years of routine raman Lidar measurements of tropospheric aerosols: backscattering, extinction, and residual layer height. *Atmos. Chem. Phys.* **2002**, *2*, 313–323. [CrossRef]

52. Matthias, V.; Bosenberg, J. Aerosol climatology for the planetary boundary layer derived from regular Lidar measurements. *Atmos. Res.* **2002**, *63*, 221–245. [CrossRef]

53. Van der Kamp, D.; McKendry, I. Diurnal and seasonal trends in convective mixed-layer heights estimated from two years of continuous ceilometer observations in Vancouver, BC. *Bound. Layer Meteorol.* **2010**, *137*, 459–475. [CrossRef]

54. Strawbridge, K.; Travis, M.; Harwood, M. Preliminary results from scanning Lidar measurements of stack plumes during winter/summer. In Proceedings of the SPIE 4546 Laser Radar: Ranging and Atmospheric Lidar Techniques III, Toulouse, France, 17 September 2017; pp. 101–110.

55. Chinese Ministry of Environmental Protection (MEP) and General Administration of Quality Supervision, Inspection, and Quarantine (AQISQ) of the People's Republic of China. *Ambient Air Quality Standards (GB 3095-2012)*; China Environmental Science Press: Beijing, China, 2012.

56. Batchvarova, E.; Gryning, S.-E. Applied model for the growth of the daytime mixed layer. *Bound. Layer Meteorol.* **1991**, *56*, 261–274. [CrossRef]

57. Deng, J.J.; Wang, T.J.; Jiang, Z.Q.; Xie, M.; Zhang, R.J.; Huang, X.X.; Zhu, J.L. Characterization of visibility and its affecting factors over Nanjing, China. *Atmos. Res.* **2011**, *101*, 681–691. [CrossRef]

58. Levin, E.J.T.; Prenni, A.J.; Palm, B.B.; Day, D.A.; Campuzano-Jost, P.; Winkler, P.M.; Kreidenweis, S.M.; DeMott, P.J.; Jimenez, J.L.; Smith, J.N. Size-resolved aerosol composition and its link to hygroscopicity at a forested site in Colorado. *Atmos. Chem. Phys.* **2014**, *14*, 2657–2667. [CrossRef]

remote sensing

MDPI

Article

Intercomparison of Ozone Vertical Profile Measurements by Differential Absorption Lidar and IASI/MetOp Satellite in the Upper Troposphere–Lower Stratosphere

Sergey I. Dolgii [1], Alexey A. Nevzorov [1], Alexey V. Nevzorov [1], Oleg A. Romanovskii [1,2,*] and Olga V. Kharchenko [1]

[1] V.E. Zuev Institute of Atmospheric Optics, 1 Academician Zuev Square, Tomsk 634055, Russia; dolgii@iao.ru (S.I.D.); naa@iao.ru (A.A.N.); nevzorov@iao.ru (A.V.N.); olya@iao.ru (O.V.K.)
[2] Department of Innovative Technologies, National Research Tomsk State University, 36 Lenin Avenue, Tomsk 634050, Russia
* Correspondence: roa@iao.ru; Tel.: +7-913-868-4294

Academic Editors: Yang Liu, Jun Wang, Omar Torres, Richard Müller and Prasad S. Thenkabail
Received: 10 February 2017; Accepted: 27 April 2017; Published: 8 May 2017

Abstract: This paper introduces the technique of retrieving the profiles of vertical distribution of ozone considering temperature and aerosol correction in DIAL sounding of the atmosphere. The authors determine wavelengths, which are promising for measurements of ozone profiles in the upper troposphere–lower stratosphere. An ozone differential absorption lidar is designed for the measurements. The results of applying the developed technique to the retrieval of the vertical profiles of ozone considering temperature and aerosol correction in the altitude range 6–15 km in DIAL sounding of the atmosphere confirm the prospects of ozone sounding at selected wavelengths of 341 and 299 nm with the proposed lidar. The 2015 ozone profiles retrieved were compared with satellite IASI data and the Kruger model.

Keywords: intercomparison; differential absorption lidar; ozone; satellite measurements; IASI; upper troposphere; lower stratosphere

1. Introduction

Laser remote sounding techniques with the use of the lidar (Light Detection and Ranging) technology are widely used for the study and monitoring of the atmosphere. Among the techniques used for measuring the spatial distribution of the concentration of an atmospheric gas, the most sensitive is the differential absorption method (DIAL). The essence of the method is that radiation is transmitted in the atmosphere simultaneously at two wavelengths: one of them (λ_{on}) is on a strong absorption line or band of the gas to be measured and another (λ_{off}) is off the absorption line and is weakly absorbed in the atmosphere or not absorbed at all. At present, DIAL sounding of trace atmospheric gases including ozone, is implemented [1–3].

Laser sounding of the ozonosphere became routine at some observatories since the second half of the 1980s [4–7]. It allows data to be received on vertical distribution of ozone (VDO), which successfully supplement similar data received by in situ methods with the use of ozonesondes, rockets, or satellites (TOMS, SAGE-II, TERRA, MetOp, etc.).

It should be noted that the profiles of IASI/MetOp-measured atmospheric ozone concentrations were earlier compared with the profiles measured in the troposphere with compact airborne lidars, ozone sondes, ground-based lidars, and Brewer–Dobson spectrophotometers [8–10].

Laser sounding of the vertical distribution of stratospheric ozone has been carried out at the Siberian Lidar Station (SLS) of Institute of Atmospheric Optics SB RAS (Tomsk, 56.5°N, 85.0°E) since 1989. A long period of lidar observations of stratospheric ozone has shown that the most important part of the ozonosphere for the study is located in the lower stratosphere, where ozone is affected by the dynamic factor. DIAL measurements of ozone profiles in different altitude ranges with different ozone content are carried out at different combinations of wavelengths [4–7].

The aims of this work are the development of an algorithm and program for VDO retrieval with temperature and aerosol correction; selection of the VDO sounding wavelengths; the design of an ozone lidar; and analysis of the measurement results, including the comparison between the ozone vertical profiles retrieved from the ground-based DIAL lidar system data and the profiles retrieved from IASI/MetOp data.

2. Methods

2.1. Selection of Wavelengths

Lidar measurements of VDO are carried out on the basis of the method of the differential absorption of the backscattered laser radiation in the ultraviolet spectral range 200–370 nm (Hartley–Huggins band) [11]. In practice, several pairs of wavelengths can be implemented in ozone lidars with the help of different lasers. Table 1 represents the specifications of some lidars used for ozone measurements and operating at different combinations of wavelengths.

Table 1. Specifications of lidars used for measurements of stratospheric ozone.

Country, Observations Site	Start of Measu-Rements	Radiation Source: Wavelength, nm/Pulse Energy, mJ/Pulse Repetition Frequency, Hz	Receiving Mirror (Diameter), m	Ref.
Russia, Tomsk (56.5°N, 85°E)	1989	XeCl + SRS (H₂) 308/100/100; 353/50/1000	2.2 0.5 0.3	[12]
USA, California (34°N, 118°E)	1986	XeCl + SRS (H₂) 308–353	0.9	[13]
France, Provance (44°N, 6°E)	1986	XeCl+ Nd:YAG 308/250/50; 355/150/50	4 mirrors of 0.53	[14]
Germany, Hohenpeißenberg (48°N, 11°E)	1987	XeCl + SRS (H₂) 308/300/20; 353/150/20	0.9	[15]
France, Italia, the Antarctic (66°S, 140°E)	1991	XeCl+ Nd:YAG 308/180/80; 355/150/10	0.8	[16]
Argentina, Buenos-Aires (35°S, 59°W)	1999	XeCl+ Nd:YAG 308/300/100; 355/255/10	0.5	[17]

More than 85% of atmospheric ozone is located in the stratosphere. To measure the much smaller concentrations of tropospheric ozone, sounding wavelengths should be selected from the short wavelength range, closer to the ozone absorption band center, to increase the concentration sensitivity of the method. In this spectral range, the absorption cross-section σ is several times larger than that for wavelengths used in stratospheric measurements (e.g., $\sigma_{299} = 4.4 \times 10^{-19}$ cm² for the wavelength $\lambda_{on} = 299$ nm and $\sigma_{308} = 1.4 \times 10^{-19}$ cm² for $\lambda_{on} = 308$ nm).

KrF laser radiation (248 nm) or the fourth harmonics of an Nd:YAG laser (266 nm) is usually used for measurements of tropospheric ozone, in combination with a technique based on the stimulated

Raman scattering (SRS) in H_2, D_2, CO_2, and other gases [4,5,7]. The most common are hydrogen and deuterium. Table 2 represents possible sets of wavelengths that correspond to the first, second, and third Stokes (C) frequencies of SRS conversion in H_2, D_2, and CO_2.

Table 2. Sets of wavelengths that correspond to Stokes (C) frequencies of SRS conversion in H_2, D_2, and CO_2.

Pumping Radiation	Wavelength, nm		
	H_2 C_1 C_2	D_2 C_1 C_2 C_3	CO_2 C_1 C_2
Nd:YAG, 266 nm	299 341	289 316	287 299
KrF, 248 nm	277 313	268 291 319	

Different wavelength combinations are used in practice in different altitude ranges in the troposphere and lower stratosphere. Thus, the wavelength pairs 289/316 and 287/299 nm allow measurements of ozone profiles up to altitudes of about 10 km [4,5]; the pair 292/319 nm, up to 14–16 km [4]; the pairs 277/313 and 292/313 nm, up to altitudes of 8–12 and 15 km, respectively [7].

We have estimated possibilities of the vertical ozone profile sounding in the upper troposphere–lower stratosphere at the wavelength pair 299/341 nm. During the calculations, actual lidar parameters were used: radiation energy of 20 mJ at both wavelengths, pulse repetition frequency of 15 Hz, receiving mirror diameter of 0.5 m, signal accumulation time of 1.5 h. To determine the transmitter–receiver efficiency, actual values of transmittance of optical elements of spectral selection and of efficiency of photomultiplier tubes (PMTs) were used; noises were taken from actual measurements. The calculations have shown that these wavelengths allow the sounding up to about 22 km (the ozone maximum is located in the altitude range 19–21 km in Tomsk) and ozone measurements in the troposphere. The ozone concentration error is within 4–10% limits up to altitudes of about 20 km.

The sounding altitude maximum is determined, first, by the range of signal detection at λ_{on}, which is always shorter than the range of signal detection at λ_{off} due to stronger ozone absorption. In view of this, λ_{on} = 299 nm is preferable to 277 or 292 nm. In addition, wavelengths of 299 and 341 nm are implemented in one sounding beam (from one laser source in one SRS cell), in contrast to, e.g., the 292/313 nm pair (see Table 2).

A system on the basis of a SRS cell filled with hydrogen is cheaper than with deuterium.

Thus, the wavelength pair 299/341 nm is of higher information content for VDO measurements in the upper troposphere–lowerstratosphere (5–22 km altitude range).

2.2. Theoretical Base of the VDO Retrieval

Initial equations for the calculation of the ozone concentration during DIAL lidar sounding of the atmosphere have the form

$$N_{on}(H) = c \cdot [\beta_{on}^a(H) + \beta_{on}^m(H)] \cdot \exp\left[-2\int_0^H \alpha_{on}^a(H) + \alpha_{on}^m + k_{on} \cdot n(H)\right] \quad (1)$$

$$N_{off}(H) = c \cdot \left[\beta_{ff}^a(H) + \beta_{off}^m(H)\right] \cdot \exp\left[-2\int_0^H \alpha_{off}^a(H) + \alpha_{off}^m + k_{off} \cdot n(H)\right] \quad (2)$$

where *N(H)* is the echo-signal recorded at corresponding wavelengths (*on* at an absorption line and *off* out of the absorption line), *C* is the instrumental constant, α^a is the aerosol extinction coefficient, β^a is the aerosol backscattering coefficient, *k* is the absorption cross-section of ozone, *n(H)* is the ozone concentration.

Let Equation (1) be divided by Equation (2):

$$\frac{N_{off}(H)}{N_{on}(H)} = \frac{\beta^a_{off}(H) + \beta^m_{off}(H)}{\beta^a_{on}(H) + \beta^m_{on}(H)} \cdot \exp\left\{-2\int_0^H \left[\alpha^a_{off}(H) - \alpha^a_{on}(H)\right] \cdot dH\right\}$$

$$\cdot \exp\left\{-2\int_0^H \left[\alpha^m_{off}(H) - \alpha^m_{on}(H)\right] \cdot dH\right\} \cdot \exp\left\{-2\int_0^H \left[k_{off}(H) - k_{on}(H)\right] \cdot n(H) \cdot dH\right\} \tag{3}$$

where $k_{on}(H)$ and $k_{off}(H)$ are the absorption coefficients and are off the ozone absorption line, dependent on the temperature.

Let us transform Equation (3) to the form

$$\ln\left\{\frac{N_{off}(H)}{N_{on}(H)} \cdot \left[\frac{\beta^a_{off}(H) + \beta^m_{off}(H)}{\beta^a_{on}(H) + \beta^m_{on}(H)}\right]\right\} = -2\int_0^H \left[\alpha^a_{off}(H) - \alpha^a_{on}(H)\right] \cdot dH - 2\int_0^H \left[\alpha^m_{off}(H) - \alpha^m_{on}(H)\right] \cdot dH$$

$$-2\int_0^H \left[k_{off}(H) - k_{on}(H)\right] \cdot n(H) \cdot dH \tag{4}$$

and derive Equation (4):

$$\frac{d}{dH} \ln\left\{\frac{N_{off}(H)}{N_{on}(H)} \cdot \left[\frac{\beta^a_{off}(H) + \beta^m_{off}(H)}{\beta^a_{on}(H) + \beta^m_{on}(H)}\right]\right\} = -2 \cdot \left[\alpha^a_{off}(H) - \alpha^a_{on}(H)\right] - 2 \cdot \left[\alpha^m_{off}(H) - \alpha^m_{on}(H)\right]$$

$$-2 \cdot \left[k_{off}(H) - k_{on}(H)\right] \cdot n(H).$$

Then

$$-2 \cdot \left[k_{off}(H) - k_{on}(H)\right] \cdot n(H) =$$

$$\frac{d}{dH} \ln\left\{\frac{N_{off}(H)}{N_{on}(H)} \cdot \left[\frac{\beta^a_{off}(H) + \beta^m_{off}(H)}{\beta^a_{on}(H) + \beta^m_{on}(H)}\right]\right\} + 2 \cdot \left[\alpha^a_{off}(H) - \alpha^a_{on}(H)\right] + 2 \cdot \left[\alpha^m_{off}(H) - \alpha^m_{on}(H)\right]. \tag{5}$$

The final equation for the ozone concentration is derived with the use of mathematical transformation

$$n(H) = \underbrace{\frac{1}{k_{on}(H) - k_{off}(H)}}_{A}$$

$$\cdot \left\{\underbrace{\frac{d}{dH} \ln\left[\frac{N_{off}(H)}{N_{on}(H)}\right]}_{B} - \underbrace{\frac{d}{dH} \ln\left[\frac{\beta^a_{off}(H) + \beta^m_{off}(H)}{\beta^a_{on}(H) + \beta^m_{on}(H)}\right]}_{C} - \underbrace{2 \cdot \left[\alpha^a_{off}(H) - \alpha^a_{on}(H)\right]}_{D} - \underbrace{2 \cdot \left[\alpha^m_{off}(H) - \alpha^m_{on}(H)\right]}_{F}\right\} \tag{6}$$

or

$$n(H) = A \cdot \{B - C - D - F\}. \tag{7}$$

Actual variations in the air temperature can cause variations in the ozone absorption cross-section, which results in systematic errors in VDO retrieval. Therefore, it is reasonable to carry out a correction to the temperature dependence $k_{on}(H,T)$, $k_{off}(H,T)$ in the algorithm for VDO retrieval.

A model of the behavior of ozone absorption cross-sections, presented in Table 3 and based on data from [18,19], was used in the technique.

The absorption coefficients $k_{on}(H)$ and $k_{off}(H)$ are used in term A of Equation (6). Real temperature variations in the atmosphere can significantly change the ozone absorption coefficient. Therefore, it is reasonable to use the correction to the temperature dependence $k_{on}(H,T)$, $k_{off}(H,T)$ in the VDO retrieval algorithm such that A takes the form

$$A = K_{299}(H,T) - K_{341}(H,T) = (5.8815E{-}16) - (1.1538E{-}17)\cdot(T(H) - 273)$$
$$+ (9.0281E{-}20)\cdot(T(H){-}273)^2 - (3.5194E{-}22)\cdot(T(H) - 273)^3 \tag{8}$$
$$+ (6.8356E{-}25)\cdot(T(H){-}273)^4 - (5.2918E{-}8)\cdot(T(H) - 273)^5,$$

where $T(H)$–is the model Kelvin temperature distributed with altitude.

Table 3. Ozone absorption cross-section (cm^2) at the ozone sounding wavelengths [18,19].

Wavelength, nm	Temperature, K				
	218	228	243	273	295
	On line				
299	4.1×10^{-19}	4.1×10^{-19}	4.25×10^{-19}	4.3×10^{-19}	4.6×10^{-19}
	Off line				
341	6×10^{-22}	6×10^{-22}	6×10^{-22}	6×10^{-22}	1.2×10^{-21}

Let us consider term C in Equation (6):

$$C = \frac{d}{dH} \ln \left[\frac{\beta^a_{off}(H) + \beta^m_{off}(H)}{\beta^a_{on}(H) + \beta^m_{on}(H)} \right] \tag{9}$$

and transform it to the form

$$\frac{\beta^a_{off}(H) + \beta^m_{off}(H)}{\beta^a_{on}(H) + \beta^m_{on}(H)} = \frac{\beta^a_{off}(H) + \beta^m_{off}(H)}{\beta^m_{off}(H) - R_{off}(H)} = \frac{1}{R_{off}(H)} \cdot \left\{ \frac{\beta^a_{on}(H)}{\beta^a_{off}(H)} + \frac{\beta^m_{on}(H)}{\beta^m_{off}(H)} \right\} =$$
$$\frac{1}{R_{off}(H)} \cdot \left\{ \left(\frac{\lambda_{off}}{\lambda_{on}}\right)^x \cdot \frac{\beta^a_{off}(H)}{\beta^m_{off}(H)} + \left(\frac{\lambda_{off}}{\lambda_{on}}\right)^4 \right\} = \frac{1}{R_{off}(H)} \cdot \left(\frac{\lambda_{off}}{\lambda_{on}}\right)^x \cdot \left[R_{off}(H) - 1\right] + \frac{1}{R_{off}(H)} \cdot \left(\frac{\lambda_{off}}{\lambda_{on}}\right)^4 \tag{10}$$
$$= \left(\frac{\lambda_{off}}{\lambda_{on}}\right)^x \cdot \left[1 - \frac{1}{R_{off}(H)}\right] + \frac{1}{R_{off}(H)} \cdot \left(\frac{\lambda_{off}}{\lambda_{on}}\right)^4,$$

where, at the corresponding wavelengths λ (*on* at an absorption line and *off* out of the absorption line), $R_{off}(H)$ is the real distribution of the scattering ratio, x is the parameter that characterizes the particle size, $\beta^a_{off}(H)$ is the aerosol backscattering coefficient, $\beta^m_{off}(H)$ is the molecular backscattering coefficient.

From the equalities $\frac{\beta^m_{off}(H)}{\beta^m_{off}(H)} = \left(\frac{\lambda_{off}}{\lambda_{on}}\right)^4$ and $\frac{\beta^a_{off}(H)}{\beta^a_{off}(H)} = \left(\frac{\lambda_{off}}{\lambda_{on}}\right)^x$, one can find

$$\beta^a_{on}(H) = \left(\frac{\lambda_{off}}{\lambda_{on}}\right)^x \cdot \beta^a_{off}(H) \tag{11}$$

Finally,

$$C = \frac{d}{dH} \left\{ \ln \left(\frac{\lambda_{off}}{\lambda_{on}}\right)^x \cdot \left[1 - \frac{1}{R_{off}(H)}\right] + \frac{1}{R_{off}(H)} \cdot \left(\frac{\lambda_{off}}{\lambda_{on}}\right)^4 \right\} \tag{12}$$

Now consider term D in Equation (6). Since

$$\alpha^a = b \cdot \beta^a, \text{ i.e., } \alpha^a(H) = b \cdot \beta^a(H) \tag{13}$$

then

$$D = 2 \cdot 0.04 \left[\beta^a_{off}(H) - \beta^a_{on}(H)\right] \tag{14}$$

Using $\beta_{on}^a(H) = \left(\frac{\lambda_{off}}{\lambda_{on}}\right)^x \cdot \beta_{off}^a(H)$ and $\beta_{off}^a(H) = \left[R_{off}(H) - 1\right] \cdot \beta_{off}^m(H)$

$$D = 2 \cdot 0.04 \cdot \left\{\beta_{off}^a(H) \cdot \left[1 - \left(\frac{\lambda_{off}}{\lambda_{on}}\right)^x\right]\right\} = 2 \cdot 0.04 \cdot \left\{\left[R_{off}(H) - 1\right] \cdot \beta_{off}^m(H) \cdot \left[1 - \left(\frac{\lambda_{off}}{\lambda_{on}}\right)^x\right]\right\} \quad (15)$$

Let us also transform term F in Equation (6) using the relations $\frac{\alpha_{on}^m(H)}{\alpha_{off}^m(H)} = \left(\frac{\lambda_{off}}{\lambda_{on}}\right)^4$ and $\alpha_{off}^m = \underbrace{\frac{3}{8\pi}}_{0.119} \cdot \beta_{\pi ff}^m(H)$:

$$F = 2 \cdot 0.119 \cdot \alpha_{off}^m \left[1 - \frac{\alpha_{on}^m(H)}{\alpha_{off}^m(H)}\right] = 2 \cdot 0.119 \cdot \beta_{off}^m(H) \left[1 - \left(\frac{\lambda_{off}}{\lambda_{on}}\right)^4\right] \quad (16)$$

Aerosol scattering exceeds molecular scattering by several times in the case of high atmospheric aerosol content, which significantly distorts ozone profiles retrieved under unconsidered scattering and attenuating properties of the atmosphere at the sounding wavelengths. In the algorithm for VDO retrieval described, the aerosol correction is considered in the Equations (9) and (15) by means of introduction of a real distribution of the scattering ratio $R_{off}(H)$, while VDO in the usual, undisturbed atmosphere can be calculated at $R_{off}(H) = 1$.

The vertical profile of the backscattering coefficient $\beta_{\pi}^a(H)$ is derived from laser sounding data. The coefficient decreases with an increase in altitude. For clearer representation of aerosol stratification, the scattering ratio

$$R(H) = \left[\beta_{\pi}^a(H) + \beta_{\pi}^M(H)\right]/\beta_{\pi}^M(H) \quad (17)$$

is used, where $\beta_{\pi}^M(H)$ is the molecular backscattering coefficient.

2.3. Software for VDO Retrieval

Software for altitude ozone profile retrieval from laser sounding data developed on the basis of the above algorithm (see the block diagram in Figure 1) allows:

1. Reading the lidar data;
2. Recording the retrieval results in ASCII format;
3. Moving average smoothing of lidar signals;
4. Temperature and aerosol correction;
5. Smoothing of the VDO retrieval results.

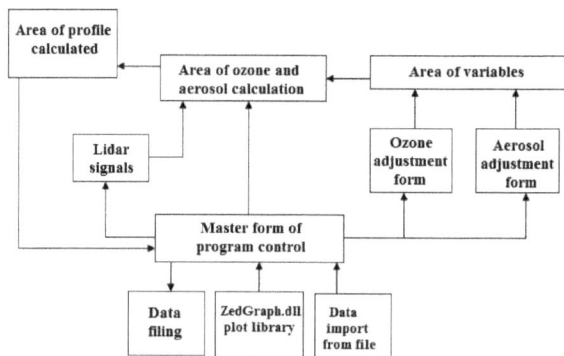

Figure 1. Structure of the software for vertical distribution of ozone (VDO) retrieval from laser sounding data.

To decrease the retrieval errors, the temperature correction of ozone absorption coefficients is used in the software. A high aerosol concentration in the 0–20 km altitude range should be considered during lidar signal retrieval when sounding at 272/289 nm and 299/341 nm wavelengths; therefore, aerosol correction is considered in the software. Model seasonal (winter and summer) midlatitude values of altitude distributions of the temperature and molecular backscattering coefficient have been introduced in the software for calculations.

Linear smoothing is used in the software for both input lidar data and retrieval results. The linear smoothing (moving average smoothing) is a well-known technique and is widely used in processing of experimental data in different fields of natural sciences. The linear smoothing is a special case of numerical filtering of signals with random errors using a rectangular window and unit weight coefficients.

The technique and software developed were used for VDO retrieval in the upper stratosphere–lower stratosphere at 299/341 nm wavelengths.

3. Materials

3.1. SLS Ozone Lidar

The sounding wavelengths selected (299/341 nm) were used in the SLS ozone lidar designed. Its block diagram is shown in Figure 2.

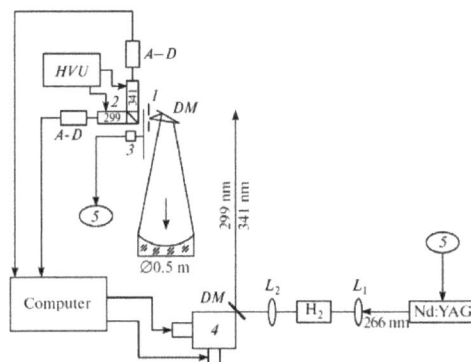

Figure 2. Block diagram of Siberian Lidar Station (SLS) ozone lidar: field diaphragm (*1*), cell for spectral selection with a PMT (*2*), mechanical shutter (*3*), rotating mirrors (RM); automated adjustment unit of an output rotating mirror (*4*); solid-state laser (Nd:YAG); SRS conversion cell with $H_2(H_2)$ amplifiers/discriminators (AD)); high-voltage power supply units for the PMT(HSU); lenses (L_1 and L_2); system for synchronizing the shutter operation time and the instant of laser pulses emission (*5*).

The fourth harmonics (266 nm) of the fundamental frequency of a Nd:YAG laser (LS-2134UT laser, LOTIS TII company, Minsk) is used as a laser radiation source, which is then

SRS converted in hydrogen in the first (299 nm) and second (341 nm) Stokes components. The receiving telescope has been designed according to the Newton scheme on the basis of a primary mirror, 0.5 m in diameter, with a focal length of 1.5 m. The recording channel of the lidar is equipped with PMTs (R7207-01) and HAMAMATSU amplifier/discriminators (C3866). Lidar signals are recorded in the photocurrent pulse counting mode. To support PMT linear modes, a mechanical shutter is used, which cuts off a high-power optical signal from the nearest sounding zone. An automated unit for the output rotary mirror adjustment has been designed on the basis of computer-driven step motors.

The SRS cell is made from a tube (stainless steel) 3 cm in inner diameter and 1 m in length. Input and output windows are made of quartz Quartz Ultraviolet (KU). The pumping pulse energy at a wavelength of 266 nm is 60 mJ. The pumping power density required for SRS conversion is provided by lens L1 with a focal length of 1 m. It is mounted before the SRS cell and focuses radiation at its center. Confocal collimating lens L2 is mounted behind the cell.

Basic specifications of laser sources and receiving optical elements of the SLS ozone lidar:

1	Transmitter	
2	Sounding wavelength, λ, nm	299, 341
3	Pulse energy, mJ (corr. λ)	25, 20
4	Frequency, Hz (corr. λ)	15
5	Divergence, mrad	0.1–0.3
6	Receiver	
7	Mirror diameter, m	0.5
8	Focal length, m	1.5

The efficiency of SRS conversion was measured versus the hydrogen pressure in the Raman cell, which was varied from 1 to 9 atm. Figure 3 shows the relative intensities of pumping radiation (266 nm), the first (299 nm) and the second (341 nm) Stokes components of SRS conversion as functions of hydrogen pressure at the SRS cell exit.

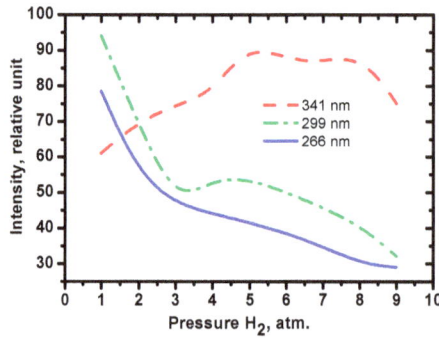

Figure 3. Relative intensities of pumping (266 nm), the first (299 nm) and the second (341 nm) Stokes conversion components of SRS conversion as functions of hydrogen pressure.

The intensities of 299 and 341 nm lines become equal at a hydrogen pressure of 2 atm, which allows ozone sounding under equal radiation energies at these wavelengths. However, to increase the upper sounding limit, a pressure of 1 atm is more efficient, since the energy is redistributed toward the 299 nm line, which is absorbed by ozone stronger than the 341 nm line.

3.2. IASI/MetOp

The IASI is mounted onboard the European Space Agency meteorological satellite MetOp. The satellite monitors CO_2, CH_4, N_2O, CO, O_3, and HNO_3 atmospheric gases; measures the temperature and humidity profiles in the troposphere and lower stratosphere in the near real time within the European Program "European Polar System". IASI provides for spectra of high radiometric quality with a resolution of 0.5 cm^{-1} in the range from 625 to 2760 cm^{-1} [20]. The ozone profiles were retrieved from satellite sounding data in the range 1025–1075 cm^{-1}.

The satellite data were received by the 2.4 XLB satellite data reception station (Orbital Systems, USA) put into operation at IAO SB RAS in 2011 [21]. The information from the station allows comparison between the satellite data and SLS lidar sounding data.

4. Results

The VDO profiles retrieved from lidar sounding data are compared with the IASI/MetOp measured profiles.

VDO was measured at SLS IAO SB RAS during 2015. Using the above described technique, the retrieved profiles of ozone were calculated for the upper troposphere–lower stratosphere and compared with the IASI/MetOpt measured profiles. The range of lidar signal detection was from 6 to 15 km. The ozone vertical profile error was from 6 to 11%. The retrieval error is estimated by the sum

$$E_{sum}{}^2 = e_1^2 + e_2^2 + e_3^2$$

where e_1 is the absorption cross-section error, e_2 is the standard error of measurements in the photon counting mode, and e_3 is the scattering ratio error. The absorption cross-section error e_2 does not exceed 2% [18].

The standard error of measurements in the photon counting mode e_2 is defined as

$$e_2^2 = 0.25 \cdot \left[\frac{1}{N_{on}(H)} + \frac{1}{N_{off}(H)} \right]$$

The scattering ratio error e_3 is defined as

$$e_3^2 = \frac{N_{off}(H)}{\left[N_{off}(H) - N_{noise}(H) \right]^2} + \frac{N_{off}(H_{calib})}{\left[N_{off}(H_{calib}) - N_{noise}(H_{calib}) \right]^2} + K$$

where H_{calib}—calibration height, N_{noise}—noise signal, K—constant value that appears due to the assumptions of the processing technique and the estimated value $3 \cdot (0.01)^2$.

The aerosol impact on the ozone profile in the troposphere and lower stratosphere is strong; therefore, the aerosol correction is to be used in the VDO retrieval algorithm.

Before the VDO retrieval, the scattering ratio is calculated by the lidar signal at the wavelength 341 nm. This allows the aerosol correction with the use of a real scattering ratio instead of its model values, and thus minimizes the aerosol impact on the ozone profile in the dynamic gaseous and aerosol medium at tropospheric and lower stratospheric altitudes. Figure 4 shows the mean error of VDO retrieval over all the measurement days.

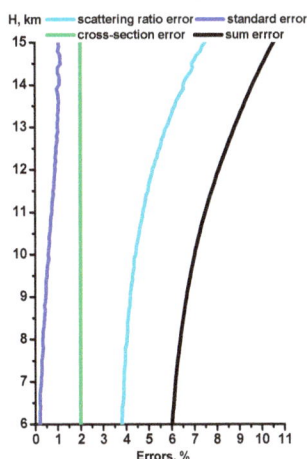

Figure 4. Mean error of ozone profile retrieval over 2015.

It should be noted that the lidar operates in the photon counting mode at a spatial resolution of 100 m, while the IASI profile has a step of 150 m and larger as the altitude increases, which is connected with the ozone profile retrieval algorithm of this interferometer [22]. IASI O_3 retrieval errors, according to the estimates given in paper [23], reach values of 30%. The IASI curve is smoother and does not describe VDO in detail; the lidar measurements show the ozone behavior in more detail.

The coordinates and time of the lidar and satellite sounding are given in Table 4.

Table 4. Coordinates and time of satellite and lidar sounding used for comparison of VDO profiles for 2015.

Date	Siberian Lidar Station		MetOp (IASI) Satellite	
	Greenwich Time	Coordinates (56.5°N, 85.0°E)	Greenwich Time	Coordinates
13 January 2015	11:53–13:45		14:17	56.472°N, 85.387°E
19 February 2015	12:39–14:13		14:53	56.681°N, 85.164°E
5 March 2015	13:05–14:56		15:02	56.472°N, 85.118°E
20 March 2015	13:32–15:24		14:53	56.691°N, 85.124°E
6 April 2015	14:25–16:17		15:41	56.254°N, 84.935°E
26 April 2015	15:11–17:03		15:26	56.585°N, 84.594°E

Figure 5 exemplifies the comparison of lidar and satellite vertical profiles of ozone measured on 26 April 2015; the VDO profiles are compared with the Krueger model [24].

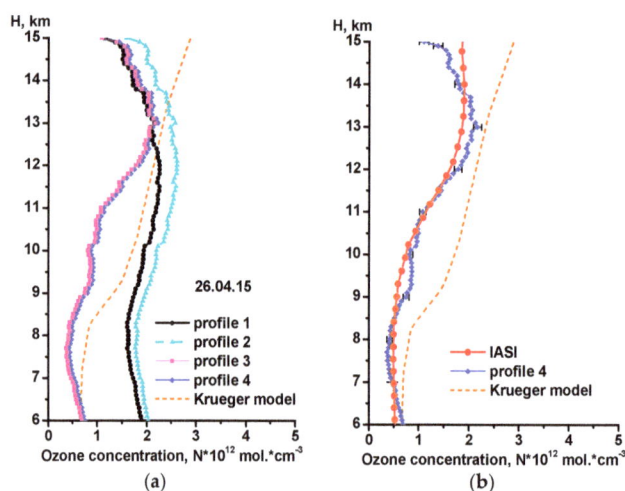

(a)   (b)

Figure 5. VDO profiles retrieved in comparison with the Kruger model: ozone profile without temperature and aerosol corrections (1), ozone profile with temperature and without aerosol corrections (2), ozone profile without temperature and with aerosol corrections (3), combination of corrections (aerosol and temperature) (4) (**a**); VDO profiles retrieved in comparison with the Kruger model and IASI satellite data (**b**).

We have analyzed the ozone profiles retrieved for April 2015 corrected to the temperature and aerosol and without corrections. One can see from Figure 5a that the efficiency of the corrections is significant. For the retrieval without temperature correction, the absorption cross-section constant $\sigma_{299} = 4.4 \times 10^{-19}$ cm^2 was used. High saturation of the aerosol component at tropospheric and lower stratospheric altitudes is corrected to both aerosol (terms C and D or Equations (12) and (15)) and temperature (term A or Equation (8)). Each correction used approaches the profiles to the IASI data.

Thus, the combination of corrections provides for a reliable ozone profile close to the IASI data (Figure 5b). The differences between profiles 3 and 4 are caused by the absorption cross-section values: the absorption cross-section in profile 4 is smaller than in profile 3; therefore, the values of profile 4 are higher than the values of profile 3 (Figure 5a).

Figure 6 shows all the considered cases of comparison of ozone profiles measured at the SLS and retrieved from IASI data (see Table 4). The measurements are reduced to the total altitude range 6–15 km for convenience. Increased ozone concentrations measured at the SLS as compared to IASI are seen in the range from 8.5 to 12.5 km throughout the whole observation period.

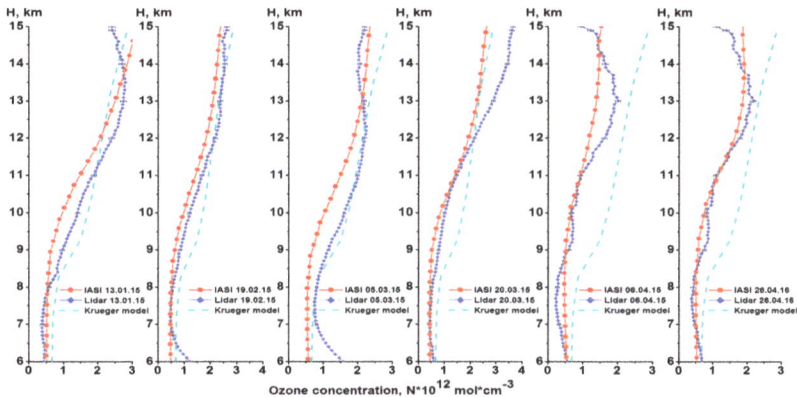

Figure 6. Intercomparison of vertical profiles of ozone for 2015.

5. Discussion

The following conclusions can be drawn from the analysis of the ozone profiles average over the period under study (Figures 7 and 8).

Figure 7a shows the mean lidar and satellite ozone profiles, and Figure 7b, the total difference between them (lidar–IASI) over all measurement days, as well as the standard deviation of this difference, minimum and maximum, and the mean. The error is calculated with the use of the standard deviation of the difference (lidar–IASI) and the ratio (lidar–IASI)/lidar over all days of measurements. Figure 7c shows the total difference between the lidar and IASI profiles normalized to the lidar profile for each measurement day, and the standard deviation with the minimum, maximum, and mean.

Figure 8a shows the mean lidar and Krueger model profiles, and Figure 8b, the total difference between them (lidar–Krueger model) over all measurement days, as well as the standard deviation of this difference, minimum and maximum, and the mean. Figure 8c shows the total difference between the lidar and Krueger model profiles normalized to the lidar profile for each measurement day, and the standard deviation with the minimum, maximum, and mean.

Thus, in Figure 7b, one can trace the variability of the difference between the lidar and satellite VDO in absolute units. The difference minima show how IASI data exceed the lidar data on VDO, and the minima, vice versa. The mean difference shows the difference between the lidar and IASI data over all measurement days. The normalized difference in Figure 7c shows more clearly the deviations between the values over all measurement days in percentage.

The difference between the lidar and IASI profiles of the ozone concentration grows with altitude. Hence, the mean difference in the ozone concentrations (lidar–IASI) varies from -1.56×10^{12} mol. cm^{-3} at an altitude of 15 km to 0.53×10^{12} mol. cm^{-3} at 13 km (Figure 6b). The maximal differences over all the profiles are from -0.01×10^{12} mol. cm^{-3} at an altitude of 7.2 km to 1.08×10^{12} mol. cm^{-3} at 14.3 km. The minimal differences over all the profiles vary from -0.77×10^{12} mol. cm^{-3} at 15 km to 0.39×10^{12} mol. cm^{-3} at 12.3 km.

Ozone shows pronounced annual variations; therefore, to find relative errors of its measurements with IASI, the difference in the concentrations was normalized to the lidar data: (Lidar–IASI)/Lidar. These data are shown in Figure 7c; it is seen that the mean relative difference is positive in the altitude ranges 6–6.5 and 8.4–14.6 km and attains 23.2% at an altitude of 12.4 km. The mean relative difference is negative in the altitude range from 6.5 to 8.4 km and 14.6–15 km: it attains the negative maximum of −33.8% at 7.5 km. The maximal relative difference changes from 3.45 to 60% in the range 6–15 km. The minimal relative difference over all the profiles in this range is −120.6% at 7.7 km; it attains 15.4% at 12.3 km. The relative difference varies in the range from −19.3 to 60% at 6 km and from −70.5 to 28.8% at 15 km.

Thus, the intercomparison performed shows that the absolute differences in the lidar and IASI measured ozone concentrations can change from −0.77 to 1.08×10^{12} mol. cm^{-3}; therefore, the relative difference is in the range from −120.6 to 60%.

It should be noted that the retrieved profiles of altitude distribution of the ozone concentration tend to IASI satellite data profiles more than to the Krueger model.

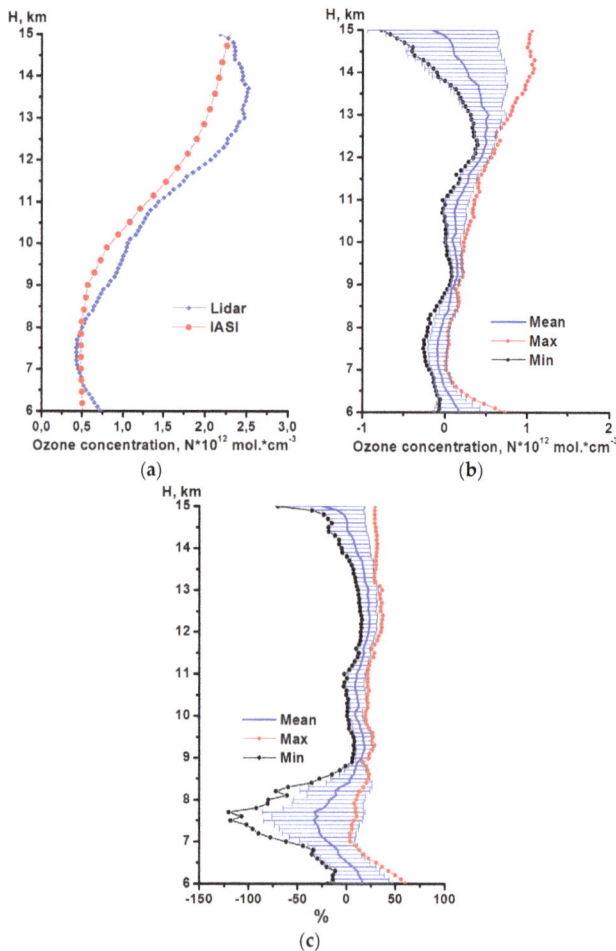

Figure 7. Mean vertical profiles of ozone (**a**), their difference (Lidar–IASI) in abs. units (**b**), and the relative difference (Lidar–IASI)/lidar (**c**).

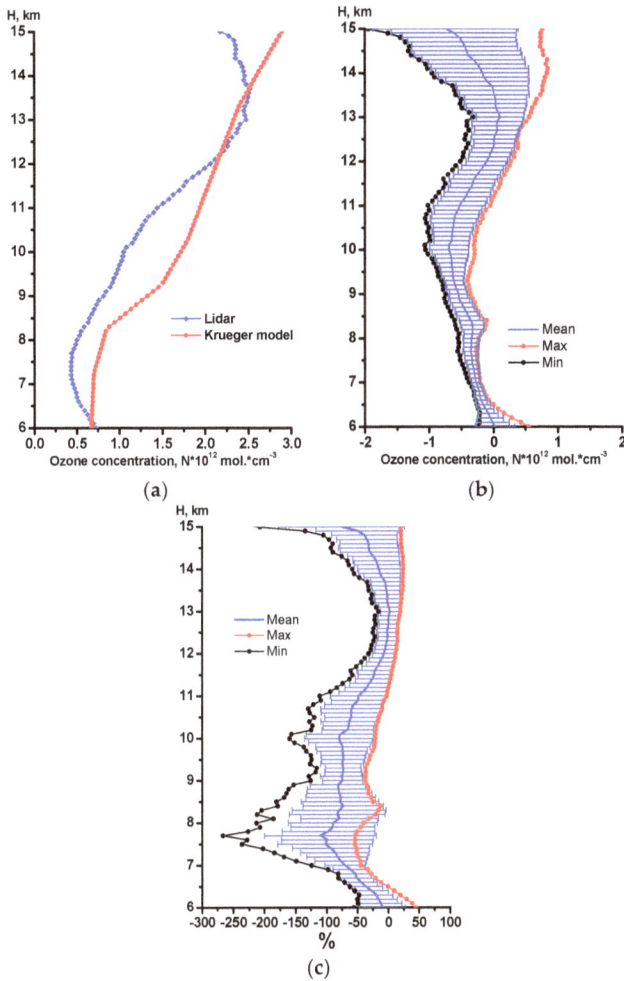

Figure 8. Mean vertical profiles of ozone (**a**), their difference (Lidar–Krueger) in abs. units (**b**), and the relative difference (Lidar–Krueger)/lidar (**c**).

6. Conclusions

The results of using the developed techniques for VDO retrieval with temperature and aerosol correction in the altitude range 6–15 km during the DIAL lidar sounding of the atmosphere confirm the prospects of the wavelengths chosen (299 and 341 nm) for ozone lidar sounding.

Results of lidar measurements at 299 and 341 nm agree with model estimates, which point towards acceptable accuracy of ozone sounding in altitudes near 6–15 km.

At present, works are being carried out on optimization of optical and photoelectronic elements of the lidar signal detection system with the aim of increasing the upper limit of the sounding and improving the measurement accuracy. A more effective and comparatively easy-to-use BaB_2O_4 (BBO) crystal has been mounted as a fourth harmonic converter.

Acknowledgments: The work was supported by the Russian Science Foundation (Agreement No. 15-17-10001 in carrying out lidar measurements of ozone) and President of the Russian Federation within the Program for Support of Leading Scientific Schools (grant No. NSh-8199.2016.5). Authors Alexey V. Nevzorov and Oleg Romanovskii defined the research theme.

Author Contributions: The work presented here was conducted in collaboration with all a experiments. This manuscript was written by Oleg Romanovskii and Alexey A. Nevzorov. Sergey Dolgii and Olga Kharchenko checked the experimental results. All authors agreed to the submission of the manuscript.

Conflicts of Interest: The authors declare no conflict of interest.

References

1. Sullivan, J.T.; McGee, T.J.; Sumnicht, G.K.; Twigg, L.W.; Hoff, R.M. A mobile differential absorption lidar to measure sub-hourly fluctuation of tropospheric ozone profiles in the Baltimore–Washington, D.C. *Atmos. Meas. Tech.* **2014**, *7*, 3529–3548. [CrossRef]
2. Repasky, K.S.; Moen, D.; Spuler, S.; Nehrir, A.R.; Carlsten, J.L. Progress towards an autonomous field deployable diode-laser-based differential absorption lidar (DIAL) for profiling water vapor in the lower troposphere. *Remote Sens.* **2013**, *6*, 6241–6259. [CrossRef]
3. Queißer, M.; Burton, M.; Fiorani, L. Differential absorption lidar for volcanic CO_2 sensing tested in an unstable atmosphere. *Opt. Exp.* **2015**, *23*. [CrossRef] [PubMed]
4. Galani, E.; Balis, D.; Zanis, P.; Zerefos, C.; Papayannis, A.; Wernli, H.; Gerasopoulo, E. Observations of stratosphere-to-troposphere transport events over the eastern Mediterranean using a ground-based lidar system. *J. Geophys. Res.* **2013**, *108*, 6634–6644. [CrossRef]
5. Nakazato, M.; Nagai, T.; Sakai, T.; Hirose, Y. Tropospheric ozone differential-absorption lidar using stimulated Raman scattering in carbon dioxide. *Appl. Opt.* **2007**, *46*, 2269–2279. [CrossRef] [PubMed]
6. Bukreev, V.S.; Vartapetov, S.K.; Veselovskii, I.A.; Galustov, A.S.; Kovalev, Y.M.; Prokhorov, A.M.; Svetogorov, E.S.; Khemelevtsov, S.S. Excimer-laser-based lidar system for stratospheric and tropospheric ozone measurements. *Quantum Electron.* **1994**, *21*, 591–596.
7. Eisele, H.; Scheel, H.E.; Sladkovic, R.; Trickl, T. High resolution lidar measurements of stratosphere-troposphere exchange. *J. Atmos. Sci.* **1999**, *56*, 319–330. [CrossRef]
8. Pommier, M.; Clerbaux, C.; Law, K.S.; Ancellet, G.; Bernath, P.; Coheur, P.F.; Hadji-Lazaro, J.; Hurtmans, D.; Nedelec, P.; Paris, J.D.; et al. Analysis of IASI tropospheric O_3 data over Arctic during POLARCAT campaigns in 2008. *Atmos. Chem. Phys.* **2012**, *12*, 7371–7389. [CrossRef]
9. Gazeaux, J.; Clerbaux, C.; George, M.; Hadji-Lazaro, J.; Kuttippurath, J.; Coheur, P.F.; Hurtmans, D.; Deshler, T.; Kovilakam, M.; Campbell, P.; et al. Intercomparison of polar ozone profiles by IASI/MetOp sounder with 2010 Concordiasiozonesonde observations. *Atmos. Meas. Tech.* **2013**, *6*, 613–620. [CrossRef]
10. Viatte, C.; Schneider, M.; Redondas, A.; Hase, F.; Eremenko, M.; Chelin, P; Flaud, J.M.; Blumenstock, T.; Orphal, J. Comparison of ground-based FTIR and Brewer O_3 total column with data from two different IASI algorithms and from OMI and GOME-2 satellite instruments. *Atmos. Meas. Tech.* **2011**, *4*, 535–546. [CrossRef]
11. Molina, L.T.; Molina, M.T. Absolute absorption cross section of ozone in the 185 nm to 350 nm wavelengthrange. *J. Geophys. Res.* **1988**, *91*, 14501–14508. [CrossRef]
12. El'nikov, A.V.; Zuev, V.V.; Marichev, V.N.; Tsaregorodtsev, S.I. First results of lidar observations of stratospheric ozone above western Siberia. *Atmos. Ocean.Opt.* **1989**, *2*, 841–842.
13. McDermid, I.S.; Walsh, T.D.; Deslis, A.; White, M.L. Optical systems design for a stratospheric lidar system. *Appl. Opt.* **1995**, *34*, 6201–6210. [CrossRef] [PubMed]
14. Godin, S.; David, C.; Lakoste, A.M. Systematic ozone and aerosol lidar measurements at OHP (44°N, 6°E) and Dumont. In Proceedings of the Abstracts of Papers: 17th International Laser Radar Conference, Sendai, Japan, 25–29 July 1994; pp. 409–412.
15. Claude, H.; Scönenborn, F.; Streinbrecht, W.; Vandersee, W. DIAL ozone measurements at the Met. Obs. Hohenpeißenberg: Climatology and trends. In Proceedings of the Abstracts of Papers: 17th International Laser Radar Conference, Sendai, Japan, 25–29 July 1994; pp. 413–415.
16. Stefanutti, L.; Castagnoli, F.; Guasta, D.M.; Morandi, M.; Sacco, V.M.; Zuccagnoli, L.; Godin, S.; Megie, G.; Porteneuve, J. A four-wavelength depolarization backscattering LIDAR for IISC monitoring. *Appl. Phys. B* **1992**, *55*, 13–17. [CrossRef]

17. Pazmino, A.F.; Lavorato, M.B.; Fochesatto, G.J. DIAL system for measurements of stratospheric ozone at Buenos Aires. In *Advances in Laser Remote Sensing, Proceedings of the Selected Papers Presented at the 20th ILRC, Vichy, France, 10–14 July 2000*; Ecole Polytechnique Impr: Paris, France, 2001.
18. Malicet, J.; Daumont, D.; Charbonnier, J.; Parisse, C.; Chakir, A.; Brion, J. Ozone UV spectroscopy. II. Absorption cross-sections and temperature dependence. *J. Atmos. Chem.* **1995**, *21*, 263–273. [CrossRef]
19. Zhu, H.; Qu, Z.W.; Grebenshchikov, S.Y.; Schinke, R.; Malicet, J.; Brion, J.; Daumont, D. Huggins band of ozone: Assignment of hot bands. *J. Chem. Phys.* **2005**, *122*. [CrossRef] [PubMed]
20. Clerbaux, C.; Boynard, A.; Clarisse, L.; George, M.; Hadji-Lazaro, J.; Herbin, H.; Hurtmans, D.; Pommier, M.; Razavi, A.; Turquety, S.; et al. Monitoring of atmospheric composition using the thermal infrared IASI/MetOp sounder. *Atmos. Chem. Phys.* **2009**, *9*, 6041–6054. [CrossRef]
21. Matvienko, G.G.; Belan, B.D.; Panchenko, M.V.; Romanovskii, O.A.; Sakerin, S.M.; Kabanov, D.M.; Turchinovich, S.A.; Turchinovich, Y.S.; Eremina, T.A.; Kozlov, V.S.; et al. Complex experiment on studying the microphysical, chemical, and optical properties of aerosol particles and estimating the contribution of atmospheric aerosol-to-earth radiation budget. *Atmos. Meas. Tech.* **2015**, *8*, 4507–4520. [CrossRef]
22. August, T.; Klaes, D.; Schlüssel, P.; Hultberg, T.; Crapeau, M.; Arriaga, A.; O'Carroll, A.; Coppens, D.; Munro, R.; Calbet, X. IASI on Metop-A: Operational Level 2 retrievals after five years in orbit. *J. Quant. Spectrosc. Radiat. Trans.* **2012**, *113*, 1340–1371. [CrossRef]
23. Keim, C.; Eremenko, M.; Orphal, J.; Dufour, G.; Flaud, J.M.; Höpfner, M.; Boynard, A.; Clerbaux, C.; Payan, S.; Coheur, P.F.; et al. Tropospheric ozone from IASI: Comparison of different inversion algorithms and validation with ozone sondes in the northern middle latitudes. *Atmos. Chem. Phys.* **2009**, *9*, 9329–9347. [CrossRef]
24. Krueger, A.J.; Minzner, R.A. Mid-latitude ozone model for the 1976 U.S. Standard Atmosphere. *J. Geophys. Res.* **1976**, *81*, 4477–4488. [CrossRef]

remote sensing

MDPI

Article

Ground Ammonia Concentrations over China Derived from Satellite and Atmospheric Transport Modeling

Lei Liu [1], Xiuying Zhang [1,*], Wen Xu [2], Xuejun Liu [2], Xuehe Lu [1], Shanqian Wang [1], Wuting Zhang [1] and Limin Zhao [1,3]

[1]   Jiangsu Provincial Key Laboratory of Geographic Information Science and Technology, International Institute for Earth System Science, Nanjing University, Nanjing 210023, China; liulei_nju_geo@163.com (L.L.); luxh43@gmail.com (X.L.); shanqianwang@163.com (S.W.); zhangwuting75@163.com (W.Z.); zlm2016nju@163.com (L.Z.)
[2]   College of Resources and Environmental Sciences, Centre for Resources, Environment and Food Security, Key Lab of Plant-Soil Interactions of MOE, China Agricultural University, Beijing 100193, China; hi.xuwen@163.com (W.X.); liu310@cau.edu.cn (X.L.)
[3]   Jiangsu Center for Collaborative Innovation in Geographical Information Resource Development and Application, Nanjing 210023, China
*   Correspondence: lzhxy77@163.com

Academic Editors: Yang Liu, Jun Wang, Omar Torres, Richard Müller and Prasad S. Thenkabail
Received: 27 March 2017; Accepted: 7 May 2017; Published: 15 May 2017

Abstract: As a primary basic gas in the atmosphere, atmospheric ammonia (NH_3) plays an important role in determining air quality, environmental degradation, and climate change. However, the limited ground observation currently presents a barrier to estimating ground NH_3 concentrations on a regional scale, thus preventing a full understanding of the atmospheric processes in which this trace gas is involved. This study estimated the ground NH_3 concentrations over China, combining the Infrared Atmospheric Sounding Interferometer (IASI) satellite NH_3 columns and NH_3 profiles from an atmospheric chemistry transport model (CTM). The estimated ground NH_3 concentrations showed agreement with the variability in annual ground NH_3 measurements from the Chinese Nationwide Nitrogen Deposition Monitoring Network (NNDMN). Great spatial heterogeneity of ground NH_3 concentrations was found across China, and high ground NH_3 concentrations were found in Northern China, Southeastern China, and some areas in Xinjiang Province. The maximum ground NH_3 concentrations over China occurred in summer, followed by spring, autumn, and winter seasons, which were in agreement with the seasonal patterns of NH_3 emissions in China. This study suggested that a combination of NH_3 profiles from CTMs and NH_3 columns from satellite obtained reliable ground NH_3 concentrations over China.

Keywords: NH_3; satellite; CTM; spatial; ground

1. Introduction

Ammonia (NH_3) is the primary form of reactive nitrogen (Nr) in the environment and a key component of the ecosystems, representing more than half of atmospheric Nr emissions [1,2]. NH_3 emissions have been increasing in recent years due to the increasing agricultural livestock numbers and the increasing application of Nr fertilization [2,3], resulting in the high NH_3 concentrations in the atmosphere. NH_3 increase has enhanced the acidification and eutrophication of the ecosystems on local and international scales [2,4]. Previous studies have shown that the lifetime of NH_3 is very short from hours to several days [5,6] converting to particulate matter (PM) as well as leading to dry and wet depositions. NH_3 reacts with acid-forming compounds such as sulfur dioxide (SO_2) and nitrogen oxides (NO_x) to form particles containing ammonium sulfate ((NH_4)$_2SO_4$) and ammonium nitrate (NH_4NO_3) in the atmosphere [7]. These processes increase the amount of atmospheric particulate matter, particularly for particles smaller than 2.5 micrometers in diameter (PM2.5), thereby reducing visibility and negatively affecting environmental and human health [8,9]. Therefore, monitoring the ground NH_3 concentrations on a regional scale is vitally important to assist in enacting effective measures to protect the eco-environments and public health, with respect to air, soil, and water quality.

Progress in the understanding of the NH_3 cycling process, flux measurements, and instrumentation have allowed advances in estimating NH_3 concentrations in the atmosphere on a local or regional scale, based on the simulation of the chemical transport models (CTM). For example, a coupled MM5-CMAQ modeling system was used for computing the ground NH_3 concentration based on the NH_3 emission developed with a spatial resolution of 27 km $\times$ 27 km in the Beijing–Tianjin–Hebei (BTH) region of China [10]. The simulation error of ground NH_3 concentration in different seasons in BTH range from -24.4% to 7.8%, indicating the ground NH_3 concentrations simulated by MM5-CMAQ are comparable with the observations; A GEOS-Chem model was used to estimate the global and seasonal NH_3 with a resolution of 2° latitude $\times$ 2.5° longitude [11], showing that the simulated ground NH_3 concentrations are biased low compared to the Tropospheric Emission Spectrometer (TES) with seasonal mean differences of -0.92 to 1.58 ppb. Similar reports on estimating ground NH_3 concentrations from CMT could also be tracked in several studies [12–14]. Although these CTMs could simulate the profiles of NH_3 concentrations in the atmosphere, the ground NH_3 concentrations over a large scale, such as on a national scale over the entire area of China, are still poorly understood due to the large pixel sizes and the relatively high uncertainties resulting from errors of the emission data and the simplification of the chemistry schemes. Fortunately, numerous studies have shown that CTMs can produce profiles for aerosol [15–18], NO_2 [19–21], NH_3 [2,22–24], and SO_2 [19,25], denoting that the vertical profiles of the NH_3 concentrations from CTM were highly beneficial in calculating the ground NH_3 concentrations.

In comparison with CTM simulations, satellite remote sensing is considered as an observational perspective and offers another way to obtain large-scale NH_3 columns with high spatial resolutions, based on advanced infrared spectroscopy (IR) sounders, such as the Infrared Atmospheric Sounding Interferometer (IASI), the Tropospheric Emission Spectrometer (TES), and the Cross-track Infrared Sounder (CrIS) [26,27]. Large-scale distributions of IASI NH_3 columns could denote the status of NH_3 levels in regions not covered by ground measurement networks, expanding insight into new NH_3 sources including industry, agriculture, and biomass burning [2,22]. However, satellite NH_3 can only provide the columns and has no information of the vertical distributions of the columns (from the ground to the top of the atmosphere), presenting a barrier in obtaining the ground NH_3 concentrations. Fortunately, as mentioned in the last paragraph, the detailed NH_3 profiles could be obtained from CTMs. Combining the advantages of CTMs (NH_3 profiles) and satellite observations (large-scale overages with high spatiotemporal resolutions), the ground NH_3 concentrations can be derived.

We aimed to generate spatiotemporal ground NH_3 concentrations with the aid of the remotely sensed NH_3 columns and vertical NH_3 profiles from a CTM. The estimated ground NH_3 concentrations were further compared with the national ground monitoring network of the Chinese Nationwide Nitrogen Deposition Monitoring Network (NNDMN). Our purpose is not to replace traditional algorithms, but to combine the advantages of satellite with high spatial and temporal resolutions, and CTMs with detailed NH_3 vertical profiles in order to obtain high spatiotemporal ground NH_3 concentrations over China, hence providing basic information for the ground status of NH_3 concentrations and guiding the monitoring plans in the future over China.

2. Materials and Methods

2.1. Ground NH₃ Concentrations in the Atmosphere

Monitoring ground-based NH_3 concentrations on a regional scale is not straightforward due to the technical limitations and great variability of the concentrations in time and space [28]. While the availability of NH_3 concentration data and the flux measurements on local scales is increasing, the measurements on a regional scale are sparser [1].

We used the monthly ground NH_3 concentrations from the Chinese Nationwide Nitrogen Deposition Monitoring Network (NNDMN, made available on request by Prof. X.J. Liu, China Agricultural University) to evaluate the accuracy of the satellite-derived ground NH_3 concentrations. Monthly NH_3 concentrations (in units of µg N m^{-3}) were measured at 44 sites from 2010 to 2013 (Figure 1). The network mainly covered farmland sites but also included some grassland (two) and forest (four) sites across China [29,30]. The ground NH_3 concentrations in NNDMN were monitored using both DEnuder for Long-Term Atmospheric (DELTA) systems as well as Adapted Low-cost, Passive High Absorption (ALPHA) samplers [30,31]. ALPHA is a passive sampling system, while DELTA is an active sampling system. Monthly ground NH_3 concentrations were mostly monitored by DELTA, and few monitoring sites were measured by ALPHA. Xu et al. [30] showed that these two methods on measuring ground NH_3 concentrations were not significantly different and can be considered consistent.

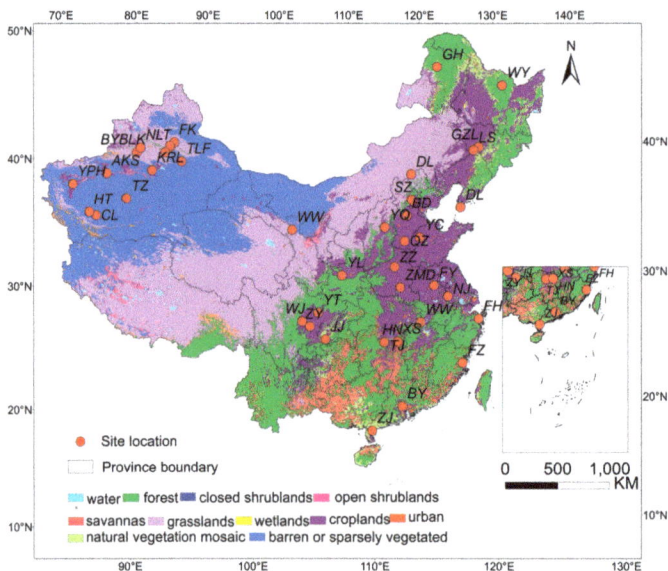

Figure 1. Spatial distribution of ground monitoring NH_3 sites in the Chinese Nationwide Nitrogen Deposition Monitoring Network (NNDMN).

2.2. IASI NH₃ Columns

The IASI instrument is on board the polar sun-synchronous MetOp platform, which crosses the equator at a mean local solar time of 9.30 a.m. and p.m. [32]. In this study, we used the measurements from the morning overpass as they are generally more sensitive to NH_3 because of higher thermal contrast at this time of day [1]. IASI has an elliptical footprint of 12 km by 12 km (at nadir) and up to 20 km by 39 km (off nadir), depending on the satellite viewing angle. The availability of measurements is mainly dependent on the cloud coverage.

The current method is based on the calculation of a spectral hyperspectral range index and subsequent conversion to a NH_3 total column using a neural network. Details on the retrieval algorithms can be found in Whitburn et al. [32]. We requested the IASI NH_3 data from Université Libre De Bruxelles, and processed the daily observation data to monthly average data for deriving the ground NH_3. In the present work, the observations with a cloud coverage lower than 25%, and relative error lower than 100% or absolute error less than 5×15 molec. cm^{-2} were processed [27].

2.3. NH₃ Profiles from MOZART-4

MOZART-4 (Model for Ozone and Related chemical Tracers, version 4) is a three-dimensional (3-D) global chemical transport model simulating the chemical and transport processes, which can be driven by essentially any meteorological dataset and with any emissions inventory [24,33]. The MOZART-4 used in this study includes detailed chemistry, an improved scheme for the determination of albedo, aerosols, online calculations of photolysis rates, dry deposition, H_2O concentration, and biogenic emissions. A comprehensive tropospheric chemistry with 85 gas-phase species, 12 bulk aerosol species, 39 photolyses, and 157 gas-phase reactions has been included in MOZART-4 [24]. The chemical initial and boundary conditions, spatially and temporally varying (6 h), are constrained by global chemical transport simulations from MOZART-4/GEOS-5 (Goddard Earth Observing System-5) with 1.9° latitude × 2.5° longitude horizontal resolution and 56 vertical levels from the surface. Details on the meteorological data and emission inventory used for driving MOZART-4 as well as related configurations can be tracked in Emmons et al. [24]. We requested the MOZART output data from NCAR (National Center for Atmospheric Research, Boulder, CO, USA). The output data are varying 6 h (daily). We calculated the monthly data by averaging the daily data, and then used the monthly data for analysis.

2.4. Satellite Derived Ground NH₃ Measurements

The fundamental thoughts of the methodology in this work were demonstrated in previous studies for aerosol [15–17], NO_2 [19–21] and SO_2 [19,25]. The recent progress in satellite NH_3 measurements also made this methodology applicable in estimating the ground NH_3 concentrations by combining the NH_3 profiles from CTM and NH_3 columns.

We had three major steps to estimate the satellite-derived ground NH_3 concentrations (Figure 2). First, we produced continuous monthly IASI NH_3 columns according to the method in previous studies [27,32]. Second, we simulated the vertical profiles from MOZART-4, and calculated the ratio of ground NH_3 to NH_3 columns. Third, we derived the satellite-derived ground NH_3 concentrations combining the IASI NH_3 columns and the ratio in the second step. Of these three steps, the second step of simulating the vertical profiles was the most important and complex one. We demonstrate here the key algorithms to simulate the vertical profiles from MOZART.

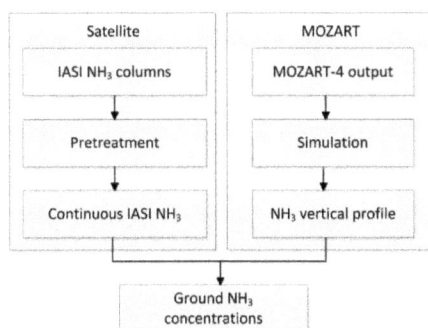

Figure 2. Schematic of the method to estimate the satellite-derived ground NH_3 concentrations.

We retrieved the NH_3 profiles from MOZART to convert the IASI NH_3 columns to ground NH_3 concentrations. The NH_3 vertical profile function was simulated by the following equation in the grid cell using the output data from MOZART-4:

$$f(h) = \sum_{i=1}^{n} a_i e^{\frac{-(h-b_i)^2}{c_i^2}} \tag{1}$$

where n ranges from 2 to 6, representing the number of Gaussian items; a_i, b_i, and c_i indicate the constants for each Gaussian item; h indicates the vertical height from the ground and $f(h)$ is the NH_3 concentration at height h. Theoretically, we can use n larger than 6 (with more Gaussian items). However, it is highly dependent on the computational time cost and computer memory limitations.

We simulated the NH_3 vertical profile using Equation (1) by each grid cell, based on the 56 vertical layers of NH_3 concentrations from MOZART. For each grid cell, we had five models ($n = 2, 3, 4, 5, 6$) and used R^2 and root-mean-square error (RMSE) to assess each model performance. We selected the best one with highest R^2 and lowest RMSE (i.e., determined the value of n).

The MOZART NH_3 columns can be gained by integration based on the simulated profile function:

$$F(h_{trop}) = \int_{0}^{h_{trop}} f(h)dh \tag{2}$$

where $F(h_{trop})$ denotes NH_3 columns and h_{trop} indicates the tropospheric height.

The satellite-derived ground NH_3 concentration is calculated as:

$$[_S NH_3]_G = [_S NH_3]_{Trop} \times \frac{f(h_G)}{F(h_{trop})} \tag{3}$$

where $[_S NH_3]_{Trop}$ indicates the IASI NH_3 columns, $f(h_G)$ denotes the ground NH_3 concentration from MOZART, and $F(h_{trop})$ represents the MOZART NH_3 columns.

We used the national ground-based NH_3 concentrations in NNDMN between 2010–2013 to validate the satellite-derived ground NH_3 concentrations. We applied the correlation coefficient (r) and relative error ((observation-estimation)/observation) at each monitoring site to assess the accuracy of the satellite-derived ground NH_3 concentrations.

3. Results and Discussion

3.1. Accuracy Assessment of the Estimated Ground NH_3 Concentrations

To convert the IASI NH_3 columns to ground NH_3 concentrations, it is essential to obtain the vertical NH_3 profiles. We retrieved the vertical NH_3 profiles from MOZART in this study (as an example, the vertical NH_3 concentrations at five locations in January 2013 from MOZART are shown in Figure A1). The NH_3 profiles were simulated by each grid cell in China (Figure A9) with determination of coefficients (R^2) larger than 0.95 accounting for 99.81% of all grid cells (Table A1 and Figure A9). Then, we estimated the ground NH_3 concentrations based on IASI NH_3 columns and the modeling MOZART NH_3 profiles.

We used 44 ground-based sites from NNDMN between 2010–2013 to assess the performance of the estimated monthly ground NH_3 concentrations. The correlation between the estimated and measured at each site is given in Table A2 in Appendix A, and the relative bias of each site as well as the yearly comparisons between the estimated and measured ground NH_3 concentration are given in Figures 3 and 4. We found 90.91% of minoring sites has a relative error within −30%–50%, showing an agreement between the estimated and measured. The seasonal absolute error by inverse-distance-weighted (IDW) interpolation is also shown in Figure A2. We found the absolute error in winter (December, January, and February) was higher than in other seasons, which can be explained by the highest relative error in IASI NH$_3$ columns in the winter season (Figure A3). In addition, Figure 4 demonstrates a comparison between the estimated and measured ground NH$_3$ concentrations before and after applying the IASI NH$_3$ data. We found a relatively higher correlation (R, 0.81 vs. 0.57) and a better consistency (slope, 0.96 vs. 0.50) between the satellite-derived ground NH$_3$ concentrations and the measured ground NH$_3$ concentrations than those from MOZART not applying the IASI NH$_3$ data.

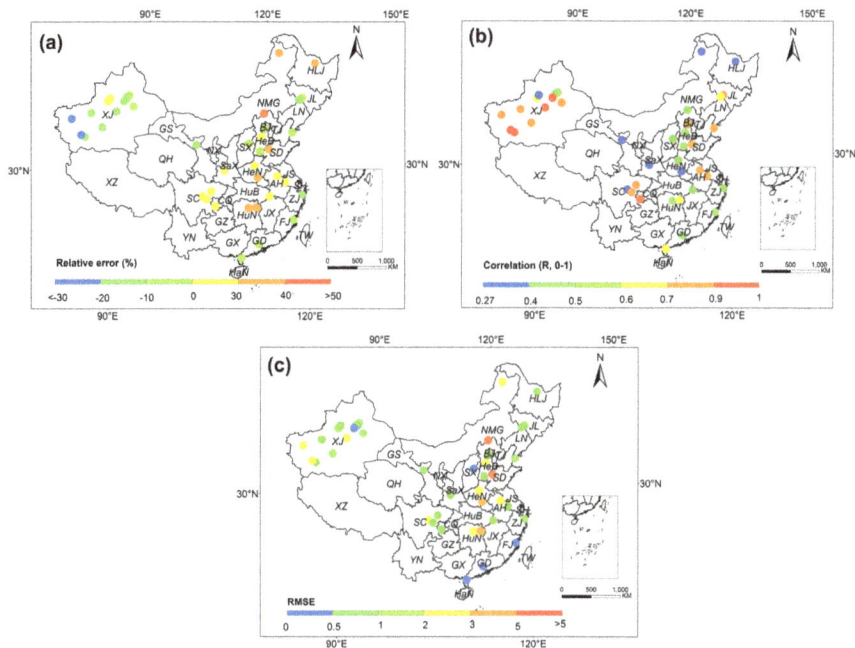

Figure 3. Spatial distribution of the relative error (**a**), correlation (**b**) and root-mean-square error (RMSE) (**c**) of the estimated ground NH$_3$ concentration (μg N m^{-3}) at 44 NNDMN sites.

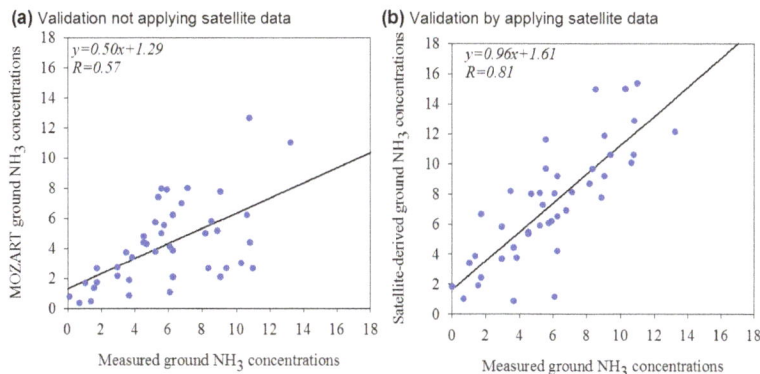

Figure 4. Yearly comparisons between the estimated and measured ground NH$_3$ concentration (μg N m^{-3}). (**a**) indicates the comparison between the measured ground NH$_3$ concentrations and the estimated ground NH$_3$ concentrations from MOZART at the lowest layer before applying the satellite data, while (**b**) represents the comparison between the measured and estimated ground NH$_3$ concentrations by applying the satellite data using the methods in Section 2.4.

3.2. Spatial Pattern of the Ground NH₃ Concentrations

Spatial distribution of ground NH_3 concentrations in 2012 over China is shown in Figure 5a. High ground NH_3 concentrations greater than 10 µg N m^{-3} were concentrated in North China and South China including Beijing–Tianjin–Hebei (BTH), Shandong, Henan, Hubei, Anhui, Sichuan and Jiangsu provinces, forming the major regions of intensive agriculture over China. Low ground NH_3 concentrations are predominantly located in TP (Tibetan Plateau), where both the synthetic fertilizers and livestock waste were the least among 32 provinces [34,35]. The spatial ground NH_3 concentrations revealed considerable spatial heterogeneity across China and were in agreement with the percent farmland area (Figure 5a,b), reflecting its unique agricultural structure and farming practice.

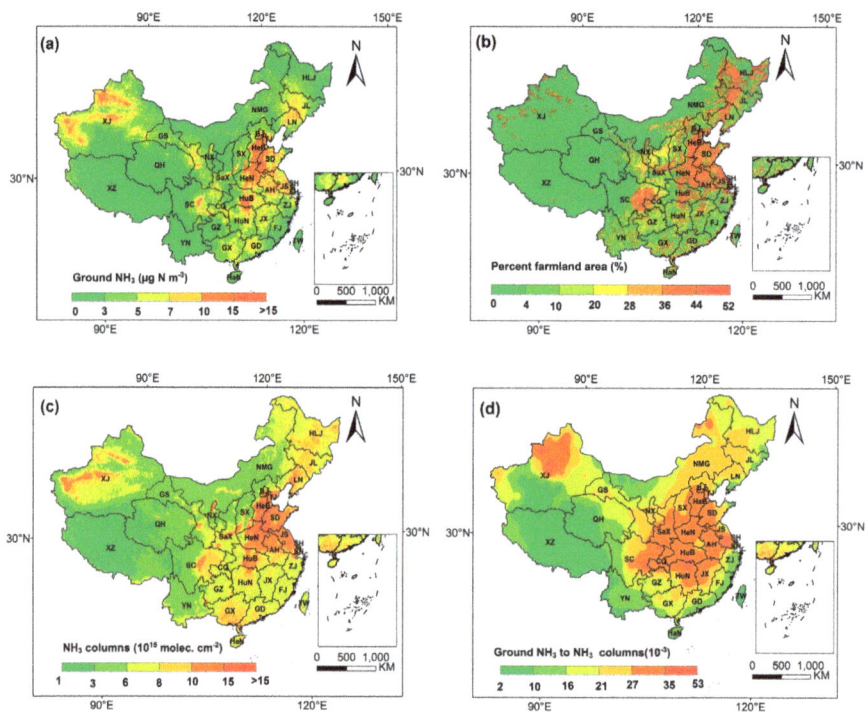

Figure 5. Spatial distribution of the ground NH_3 concentration (µg N m^{-3}). (**a**) represents the yearly estimated ground NH_3 concentrations; (**b**) denotes the percent farmland area; (**c**) denotes the Infrared Atmospheric Sounding Interferometer (IASI) NH_3 columns and (**d**) indicates the ratio of ground NH_3 concentration to NH_3 columns from MOZART.

High ground NH_3 concentrations were also observed in some areas in Xinjiang province (Figure 5a), where our estimation were about −30% to −10% underestimation compared with measurements in NNDMN (Figure 3). Moreover, relatively high NH_3 columns could be observed by satellite IASI instrument (Figure 5c). Synthetic N fertilizers and livestock waste both dominated the spatial distribution of the total emissions [34,35], hence determining the spatial patterns of the ground NH_3 concentrations. Previous studies reported that the NH_3 emissions from livestock exceeded those from the farmland in China, and NH_3 emissions from livestock accounted for about 54% of the total NH_3 emissions over China [35]. The contribution of livestock to the total NH_3 emissions in Xinjiang (where sheep are widely raised) accounted for higher than 60% [10,35]. Thus, due to the combining influence of both synthetic N fertilizers and livestock waste, the spatial distributions of ground NH_3

concentrations and percent farmland differed, especially in regions where the livestock dominated the NH_3 emissions. In addition, most of the ground NH_3 emissions were more concentrated on the ground and relatively hard to transport vertically compared with other regions in China, which can be clearly seen by the ratio of ground NH_3 concentrations to NH_3 columns from MOZART (Figure 5d).

3.3. Seasonal Variations of the Ground NH_3 Concentrations in China

To demonstrate the seasonal variations of the ground NH_3 concentrations in China, we calculated the monthly average values throughout China (Figure 6a). We found the maximum ground NH_3 concentrations over China occurred in summer (June, July, and August), followed by spring (March, April, and May), autumn (September, October, and November) and winter (December, January, and February) seasons. It is interesting that the seasonal ground NH_3 concentrations were in agreement with the seasonal patterns of NH_3 emissions in China conducted by Kang et al. [36], Huang et al. [35], and Xu et al. [37] (Figure 6b–d), indicating that the NH_3 emissions are the key factor influencing seasonal pattern of the ground NH_3 concentrations. The maximum NH_3 emissions in summer is reasonable due to more than 40% of the fertilization and more than 25% of livestock emissions occurring in summer [36,37]. In addition, high temperature in summer in China may also accelerate the NH_3 volatilization ($NH_4^+ \rightarrow NH_3 + H^+$) from fertilizer, animal waste, city garbage or vehicles [6,38–40], and hence cause high ground NH_3 concentrations. In contrast, in winter, temperature frequently below freezing leads to reduced NH_3 volatilization and lower NH_3 concentrations than in other seasons.

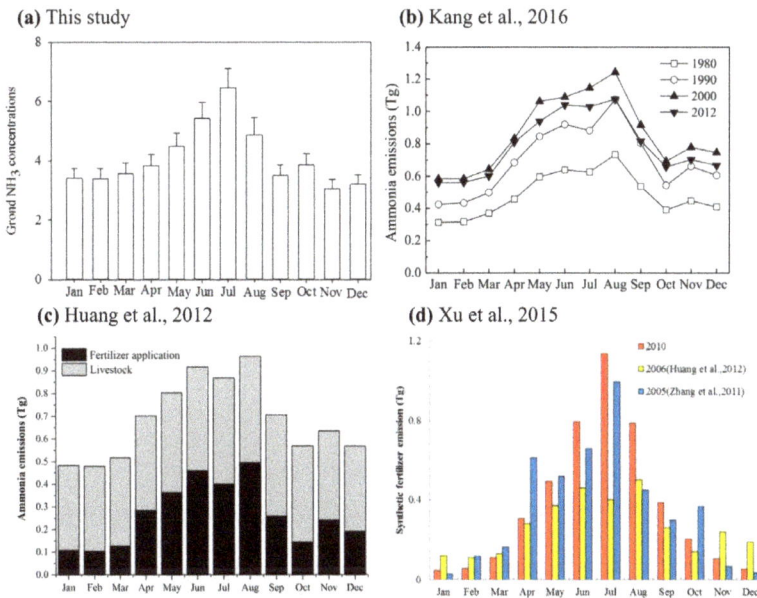

Figure 6. Seasonal patterns of ground NH_3 concentrations in China. (**a**) indicates the monthly variations of ground NH_3 concentrations (μg N m^{-3}) in China; (**b**) represents the monthly variations of the total NH_3 emissions (Tg, 10^{12} g) in China conducted by Kang et al. [36]; (**c**) shows the the monthly variations of the sum of fertilizer and livestock NH_3 emissions (Tg) in China conducted by Huang et al. [35] and (**d**) denotes the monthly variations of the fertilizer NH_3 emissions (Tg) in China conducted by Xu et al. [37].

To more accurately quantify the effects of meteorological parameters on the seasonal trends of the ground NH_3 concentrations, we selected the five best-simulated ground sites with n >30

(Table A2) for demonstrating meteorological parameters, such as temperature, wind speed, humidity, and precipitation on the seasonal variations of the ground NH_3 concentrations (Figures 7 and A4–A8). The monthly wind speed, temperature, relative humidity, and precipitation for each site were taken from the China Meteorological Administration. A positive correlation ($R = 0.6$, $p = 0.00$) was found between the ground NH_3 concentrations and temperature. An inverse relationship between the ground NH_3 concentrations and humidity (Figure 7), indicated that higher relative humidity may contribute to more NH_3 loss rates ($NH_3 \rightarrow NH_4^+$). In addition, we also conducted a partial correlation analysis [41] regarding ground NH_3 concentrations, temperature, and humidity by considering their interactions using the function "partialcorr" in Matlab. We found the partial correlation between ground NH_3 concentrations and humidity was -0.10 ($p = 0.03$), showing a significant inverse relationship between the ground NH_3 concentrations and humidity. Significant effects of air humidity on NH_3 loss were also demonstrated previously [42,43]. However, precipitation and wind speed were not significantly correlated with ground NH_3 concentrations ($p = 0.632$, precipitation vs. NH_3; $p = 0.156$, wind speed vs. NH_3) as shown in Figures A4–A8.

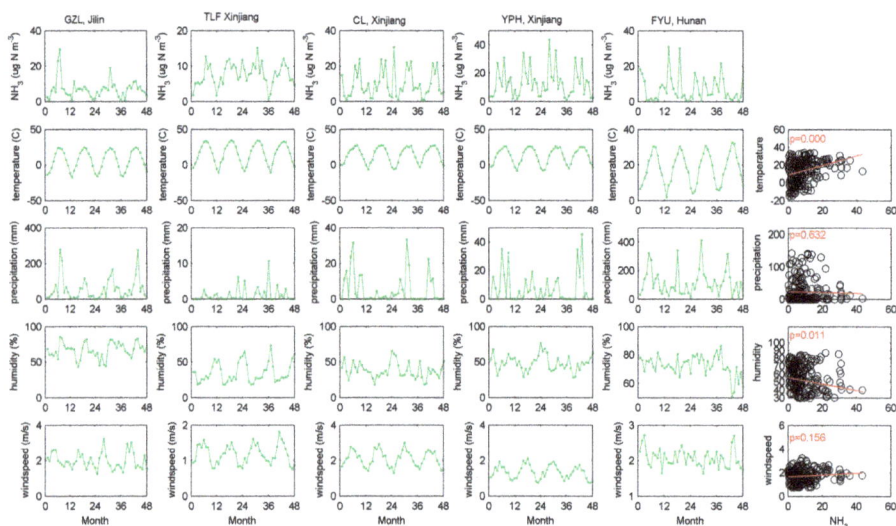

Figure 7. The seasonal variations of ground NH_3 concentrations (μg N m^{-3}), temperature ($^\circ$C), precipitation (mm), humidity (%), and wind speed (m/s) at five sites with best-simulated ground NH_3 concentrations from January 2010 to December 2013 (0–12, 2010; 13–24, 2011; 25–36, 2012; 37–48, 2013). The relationship between the ground NH_3 concentrations and precipitation (mm), humidity (%), and wind speed (m/s) at each site is provided in Figures A4–A8.

3.4. Comparison with Previous Studies

The first relatively complete work on the national ground measurements of NH_3 concentrations in China is NNDMN, and the results of ground measurements were published by Xu et al. [30], which we considered as a truly comprehensive and valuable work on the national status of the ground NH_3 concentrations, and which shed some light on the actual status of ground NH_3 concentrations. The national measurements in NNDMN provide the best accurate datasets for validating the modeling ground NH_3 concentrations. In the previous studies, due to very limited ground measurements (not to mention the national monitoring measurements), it was difficult to validate the accuracy of the modeling ground NH_3 concentrations in China. The lack of measurements makes it necessary to assess the modeling ground NH_3 concentrations in China [44]. Recently, Zhao et al. [45] presented a comprehensive work on the national-scale model validation of ground NH_3 concentrations with $1/2°$ longitude by $1/3°$ latitude horizontal resolution using the GEOS-Chem model, showing the correlation coefficient with NNDMN between 2011–2012 which was about 0.65 on the annual scale [45]. Compared with Zhao et al. [45], we used the same datasets from NNDMN while having a longer time period (2010–2013) to validate our estimated ground NH_3 concentrations, and found the correlation coefficient was about 0.81 (slope = 0.96 and intercept = 1.31) on the annual scale as shown in Figure 4, demonstrating better agreement with the ground measurements. The relatively higher accuracy in estimating ground NH_3 concentrations may result from different datasets used for estimation, where we used the satellite observation and Zhao et al. [45] used the NH_3 emission data used for modeling. Uncertainties existed in the estimation of NH_3 emission resulting from the methodology of calculation, which simplified the complexity of the real status of emission process [36]. For example, N-fertilizer NH_3 emission in BTH between different studies varied greatly as 256.5 Gg [35], 502.5 Gg [46], 432.7 Gg [10]; livestock NH_3 emission in BTH between different studies varied as 556.6 Gg [35], 675.2 Gg [46], and 891.6 Gg [10]. The estimation of NH_3 emissions by Zhou et al. [10] even nearly doubled that by Huang et al. [35] and Dong et al. [46]. The actual local emission factors in different regions differed from each other greatly, due to the difference of the local meteorological conditions, fertilizing time, and fertilizer kinds [37]. The NH_3 emissions are mainly based on statistical NH_3 emissions at a city or county level, and the accuracy is strongly dependent on both the limited spatial and temporal resolutions of the coarse statistical data [35–37,44,47].

The present study derived ground NH_3 concentrations from IASI NH_3 columns and the profiles from MOZART-4, implying that a combination of CTM modeling and satellite monitoring obtained a reliable ground NH_3 estimation over China. More generally, this attempt to generate the ground NH_3 measurements with a relative high resolution from IASI and MOZART has highlighted known limitations in the ground NH_3 monitoring measurements, which may in some cases not be representative of the estimated NH_3 concentrations horizontally and vertically. Here we highlight the need to acquire more comprehensive datasets of ground NH_3 concentrations, and dedicated measurement campaigns focusing on the ground NH_3 measurement will no doubt allow improvements in the validation of estimated NH_3 in the future. In addition, we focused on the spatial pattern of ground NH_3 concentrations derived from satellite and a CTM, which is based on the monthly average and may be limited for the specific analysis such as secondary aerosol formation, photochemistry, and consideration of regulation. It is also beneficial and even essential to gain higher temporal resolution of ground NH_3 concentrations in the future.

4. Conclusions

We critically estimated the ground NH_3 concentrations over China, combining IASI NH_3 columns and NH_3 profiles from MOZART. We aimed to generate ground NH_3 concentrations over China, and hence provide potential to understand both the spatial and temporal variations of ground NH_3 concentrations in order to guide future ground NH_3 monitoring plans. The intention was not to replace traditional algorithms but to provide new insight on the current status of ground NH_3 over China, and to generate more reliable ground NH_3 concentrations. The IASI NH_3 columns and NH_3 profiles from the atmospheric chemistry transport model are encouraged to be combined to generate ground NH_3 concentrations at local or regional scales, and the estimated results should be further improved.

This study introduced methods to estimate ground NH_3 concentrations over China using IASI NH_3 columns and NH_3 profiles. The estimated ground NH_3 concentrations were validated by 44 sites from NNDMN, showing promising results between the estimated and measured, and then the spatial and temporal variations of ground NH_3 concentrations were demonstrated. High ground NH_3 concentrations greater than 10 μg N m^{-3} were mainly located in Beijing, Hebei, Shandong, Henan, Jiangsu, eastern Sichuan, and some regions in Xinjiang provinces, while low ground NH_3 concentrations were concentrated in the Tibet-Plateau area. The maximum ground NH_3 concentrations over China occurred in summer, followed by spring, autumn, and winter seasons, which are in agreement with the seasonal patterns of NH_3 emissions in China.

Acknowledgments: This study was supported by the National Natural Science Foundation of China (No. 41471343, 40425007 and 41101315) and Doctoral Research Innovation Fund (2016CL07). We also much appreciate the free use of the IASI NH_3 data provided by Université Libre de Bruxelles (ULB) (http://www.ulb.ac.be/cpm/atmosphere.html).

Author Contributions: L.L. and X.Z conceived the idea; L.L. and S.W. conducted the analyses; L.L. and S.W. processed the data; X.L. and W.X. provided the observation data for validation; X.Z, X.L., L.Z., and W.Z. contributed to the writing and revisions.

Conflicts of Interest: The authors declare no competing financial interest.

Appendix A

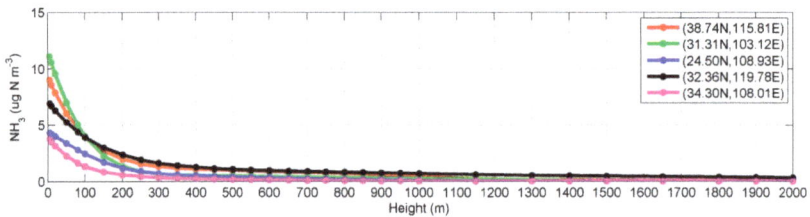

Figure A1. Vertical NH_3 concentrations (μg N m^{-3}) simulated by Mozart at five locations in January 2013.

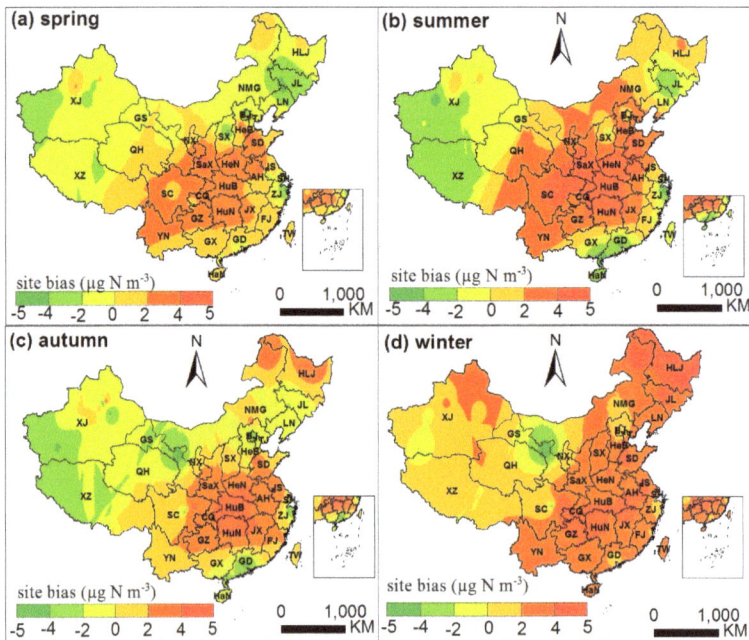

Figure A2. A quick illustration of the site bias of ground NH_3 concentrations across China by interpolating the residuals between the measured and estimated using the inverse-distance-weighted (IDW) interpolation. The figures were generated using ArcGIS 12.0 software (https://www.arcgis.com/).

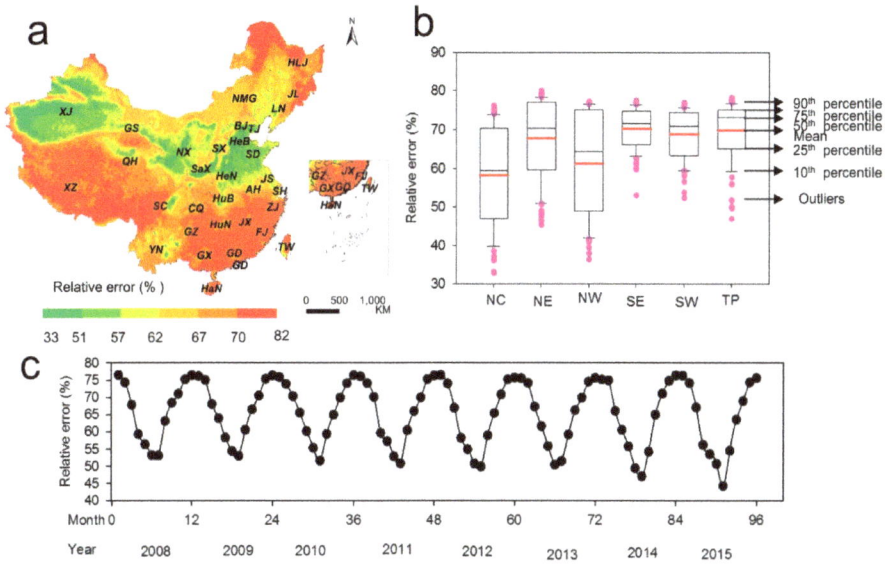

Figure A3. Relative error (%) of IASI NH$_3$ columns. (**a**) indicates the annual IASI NH$_3$ error (with a cloud coverage lower than 25%) averaged from 2008 to 2015; (**b**) indicates the averaged monthly relative error from 2008 to 2015 in different regions (every dot indicates the relative error at a month in a region); (**c**) indicates the temporal variations of relative error over China at a monthly scale.

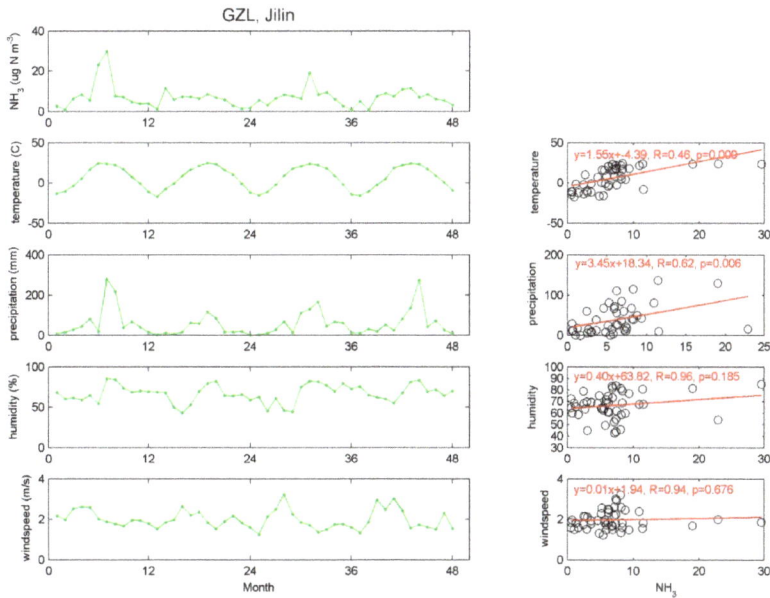

Figure A4. The seasonal variations of ground NH$_3$ concentrations (µg N m^{-3}), temperature (°C), precipitation (mm), humidity (%), and wind speed (m/s) at GZL from January 2010 to December 2013 (0–12, 2010; 13–24, 2011; 25–36, 2012; 37–48, 2013).

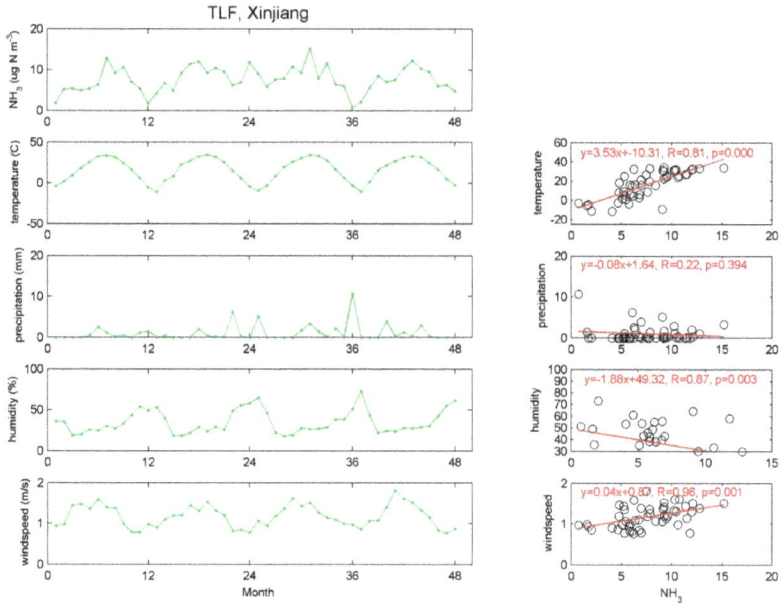

Figure A5. The seasonal variations of ground NH_3 concentrations (μg N m^{-3}), temperature (°C), precipitation (mm), humidity (%), and wind speed (m/s) at TLF from January 2010 to December 2013 (0–12, 2010; 13–24, 2011; 25–36, 2012; 37–48, 2013).

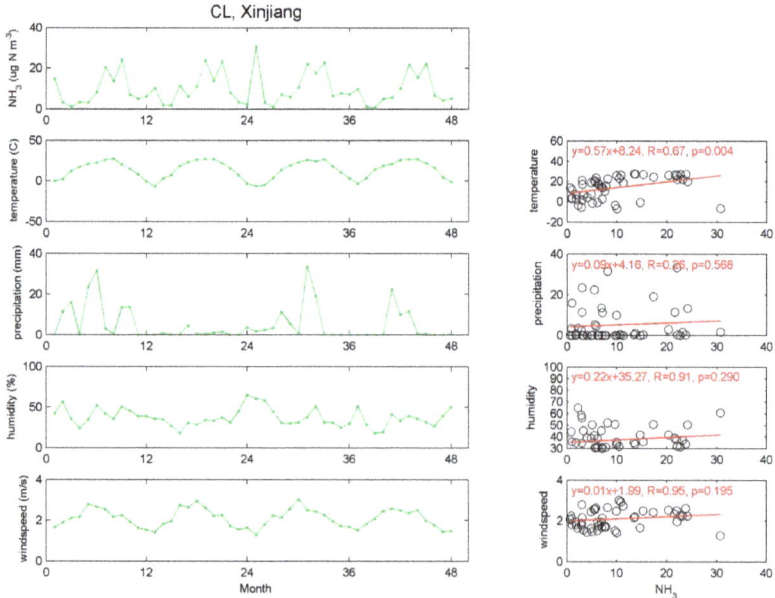

Figure A6. The seasonal variations of ground NH_3 concentrations (μg N m^{-3}), temperature (°C), precipitation (mm), humidity (%), and wind speed (m/s) at CL from January 2010 to December 2013 (0–12, 2010; 13–24, 2011; 25–36, 2012; 37–48, 2013).

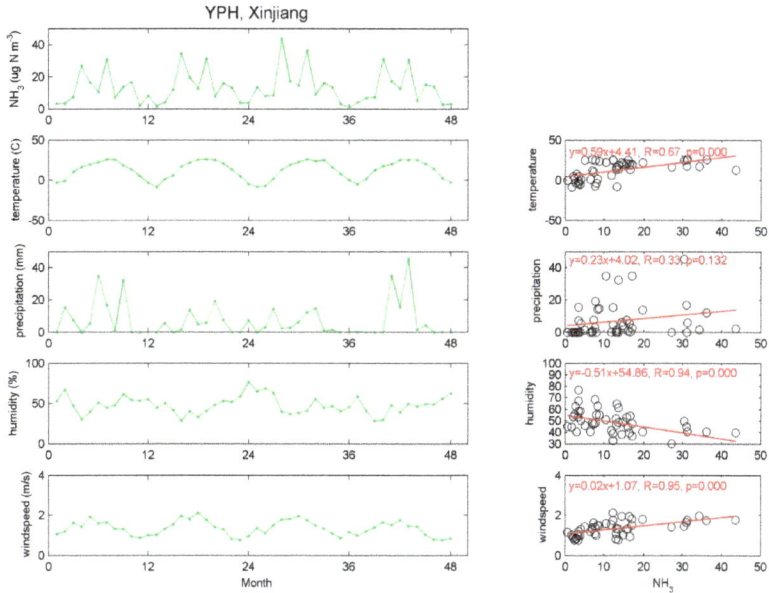

Figure A7. The seasonal variations of ground NH_3 concentrations (μg N m^{-3}), temperature ($^\circ$C), precipitation (mm), humidity (%), and wind speed (m/s) at YPH from January 2010 to December 2013 (0–12, 2010; 13–24, 2011; 25–36, 2012; 37–48, 2013).

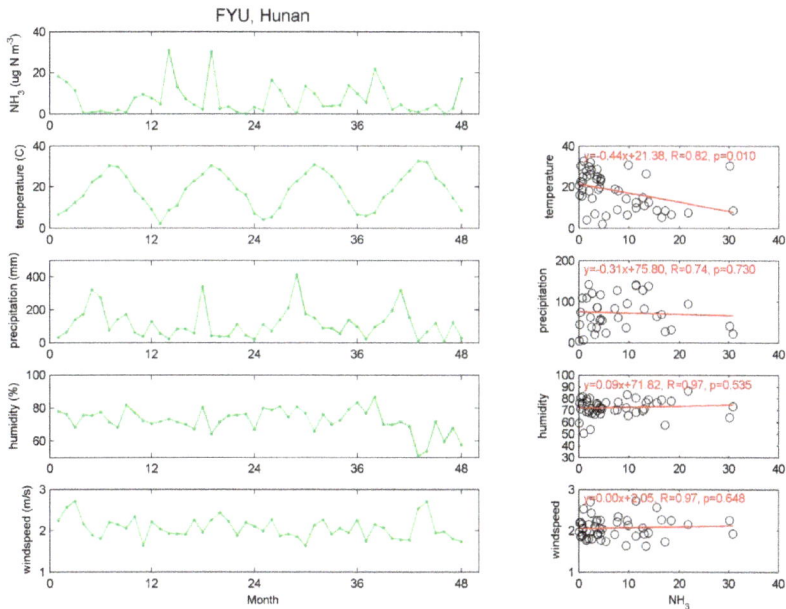

Figure A8. The seasonal variations of ground NH_3 concentrations (μg N m^{-3}), temperature ($^\circ$C), precipitation (mm), humidity (%), and wind speed (m/s) at FYU from January 2010 to December 2013 (0–12, 2010; 13–24, 2011; 25–36, 2012; 37–48, 2013).

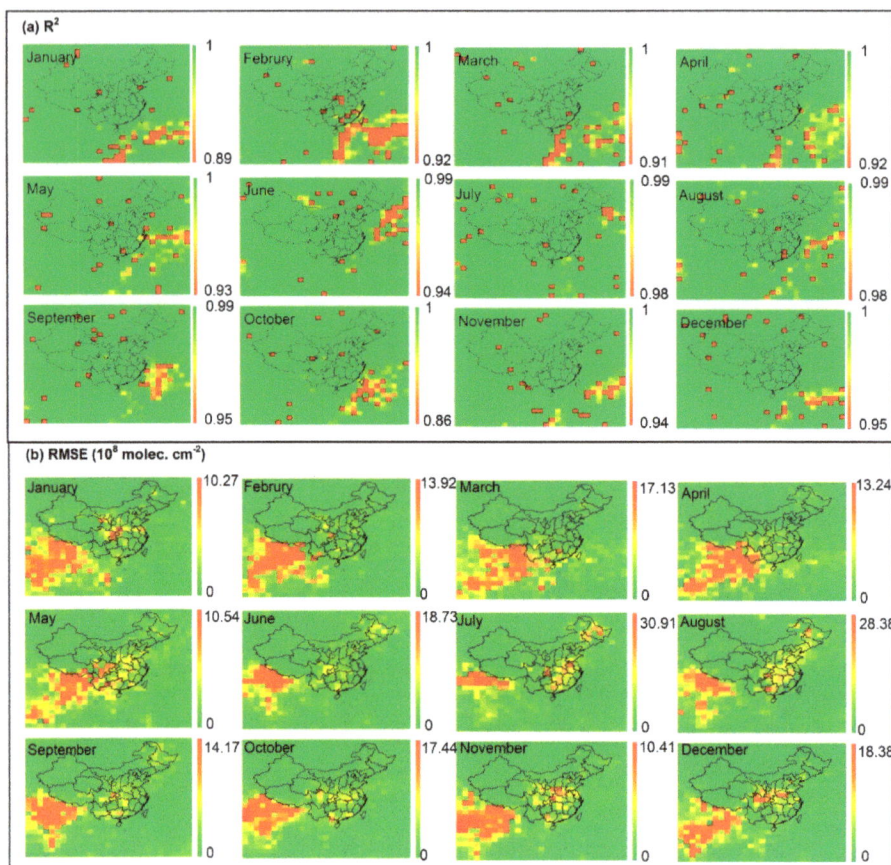

Figure A9. (**a**,**b**) R^2 and RMSE (molec./cm^2) for the Gaussian simulation of the NH$_3$ profiles (68~142°E, 5~55°N) in 2013.

Table A1. Descriptive statistics for results of Gaussian simulation.

Season (%)	N = 2	N = 3	N = 4	N = 5	N = 6	$R^2 > 0.95$	$R^2 > 0.99$
Spring	0.70	12.02	33.33	34.61	19.31	99.86	96.94
Summer	0.79	10.47	28.24	37.09	23.38	99.86	97.52
Autumn	0.48	7.60	24.58	37.93	29.39	99.86	98.89
Winter	0.92	10.25	31.03	35.80	21.97	99.64	96.46
All	0.72	10.09	29.29	36.36	23.51	99.81	97.45

Note: Spring includes March, April, and May; Summer includes June, July, and August; Autumn includes September, October, and November; Winter includes December, January, and February. *N* indicates the numbers of the Gaussian items. For details, please refer to the methods part.

Table A2. Comparison between monthly IASI satellite-derived ground NH$_3$ concentrations and the NNDMN monitoring sites from 2010 to 2013.

Site	Landuse	Long (°E)	Lat (°N)	n	R ($\pm std$) This Study
BYBLK	Alpine grassland	83.71	42.88	22	0.68 (0.05)
FK	Desert-oasis ecotone	87.93	44.29	32	0.49 (0.04)
TLF	Desert in an oasis	89.19	42.85	28	0.84 (0.07)
SDS	Urban	87.56	43.85	38	0.69 (0.06)
TFS	Suburban	87.47	43.94	35	0.56 (0.05)
CL	Desert-oasis ecotone	80.73	37.02	12	0.94 (0.08)
TZ	Desert	83.66	38.97	12	0.89 (0.07)
YPH	Farmland	77.27	39	12	0.83 (0.05)
HT	Farmland	79.89	37.15	5	0.99 (0.08)
AKS	Farmland	80.83	40.62	17	0.72 (0.06)
KRL	Farmland	85.86	41.68	6	0.94 (0.08)
NLT	Forest	84.03	43.31	4	0.33 (0.03)
NSXC	Forest	87.04	43.35	7	0.98 (0.09)
CAU	Urban	116.28	40.02	45	0.57 (0.05)
ZZ	Urban	113.63	34.75	44	0.55 (0.04)
SZ	Farmland	116.2	40.11	45	0.86 (0.07)
BD	Farmland	115.48	38.85	12	0.44 (0.04)
QZ	Farmland	114.94	36.78	45	0.50 (0.04)
YQ	Farmland	112.89	38.05	45	0.57 (0.05)
ZMD	Farmland	114.05	33.02	45	0.27 (0.02)
YL	Farmland	108.01	34.31	45	0.27 (0.02)
YC	Farmland	116.63	36.94	35	0.77 (0.06)
GZL	Farmland	124.83	43.53	42	0.82 (0.06)
LS	Farmland	124.17	43.36	42	0.62 (0.05)
DL	Coastal	121.58	38.92	40	0.73 (0.05)
WY	Forest	129.25	48.11	12	0.31 (0.02)
GH	Forest	121.52	50.78	12	0.38 (0.03)
WW	Farmland	102.6	38.07	39	0.32 (0.02)
DL	Grassland	116.49	42.2	6	0.52 (0.04)
WX	Farmland	115.79	30.01	29	0.56 (0.05)
BY	Farmland	113.27	23.16	44	0.47 (0.04)
TJ	Farmland	111.97	28.61	39	0.42 (0.03)
FYU	Farmland	113.34	28.56	40	0.76 (0.06)
HN	Farmland	113.41	28.52	40	0.36 (0.03)
NJ	Farmland	118.85	31.84	18	0.82 (0.06)
FY	Farmland	117.56	32.88	11	0.79 (0.06)
ZJ	Coastal	110.33	21.26	41	0.63 (0.05)
FZ	Coastal	119.36	26.17	45	0.49 (0.03)
FH	Coastal	121.53	29.61	41	0.57 (0.04)
XS	Forest	113.31	28.61	40	0.67 (0.06)
WJ	Farmland	103.84	30.55	39	0.28 (0.02)
ZY	Farmland	104.63	30.13	42	0.74 (0.06)
YT	Farmland	105.47	31.28	30	0.78 (0.06)
JJ	Farmland	106.18	29.06	12	0.94 (0.08)

References

1. Van Damme, M.; Clarisse, L.; Dammers, E.; Liu, X.; Nowak, J.; Clerbaux, C.; Flechard, C.; Galy-Lacaux, C.; Xu, W.; Neuman, J.; et al. Towards validation of ammonia (NH$_3$) measurements from the IASI satellite. *Atmos. Meas. Tech.* **2014**, *7*, 12125–12172. [CrossRef]
2. Warner, J.; Wei, Z.; Strow, L.; Dickerson, R.; Nowak, J. The global tropospheric ammonia distribution as seen in the 13 year AIRS measurement record. *Atmos. Chem. Phys. Discuss.* **2015**, *15*, 35823–35856. [CrossRef]

3. Van Damme, M.; Wichink Kruit, R.; Schaap, M.; Clarisse, L.; Clerbaux, C.; Coheur, P.F.; Dammers, E.; Dolman, A.; Erisman, J. Evaluating 4 years of atmospheric ammonia (NH₃) over Europe using IASI satellite observations and LOTOS-EUROS model results. *J. Geophys. Res. Atmos.* **2014**, *119*, 9549–9566. [CrossRef]

4. Paulot, F.; Jacob, D.J.; Pinder, R.; Bash, J.; Travis, K.; Henze, D. Ammonia emissions in the United States, European Union, and China derived by high-resolution inversion of ammonium wet deposition data: Interpretation with a new agricultural emissions inventory (MASAGE_NH₃). *J. Geophys. Res. Atmos.* **2014**, *119*, 4343–4364. [CrossRef]

5. Kruit, R.J.W.; Schaap, M.; Sauter, F.J.; Zanten, M.C.V. Modeling the distribution of ammonia across Europe including bi-directional surface-atmosphere exchange. *Biogeosciences* **2012**, *9*, 5261–5277. [CrossRef]

6. Sutton, M.A.; Reis, S.; Riddick, S.N.; Dragosits, U.; Nemitz, E.; Theobald, M.R.; Tang, Y.S.; Braban, C.F.; Vieno, M.; Dore, A.J.; et al. Towards a climate-dependent paradigm of ammonia emission and deposition. *Philos. Trans. R. Soc. B Biol. Sci.* **2013**, *368*. [CrossRef] [PubMed]

7. Gu, B.; Sutton, M.A.; Chang, S.X.; Ge, Y.; Chang, J. Agricultural ammonia emissions contribute to China's urban air pollution. *Front. Ecol. Environ.* **2014**, *12*, 265–266. [CrossRef]

8. Gu, B.; Ge, Y.; Ren, Y.; Xu, B.; Luo, W.; Jiang, H.; Gu, B.; Chang, J. Atmospheric reactive nitrogen in China: Sources, recent trends, and damage costs. *Environ. Sci. Technol.* **2012**, *46*, 9420–9427. [CrossRef] [PubMed]

9. Pope, C.A., III; Burnett, R.T.; Thun, M.J.; Calle, E.E.; Krewski, D.; Ito, K.; Thurston, G.D. Lung cancer, cardiopulmonary mortality, and long-term exposure to fine particulate air pollution. *JAMA* **2002**, *287*, 1132–1141. [CrossRef] [PubMed]

10. Zhou, Y.; Shuiyuan, C.; Lang, J.; Chen, D.; Zhao, B.; Liu, C.; Xu, R.; Li, T. A comprehensive ammonia emission inventory with high-resolution and its evaluation in the Beijing–Tianjin–Hebei (BTH) region, China. *Atmos. Environ.* **2015**, *106*, 305–317. [CrossRef]

11. Luo, M.; Shephard, M.W.; Cady-Pereira, K.E.; Henze, D.K.; Zhu, L.; Bash, J.O.; Pinder, R.W.; Capps, S.L.; Walker, J.T.; Jones, M.R. Satellite observations of tropospheric ammonia and carbon monoxide: Global distributions, regional correlations and comparisons to model simulations. *Atmos. Environ.* **2015**, *106*, 262–277. [CrossRef]

12. Hamaoui-Laguel, L.; Meleux, F.; Beekmann, M.; Bessagnet, B.; Génermont, S.; Cellier, P.; Létinois, L. Improving ammonia emissions in air quality modelling for France. *Atmos. Environ.* **2014**, *92*, 584–595. [CrossRef]

13. Wen, D.; Zhang, L.; Lin, J.; Vet, R.; Moran, M. An evaluation of ambient ammonia concentrations over southern Ontario simulated with different dry deposition schemes within STILT-Chem v0.8. *Geosci. Model Dev.* **2014**, *7*, 1037–1050. [CrossRef]

14. Wen, D.; Lin, J.; Zhang, L.; Vet, R.; Moran, M. Modeling atmospheric ammonia and ammonium using a stochastic Lagrangian air quality model (STILT-Chem v0.7). *Geosci. Model Dev.* **2013**, *6*, 327–344. [CrossRef]

15. Van Donkelaar, A.; Martin, R.V.; Park, R.J. Estimating ground-level PM2.5 using aerosol optical depth determined from satellite remote sensing. *J. Geophys. Res. Atmos.* **2006**, *111*. [CrossRef]

16. Liu, Y.; Park, R.J.; Jacob, D.J.; Li, Q.; Kilaru, V.; Sarnat, J.A. Mapping annual mean groun-level PM2.5 concentrations using Multiangle Imaging Spectroradiometer aerosol optical thickness over the contiguous United States. *J. Geophys. Res. Atmos.* **2004**, *109*, 1–10.

17. Van Donkelaar, A.; Martin, R.V.; Brauer, M.; Kahn, R.; Levy, R.; Verduzco, C.; Villeneuve, P.J. Global estimates of ambient fine particulate matter concentrations from satellite-based aerosol optical depth: Development and application. *Environ. Health Perspect.* **2010**, *118*, 847. [CrossRef] [PubMed]

18. Wang, J.; Xu, X.; Spurr, R.; Wang, Y.; Drury, E. Improved algorithm for MODIS satellite retrievals of aerosol optical thickness over land in dusty atmosphere: Implications for air quality monitoring in China. *Remote Sens. Environ.* **2010**, *114*, 2575–2583. [CrossRef]

19. Nowlan, C.; Martin, R.; Philip, S.; Lamsal, L.; Krotkov, N.; Marais, E.; Wang, S.; Zhang, Q. Global dry deposition of nitrogen dioxide and sulfur dioxide inferred from space-based measurements. *Glob. Biogeochem. Cycles* **2014**, *28*, 1025–1043. [CrossRef]

20. Lamsal, L.N.; Martin, R.V.; van Donkelaar, A.; Steinbacher, M.; Celarier, E.A.; Bucsela, E.; Dunlea, E.J.; Pinto, J.P. Ground-level nitrogen dioxide concentrations inferred from the satellite-borne Ozone Monitoring Instrument. *J. Geophys. Res. Atmos.* **2008**, *113*, 1–15. [CrossRef]

21. Hudman, R.C.; Jacob, D.J.; Turquety, S.; Leibensperger, E.M.; Murray, L.T.; Wu, S.; Gilliland, A.B.; Avery, M.; Bertram, T.H.; Brune, W.; et al. Surface and lightning sources of nitrogen oxides over the United States: Magnitudes, chemical evolution, and outflow. *J. Geophys. Res. Atmos.* **2007**, *112*, 1–14. [CrossRef]
22. Dammers, E.; Vigouroux, C.; Palm, M.; Mahieu, E.; Warneke, T.; Smale, D.; Langerock, B.; Franco, B.; Damme, M.V.; Schaap, M.; et al. Retrieval of ammonia from ground-based FTIR solar spectra. *Atmos. Chem. Phys.* **2015**, *15*, 12789–12803. [CrossRef]
23. Van Damme, M.; Erisman, J.W.; Clarisse, L.; Dammers, E.; Whitburn, S.; Clerbaux, C.; Dolman, A.J.; Coheur, P.F. Worldwide spatiotemporal atmospheric ammonia (NH$_3$) columns variability revealed by satellite. *Geophys. Res. Lett.* **2015**, *42*, 8660–8668. [CrossRef]
24. Emmons, L.; Walters, S.; Hess, P.; Lamarque, J.-F.; Pfister, G.; Fillmore, D.; Granier, C.; Guenther, A.; Kinnison, D.; Laepple, T.; et al. Description and evaluation of the Model for Ozone and Related chemical Tracers, version 4 (MOZART-4). *Geosci. Model Dev.* **2010**, *3*, 43–67. [CrossRef]
25. Lee, C.; Martin, R.V.; van Donkelaar, A.; Lee, H.; Dickerson, R.R.; Hains, J.C.; Krotkov, N.; Richter, A.; Vinnikov, K.; Schwab, J.J. SO$_2$ emissions and lifetimes: Estimates from inverse modeling using in situ and global, space-based (SCIAMACHY and OMI) observations. *J. Geophys. Res. Atmos.* **2011**, *116*, 1–13. [CrossRef]
26. Coheur, P.-F.; Clarisse, L.; Turquety, S.; Hurtmans, D.; Clerbaux, C. IASI measurements of reactive trace species in biomass burning plumes. *Atmos. Chem. Phys.* **2009**, *9*, 5655–5667. [CrossRef]
27. Liu, L.; Zhang, X.; Xu, W.; Liu, X.; Li, Y.; Lu, X.; Zhang, Y.; Zhang, W. Temporal characteristics of atmospheric ammonia and nitrogen dioxide over China based on emission data, satellite observations and atmospheric transport modeling since 1980. *Atmos. Chem. Phys. Discuss.* **2017**, *2017*, 1–32. [CrossRef]
28. Hertel, O.; Skjøth, C.A.; Reis, S.; Bleeker, A.; Harrison, R.; Cape, J.N.; Fowler, D.; Skiba, U.; Simpson, D.; Jickells, T.; et al. Governing processes for reactive nitrogen compounds in the European atmosphere. *Biogeosciences* **2012**, *9*, 4921–4954. [CrossRef]
29. Liu, X.; Duan, L.; Mo, J.; Du, E.; Shen, J.; Lu, X.; Zhang, Y.; Zhou, X.; He, C.; Zhang, F. Nitrogen deposition and its ecological impact in China: An overview. *Environ. Pollut.* **2011**, *159*, 2251–2264. [CrossRef] [PubMed]
30. Xu, W.; Luo, X.S.; Pan, Y.P.; Zhang, L.; Tang, A.H.; Shen, J.L.; Zhang, Y.; Li, K.H.; Wu, Q.H.; Yang, D.W.; et al. Quantifying atmospheric nitrogen deposition through a nationwide monitoring network across China. *Atmos. Chem. Phys.* **2015**, *15*, 12345–12360. [CrossRef]
31. Flechard, C.R.; Nemitz, E.; Smith, R.I.; Fowler, D.; Vermeulen, A.T.; Bleeker, A.; Erisman, J.W.; Simpson, D.; Zhang, L.; Tang, Y.S.; et al. Dry deposition of reactive nitrogen to European ecosystems: A comparison of inferential models across the NitroEurope network. *Atmos. Chem. Phys.* **2011**, *2011*, 2703–2728. [CrossRef]
32. Whitburn, S.; Van Damme, M.; Clarisse, L.; Bauduin, S.; Heald, C.L.; Hadji-Lazaro, J.; Hurtmans, D.; Zondlo, M.A.; Clerbaux, C.; Coheur, P.F. A flexible and robust neural network IASI-NH$_3$ retrieval algorithm. *J. Geophys. Res. Atmos.* **2016**, *121*, 6581–6599. [CrossRef]
33. Sahu, L.; Sheel, V.; Kajino, M.; Gunthe, S.S.; Thouret, V.; Nedelec, P.; Smit, H.G. Characteristics of tropospheric ozone variability over an urban site in Southeast Asia: A study based on MOZAIC and MOZART vertical profiles. *J. Geophys. Res. Atmos.* **2013**, *118*, 8729–8747. [CrossRef]
34. Xu, P.; Liao, Y.J.; Lin, Y.H.; Zhao, C.X.; Yan, C.H.; Cao, M.N.; Wang, G.S.; Luan, S.J. High-resolution inventory of ammonia emissions from agricultural fertilizer in China from 1978 to 2008. *Atmos. Chem. Phys.* **2016**, *16*, 1207–1218. [CrossRef]
35. Huang, X.; Song, Y.; Li, M.; Li, J.; Huo, Q.; Cai, X.; Zhu, T.; Hu, M.; Zhang, H. A high resolution ammonia emission inventory in China. *Glob. Biogeochem. Cycles* **2012**, *26*, 1–14. [CrossRef]
36. Kang, Y.; Liu, M.; Song, Y.; Huang, X.; Yao, H.; Cai, X.; Zhang, H.; Kang, L.; Liu, X.; Yan, X.; et al. High-resolution ammonia emissions inventories in China from 1980 to 2012. *Atmos. Chem. Phys.* **2016**, *16*, 2043–2058. [CrossRef]
37. Xu, P.; Zhang, Y.; Gong, W.; Hou, X.; Kroeze, C.; Gao, W.; Luan, S. An inventory of the emission of ammonia from agricultural fertilizer application in China for 2010 and its high-resolution spatial distribution. *Atmos. Environ.* **2015**, *115*, 141–148. [CrossRef]
38. Zhang, Y.; Dore, A.; Ma, L.; Liu, X.; Ma, W.; Cape, J.; Zhang, F. Agricultural ammonia emissions inventory and spatial distribution in the North China Plain. *Environ. Pollut.* **2010**, *158*, 490–501. [CrossRef] [PubMed]
39. Aneja, V.P.; Chauhan, J.; Walker, J. Characterization of atmospheric ammonia emissions from swine waste storage and treatment lagoons. *J. Geophys. Res. Atmos.* **2000**, *105*, 11535–11545. [CrossRef]

40. Pan, Y.; Wang, Y.; Tang, G.; Wu, D. Spatial distribution and temporal variations of atmospheric sulfur deposition in Northern China: Insights into the potential acidification risks. *Atmos. Chem. Phys.* **2013**, *13*, 1675–1688. [CrossRef]

41. Fuente, A.D.L.; Bing, N.; Hoeschele, I.; Mendes, P. Discovery of meaningful associations in genomic data using partial correlation coefficients. *Bioinformatics* **2004**, *20*, 3565. [CrossRef] [PubMed]

42. Sommer, S.G.; Olesen, J.E.; Christensen, B.T. Effects of temperature, wind speed and air humidity on ammonia volatilization from surface applied cattle slurry. *J. Agric. Sci.* **1991**, *117*, 91–100. [CrossRef]

43. Cassity-Duffey, K.; Cabrera, M.; Rema, J. Ammonia Volatilization from Broiler Litter: Effect of Soil Water Content and Humidity. *Soil Sci. Soc. Am. J.* **2015**, *79*, 543–550. [CrossRef]

44. Vet, R.; Artz, R.S.; Carou, S.; Shaw, M.; Ro, C.-U.; Aas, W.; Baker, A.; Bowersox, V.C.; Dentener, F.; Galy-Lacaux, C.; et al. A global assessment of precipitation chemistry and deposition of sulfur, nitrogen, sea salt, base cations, organic acids, acidity and pH, and phosphorus. *Atmos. Environ.* **2014**, *93*, 3–100. [CrossRef]

45. Zhao, Y.; Zhang, L.; Chen, Y.; Liu, X.; Xu, W.; Pan, Y.; Duan, L. Atmospheric nitrogen deposition to China: A model analysis on nitrogen budget and critical load exceedance. *Atmos. Environ.* **2017**, *153*, 32–40. [CrossRef]

46. Dong, W.X.; Xing, J.; Wang, S.X. Temporal and spatial distribution of anthropogenic ammonia emissions in China: 1994–2006. *Huanjing Kexue Environ. Sci.* **2010**, *31*, 1457–1463.

47. Zhang, L.; Jacob, D.J.; Knipping, E.M.; Kumar, N.; Munger, J.W.; Carouge, C.; Van Donkelaar, A.; Wang, Y.; Chen, D. Nitrogen deposition to the United States: Distribution, sources, and processes. *Atmos. Chem. Phys.* **2012**, *12*, 4539–4554. [CrossRef]

remote sensing

MDPI

Article

Ground-Level NO$_2$ Concentrations over China Inferred from the Satellite OMI and CMAQ Model Simulations

Jianbin Gu [1,2], Liangfu Chen [1,*], Chao Yu [1,3,*], Shenshen Li [1], Jinhua Tao [1], Meng Fan [1], Xiaozhen Xiong [4], Zifeng Wang [1], Huazhe Shang [1] and Lin Su [1]

[1] State Key Laboratory of Remote Sensing Science, Institute of Remote Sensing and Digital Earth, Chinese Academy of Sciences, Beijing 100101, China; gujianbin110@163.com (J.G.); liss01@radi.ac.cn (S.L.); taojh@radi.ac.cn (J.T.); fanmeng@radi.ac.cn (M.F.); wangzf@radi.ac.cn (Z.W.); huazhe_zhang@126.com (H.S.); sulin@irsa.ac.cn (L.S.)
[2] University of the Chinese Academy of Sciences, Beijing 100049, China
[3] State Key Joint Laboratory of Environment Simulation and Pollution Control, School of Environment, Tsinghua University, Beijing 100101, China
[4] NOAA/NESDIS/Center for Satellite Applications and Research, College Park, MD 20740, USA; xiaozhen.xiong@noaa.gov
* Correspondence: chenlf@radi.ac.cn (L.C.); yuchao@radi.ac.cn (C.Y.); Tel.: +86-10-6483-6589 (L.C.)

Academic Editors: Yang Liu, Jun Wang, Omar Torres, Richard Müller and Prasad S. Thenkabail
Received: 9 March 2017; Accepted: 19 May 2017; Published: 24 May 2017

Abstract: In the past decades, continuous efforts have been made at a national level to reduce Nitrogen Dioxide (NO$_2$) emissions in the atmosphere over China. However, public concern and related research mostly deal with tropospheric NO$_2$ columns rather than ground-level NO$_2$ concentrations, but actually ground-level NO$_2$ concentrations are more closely related to anthropogenic emissions, and directly affect human health. This paper presents one method to derive the ground-level NO$_2$ concentrations using the total column of NO$_2$ observed from the Ozone Monitoring Instrument (OMI) and the simulations from the Community Multi-scale Air Quality (CMAQ) model in China. One year's worth of data from 2014 was processed and the results compared with ground-based NO$_2$ measurements from a network of China's National Environmental Monitoring Centre (CNEMC). The standard deviation between ground-level NO$_2$ concentrations over China, the CMAQ simulated measurements and in-situ measurements by CNEMC for January was 21.79 µg/m^3, which was improved to a standard deviation of 18.90 µg/m^3 between our method and CNEMC data. Correlation coefficients between the CMAQ simulation and in-situ measurements were 0.75 for January and July, and they were improved to 0.80 and 0.78, respectively. Our results revealed that the method presented in this paper can be used to better measure ground-level NO$_2$ concentrations over China.

Keywords: NO$_2$; ground-level concentrations; OMI; CMAQ; profile shape

1. Introduction

Nitrogen dioxide (NO$_2$) is a pollutant trace gas in the atmosphere that plays an important role in atmospheric tropospheric chemistry and radiative heating [1–3]. Atmospheric ozone chemistry is affected by NO$_2$ in terms of ozone formation, whereas in the troposphere, NO$_2$ regulates the surface ozone level and maintains oxidizing capacities [4]; furthermore, exposure to ozone leads to adverse health effects for humans [5]. At high concentrations, NO$_2$ is toxic to humans [6,7]. Some epidemiological studies have shown that long-term NO$_2$ exposure is consistently associated with decreased lung function and with increased risks of respiratory symptoms [8–13], and daily time-series research results show that NO$_2$ and non-accidental mortality are strongly correlated [14–16].

In addition, NO_2 can initiate the formation of acid rain and can indirectly affect the global climate by perturbing greenhouse gas, ozone and methane levels [3]. Furthermore, NO_2 is a precursor of ammonium nitrate, which is an important component of atmospheric particulate matter pollution [17].

Concentrations of NO_2 columns are traditionally monitored through in-situ measurement networks [18]; however, these in-situ measurements are sparse in many parts of the world. Since 1995, satellite retrievals of NO_2 columns have provided more measurements than the ground-based and aircraft measurements. Some attempts to estimate emission levels have been made using a top-down approach [19] and satellite measurements. These studies show that satellite remote sensing can be used to monitor NO_2 columns at regional to global scales [20–22]. Satellite observations of global NO_2 columns began in 1995 with the development of the Global Ozone Monitoring Experiment (GOME) [23], followed by launch of the Scanning Imaging Absorption Spectrometer for Atmospheric Chartography (SCIAMACHY) [24], the Ozone Monitoring Instrument (OMI) onboard Earth Observing System (EOS)/Aura [25,26], and the GOME-2 [27]. The retrieval of tropospheric NO_2 columns is especially relevant to the state of the atmosphere (e.g., NO_2 profile shape). The uncertainties of GOME, SCIAMACHY and OMI observations are estimated to be on the order of 30–60% for individual measurements [28–31].

Over the last decades, the incredible economic growth of China has led to serious atmosphere pollution problems, continuous efforts have been made at national levels to reduce NO_2 emissions in the atmosphere. The monitoring of long-term pollutant emissions and the trend of concentration has been a key aspect of the evaluation of NO_2 emission abatement strategy effects. However, public concern and related research based on satellite observations of NO_2 columns have been focusing mainly on tropospheric NO_2 columns instead of ground-level NO_2 concentrations; when actually, ground-level NO_2 concentrations are more closely linked to the air pollution and impact on human health. This paper presents a method of estimating ground-level NO_2 concentrations over China based on tropospheric NO_2 columns retrieved from the OMI and model simulations. Section 2 provides a brief introduction to OMI and its retrieval of tropospheric NO_2 columns, the CMAQ model, ground-level in-situ measurements, and a method to derive the ground-level NO_2 concentrations by combining OMI tropospheric NO_2 columns and the CMAQ model. Comparisons of the derived NO_2 with ground-based NO_2 concentrations and model simulations in China are given in Section 3. A discussion and conclusion are given in Sections 4 and 5, respectively.

2. Materials and Methods

2.1. Measurement of OMI Tropospheric NO_2 Columns

The Dutch-Finnish OMI installed on NASA's EOS Aura satellite is a nadir-viewing imaging spectrograph that measures direct and atmosphere-backscattered sunlight within an ultraviolet-visible (UV-VIS) range of 270 nm to 500 nm [26]. EOS Aura was launched on 15 July 2004, and it traces a sun-synchronous polar orbit at approximately 705 km altitude over a period of 100 min and with a local equator crossing time of between 13:40 and 13:50, local time [31]. The OMI instrument is equipped with two two-dimensional Charge Coupled Device (CCD) detectors. The CCDs record the complete 270–500 nm spectrum in one direction and observe the Earth's atmosphere with a 114° field of view that is distributed over 60 discrete viewing angles, and which is perpendicular to the flight direction. The OMI's wide field of view corresponds to a 2600 km-wide spatial swath across the Earth's surface, which is large enough to achieve complete global coverage once a day. The exposure time of the CCD-camera is 2 s, corresponding to a spatial sampling of 13 km along the track (2 s × 6.5 km/s, with the latter being the orbital velocity projected onto the Earth's surface). Along the cross track, OMI pixel sizes vary with viewing zenith angles from 24 km in the nadir to approximately 128 km in extreme viewing angles of 57° along the edges of the swath [31].

Detailed descriptions of the NO_2 retrieval algorithm were provided by Boersma et al. [32], Bucsela et al. [33] and Celarier et al. [34]. The NO_2 retrieval algorithm involves a two-step procedure [31].

The first step employs a standard Differential Optical Absorption Spectroscopy (DOAS) technique [35] to determine slant column densities with a nonlinear least squares fitting within the 415–465 nm windows. The slant column represents the integrated abundance of NO_2 along the average photon path through the atmosphere. The second step is to derive initial vertical column densities by dividing slant column densities with an unpolluted air mass factor (AMF), which is defined as the ratio of the observed slant column to the vertical column. The AMF can be calculated using a single mean unpolluted NO_2 profile, and it estimates the stratospheric contributions to slant columns, which can be made by assimilating slant columns into the Thematic Mapper 4 (TM4) atmospheric Chemical Transport Model (CTM), including stratospheric chemistry and meteorological fields.

Major errors in the retrieval of tropospheric NO_2 columns have been estimated at ~0.7×10^{15} mol cm^{-2} from the slant column fitting (~0.15×10^{15} mol cm^{-2} in the stratospheric slant column and ~0.5×10^{15}–1.5×10^{15} mol cm^{-2} in the tropospheric AMF for individual cloud-free pixels (with an effective cloud fraction of <0.2)) [31]. AMF errors are primarily caused by cloud interference, surface albedo, aerosol, and profile shape uncertainties [29,30,32,36,37]. Error contributions to relative tropospheric AMF uncertainties (31%) are reported to include the following: 15% from surface albedo, 30% from cloud fractions, 15% from cloud top pressure levels and 9% from profile shapes [31]. The separation between the stratosphere and troposphere also serves as a source of error, and while the overall error in the OMI vertical column density under clear and unpolluted conditions is estimated at 5%, it can reach up to 50% in the presence of pollution and clouds [32]. Stripes affecting slant columns in the swath direction in Version 1.0.0 have been greatly reduced in Version 1.0.5, largely due to the improved dark current correction mechanisms that are available through Collection 3 Level 1B processing [38]. In this study we used the OMI standard tropospheric NO_2 product (version 3.0) available from the NASA Goddard Earth Sciences (GES) Data Active Archive Center (http://disc.sci.gsfc.nasa.gov/Aura/overview/data-holdings/OMI/). One year's worth of OMI NO_2 tropospheric columns data in 2014 in China were used because the ground-based NO_2 measurements are available for validation. We used here the data taken at an effective cloud fraction of <0.2. We used OMI tropospheric NO_2 columns covering an area of 18°N–55°N and 70°E–138°E.

2.2. Model Description

The two main components of the modeling system are Community Multi-scale Air Quality (CMAQ), developed by the US Environmental Protection Agency (US EPA) to simulate multiple atmosphere quality issues with multi-scale capabilities [39], and Regional Atmospheric Modeling System (RAMS). CMAQ is a multi-scale and multi-pollutant air quality model developed for depicting the detail processes about dust formation, transport, deposition, and other important characteristics [40]. The comprehensive suite aerosol composition (sulfate, nitrate, ammonium, black carbon, organic mass, dust and sea salt) is taken into consideration. The aerosol particle size distribution is comprised of three modes: Aitken mode, accumulation mode, and coarse mode. In this study the chemical mechanism CB05 [41] and aerosol evaluation processes of CMAQ Version 4.7 is used.

RAMS is a multifunctional numerical code for simulating and forecasting meteorological phenomena, and has good capacity to depict the boundary layer, which is important for simulating the dust formation. In this study, RAMS is used to provide the three-dimensional meteorological field for CMAQ, including boundary-layer turbulence, cloud, precipitation, and other meteorological elements. The meteorological fields from RAMS are used instead of the CMAQ default meteorological driver. In this study, the RAMS was run in a four-dimensional data assimilation mode along with re-initialization every 4 days, with the first 24 h designated as the initialization period. The three-dimensional meteorological fields of the RAMS were obtained from the European Center for Medium-Range Weather Forecast (ECMWF) datasets, which were available every 6 h with a spatial resolution of 1° × 1°. Many previous works have shown the successful use of the RAMS-CMAQ modeling system by comparing the simulation results with diverse measurement data [42–45].

In this study, the emission inventory by the RAMS-CMAQ modeling system is introduced as follows. The anthropogenic emissions of aerosols and their precursors (CO, NOx, SO_2, volatile organic compounds (VOCs), black carbon, organic carbon, $PM_{2.5}$, and PM_{10}) are obtained from the monthly-based emission inventory [46–48], updated from the previous version [49], over East Asia. This emission inventory has a spatial resolution of $0.25° \times 0.25°$ and involves four emission categories, including industry, power, transport and residential. The model domain (Figure 1) is on a rotated polar stereographic map projection centered at (35°N, 116°E) with a 64 km grid cell. The modeling system has 15 vertical layers in the coordinates system unequally spaced from the ground to ~23 km, and approximately half of them are concentrated in the lowest 2 km to improve the simulation of the atmospheric boundary layer. Research has shown that the NO_2 concentrations in China modeled using the RAMS-CMAQ modeling system are generally in good agreement with surface observations and satellite measurements [50–53].

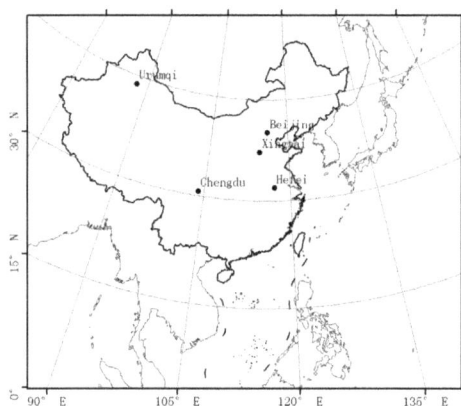

Figure 1. Model domain for RAMS-CMAQ used in this study is on a rotated polar stereographic map projection centered at (35°N, 116°E) with a 64 km grid cell.

2.3. Ground-Level In Situ Measurement

Along with the rapid economic growth that has occurred over the past two decades, environmental pollution has emerged as a severe issue in China. The Chinese government has established the China National Environmental Monitoring Centre (CNEMC), which is directly affiliated with the ministry of environmental protection of the People's Republic of China. CNEMC's main functions are to undertake state environmental monitoring, develop state environmental monitoring technologies, and provide monitoring information and technical support to the country's environmental management and decision-making bodies. Since the beginning of 2013, CNEMC has begun to establish a network for monitoring ground-level NO_2 concentrations over China. Currently, there are more than 800 atmospheric pollution-monitoring stations in this network, hourly ground-level NO_2 concentrations released by these monitoring stations were measured with the standard methods (http://www.cnemc.cn/publish/totalWebSite/0493/187/newList_1.html). The Thermo Scientific Model 42i, which is used to monitor ground-level NO_2 concentrations in these monitoring stations, is designated by the United States Environmental Protection Agency (US EPA) as a Reference Method for the measurement of ambient concentrations of NO_2 pursuant with the requirements defined in the Code of Federal Regulations. The Model 42i Chemiluminescence Analyzer combines proven detection technology, easy-to-use menu-driven software, and advanced diagnostics to offer unsurpassed flexibility and reliability. We found that the 2014 annual observation data is the most abundant, and has the minimum discontinuity after statistics.

In this study, we used ground-level NO_2 concentration observation data released by CNEMC-monitored stations in 2014 to compare. We first eliminated the data released by the CNEMC stations that had a monitoring time of less than 20 days a month. Then, we analyzed the longitude and latitude of each monitored station, computed the mean of the data released by stations distributed in the same 64 km grid cell according to latitude and longitude. In addition, because the OMI was launched into a Sun-synchronous orbit crossing the equator at approximately 13:45, we computed the mean of the CNEMC's ground-level NO_2 concentration observation data between 13:00 and 14:00 as daily mean values to compare. In this study we obtained monthly mean values of ground-level NO_2 concentrations released by 100 CNEMC monitored stations. Distribution of these 100 monitored stations is shown in Figure 2.

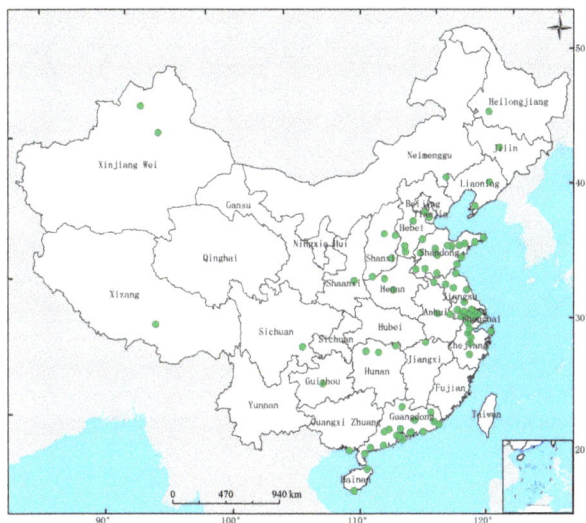

Figure 2. Distribution of the 100 ground-level NO_2 concentrations CNEMC monitored stations.

2.4. Determination of Ground-Level NO$_2$ Concentrations

Airborne measurements of the southeastern United States show that NO_2 in the boundary layer can greatly contribute to NO_2 tropospheric columns over polluted regions [54]. Retrievals based on satellite observations have revealed a close relationship between land surface NO_2 emissions and tropospheric NO_2 columns [36,55–59]. These studies clearly suggest that tropospheric NO_2 columns retrieved from satellite observations can be used to derive the ground-level NO_2 concentrations. In this study, we used the RAMS-CMAQ modeling system to simulate the relationship between satellite observations of tropospheric NO_2 columns and the ground-level NO_2 concentrations over China following the method described by Lamsal et al. [60,61], who conducted a simulation of tropospheric NO_2 profiles over the United States and Canada using the Goddard Earth Observing System (GEOS)-Chem global three-dimensional model of tropospheric chemistry at $2° \times 2.5°$, version 7-03-06. However, previous research on ground-level NO_2 concentrations by combining with satellite observations and model simulations has failed to consider the influence of China's high atmospheric pollution on obtaining the vertical distribution of tropospheric NO_2 profiles over China. In addition, compared to the GEOS-Chem global model, the RAMS-CMAQ modeling system, with its higher spatial resolution of a 64 km grid cell, is more appropriate to simulating tropospheric NO_2 profiles over China. Many works have shown that the modeled NO_2 concentrations over China by the RAMS-CMAQ modeling system are more appropriately suited to China's high atmospheric pollution [51–53]. In this study we compensated for this shortcoming by estimating ground-level NO_2

concentrations over China using the OMI standard NO_2 product combined with simulations from the RAMS-CMAQ model, after obtaining the more appropriate vertical distribution of tropospheric NO_2 profiles.

In this study, the different spatial resolution from the OMI standard NO_2 product and simulation from the RAMS-CMAQ modeling system has hindered the estimating of ground-level NO_2 concentrations over China. Our approach to reconstructing consistent spatial resolution was to degrade higher spatial resolution data to a single consistent coarse spatial resolution. Here we first calculated the distance of latitude and longitude from each grid cell of the RAMS-CMAQ model with relatively coarse spatial resolution of 64×64 km^2 grid cells to all grid cells of the OMI tropospheric NO_2 columns with higher spatial resolution of 13×24 km^2, and considered the two grid cells of the different data with minimum distance of latitude and longitude to correspond to each other. Then we estimated ground-level NO_2 concentrations over China using the OMI standard NO_2 product combined with simulation from the RAMS-CMAQ model after reconstructing a consistent spatial resolution of 64×64 km^2.

3. Results

3.1. Verification of Distributions of Tropospheric NO_2 Profiles from RAMS-CMAQ

In this section, verification of vertical distributions of tropospheric NO_2 profiles from the RAMS-CMAQ model is presented. We first counted the monthly mean of tropospheric NO_2 columns from OMI observations over China in 2014. After the corresponding unit conversion, we calculated the ratio of ground-level NO_2 concentrations released by CNEMC to tropospheric NO_2 columns by OMI, then compared the ratio of simulated NO_2 concentrations distributed within the atmosphere from the ground to a height of 100 m to simulated concentrations distributed within the atmosphere from the ground to ~23 km by the RAMS-CMAQ model. We verified the accuracy of the vertical distribution of tropospheric NO_2 profiles over China from the RAMS-CMAQ model by comparing the correlation of these two sets of ratios, the results of which are shown in Figure 3.

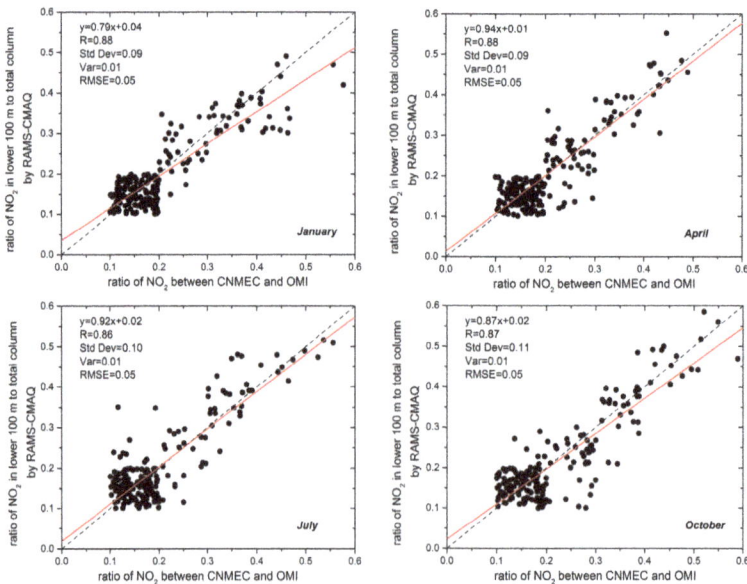

Figure 3. Scatter plots to verify accuracy of vertical distribution of tropospheric NO_2 profiles over China from the RAMS-CMAQ model in January, April, July and October in 2014.

It is found that the vertical distribution of tropospheric NO_2 profiles over China has a large spatial and temporal variation. To illustrate these spatial-temporal variations in tropospheric NO_2 profiles over China, we chose to analyze the vertical distributions of tropospheric NO_2 profiles in five representative cities: Beijing, Xingtai, Chengdu, Urumqi and Hefei. Information on these five cities is listed in Table 1. The vertical distributions of tropospheric NO_2 profiles in these five cities are shown in Figure 4. The *x*-axis shows NO_2 concentrations simulated by the RAMS-CMAQ modeling system, and the *y*-axis shows the corresponding heights expressed by the natural logarithm. Natural logarithm height was used because the vertical layers by the RAMS-CMAQ modeling system were unequally spaced in the coordinates system, and heights corresponding to the different vertical layers varied dramatically. It is evident from Figure 4 that the tropospheric NO_2 columns are mainly distributed within the atmosphere from the ground to a height of 100–150 m.

Table 1. The information of the selected five cities.

City Name	Latitude	Longitude	City Condition
Beijing	40.00°	116.00°	a megalopolis located in northeastern China that presents relatively high levels of air pollution
Xingtai	37.05°	114.48°	one of the most air-polluted cities in China, and an important energy base in the North China area
Chengdu	30.67°	104.06°	a large city located in southwestern China with relatively high air pollution compared to other southwestern cities
Urumqi	43.77°	87.68°	a large city located in northwestern China but with lower levels of air pollution compared to other cities located in the east
Hefei	31.86°	117.27°	a large city located in southeastern China with relatively low levels of air pollution compared to other cities located in the north

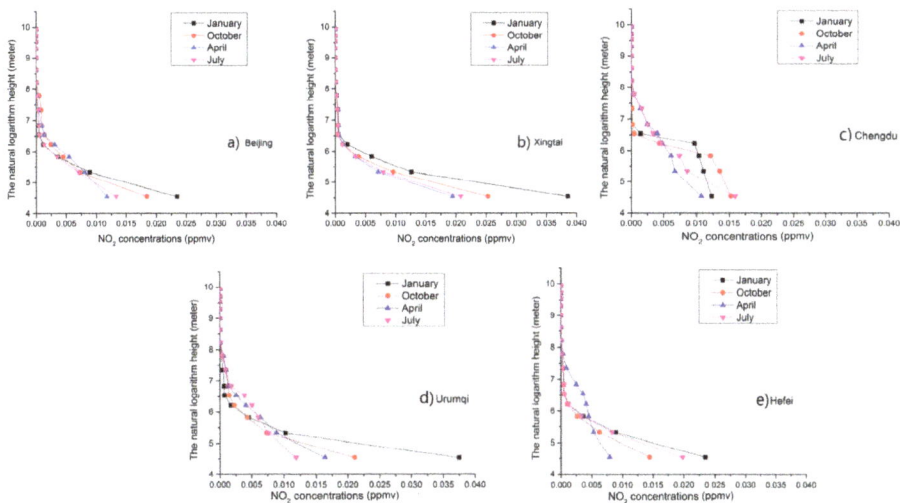

Figure 4. The simulated NO_2 vertical profiles over Beijing (**a**), Xingtai (**b**), Chengdu (**c**), Urumqi (**d**) and Hefei (**e**) in January, April, July and October in 2014.

3.2. Spatial-Temporal Variations of Derived Ground-Level NO_2 Concentrations

In this section, spatio-temporal variation trends of derived ground-level NO_2 concentrations combining OMI observations with the RAMS-CMAQ model are presented. Figure 5 shows the monthly mean values of the derived ground-level NO_2 concentrations over China in January, April,

July, and October, 2014. From Figure 5, we can see that among different seasons the ground-level NO_2 concentration levels over China were high in the winter/spring and low in the summer/fall, and for different regions they were high in the eastern, developed areas, but low in the western, developing areas of China, due to there being more anthropogenic emissions in the eastern areas (e.g., ground-level NO_2 concentrations in the developed North China Plain reached 60.00–100.00 μg/m^3 whereas concentrations in the less developed western areas reached 5.00–20.00 μg/m^3). We analyzed these spatial distribution characteristics of ground-level NO_2 concentrations for China in combination with industrial development and anthropogenic emissions. A significant positive correlation was found between the magnitude of ground-level NO_2 concentrations and levels of industrial development and anthropogenic emissions. Of the different seasons in China, winter ground-level NO_2 concentration values are the highest of the year, which is mainly attributable to winter coal use for heating.

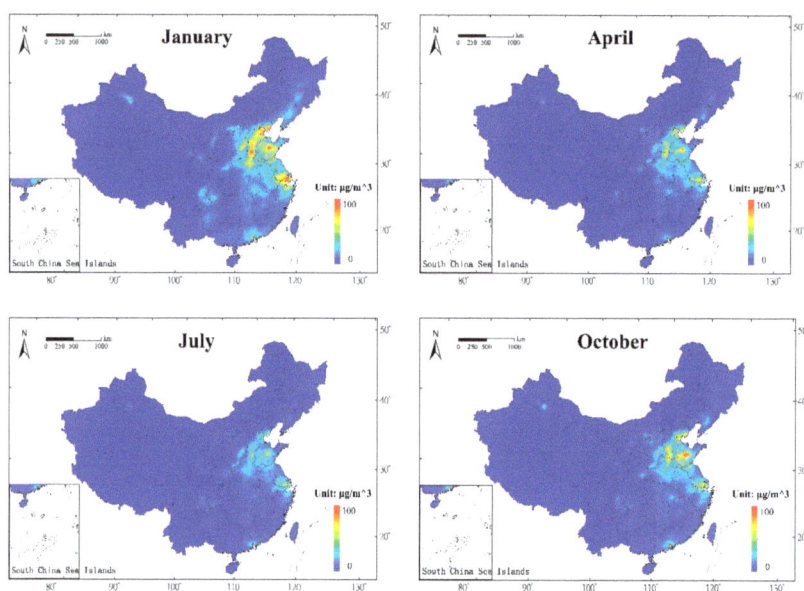

Figure 5. Monthly derived ground-level NO_2 concentrations over China in January, April, July, and October in 2014.

3.3. Comparisons of the Derived NO_2 with Ground-Based Measurements for Different Regions

A comparison between monthly mean ground-level NO_2 concentrations for different regions from in-situ measurements and the concentrations derived from OMI satellite data in conjunction with the RAMS-CMAQ modeling results is shown in Figure 6, the x-axis is the month, and the y-axis is the ground-level NO_2 concentrations. We analyzed the ground-level NO_2 concentrations in the five representative cities, i.e., Beijing, Xingtai, Chengdu, Urumqi and Hefei. As the CNEMC atmospheric pollution monitoring stations are unevenly distributed across the country, we first computed the average values of ground-level NO_2 concentrations released by all monitoring stations distributed within a city, calculated the average value of OMI tropospheric NO_2 columns over the same city, and then we obtained the ground-level NO_2 concentrations of the same city. Figure 6 shows that the average correlation coefficients largely fall within a range of 0.70~0.80 for the selected five representative cities of Beijing, Xingtai, Chengdu, Urumqi and Hefei, and that the correlation coefficients for Beijing, Xingtai and Hefei are, relatively, better. Beijing is one of China's mega cities and has, relatively, more ground monitoring stations. The data quality in Beijing is better, creating a more solid foundation for air pollution monitoring research and long-term monitoring. Xingtai is currently one of the most heavily air-polluted cities in China, and the air quality levels in Xingtai were ranked last. Hefei's air is similar to Beijing. Figure 6 shows that our method is more precise at determining levels for cities with heavy air pollution, and for those cities with more ground monitoring stations. In general the vertical distributions of tropospheric NO_2 profiles for these cities based on the RAMS-CMAQ model are, relatively, more accurate representations of air quality conditions; therefore, ground-level NO_2 concentrations retrieved are more accurate.

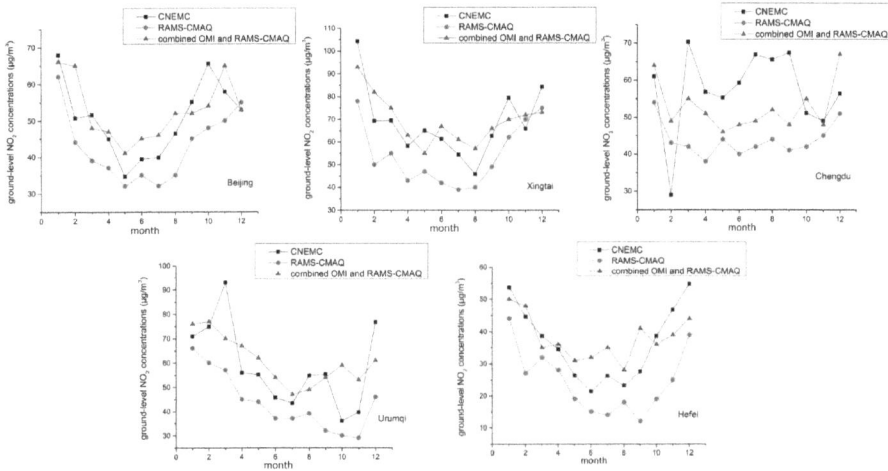

Figure 6. Line charts of the derived ground-level NO_2 concentrations with in-situ measurements from ground network and the RAMS-CMAQ simulation over Beijing, Xingtai, Chengdu, Urumqi and Hefei cities in 2014.

From Figure 6, we also found that, compared to the in-situ ground measurements, the RAMS-CMAQ simulated values are underestimated, and in areas with high NO_2 concentrations, the underestimations are more obvious (e.g., in Xingtai City). Overall, the results of our method are similar to the in-situ measurements. The standard deviation between ground-level NO_2 concentrations, the RAMS-CMAQ simulated measurements and those measurements released by CNEMC in Beijing was 10.25 $\mu g/m^3$, and the standard deviation between our method and CNEMC's was 9.03 $\mu g/m^3$. The variance between the RAMS-CMAQ and CNEMC data for Beijing was 105.01, and the variance between our method and that of CNEMC was 81.51. The standard deviation between the RAMS-CMAQ simulated measurements and those released by CNEMC for Xingtai was 15.68 $\mu g/m^3$, and the standard deviation between our method and that released by CNEMC was 12.65 $\mu g/m^3$. The variance between the RAMS-CMAQ model and CNEMC results for Xingtai was 246.00, and the variance between our method and that of CNEMC was 160.07. The analysis shows that ground-level NO_2 concentrations derived using OMI tropospheric NO_2 columns together with vertical distributions of tropospheric NO_2 profiles from RAMS-CMAQ model simulations were more accurate than the simulations from RAMS-CMAQ model only.

Figure 6 shows the derived ground-level NO_2 concentrations by OMI and RAMS-CMAQ model are generally lower than the in-situ ground measurements by CNEMC in these cities in winter, except for Chengdu, this result is similar to Lamsal's result that the derived surface NO_2 by OMI and GEOS-chem over western North America are generally lower than the local in-situ measurements in winter [60]. Larger differences between the in-situ measurements and the derived ground-level NO_2 concentrations in China likely reflect a combination of enhanced spatial variations in polluted regions and preferential placement of in-situ monitors in polluted locations. The ground-level

NO_2 concentrations estimated from our approach are generally consistent with the in-situ ground measurements for these five cities, and the occasional large discrepancies may reflect local variation processes. In addition, the consumption of bulk coal and household coal for heating, cooking and other uses in winter is common in China, especially in rural areas. However, information about these sources of NO_2 emissions are generally not grasped by the government. The emission inventory by the RAMS-CMAQ model generally also overlooks wintertime NO_2 emission caused by bulk coal and household coal in China, which led to the underestimation of the derived ground-level NO_2 concentrations in winter from our approach. The vertical distribution of tropospheric NO_2 profiles over China from the RAMS-CMAQ model as shown in Figure 3 also verified this underestimation of our results. Chengdu is surrounded by mountains. From west to east, the terrain of Chengdu is divided into three parts, comprising mountains, plains and hills. It is difficult to operate the local in-situ NO_2 measurements such as the elevation of the western part of Chengdu is mainly over 3 km while the elevation of the central part is about 400 m to 700 m. However, in order to ensure the integrity of this study, it is necessary to study ground-level NO_2 concentration in Chengdu, which is a representative metropolis in Southwest China. Meanwhile, the worse result in Chengdu reflects that our approach has to be improved when considering complicated topographies, which is a future work we will conduct. Additionally, more ground-based monitoring data will be helpful for future analysis.

3.4. Comparisons of the Derived NO_2 with Ground-Based Measurements for Different Seasons

In this section, we compared the ground-level NO_2 concentrations retrieved from our method with in-situ measurements for different seasons across China. As satellite data offer broader observational data coverage than in-situ measurements, data in a large area of China and for different seasons were used. We used mean monthly values from the 100 ground-level NO_2 concentration CNEMC-monitored stations mentioned in Section 2.3 to compare. Figure 7 shows good correlations between the derived ground-level NO_2 concentrations and the monthly mean ground-level NO_2 over China for 2014. The correlation coefficient, R, was 0.80 for January and was 0.78 for July; compared to ground-level NO_2 concentrations simulated using the RAMS-CMAQ model only. The standard deviation between our method and those measurements released by CNEMC for January was 18.90 $\mu g/m^3$, and the standard deviation between the RAMS-CMAQ simulated measurements and CNEMC's was 21.79 $\mu g/m^3$. The variance between our method and that released by CNEMC for January was 257.19, and the variance between the RAMS-CMAQ and CNEMC data was 353.09. The standard deviation between our method and that released by CNEMC for July was 11.31 $\mu g/m^3$, and the standard deviation between the RAMS-CMAQ simulated concentrations and CNEMC's was 12.11 $\mu g/m^3$. The variance between our method and that released by CNEMC for July was 127.96, and the variance between the RAMS-CMAQ and CNEMC data was 146.59. Analyzing the ground-level NO_2 concentrations retrieved from our method for different seasons, we found clear seasonal variations in the derived ground-level NO_2 concentrations, with the largest variations occurring in the winter and the least pronounced in the summer. This is mainly attributable to increases in burning and heating emissions in the winter. Therefore, anthropogenic emission is the main factor that is impacting changes in ground-level NO_2 concentrations. The ground-level NO_2 concentrations calculated in this paper can be used to measure the influence of anthropogenic emissions on atmospheric quality.

Figure 7 also shows that the ground-level NO_2 concentrations estimated by OMI and RAMS-CMAQ model are generally lower than the in-situ measurements by CNEMC over China in January and July, the underestimation of the derived ground-level NO_2 concentrations in January from our approach may be caused by the wintertime consumption of bulk coal and household coal in China, which are not considered in the emission inventory by the RAMS-CMAQ model. The underestimation level of the derived ground-level NO_2 concentrations in July is lower than in January, which corresponds to the underestimation level of the vertical distributions of tropospheric NO_2 profiles as shown in Figure 3. OMI-derived ground-level NO_2 concentration represents the mean concentration over several hundred square kilometers, while in-situ measurements are point observations in general, which led to a slope of linear regression line of less than 1 in Figure 7. Possible explanations for the seasonal discrepancy between the derived ground-level NO_2 concentrations and in-situ measurements include errors in the in-situ NO_2 concentrations, in the RMAS-CMAQ simulated vertical distributions of tropospheric NO_2 profiles, and in the OMI tropospheric NO_2 column retrieval. Another likely contributor to the seasonal discrepancy is the use of mean NO_2 profiles in the OMI air mass factor calculation [60]. Seasonal variation would yield an underestimate in retrieved NO_2 columns in winter versus summer. In addition, seasonal variation in surface reflectivity could also play a part.

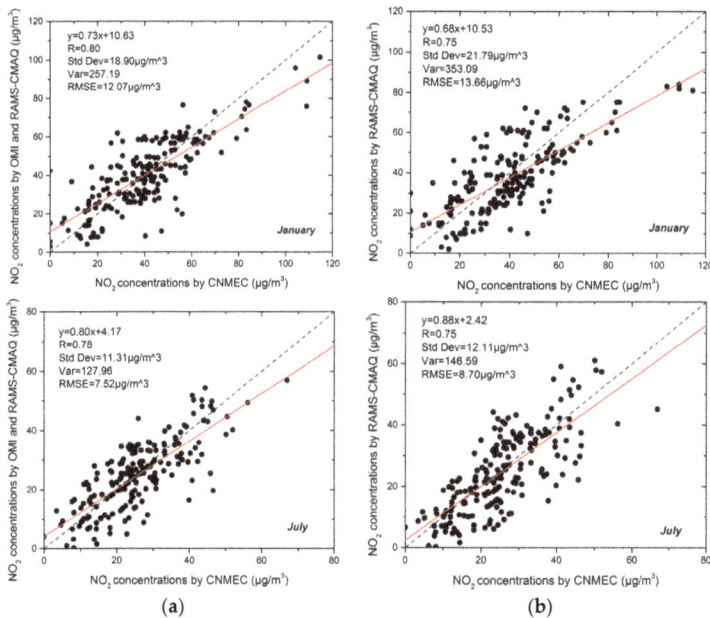

Figure 7. Scatter plots of the derived ground-level NO_2 concentrations with in-situ measurements from ground network (**a**) and the RAMS-CMAQ simulation (**b**) over China in January and July 2014.

4. Discussion

In this study, we have used the OMI tropospheric NO_2 columns and the RAMS-CMAQ modeling system to infer ground-level NO_2 concentrations. Several previous studies reported that the simulation of tropospheric NO_2 profiles over the United States and Canada using the GEOS-Chem global three-dimensional model of tropospheric chemistry at $2° \times 2.5°$ were obtained, and ground-level NO_2 concentrations over the United States and Canada were measured by the Ozone Monitoring Instrument (OMI) [60,61]. However, these previous studies on ground-level NO_2 concentrations by combining satellite observations and model simulations, on the one hand, has not been implemented in China, and the influence of severe atmospheric pollution also increases the level of difficulty of obtaining the vertical distribution of tropospheric NO_2 profiles over China. On the other hand, compared to the GEOS-Chem global model used in the previous studies, the RAMS-CMAQ modeling system with the higher spatial resolution of 64 km grid cell is more appropriate for simulating tropospheric NO_2 profiles over China. Much research has shown that the modeled NO_2 concentrations over China by the RAMS-CMAQ model are more appropriate to China's high atmospheric pollution [51–53]. In this study we inferred the ground-level NO_2 concentrations over China using the OMI NO_2 product combined with simulation from the RAMS-CMAQ model after obtaining the more appropriately vertical distribution of tropospheric NO_2 profiles.

We derived the ground-level NO_2 concentrations using the total column of NO_2 observed from the OMI and the simulations from the RAMS-CMAQ model in China. Data for 2014 were processed and compared to in-situ measurements derived from the CNEMC monitoring network. The derived ground-level NO_2 concentrations were also compared with the simulated ground-level NO_2 concentrations by the RAMS-CMAQ. Overall, ground level NO_2 concentrations were underestimated by the RAMS-CMAQ model. Using observed data corresponding to in-situ CNEMC measurements, the standard deviation between the RAMS-CMAQ simulated measurements and in-situ measurements by CNEMC for January was 21.79 $\mu g/m^3$, it was improved to 18.90 $\mu g/m^3$ between our method and CNEMC data. The variance between our method and that released by CNEMC for January was 257.19, and the variance between the RAMS-CMAQ and CNEMC data was 353.09, the root mean square error between the RAMS-CMAQ simulations and in-situ measurements by CNEMC for January was 13.66 $\mu g/m^3$, which was improved to 12.07 $\mu g/m^3$ between our method and CNEMC data. The standard deviation between the RAMS-CMAQ and CNEMC data for July was 12.11 $\mu g/m^3$, and it was improved to 11.31 $\mu g/m^3$ between our method and CNEMC data. The variance between our method and that released by CNEMC for July was 127.96, and the variance between the RAMS-CMAQ data and CNEMC data was 146.59. The root mean square error between the RAMS-CMAQ simulations and in-situ measurements by CNEMC for July was 8.70 $\mu g/m^3$, which was improved to 7.52 $\mu g/m^3$ between our method and CNEMC data. Correlation coefficients between the RAMS-CMAQ simulation and in-situ measurements were 0.75 for January and July, and they were improved to 0.80 and 0.78, respectively, from our approach. Compared to that of the in-situ measurements in different regions, the standard deviation between ground-level NO_2 concentrations, the RAMS-CMAQ simulated measurements, and in-situ measurements released by CNEMC in Beijing was 10.25 $\mu g/m^3$, which was improved to 9.03 $\mu g/m^3$ between our method and CNEMC data33; the standard deviation between the RAMS-CMAQ and CNEMC data in Xingtai was 15.68 $\mu g/m^3$, which was improved to 12.65 $\mu g/m^3$ between our method and CNEMC data.

Major errors in the retrieval of ground-level NO_2 concentrations using the total column of NO_2 observed from the OMI and the simulations from the RAMS-CMAQ model have been estimated in the following aspects. First, the errors in the retrieval of OMI tropospheric NO_2 columns have been estimated at ~0.7 × 10^{15} mol cm^{-2} from the slant column fitting (~0.15 × 10^{15} mol cm^{-2} in the stratospheric slant column and ~0.5 × 10^{15}–1.5 × 10^{15} mol cm^{-2} in the tropospheric AMF for individual cloud-free pixels (with an effective cloud fraction of <0.2)) [31]. AMF errors are mainly from cloud interference, surface albedo, aerosol, and profile shape uncertainties [29,30,32,36,37]. Separation between the stratosphere and troposphere is also a source of error; and while the overall error in the OMI vertical NO_2 columns under clear and unpolluted conditions is estimated at 5%, it can reach up to 50% in the presence of pollution and clouds [32]. Second, in this study, we used the RAMS-CMAQ modeling system to simulate of tropospheric NO_2 profiles over China. The emission inventory by the RAMS-CMAQ model, such as the anthropogenic emissions of aerosols and their precursors (CO, NOx, SO_2, volatile organic compounds (VOCs), black carbon, and organic carbon, $PM_{2.5}$, and PM_{10}), are obtained from the monthly emission inventory [46–48] updated from the previous version [49] over East Asia. The errors in the retrieval of tropospheric NO_2 profiles over China by the RAMS-CMAQ model are mainly due to the emission inventory and the meteorology field. The meteorology field is important to the modeled mass concentrations, aerosols and their precursor simulations. The accuracy of wind field and relative humidity simulation could obviously affect the dust particle transport and optical properties calculation [50]. Last, because the different spatial resolution from the OMI standard NO_2 product and simulation from the RAMS-CMAQ model has hindered the estimating ground-level NO_2 concentrations, another source of error was the reconstructed consistent spatial resolution, for which we degraded the higher spatial resolution data to a single consistent coarse spatial resolution.

In this study, we did not take into account the averaging kernel (AK). AK is a well-established concept in the retrieval of remote sensing observations as the link between the retrieved quantities and reality. It is proportional to the height-dependent sensitivity of satellite observation to changes in tracer concentration, and provides the interpretation of the value of the air mass factor. The AK provides important information needed for quantitative analysis of the satellite data, especially for interpreting the satellite retrieval of trace gases to users. It's very useful to remove the dependence on a priori assumptions about the profile shape for inter-comparison between model simulations and satellite measurements. In this paper, we focused on estimating ground-level NO_2 by combining the OMI standard product OMNO2 from NASA (Version 3) and the RAMS-CMAQ modeled NO_2 profiles, rather than on satellite-model comparison, so we didn't consider the AK in this study. However, the OMI NO_2 standard product is not directly suitable to studying VCDs on a scale below the resolution of the ancillary parameters (such as $2° \times 2.5°$ for priori NO_2 profiles), and the AK can be used to correct the NO_2 VCDs when more accurate NO_2 profiles are available, which will improve satellite NO_2 VCD products and also the ground-level nitrogen dioxide concentrations inferred from satellites. While this correction is beyond the scope of this study, we are developing a customized OMI NO_2 retrieval by recalculating AKs using the high-resolution RAMS-CMAQ modeled NO_2 profiles, and will apply it to the estimation of ground-level NO_2 in future study. We did not take into account NO_2 emissions from natural sources such as biomass burning, soil, and lightning, because they are negligible compared to anthropogenic emissions over urban areas. In addition, the contribution of free tropospheric NO_2 produced from lightning mainly occur in low latitude areas, while the NO_2 high concentration areas, such as the Beijing Tianjin Hebei region, were mainly distributed in the middle and high latitude areas.

5. Conclusions

Most works using satellite observations of NO_2 focused on tropospheric NO_2 columns rather than on ground-level NO_2 concentrations. However, ground-level NO_2 concentrations are more closely related to anthropogenic emissions and directly affect human health. This paper presents a means of estimating ground-level NO_2 concentrations based on OMI tropospheric NO_2 columns and the vertical distribution of tropospheric NO_2 profiles simulated using the RAMS-CMAQ model. One year's worth of data from 2014 was processed and the results were compared to ground-based NO_2 measurements from a network of CNEMC, and the simulated ground-level NO_2 concentrations by the RAMS-CMAQ model. Our results revealed that the method presented in this paper can be used to better measure ground-level NO_2 concentrations over China.

Further analysis of the ground-level NO_2 concentrations retrieved from this algorithm shows that seasonal variations of the ground-level NO_2 concentrations are pronounced, with the largest occurring in the winter and the lowest in the summer. Such variations are mainly due to significant increases in burning and heating emissions levels in winter. The ground-level NO_2 concentrations presented in this paper can be used to better measure distribution of NO_2 in the atmosphere and study the effects of anthropogenic emissions on atmospheric pollution conditions. Further validations and improvements of this method are ongoing, and include the quality control of in-situ measurements, whose errors have not been taken into account. We plan to develop a simple empirical formula based on the CMAQ model for directly converting OMI column values to ground-level NO_2. Such an algorithm may be used to more accurately monitor ground-level NO_2 using satellite data and to generate measurements with better spatial and temporal coverage than surface measurements.

Acknowledgments: This work was supported by the National Natural Science Foundation of China (Grant No. 41501476) and the National Science and Technology Ministry (Grant No. 2014BAC21B03), this work was partially supported by China Postdoctoral Science Foundation funded project (Grant No. 2015M580084, 2016T90083). We greatly acknowledge the free use of tropospheric NO_2 column data from the OMI sensor. We also thank State Key Laboratory of Atmospheric Boundary Layer Physics and Atmospheric Chemistry, Chinese Academy of Sciences for the RAMS-CMAQ simulation data. We also thank China's National Environmental Monitoring Centre for in-situ measurements.

Author Contributions: The authors contributed equally to the manuscript.

Conflicts of Interest: The authors declare no conflict of interest.

References

1. Logan, J.A. Nitrogen Oxides in the Troposphere: Global and Regional Budgets. *J. Geophys. Res.* **1983**, *88*, 10785–10807. [CrossRef]
2. Finlayson-Pitts, F.C.; Pitts, J.N., Jr. (Eds.) *Atmospheric Chemistry: Fundamentals and Experimental Techniques*; John Wiley: Hoboken, NJ, USA, 1986.
3. Solomon, S.; Portmann, R.W.; Sanders, R.W.; Daniel, J.S.; Madsen, W.; Bartram, B.; Dutton, E.G. On the Role of Nitrogen Dioxide in the Absorption of Solar Radiation. *J. Geophys. Res.* **1999**, *1041*, 12047–12058. [CrossRef]
4. Crutzen, P.J. The Role of NO and NO_2 in the Chemistry of the Troposphere and Stratosphere. *Annu. Rev. Earth Planet. Sci.* **1979**, *7*, 443–472. [CrossRef]
5. Sunyer, J.; Spix, C.; Quénel, P.; Ponce-de-León, A.; Pönka, A.; Barumandzadeh, T.; Touloumi, G.; Bacharova, L.; Wojtyniak, B.; Vonk, J.; et al. Urban Air Pollution and Emergency Admissions for Asthma in Four European Cities: the APHEA Project. *Thorax* **1997**, *52*, 760–765. [CrossRef] [PubMed]
6. Latza, U.; Gerdes, S.; Baur, X. Effects of Nitrogen Dioxide on Human Health: Systematic Review of Experimental and Epidemiological Studies Conducted between 2002 and 2006. *Int. J. Hyg. Environ. Health* **2009**, *212*, 271–287. [CrossRef] [PubMed]
7. Chen, R.; Samoli, E.; Wong, C.M.; Huang, W.; Wang, Z.; Chen, B.; Kan, H. CAPES Collaborative Group. Associations between Short-Term Exposure to Nitrogen Dioxide and Mortality in 17 Chinese Cities: the China Air Pollution and Health Effects Study (CAPES). *Environ. Int.* **2012**, *45*, 32–38. [CrossRef] [PubMed]
8. Ackermann-Liebrich, U.; Leuenberger, P.; Schwartz, J.; Schindler, C.; Monn, C.; Bolognini, G.; Bongard, J.P.; Brändli, O.; Domenighetti, G.; Elsasser, S.; et al. Lung Function and Long Term Exposure to Air Pollutants in Switzerland. Study on Air Pollution and Lung Diseases in Adults (SAPALDIA) Team. *Am. J. Respir. Crit. Care Med.* **1997**, *155*, 122–129. [CrossRef] [PubMed]
9. Schindler, C.; Ackermann-Liebrich, U.; Leuenberger, P.; Monn, C.; Rapp, R.; Bolognini, G.; Bongard, J.P.; Brändli, O.; Domenighetti, G.; Karrer, W.; et al. Associations between Lung Function and Estimated Average Exposure to NO_2 in Eight Areas of Switzerland. The SAPALDIA Team. Swiss Study of Air Pollution and Lung Diseases in Adults. *Epidemiology* **1998**, *9*, 405–411. [CrossRef] [PubMed]
10. Panella, M.; Tommasini, V.; Binotti, M.; Palin, L.; Bona, G. Monitoring Nitrogen Dioxide and Its Effects on Asthmatic Patients: Two Different Strategies Compared. *Environ. Monit. Assess.* **2000**, *63*, 447–458. [CrossRef]
11. Smith, B.J.; Nitschke, M.; Pilotto, L.S.; Ruffin, R.E.; Pisaniello, D.L.; Willson, K.J. Health Effects of Daily Indoor Nitrogen Dioxide Exposure in People with Asthma. *Eur. Respir. J.* **2000**, *16*, 879–885. [CrossRef] [PubMed]
12. Gauderman, W.J.; McConnell, R.; Gilliland, F.; London, S.; Thomas, D.; Avol, E.; Vora, H.; Berhane, K.; Rappaport, E.B.; Lurmann, F.; et al. Association between Air Pollution and Lung Function Growth in Southern California Children. *Am. J. Respir. Crit. Care Med.* **2000**, *162*, 1383–1390. [CrossRef] [PubMed]
13. Gauderman, W.J.; Gilliland, G.F.; Vora, H.; Avol, E.; Stram, D.; McConnell, R.; Thomas, D.; Lurmann, F.; Margolis, H.G.; Rappaport, E.B.; et al. Association between Air Pollution and Lung Function Growth in Southern California Children: Results from a Second Cohort. *Am. J. Respir. Crit. Care Med.* **2002**, *166*, 76–84. [CrossRef] [PubMed]
14. Stieb, D.M.; Judek, S.; Burnett, R.T. Meta-Analysis of Time-Series Studies of Air Pollution and Mortality: Update in Relation to the Use of Generalized Additive Models. *J. Air Waste Manag. Assoc.* **2003**, *53*, 258–261. [CrossRef] [PubMed]
15. Burnett, R.T.; Stieb, D.; Brook, J.R.; Cakmak, S.; Dales, R.; Raizenne, M.; Vincent, R.; Dann, T. The Short-Term Effects of Nitrogen Dioxide on Mortality in Canadian Cities. *Arch. Environ. Health* **2004**, *59*, 228–237. [CrossRef] [PubMed]
16. Samoli, E.; Aga, E.; Touloumi, G.; Nisiotis, K.; Forsberg, B.; Lefranc, A.; Pekkanen, J.; Wojtyniak, B.; Schindler, C.; Niciu, E.; et al. Short Term Effects of Nitrogen Dioxide and Mortality: An Analysis within the APHEA Project. *Eur. Respir. J.* **2006**, *27*, 1129–1137. [CrossRef] [PubMed]
17. Schaap, M.; Müller, K.; ten Brink, H.M. Constructing the European Aerosol Nitrate Concentration Field from Quality Analysed Data. *Atmos. Environ.* **2002**, *36*, 1323–1335. [CrossRef]

18. Tørseth, K.; Aas, W.; Breivik, K.; Fjæraa, A.M.; Fiebig, M.; Hjellbrekke, A.G.; Lund Myhre, C.; Solberg, S.; Yttri, K.E. Introduction to the European Monitoring and Evaluation Programme (EMEP) and Observed Atmospheric Composition Change During 1972–2009. *Atmos. Chem. Phys.* **2012**, *12*, 5447–5481. [CrossRef]

19. Zhou, Y.; Brunner, D.; Hueglin, C.; Henne, S.; Staehelin, J. Changes in OMI Tropospheric NO_2 Columns over Europe from 2004 to 2009 and the Influence of Meteorological Variability. *Atmos. Environ.* **2012**, *46*, 482–495. [CrossRef]

20. Castellanos, P.; Boersma, K.F. Reductions in Nitrogen Oxides over Europe Driven by Environmental Policy and Economic Recession. *Sci. Rep.* **2012**, *2*, 265. [CrossRef] [PubMed]

21. Hilboll, A.; Richter, A.; Burrows, J.P. Long-Term Changes of Tropospheric NO_2 over Megacities Derived from Multiple Satellite Instruments. *Atmos. Chem. Phys.* **2013**, *13*, 4145–4169. [CrossRef]

22. Curier, R.L.; Kranenburg, R.; Segers, A.J.S.; Timmermans, R.M.A.; Schaap, M. Synergistic Use of OMI NO_2 Tropospheric Columns and LOTOS-EUROS to Evaluate the NOx Emission Trends across Europe. *Remote Sens. Environ.* **2014**, *149*, 58–69. [CrossRef]

23. Burrows, J.P.; Weber, M.; Buchwitz, M.; Rozanov, V.; Ladstätter-Weißenmayer, A.; Richter, A.; DeBeek, R.; Hoogen, R.; Bramstedt, K.; Eichmann, K.; et al. The Global Ozone Monitoring Experiment (GOME): Mission Concept and First Scientific Results. *J. Atmos. Sci.* **1999**, *56*, 151–175. [CrossRef]

24. Bovensmann, H.; Burrows, J.P.; Buchwitz, M.; Frerick, J.; Noël, S.; Rozanov, V.V.; Chance, K.V.; Goede, A.P.H. SCIAMACHY: Mission Objectives and Measurement Modes. *J. Atmos. Sci.* **1999**, *56*, 127–150. [CrossRef]

25. Levelt, P.F.; Hilsenrath, E.; Leppelmeier, G.W.; van den Oord, G.H.J.; Bhartia, P.K.; Tamminen, J.; de Haan, J.F.; Veefkind, J.P. Science Objectives of the Ozone Monitoring Instrument. *IEEE Trans. Geosci. Remote Sens.* **2006**, *44*, 1199–1208. [CrossRef]

26. Levelt, P.F.; van den Oord, G.H.J.; Dobber, M.R.; Malkki, A.; Visser, H.; de Vries, J.; Stammes, P.; Lundell, J.O.V.; Saari, H. The Ozone Monitoring Instrument. *IEEE Trans. Geosci. Remote Sens.* **2006**, *44*, 1093–1101. [CrossRef]

27. Callies, J.; Corpaccioli, E.; Eisinger, M.; Hahne, A.; Lefebvre, A. GOME-2, Metop's Second Generation Sensor for Operational Ozone Monitoring. *ESA Bull.* **2000**, *102*, 28–36.

28. Richter, A.; Burrows, J.P. Retrieval of Tropospheric NO_2 from GOME Measurements. *Adv. Space Res.* **2002**, *29*, 1673–1683. [CrossRef]

29. Martin, R.V.; Chance, K.; Jacob, D.J.; Kurosu, T.P.; Spurr, R.J.D.; Bucsela, E.; Gleason, J.F.; Palmer, P.I.; Bey, I.; Fiore, A.M.; et al. An Improved Retrieval of Tropospheric Nitrogen Dioxide from GOME. *J. Geophys. Res.* **2002**, *107*, 4437. [CrossRef]

30. Boersma, K.F.; Eskes, H.J.; Brinksma, E.J. Error Analysis for Tropospheric NO_2 Retrieval from Space. *J. Geophys. Res.* **2004**, *109*, D04311. [CrossRef]

31. Boersma, K.F.; Eskes, H.J.; Veefkind, J.P.; Brinksma, E.J.; van der A, R.J.; Sneep, M.; van den Oord, G.H.J.; Levelt, P.; Stammes, P.; Gleason, J.F.; et al. Near-Real Time Retrieval of Tropospheric NO_2 from OMI. *Atmos. Chem. Phys.* **2007**, *7*, 2103–2118. [CrossRef]

32. Boersma, K.F.; Bucsela, E.J.; Brinksma, E.J.; Gleason, J.F. NO_2. In *OMI Algorithm Theoretical Basis Document, OMI Trace Gas Algorithms*; ATB-OMI-04, Version 2.0; Chance, K., Ed.; NASA Distributed Active Archive Centers: Greenbelt, MD, USA, 2002; Volume 4.

33. Bucsela, E.J.; Celarier, E.A.; Wenig, M.O.; Gleason, J.F.; Veefkind, J.P.; Boersma, K.F.; Brinksma, E.J. Algorithm for NO_2 Vertical Column Retrieval from the Ozone Monitoring Instrument. *IEEE Trans. Geosci. Remote Sens.* **2006**, *44*, 1245–1258. [CrossRef]

34. Celarier, E.A.; Brinksma, E.J.; Gleason, J.F.; Veefkind, J.P.; Cede, A.; Herman, J.R.; Ionov, D.; Goutail, F.; Pommereau, J.; Lambert, J.; et al. Validation of Ozone Monitoring Instrument Nitrogen Dioxide Columns. *J. Geophys. Res.* **2008**, *113*, D15S15. [CrossRef]

35. Platt, U. Differential Optical Absorption Spectroscopy (DOAS). In *Air Monitoring by Spectroscopic Techniques*; Sigrist, M., Ed.; John Wiley: Hoboken, NJ, USA, 1994; pp. 27–84.

36. Martin, R.V.; Jacob, D.J.; Chance, K.; Kurosu, T.P.; Perner, P.I.; Evans, M.J. Global Inventory of Nitrogen Oxide Emission Constrained by Space-Based Observations of NO_2 Columns. *J. Geophys. Res.* **2003**, *108*, 4537. [CrossRef]

37. Hong, H.; Lee, H.; Kim, J.; Jeong, U.; Ryu, J.; Lee, D.S. Investigation of Simultaneous Effects of Aerosol Properties and Aerosol Peak Height on the Air Mass Factors for Space-Borne NO_2 Retrievals. *Remote Sens.* **2017**, *9*, 208. [CrossRef]

38. Dobber, M.; Kleipool, Q.; Dirksen, R.; Levelt, P.; Jaross, G.; Taylor, S.; Kelly, T.; Flynn, L.; Leppelmeier, G.; Rozemeijer, N. Validation of Ozone Monitoring Instrument Level 1b Data Products. *J. Geophys. Res.* **2008**, *113*, D15S06. [CrossRef]

39. Byun, D.; Schere, K. Review of the governing equations, computation algorithms, and other components of the Models-3 Community Multiscale Air Quality (CMAQ) modeling system. *Appl. Mech. Rev.* **2006**, *59*, 51–77. [CrossRef]

40. Byun, D.W.; Ching, J. *Science Algorithms of the EPA Models-3 Community Multi-Scale Air Quality (CMAQ) Modeling System*; NERL: Research Triangle Park, NC, USA, 1999; p. 425.

41. Sarwar, G.; Luecken, D.; Yarwood, G.; Whitten, G.; Carter, W. Impact of an Updated Carbon Bond Mechanism on Predictions from the CMAQ Modeling System: Preliminary Assessment. *J. Appl. Meteorol. Climatol.* **2008**, *47*, 3–14. [CrossRef]

42. Zhang, M.; Uno, I.; Sugata, S.; Wang, Z.; Byun, D.; Akimoto, H. Numerical Study of Boundary Layer Ozone Transport and Photochemical Production in East Asia in the Wintertime. *Geophys. Res. Lett.* **2002**, *29*. [CrossRef]

43. Zhang, M. Large-Scale Structure of Trace Gas and Aerosol Distributions over the Western Pacific Ocean During the Transport and Chemical Evolution Over the Pacific (TRACE-P) Experiment. *J. Geophys. Res.* **2003**, *108*, 8820. [CrossRef]

44. Zhang, M.G.; Uno, I.; Yoshida, Y.; Xu, Y.; Wang, Z.; Akimoto, H.; Bates, T.; Quinn, T.; Bandy, A.; Blomquist, B. Transport and Transformation of Sulfur Compounds over East Asia during the TRACE-P and ACE-Asia Campaigns. *Atmos. Environ.* **2004**, *38*, 6947–6959. [CrossRef]

45. Zhang, M.G. Modeling of Organic Carbon Aerosol Distributions over East Asia in the Springtime. *China Part.* **2004**, *2*, 192–195. [CrossRef]

46. Lu, Z.; Streets, D.G.; Zhang, Q.; Wang, S.; Carmichael, G.R.; Cheng, Y.F.; Wei, C.; Chin, M.; Diehl, T.; Tan, Q. Sulfur dioxide emissions in China and sulfur trends in East Asia since. *Atmos. Chem. Phys.* **2000**, *10*, 6311–6331. [CrossRef]

47. Lu, Z.; Zhang, Q.; Streets, D.G. Sulfur dioxide and primary carbonaceous aerosol emissions in China and India, 1996–2010. *Atmos. Chem. Phys.* **2011**, *11*, 9839–9864. [CrossRef]

48. Lei, Y.; Zhang, Q.; He, K.; Streets, D. Primary anthropogenic aerosol emission trends for China, 1990–2005. *Atmos. Chem. Phys.* **2011**, *11*, 931–954. [CrossRef]

49. Zhang, Q.; Streets, D.; Carmichael, G.; He, K.; Huo, H.; Kannari, A.; He, K.; Huo, H.; Kannari, A.; Klimont, Z.; et al. Asian emissions in 2006 for the NASA INTEX-B mission. *Atmos. Chem. Phys.* **2009**, *9*, 5131–5153. [CrossRef]

50. Han, X.; Ge, C.; Tao, J.; Zhang, M.; Zhang, R. Air quality modeling for a strong dust event in East Asia in March 2010. *Aerosol Air Qual. Res.* **2012**, *12*, 615–628. [CrossRef]

51. Han, X.; Zhang, M.; Tao, J.; Wang, L.; Gao, J.; Wang, S.; Chai, F. Modeling aerosol impacts on atmospheric visibility in Beijing with RAMS-CMAQ. *Atmos. Environ.* **2013**, *72*, 177–191. [CrossRef]

52. Han, X.; Zhang, M.; Gao, J.; Wang, S.; Chai, F. Modeling analysis of the seasonal characteristics of haze formation in Beijing. *Atmos. Chem. Phys.* **2014**, *14*, 10231–10248. [CrossRef]

53. Han, X.; Zhang, M.; Zhu, L.; Skorokhod, A. Assessment of the impact of emissions reductions on air quality over North China Plain. *Atmos. Pollut. Res.* **2016**, *7*, 249–259. [CrossRef]

54. Bucsela, E.J.; Perring, A.E.; Cohen, R.C.; Boersma, K.F.; Celarier, E.A.; Gleason, J.F.; Wenig, M.O.; Bertram, T.H.; Wooldridge, P.J.; Dirksen, R.; et al. Comparison of Tropospheric NO$_2$ In Situ Aircraft Measurements with near-Real-Time and Standard Product Data from the Ozone Monitoring Instrument. *J. Geophys. Res.* **2008**, *113*, D16S31. [CrossRef]

55. Leue, C.; Wenig, M.; Wagner, T.; Klimm, O.; Platt, U.; Jähne, B. Quantitative Analysis of NO$_2$ Emissions from Global Ozone Monitoring Experiment Satellite Image Sequences. *J. Geophys. Res.* **2001**, *106*, 5493–5505. [CrossRef]

56. Martin, R.V.; Sioris, C.E.; Chance, K.; Ryerson, T.B.; Bertram, T.H.; Wooldridge, P.J.; Cohen, R.C.; Neuman, J.A.; Swanson, A.; Flocke, F.M. Evaluation of Space-Based Constraints on Global Nitrogen Oxide Emissions with Regional Aircraft Measurements over and Downwind of Eastern North America. *J. Geophys. Res.* **2006**, *111*. [CrossRef]

57. Jaeglé, L.; Steinberger, L.; Martin, R.V.; Chance, K. Global Partitioning of NOx Sources Using Satellite Observations: Relative Roles of Fossil Fuel Combustion, Biomass Burning and Soil Emissions. *Faraday Discuss.* **2005**, *130*, 407–423. [CrossRef] [PubMed]

58. Zhang, Q.; Streets, D.G.; He, K.; Wang, Y.; Richter, A.; Burrows, J.P.; Uno, I.; Jang, C.J.; Chen, D.; Yao, Z.; et al. NOx Emission Trends for China, 1995–2004, The View from the Ground and the View from Space. *J. Geophys. Res.* **2007**, *112*, D22306. [CrossRef]

59. Schaap, M.; Kranenburg, R.; Curier, L.; Jozwicka, M.; Dammers, E.; Timmermans, R. Assessing the Sensitivity of the OMI-NO$_2$ Product to Emission Changes across Europe. *Remote Sens.* **2013**, *5*, 4187–4208. [CrossRef]

60. Lamsal, L.N.; Martin, R.V.; van Donkelaar, A.; Steinbacher, M.; Celarier, E.A.; Bucsela, E.; Dunlea, E.; Pinto, J.P.; Lamsal, C. Ground-level nitrogen dioxide concentrations inferred from the satellite-borne Ozone Monitoring Instrument. *J. Geophys. Res.* **2008**, *113*. [CrossRef]

61. Lamsal, L.N.; Martin, R.V.; van Donkelaar, A.; Celarier, E.A.; Bucsela, E.J.; Boersma, K.F.; Dirksen, R.; Luo, C.; Wang, Y. Indirect validation of tropospheric nitrogen dioxide retrieved from the OMI satellite instrument: insight into the seasonal variation of nitrogen oxides at northern midlatitudes. *J. Geophys. Res.* **2010**, *115*. [CrossRef]

remote sensing

[MDPI]

Article

Estimation of Surface NO$_2$ Volume Mixing Ratio in Four Metropolitan Cities in Korea Using Multiple Regression Models with OMI and AIRS Data

Daewon Kim [1], Hanlim Lee [1,*], Hyunkee Hong [1], Wonei Choi [1], Yun Gon Lee [2] and Junsung Park [1]

[1] Division of Earth Environmental System Science Major of Spatial Information Engineering, Pukyong National University, Busan 608-737, Korea; k.daewon91@gmail.com (D.K.); brunhilt77@gmail.com (H.H.); cwyh3338@gmail.com (W.C.); junsung2ek@gmail.com (J.P.)

[2] Department of Atmospheric Sciences, Chungnam National University, Daejeon 34134, Korea; yungonlee@gmail.com

* Correspondence: hllee@pknu.ac.kr; Tel.: +82-51-629-6688

Academic Editors: Yang Liu, Jun Wang, Omar Torres and Richard Müller
Received: 25 April 2017; Accepted: 15 June 2017; Published: 18 June 2017

Abstract: Surface NO$_2$ volume mixing ratio (VMR) at a specific time (13:45 Local time) (NO$_2$ VMR$_{ST}$) and monthly mean surface NO$_2$ VMR (NO$_2$ VMR$_M$) are estimated for the first time using three regression models with Ozone Monitoring Instrument (OMI) data in four metropolitan cities in South Korea: Seoul, Gyeonggi, Daejeon, and Gwangju. Relationships between the surface NO$_2$ VMR obtained from in situ measurements (NO$_2$ VMR$_{In\text{-}situ}$) and tropospheric NO$_2$ vertical column density obtained from OMI from 2007 to 2013 were developed using regression models that also include boundary layer height (BLH) from Atmospheric Infrared Sounder (AIRS) and surface pressure, temperature, dew point, and wind speed and direction. The performance of the regression models is evaluated via comparison with the NO$_2$ VMR$_{In\text{-}situ}$ for two validation years (2006 and 2014). Of the three regression models, a multiple regression model shows the best performance in estimating NO$_2$ VMR$_{ST}$ and NO$_2$ VMR$_M$. In the validation period, the average correlation coefficient (R), slope, mean bias (MB), mean absolute error (MAE), root mean square error (RMSE), and percent difference between NO$_2$ VMR$_{In\text{-}situ}$ and NO$_2$ VMR$_{ST}$ estimated by the multiple regression model are 0.66, 0.41, -1.36 ppbv, 6.89 ppbv, 8.98 ppbv, and 31.50%, respectively, while the average corresponding values for the other two models are 0.75, 0.41, -1.40 ppbv, 3.59 ppbv, 4.72 ppbv, and 16.59%, respectively. All three models have similar performance for NO$_2$ VMR$_M$, with average R, slope, MB, MAE, RMSE, and percent difference between NO$_2$ VMR$_{In\text{-}situ}$ and NO$_2$ VMR$_M$ of 0.74, 0.49, -1.90 ppbv, 3.93 ppbv, 5.05 ppbv, and 18.76%, respectively.

Keywords: surface NO$_2$ volume mixing ratio; NO$_2$; OMI; multiple regression

1. Introduction

The main anthropogenic source of nitrogen dioxide (NO$_2$) is fossil fuel combustion, while natural sources of NO$_2$ include lightning, forest fires, and soil emissions [1,2]. In particular, since NO$_2$ is emitted in large quantities in automobile exhaust gas, NO$_2$ is often used as an indicator of traffic-related air pollution in urban areas [3]. In terms of its effect on human health, long-term NO$_2$ exposure can lead to respiratory depression and respiratory illness [4–8]. In addition, it is a precursor of aerosol nitrate, tropospheric ozone, and the hydroxyl radical (OH), the main atmospheric oxidant [9]. It is therefore important to measure NO$_2$ and various methods are used, with chemiluminescence, a well-known technique for measuring surface NO$_2$ volume mixing ratio (VMR) [10]. In situ measurements such as the chemiluminescence method are, in general, more accurate than remote sensing techniques, but require a large number of in situ instruments to provide the spatial distribution of the NO_2 VMR at high resolution [11]. In recent years, NO_2 vertical column density (VCD) has been measured from satellites that can monitor NO_2 at global scale over a short time scale. Space-borne sensors that have observed global distributions of NO_2 are the Global Ozone Monitoring Experiment (GOME) aboard European Remote Sensing-2 (ERS-2) (1995–2003), Scanning Imaging Absorption Spectrometer for Atmospheric Chartography/Chemistry (SCIAMACHY) aboard Environmental Satellite (Envisat) (2002–2012), the Ozone Monitoring Instrument (OMI) aboard EOS-AURA (2004–present), and GOME-2 aboard the Meteorological Operational satellite (MetOp)-A (2007–present) and MetOp-B (2012–present) [12–17]. In many countries, air quality regulation requires surface NO_2 VMR so the NO_2 VCD obtained from satellites cannot be used directly. In recent years, studies have been conducted to investigate the feasibility of estimating the surface NO_2 VMR using the NO_2 VCD obtained from satellite measurements and, in particular, the correlation between the NO_2 VCD obtained from satellite measurements and the surface NO_2 VMR.

Ordóñez et al. [18] reported the correlation between tropospheric NO_2 VCD and the NO_2 VCD measured by GOME and ground based in situ devices in Milan. Kharol et al. [3] estimated the annual average ground-level NO_2 concentrations in North America using chemical transport model (GEOS-Chem) data and OMI NO_2 columns and also reported the annual trend of the estimated ground-level NO_2 concentrations. However, no studies have attempted to estimate the surface NO_2 VMR at higher temporal resolutions such as hourly and monthly using the NO_2 VCD measured by satellites.

In this present study, we estimate for the first time the surface NO_2 VMR at a specific time (13:45 Local time (LT)) (NO_2 VMRST) and the monthly mean surface NO_2 VMR (NO_2 VMRM) using two linear regression models and a multiple regression model with the tropospheric NO_2 VCD obtained from OMI (Trop NO_2 VCDOMI) in five metropolitan cities. In addition, the performance of each regression method is evaluated by comparing the estimated surface NO_2 VMRs with those obtained from in situ measurement (NO_2 VMRIn-situ).

2. Study Area and Period

A large amount of anthropogenic NO_X is emitted in Northeast Asia including China, Korea and Japan [19]. Especially, the annual mean NO_2 tended to increase in Seoul from 1995 to 2009 [20]. The study areas were selected where the surface NO_2 VMR is continuously measured in Korean metropolitan cities (Figure 1). Metropolitan cities such as Busan and Incheon where the OMI pixel covers both sea and land are excluded since there are no surface NO_2 data available over the sea. Therefore, the selected areas are Seoul, Gyeonggi, Daejeon, and Gwangju. Seoul is covered by four OMI pixels and is divided into eastern and western areas (West Seoul and East Seoul). The study period is the nine years from 2006 to 2014. This is split into a seven-year training period (2007–2013) to determine the coefficients of the regression models used in this study, and two years of validation (2006 and 2014) when the surface NO_2 VMRs estimated from the resulting three regression models are evaluated by comparison with the in situ data. The three regression models used in this study are described in detail in Section 3.

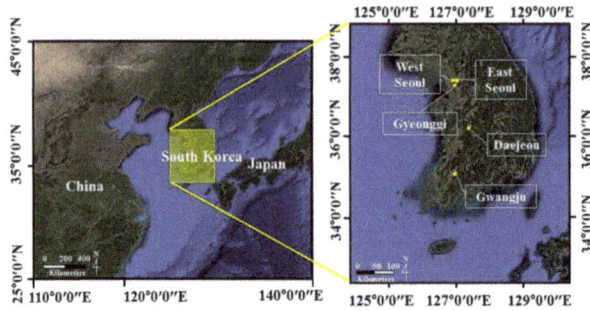

Figure 1. Study areas in South Korea.

2.1. Data

The data used in this study are Trop NO_2 VCD_{OMI} and Atmospheric Infrared Sounder (AIRS) boundary layer height (BLH_{AIRS}), atmospheric temperature ($Temp_{AIRS}$) and pressure ($Press_{AIRS}$), together with in situ measurements of NO_2 $VMR_{In-situ}$, surface temperature ($Temp_{In-situ}$), surface pressure ($Press_{In-situ}$), surface dew point ($Dewpoint_{In-situ}$), surface wind speed ($WS_{In-situ}$), and surface wind direction ($WD_{In-situ}$) (see Table 1).

Table 1. Satellite and in situ data used in this study.

	Data		Time (LT)
Satellite	Trop NO_2 VCD	OMI Level3 NO_2 Daily data (OMNO2d)	13:45
	BLH, Temperature, Pressure	AIRS/Aqua L3 Daily Support Product (AIRS + AMSU) V006 (AIRX3SPD)	13:30
In situ	Surface NO_2 VMR	Air Korea	13:00 and 14:00
	Surface Temperature, Surface Pressure, Surface Dew point, Surface Wind Speed, Surface Wind Data	AWS (Automatic Weather System)	

2.1.1. Ozone Monitoring Instrument (OMI) Data

The Trop NO_2 VCD_{OMI} data were obtained from OMI Level3 NO_2 Daily Data (OMNO2d) provided by the NASA Goddard Earth Sciences Data and Information Services Center (http://disc.sci.gsfc.nasa.gov/Aura/data-holdings/OMI) [17,21,22]. OMI is a nadir-viewing UV–visible (270–500 nm) spectrometer aboard the Aura platform launched in July 2004 [23]. Aura is a polar orbiting satellite with an overpass time of 13:45 LT. The spectral resolution of the OMI is about 0.5 nm and the spatial resolution is 13 × 24 km at nadir. Cloud-screened NO_2 data (Level-3 OMI NO_2 Cloud-Screened Total and Tropospheric Column NO_2 (V003)) are used in the present study (Cloud Fraction <30%).

2.1.2. Atmospheric Infrared Sounder (AIRS) Data

The BLH_{AIRS}, $Temp_{AIRS}$, and $Press_{AIRS}$ used in this study were obtained from the AIRS/Aqua L3 Daily Support Product (AIRS + AMSU) 1 degree × 1 degree V006 (AIRX3SPD.00) from NASA Goddard Earth Sciences Data and Information Services Center (http://disc.sci.gsfc.nasa.gov/uui/datasets/AIRX3SPD_V006/summary?keywords=%22AIRS%22) [24–26]. The AIRS/Advanced Microwave Sounding Unit (AMSU) is a sounding suite launched in May 2002 aboard Aqua [26,27]. Aqua is a polar orbiting satellite with an overpass time of 13:30 LT and a horizontal spatial resolution of 40 km at nadir.

2.1.3. In Situ NO_2 Data

The NO_2 $VMR_{In\text{-}situ}$ data were obtained from Air Korea (http://www.airkorea.or.kr/last_amb_hour_data). Since NO_2 $VMR_{In\text{-}situ}$ is available hourly, the average of the values at 13:00 and 14:00 LT is used to be closer to the OMI overpass time. In a previous study [18], the in situ measurements were grouped into five different NO_2 levels: clean, slightly polluted, averagely polluted, polluted, and heavily polluted. Many stations are located close to roads and are exposed to emissions. In addition, the in situ NO_2 data from stations within GOME pixels (320 × 40 km) were averaged, since in situ measurements are only representative of a small fraction of the satellite ground scene. In the present study, the NO_2 $VMR_{In\text{-}situ}$ obtained from in situ measurements located close to streets were excluded in this study. We used the average of three or more NO_2 $VMR_{In\text{-}situ}$ from stations located at least 2 km from each other.

2.1.4. In Situ Meteorological Data

The $Temp_{In\text{-}situ}$, $Press_{In\text{-}situ}$, $Dewpoint_{In\text{-}situ}$, $WS_{In\text{-}situ}$, and $WD_{In\text{-}situ}$ used in this study are Automatic Weather System (AWS) data provided by the Korea Meteorological Administration (http://sts.kma.go.kr/jsp/home/contents/statistics/newStatisticsSearch.do?menu=SFC&MNU=MNU). Since meteorological data are available hourly, the average of the data at 13:00 LT and 14:00 LT is used. The surface wind data, especially wind direction can be impacted by local topography and interferences.

3. Methodology

In this study, NO_2 VMR_{ST} and NO_2 VMR_M were estimated using three regression models with Trop NO_2 VCD_{OMI}. Table 2 summarizes the three models.

Table 2. Regression models used for surface NO_2 VMR estimation in this study.

Model		Equation
M1	13:45 LT and Monthly	$NO_2\ VMR_{in\ situ} = a Trop\ NO_2\ VCD_{OMI}^{(a)} + b$
M2	13:45 LT and Monthly	$NO_2\ VMR_{in\ situ} = a BLH\ NO_2\ VMR_{OMI}^{(b)} + b$
M3	13:45 LT	Section 3, Multiple regression Equation (1)
M4	Monthly	

Notes: (a) NO_2 tropospheric vertical column density obtained from OMI; and (b) BLH NO_2 $VMR_{OMI} = \frac{Trop\ NO_2\ VCD_{OMI}\ Gas\ constant\ R\ Temp_{AIRS} \times 10^{13}}{Avogadro\ constant\ NA\ BLH_{AIRS}\ Press_{AIRS}}$, where the AIRS pressure and temperature are boundary layer mean values, Gas constant R = 8.314472 m^3 pa K^{-1} mol^{-1} and Avogadro constant NA = 6.022 × 10^{23} mol^{-1}.

3.1. M1

M1 is the linear regression equation where Trop NO_2 VCD_{OMI} is used as the independent variable. Figure 2 shows the linear regression between Trop NO_2 VCD_{OMI} and NO_2 $VMR_{In\text{-}situ}$ at 13:45 LT during the training period, with R^2 (coefficient of determination), slope, and intercept of 0.47, 0.80 and 11.47, respectively. Figure 3 shows the linear regression between monthly mean Trop NO_2 VCD_{OMI} and monthly mean NO_2 $VMR_{In\text{-}situ}$ during the training period, with R^2, slope, and intercept of 0.62, 0.77, and 10.95, respectively. The final form of the M1 equation for estimating NO_2 VMR_{ST} is shown in Table 3, and that for estimating NO_2 VMR_M in Table 4.

Tables 3 and 4 show the equations M1, M2, M3, and M4 with the regression coefficients determined from the training period.

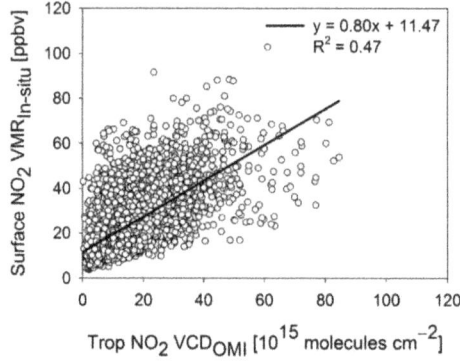

Figure 2. Scatter plot between Trop NO_2 VCD_{OMI} at 13.45 LT and NO_2 $VMR_{In-situ}$ to determine the regression coefficient for M1 for the training period 2007–2013.

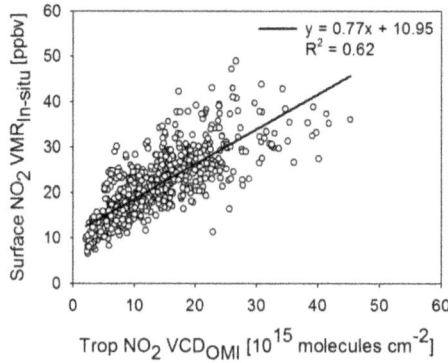

Figure 3. As Figure 2 but for the monthly mean values.

Table 3. Final form of the regression models used for estimating surface NO_2 VMR at a specific time and R^2 obtained from the regression between NO_2 $VMR_{In-situ}$ and the corresponding independent variable for the training period.

		Equation	R^2
13:45 LT	M1	$NO_2\ VMR_{ST} = 1.71 \times Trop\ NO_2\ VCD_{OMI} - 0.68$	0.47
	M2	$NO_2\ VMR_{ST} = 4.19 \times BLH\ NO_2\ VMR_{OMI} + 1.57$	0.38
	M3	$NO_2\ VMR_{ST} = 0.000602 \times Trop\ NO_2\ VCD_{OMI} - 0.000107 \times Temp_{In-situ}$ $-0.000083 \times Dewpoint_{In-situ} + 0.000061 \times Press_{In-situ}$ $-0.000002 \times BLH_{AIRS} - 0.002435 \times WS_{In-situ}$ $+0.001190 \times WD_{In-situ} - 0.039996$	0.47

Table 4. As Table 3 but for monthly mean surface NO_2 VMR.

		Equation	R^2
Monthly mean	M1	$NO_2\ VMR_M = 1.23 \times Trop\ NO_2\ VCD_{OMI} + 4.74$	0.62
	M2	$NO_2\ VMR_M = 2.92 \times BLH\ NO_2\ VMR_{OMI} + 6.74$	0.59
	M4	$NO_2\ VMR_M = 0.657241 \times Trop\ NO_2\ VCD_{OMI} - 0.137334 \times Dewpoint_{In-situ}$ $-0.136096 \times Press_{In-situ} - 0.004331 \times BLH_{AIRS} - 0.770356 \times WS_{In-situ}$ $+2.370956 \times WD\ (west)_{In-situ} + 157.361668$	0.63

3.2. M2

There might exist a minor fraction of the tropospheric NO_2 column in upper troposphere particularly because of lightning. However, the NO_2 amount in upper troposphere could be considered negligible in metropolitan cities, where a significant amount of NO_X is emitted. Therefore, assuming Trop NO_2 VCD_{OMI} is mostly present within the PBL, the relationship between Trop NO_2 VCD_{OMI} and the surface NO_2 VMR may change as the PBL varies. However, a minor fraction of the tropospheric NO_2 column can also be in the upper tropospheric, particularly because of lightning. This NO_2 fraction in upper tropospheric might cause either small or negligible reduction in correlations of the OMI NO_2 VCD between and surface NO_2 VMR as the upper part of the troposphere (free troposphere) contribution is assumed to be negligible [28]. To reflect the BLH in the regression equation, Trop NO_2 VCD_{OMI} is first divided by BLH_{AIRS} to calculate the NO_2 concentration in the PBL and then converted to the NO_2 mixing ratio in the PBL (BLH NO_2 VMR_{OMI}) using $Temp_{AIRS}$ and $Press_{AIRS}$ [29] as shown Table 2. Only a single OMI pixel contained completely within an AIRS pixel was used. Figure 4 shows the linear regression between BLH NO_2 VMR_{OMI} and NO_2 $VMR_{In\text{-}situ}$ at 13:45 LT during the training period. Here R^2, slope and intercept are 0.38, 1.58, and 14.30, respectively. Figure 5 shows the corresponding linear regression for the monthly mean data, with R^2, slope and intercept of 0.59, 1.71, and 12.75, respectively. The final form of equation M2 to estimate NO_2 VMR_{ST} is shown in Table 3, and for the monthly values in Table 4.

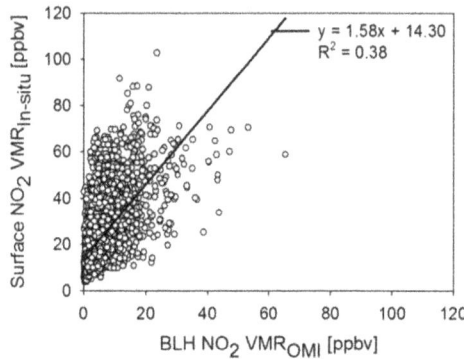

Figure 4. Scatter plot between BLH NO_2 VMR_{OMI} at a specific time (13:45 LT) and NO_2 $VMR_{In\text{-}situ}$ to determine the regression coefficient for M1 for the training period 2007–2013.

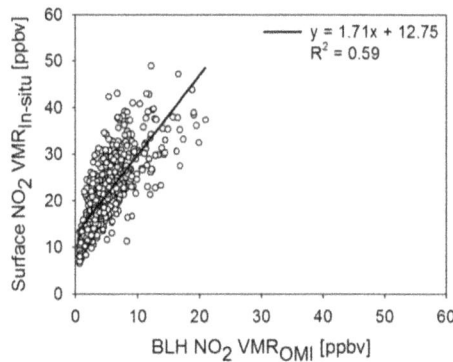

Figure 5. As Figure 4 but for the monthly mean values.

3.3. M3 and M4

M3 and M4 are multiple regression equations for estimating NO_2 VMR_{ST} and NO_2 VMR_M. Multiple regression equations consist of a dependent variable, independent variables, and their regression coefficients. In addition to Trop NO_2 VCD_{OMI} and BLH_{AIRS}, meteorological factors (surface temperature, dew point, atmospheric pressure, wind direction, and wind speed) are used as candidate independent variables for the multiple regression equation in the present study. In a previous study [30], these meteorological factors were also used as candidate independent variables to estimate surface SO_2 concentration in Shanghai, China. Temperature, pressure, boundary layer height, wind speed, and wind direction were selected as the candidates for independent variables since they are known to either directly or indirectly affect the spatial mixing of NO_2 molecules in boundary layer. Furthermore, temperature and dewpoint were selected as candidates for independent variables as they affect the boundary layer height [31].

The multiple regression equation can be defined by the following equations:

$$\hat{y} = \beta_0 + \beta_1 x_1 + \beta_2 x_2 + \dots + \beta_n x_n + \varepsilon \tag{1}$$

where $\hat{y}$ and β_0 are the dependent variable (NO_2 $VMR_{In-situ}$) and regression coefficient, respectively; $x_1, x_2,..., x_n$ are the candidate independent variables (Trop NO_2 VCD_{OMI}, $Dewpoint_{In-situ}$, $Press_{In-situ}$, $Temp_{In-situ}$, BLH_{AIRS}, $WS_{In-situ}$, and $WD_{In-situ}$); $\beta_1, \beta_2, \dots, \beta_n$ are the regression coefficients of the independent variables; and ε is the difference between observations (NO_2 $VMR_{In-situ}$) and estimated values (NO_2 $VMR_{estimate}$). The regression coefficients can be estimated by least square fitting:

$$\sum_{j=1}^{m} \varepsilon_j^2 = \sum_{j=1}^{m} (y_j - \hat{y}_j)^2 \tag{2}$$

where y_j is the observed value with m data points. By minimizing the sum of ε^2, regression coefficients can be derived. These least square fitting techniques are based on the following assumptions: the linear relationship, a normal distribution and equal variance in the residuals. The least squares regression is sensitive to the presence of some points that are excessively large or small values in the training data [32]. To determine the independent variables (x_n) and regression coefficients (β_n) included in the final form of equations M3 and M4, we considered the variation inflation factor (VIF) and *p*-value to ensure their statistical significance. First, we examined the VIF that explains the multicollinearity of a candidate independent variable with regard to other candidate independent variables. The VIF of the *j*-th independent variable is expressed as:

$$VIF(x_j) = \frac{1}{1 - R_j^2} \tag{3}$$

where R_j^2 is the coefficient of determination for the regression of x_j against another independent variable (a regression that does not involve the dependent variable *j*). The VIF indicates how much x_j is correlated with the other candidate variables. A candidate independent variable with a very high VIF can be considered redundant and should be removed from the multiple regression equations. Candidate independent variables that do not satisfy the criterion VIF < 10 [33], were excluded from the independent variables. The *p*-value was also used to select independent variables. The highest still statistically significant *p*-level was shown by Sellke et al. [34] to be 5%. Among the independent variables that satisfy the VIF criterion, those that also satisfy *p*-value <0.05 are selected as final independent variables in the multiple regression equations. The independent variables selected for equations M3 and M4 are shown in Table 5. The final form of equation M3 to estimate NO_2 VMR_{ST} is shown Table 3, and that for NO_2 VMR_M in Table 4.

Table 5. Final independent variables included in multiple regression equations (M3 and M4).

	Final Selected Independent Variables	*p*-Value	VIF
M3	Trop NO_2 VCD_{OMI}	0	1.26
	$Temp_{In\text{-}situ}$	0.000032	7.02
	$Dewpoint_{In\text{-}situ}$	0.000306	7.16
	$Press_{In\text{-}situ}$	0.009981	3.14
	BLH_{AIRS}	1.73×10^{-15}	1.12
	$WS_{In\text{-}situ}$	3.86×10^{-133}	1.33
	$WD_{In\text{-}situ}$	1.7493×10^{-38}	1.07
M4	Trop NO_2 VCD_{OMI}	2.4832×10^{-89}	1.64
	$Dewpoint_{In\text{-}situ}$	0.000421	6.47
	$Press_{In\text{-}situ}$	0.034582	6.65
	BLH_{AIRS}	0.000834	2.32
	$WS_{In\text{-}situ}$	3.86×10^{-133}	1.59
	$WD_{In\text{-}situ}$	1.699×10^{-7}	1.25

4. Results

4.1. Daily Estimates

Figure 6 shows the day-to-day variations of NO_2 $VMR_{In\text{-}situ}$ and NO_2 VMR_{ST} estimated at 13:45 LT in West Seoul and East Seoul using M1, M2 and M3 in Table 3 for 2006 and 2014. A slightly larger difference in magnitude is found between NO_2 $VMR_{In\text{-}situ}$ and NO_2 VMR_{ST} obtained with M3 compared to those between NO_2 $VMR_{In\text{-}situ}$ and NO_2 VMR_{ST} obtained with M1 and M2. However, NO_2 obtained from M3 showed moderate agreement with NO_2 $VMR_{In\text{-}situ}$ in the form of the day-to-day variation. Results for Daejeon, Gwangju, and Gyeonggi are included in the Supplementary Materials.

Figure 7 shows the *R*, slope, mean bias (MB), mean absolute error (MAE), root mean square error (RMSE) and percent difference between NO_2 VMR_{ST} and NO_2 $VMR_{In\text{-}situ}$ for the validation period (2006 and 2014). The R obtained with M1 ranges from 0.49 to 0.71, showing better agreement than that with M2 ($0.47 < R < 0.65$). M3 showed the best correlation with NO_2 $VMR_{In\text{-}situ}$ ($0.67 < R < 0.90$). The slopes from both M1 and M2 are close to one in East Seoul, whereas they are lower in the other cities. The MB from M1, M2, and M3 ranges from -7.74 to 5.80 ppbv. In all study areas, the MAE (5.79 ppbv $<$ MAE $<$ 8.25 ppbv) of M3 is lower than those (6.58 ppbv $<$ MAE $<$ 11.41 ppbv) of M1 and M2, which means that NO_2 VMR_{ST} estimated from M3 show moderate agreement with NO_2 $VMR_{In\text{-}situ}$ in terms of magnitude. The RMSE from M3 is found to be lower than those from M1 and M2. The NO_2 VMR_{ST} from M3 have the lowest RMSE in all study areas (7.21 ppbv $<$ RMSE $<$ 11.37 ppbv). In addition, percent differences estimated from M3 and NO_2 $VMR_{In\text{-}situ}$ are lower in all study areas than from M1 and M2. In estimating NO_2 VMR_{ST}, M3, which is a multiple regression method with various independent variables as inputs, generally showed good statistical performance except for MB.

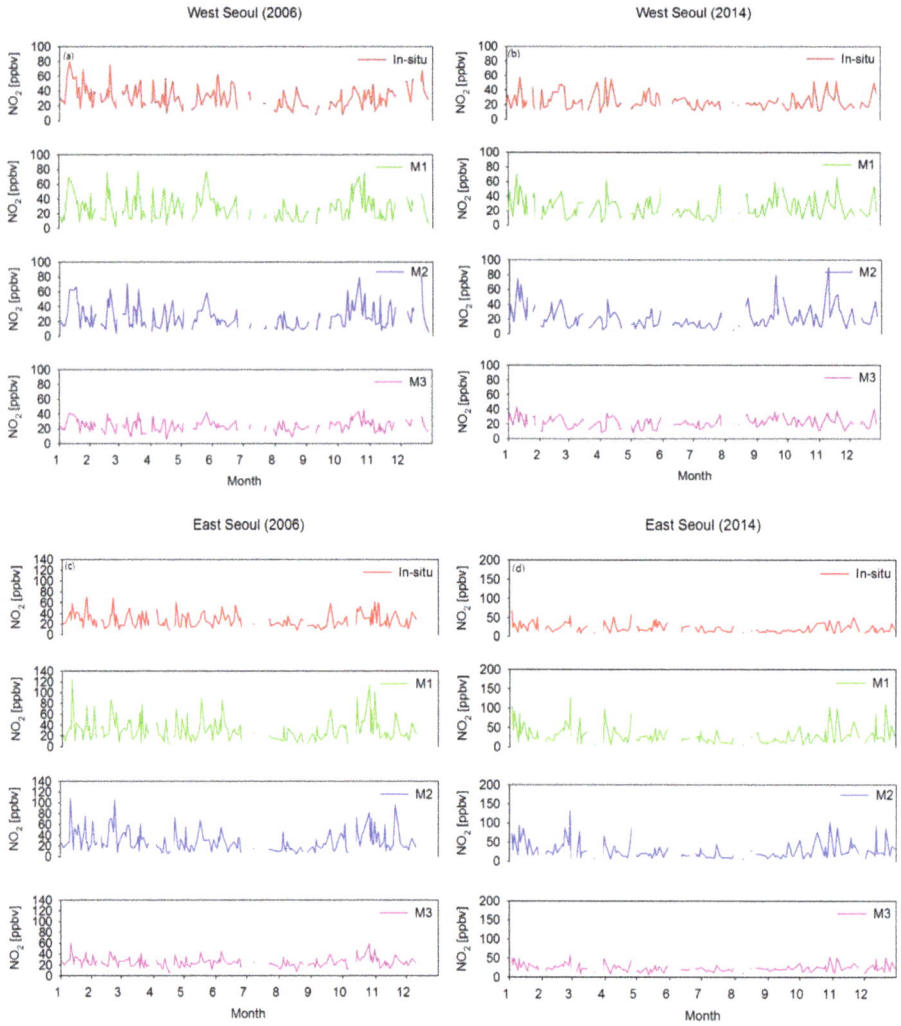

Figure 6. Time series of NO$_2$ VMR$_{In-situ}$ and NO$_2$ VMR$_{ST}$ at 13:45 LT estimated by M1, M2 and M3 in East Seoul and West Seoul for: 2006 (**a,c**); and 2014 (**b,d**).

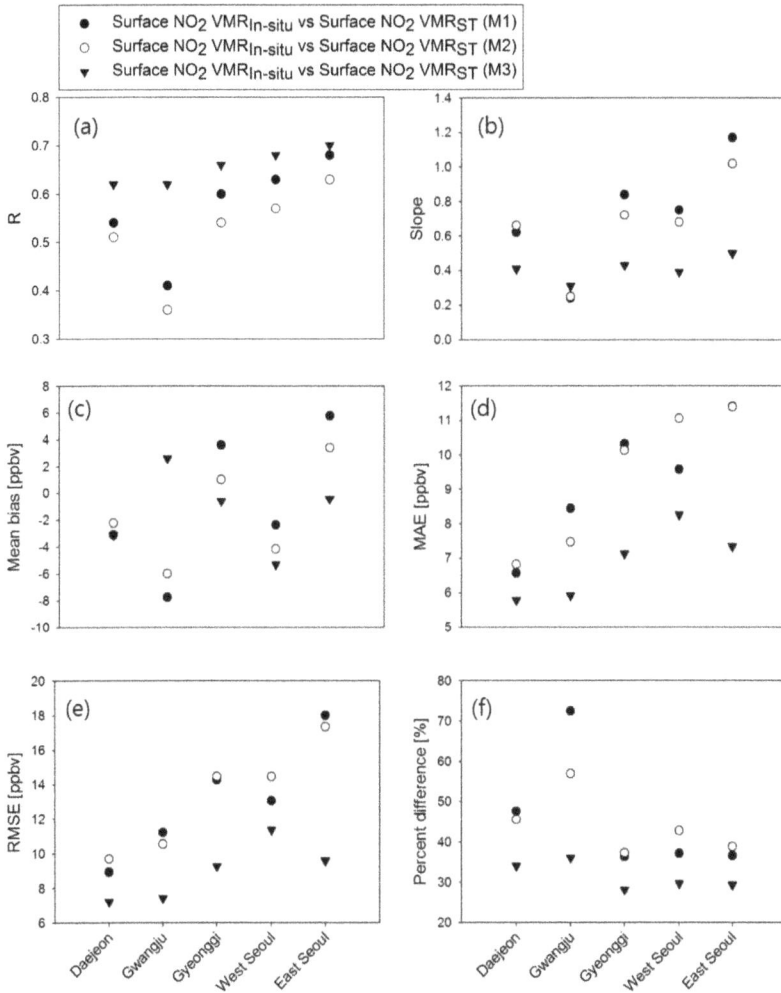

Figure 7. (a) R; (b) slope; (c) MB; (d) MAE; (e) RMSE; and (f) percent difference between NO_2 VMR_{ST} against NO_2 $VMR_{In-situ}$ in 2006 and 2014 for M1, M2, and M3.

4.2. Monthly Estimates

Figure 8 shows the temporal variation of monthly mean NO_2 $VMR_{In-situ}$ and NO_2 VMR_M estimated using M1, M2 and M4 of Table 4 in West Seoul and East Seoul using monthly mean independent variables during the validation period (see the detailed input data in Section 2.1). Figure 8 shows good agreement in terms of the temporal pattern between the estimated NO_2 VMR_M and monthly mean NO_2 $VMR_{In-situ}$. However, we found a large difference between NO_2 $VMR_{In-situ}$ and NO_2 VMR_M in periods when there was a jump in NO_2 $VMR_{In-situ}$ between successive months. For example, no models calculated NO_2 VMR_M that were similar to NO_2 $VMR_{In-situ}$ in December 2006, which is very different from that in November 2006. NO_2 $VMR_{In-situ}$ (NO_2 VMR_M from M1, M2, and M4) in November and December in 2006 are 19.32 ppbv (15.94, 17.96, and 17.62 ppbv) and 30.30 ppbv (15.94, 17.96, and 17.62 ppbv) in Daejeon, 15.26 ppbv (12.29, 13.87, and 18.09 ppbv) and 32.55 ppbv (12.73, 14.57, and 18.46 ppbv) in Gwangju, 29.31 ppbv (25.86, 25.97, and 22.35 ppbv) and

40.64 ppbv (29.91, 29.15, and 26.85 ppbv) in Gyeonggi, and 31.25 ppbv (22.80, 24.55, and 23.64 ppbv) and 45.93 ppbv (28.65, 28.92, and 26.49 ppbv) in West Seoul. Especially in West Seoul, there are several periods when NO_2 $VMR_{In\text{-}situ}$ changes rapidly compared with the previous month. The NO_2 VMR_M obtained from the three models at these times are in poor agreement with the pattern of monthly NO_2 $VMR_{In\text{-}situ}$. As described in Section 2, despite the use of NO_2 $VMR_{In\text{-}situ}$ located away from the streets, the in situ measurement sites in West Seoul are located closer to the streets than the in situ measurement sites in Daejeon and Gwangju. This may explain why there are more periods when NO_2 $VMR_{In\text{-}situ}$ changes rapidly from one month to the next. It is difficult to estimate the rapid change of NO_2 VMR near the NO_2 source using regression models that reflect the relationship between the in situ measurements and the OMI sensor covering both source and non-source areas in a single pixel.

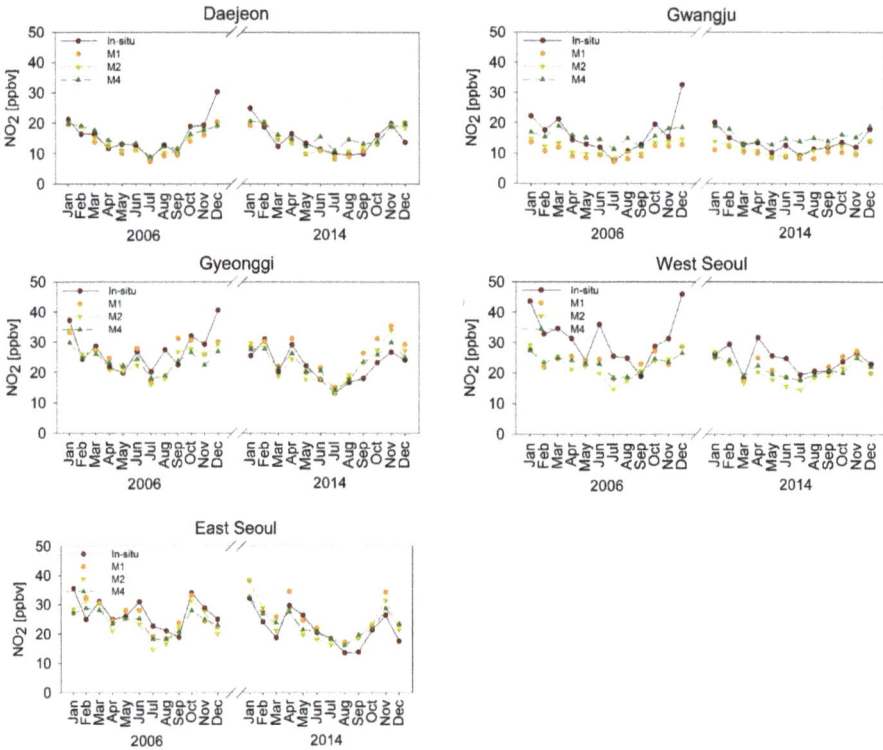

Figure 8. Time series of NO_2 $VMR_{In\text{-}situ}$ and NO_2 VMR_M estimated by M1, M2, and M4 for 2006 and 2014.

Figure 9 shows the *R*, slope, MB, MAE, RMSE and percent difference between NO_2 VMR_M and monthly mean NO_2 $VMR_{In\text{-}situ}$ in 2006 and 2014. In general, NO_2 VMR_M agreed better with NO_2 $VMR_{In\text{-}situ}$ than did the NO_2 VMR_{ST}. The value of R from M1, M2 and M4 and monthly mean NO_2 $VMR_{In\text{-}situ}$ ranged from 0.68 to 0.82 in all areas. MB was close to 0 in most study areas. MAE was less than 5 ppbv in Daejeon, Gwangju, Gyeonggi, and East Seoul where there is good agreement between NO_2 VMR_M from M1, M2, and M4 and monthly mean NO_2 $VMR_{In\text{-}situ}$, whereas MAEs in West Seoul ranged from 5.66 to 6.79. RMSEs between NO_2 $VMR_{In\text{-}situ}$ and NO_2 VMR_M from M1, M2, and M3 are found to be lower than 7 ppbv in the study areas except for West Seoul. In addition, the three models showed percent differences of less than 30% except for the value estimated from M1 in Gwangju.

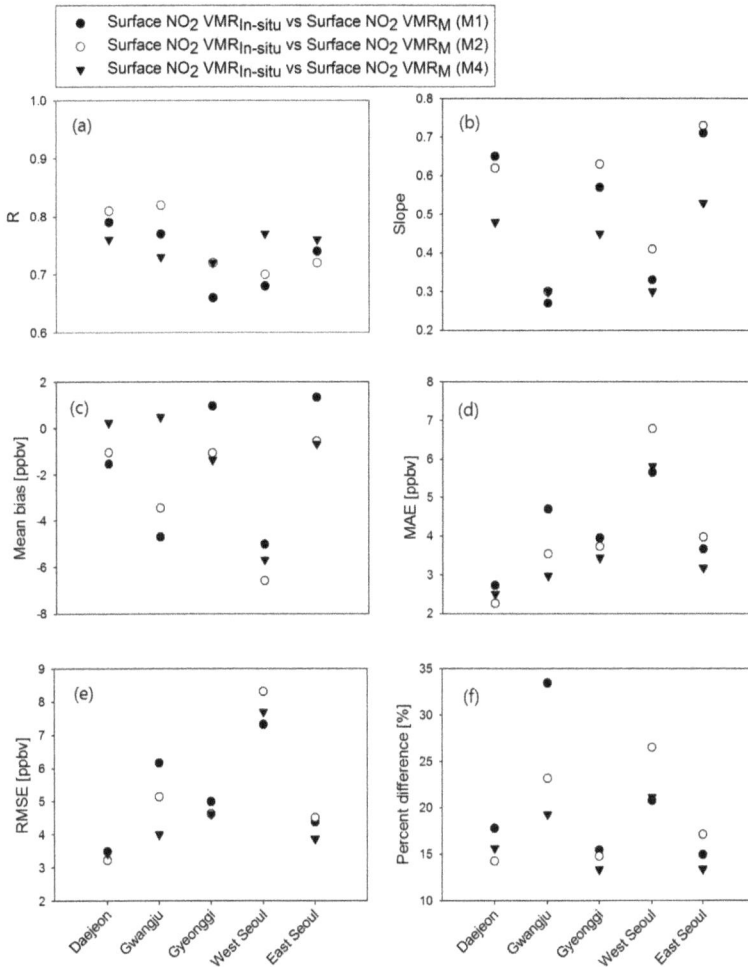

Figure 9. (**a**) *R*; (**b**) slope; (**c**) MB; (**d**) MAE; (**e**) RMSE; and (**f**) percent difference between NO$_2$ VMR$_M$ and monthly mean NO$_2$ VMR$_{In\text{-}situ}$ in 2006 and 2014.

5. Discussion

In a previous study [18], tropospheric NO$_2$ VCDs obtained from GOME were compared with tropospheric NO$_2$ VCDs calculated using NO$_2$ concentrations obtained from both in situ measurements and the Model of Ozone and Related Tracers 2 (MOZART-2). There are also several previous studies estimating surface NO$_2$ VMR using satellite data [3,35]. Among them, Kharol et al. [3] estimated the annual variation of ground-level NO$_2$ concentrations using both GEOS-Chem data and OMI data. However, in the present study, NO$_2$ VMR$_{ST}$ and NO$_2$ VMR$_M$ were estimated for the first time at higher temporal resolution using three regression models with Trop NO$_2$ VCD$_{OMI}$ as input.

5.1. Estimation of Surface NO$_2$ VMRs at a Specific Time (13:45 LT)

- Among the three regression models, the multiple regression model M3 performed best in estimating NO$_2$ VMR$_{ST}$. The linear regression model (M2), in which BLH is used as an independent variable in addition to Trop NO$_2$ VCD$_{OMI}$, has comparable performance to that of the model (M1) which uses Trop NO_2 VCD_{OMI} as the only independent variable.The BLH varies with latitude [36], but the latitudinal variation of BLH is not well represented since the spatial resolution of the AIRS used in this study is coarser than the spatial resolution of OMI. It might also be associate the BLH_{AIRS} data quality. We expect better results using BLH data obtained from LIDAR.

- The average difference was found to be 46.04% between NO_2 $VMR_{In\text{-}situ}$ and NO_2 VMR_{ST} obtained from M1, 44.29% between NO_2 $VMR_{In\text{-}situ}$ and NO_2 VMR_{ST} obtained from M2, and 31.50% between NO_2 $VMR_{In\text{-}situ}$ and NO_2 VMR_{ST} obtained from M3 in all cities, while there was moderate agreement in the temporal pattern of NO_2 variation between NO_2 $VMR_{In\text{-}situ}$ and NO_2 VMR_{ST} obtained from M1, M2, and M3 (Figure 6).
- In terms of statistical evaluation with respect to the in situ data, M3 showed the best performance in general.
- The results produced by M2 are not improved compared to those by M1 which may imply that surface NO_2 VMR is dominantly affected by tropospheric NO_2 column while the BLH effect could be negligible in areas of the present study. It might also be associate the AIRS BLH data quality.

5.2. Estimation of Monthly Mean Surface NO_2 VMRs of a Specific Time (13:45 LT)

- We found good agreement in the temporal pattern between the estimated NO_2 VMR_M and monthly mean NO_2 $VMR_{In\text{-}situ}$ (Figure 8). However, there was a large difference between NO_2 $VMR_{In\text{-}situ}$ and NO_2 VMR_M in the period when there was a clear change in NO_2 VMR_M between one month and the next. Despite the use of NO_2 $VMR_{In\text{-}situ}$ located away from streets, the in situ measurement sites in West Seoul are located closer to streets than the in situ measurement sites in Daejeon and Gwangju. This may explain why there are more periods when NO_2 $VMR_{In\text{-}situ}$ changes rapidly in successive months. It is difficult to estimate the rapid change of NO_2 VMR near NO_2 sources with regression models that reflect the relationship between the in situ measurements and the OMI sensor covering both source and non-source areas in a single pixel.
- In terms of statistical evaluation, the three regression models (M1, M2, and M4) were found to be similar (Figure 9).
- NO_2 VMR_M shows better agreement with the NO_2 $VMR_{In\text{-}situ}$ than does NO_2 VMR_{ST}. The reason for the better performance in the monthly mean estimation could be attributed to reduced errors in the monthly mean OMI data [37] as well as fewer occasions with sudden monthly changes in NO_2 $VMR_{In\text{-}situ}$ than rapid day-to-day changes in NO_2 $VMR_{In\text{-}situ}$.

This present study provides the results in the condition of 2 km distance between the in situ NO_2 measurement location and NO_X point source. For a future study, performances of the models need to be investigated depending on the distance between the in situ NO_2 data and point sources. We expect that the regression methods used to estimate the surface NO_2 VMR using Trop NO_2 VCD_{OMI} will be useful in providing information on surface NO_2 VMR in metropolitan cites on a monthly timescale. In future research, the estimation of surface NO_2 VMR may be attempted at higher time resolution with geostationary satellite sensors (e.g., geostationary environmental monitoring spectrometer (GEMS), tropospheric emissions: monitoring of pollution (TEMPO), and Sentinel-4). In further work, improvements are needed in the input data or the model formulation before the surface NO_2 can be estimated on a daily basis.

6. Conclusions

In this study, monthly and specific time estimates of NO_2 VMR were obtained for the first time using three regression models in four metropolitan cities for two years, 2006 and 2014. The multiple regression model (M3) was found to perform best in estimating NO_2 VMR_{ST} in all cities. For surface NO_2 estimates at the specific time (13:45 LT), M3 generally gives better R, MAE, RMSE, and percent difference than the other two models (M1 and M2). A comparison between monthly surface NO_2

VMR estimates and those at the specific time showed that agreement with NO_2 $VMR_{In\text{-}situ}$ was better for monthly estimates. In estimating NO_2 VMR_M, three regression models (M1, M2, and M4) showed similar performance. In estimating daily and monthly surface NO_2 VMR variations, when the surface NO_2 VMR changes rapidly, the difference between surface NO_2 VMR estimated from all models and NO_2 $VMR_{In\text{-}situ}$ is found to be large. In future studies, using higher spatial resolution satellites is expected to improve the relationship with in situ measurements. In addition, the use of other independent variables that may co-vary with rapid changes of surface NO_2 VMR should be investigated.

Supplementary Materials: The Supplementary Materials are available online at http://www.mdpi.com/2072-4292/9/6/627/s1.

Acknowledgments: This subject is supported by Korea Ministry of Environment (MOE) as "K-COSEM Research Program". This work was financially supported by the BK21 plus Project of the Graduate School of Earth Environmental Hazard System.

Author Contributions: Ozone Monitoring Instrument (OMI) and Atmospheric Infrared Sounder (AIRS) data collection and analysis were done by Wonei Choi. In situ data collection and analysis were done by Yun Gon Lee and Junsung Park. Estimation of surface NO_2 mixing ratios using three regression models was conducted by Daewon Kim, Hanlim Lee and Hyunkee Hong.

Conflicts of Interest: The authors declare no conflict of interest.

References

1. IPCC. Climate Change 2007: The Physical Science Basis. In *Contribution of Working Group I to the Fourth Assessment Report of the Intergovernmental Panel on Climate Change*; Solomon, S., Qin, D., Manning, M., Chen, Z., Marquis, M., Averyt, K.B., Tignor, M., Miller, H.L., Eds.; Cambridge University Press: Cambridge, UK; New York, NY, USA, 2007; p. 2007.
2. Van der A, R.J.; Eskes, H.J.; Boersma, K.F.; Van Noije, T.P.C.; Van Roozendael, M.; De Smedt, I.; Peters, D.H.M.U.; Meijer, E.W. Trends, seasonal variability and dominant NOx source derived from a ten year record of NO_2 measured from space. *J. Geophys. Res. Atmos.* **2008**, *113*. [CrossRef]
3. Kharol, S.K.; Martin, R.V.; Philip, S.; Boys, B.; Lamsal, L.N.; Jerrett, M.; Brauer, M.; Crouse, D.L.; McLinden, C.; Burnett, R.T.; et al. Assessment of the magnitude and recent trends in satellite-derived ground-level nitrogen dioxide over North America. *Atmos. Environ.* **2015**, *118*, 236–245.
4. Ackermann-Liebrich, U.; Leuenberger, P.; Schwartz, J.; Schindler, C.; Monn, C.; Bolognini, G.; Bongard, J.P.; Brändli, O.; Domenighetti, G.; Elsasser, S.; et al. Lung function and long term exposure to air pollutants in Switzerland. Study on Air Pollution and Lung Diseases in Adults (SAPALDIA) Team. *Am. J. Respir. Crit. Care Med.* **1997**, *155*, 122–129. [PubMed]
5. Schindler, C.; Ackermann-Liebrich, U.; Leuenberger, P.; Monn, C.; Rapp, R.; Bolognini, G.; Bongard, J.P.; Brändli, O.; Domenighetti, G.; Karrer, W.; et al. Associations between Lung Function and Estimated Average Exposure to NO_2 in Eight Areas of Switzerland. *Epidemiology* **1998**, *9*, 405–411. [CrossRef] [PubMed]
6. James Gauderman, W.; McConnell, R.O.B.; Gilliland, F.; London, S.; Thomas, D.; Avol, E.; Vora, H.; Berhane, K.; Rappaport, E.B.; Lurmann, F.; et al. Association between air pollution and lung function growth in southern California children. *Am. J. Respir. Crit. Care Med.* **2000**, *162*, 1383–1390. [CrossRef] [PubMed]
7. Panella, M.; Tommasini, V.; Binotti, M.; Palin, L.; Bona, G. Monitoring nitrogen dioxide and its effects on asthmatic patients: Two different strategies compared. *Environ. Monit. Assess.* **2000**, *63*, 447–458.
8. Smith, B.J.; Nitschke, M.; Pilotto, L.S.; Ruffin, R.E.; Pisaniello, D.L.; Willson, K.J. Health effects of daily indoor nitrogen dioxide exposure in people with asthma. *Eur. Respir. J.* **2000**, *16*, 879–885. [PubMed]
9. Boersma, K.F.; Jacob, D.J.; Trainic, M.; Rudich, Y.; DeSmedt, I.; Dirksen, R.; Eskes, H.J. Validation of urban NO_2 concentrations and their diurnal and seasonal variations observed from the SCIAMACHY and OMI sensors using in situ surface measurements in Israeli cities. *Atmos. Chem. Phys.* **2009**, *9*, 3867–3879. [CrossRef]
10. Demerjian, K.L. A review of national monitoring networks in North America. *Atmos. Environ.* **2000**, *34*, 1861–1884. [CrossRef]
11. Bechle, M.J.; Millet, D.B.; Marshall, J.D. Remote sensing of exposure to NO_2: Satellite versus ground-based measurement in a large urban area. *Atmos. Environ.* **2013**, *69*, 345–353.

12. Leue, C.; Wenig, M.; Wagner, T.; Klimm, O.; Platt, U.; Jähne, B. Quantitative analysis of NO$_x$ emissions from Global Ozone Monitoring Experiment satellite image sequences. *J. Eophys. Res. Atmos.* **2001**, *106*, 5493–5505. [CrossRef]

13. Richter, A.; Burrows, J.P. Tropospheric NO$_2$ from GOME measurements. *Adv. Space Res.* **2002**, *29*, 1673–1683. [CrossRef]

14. Martin, R.V.; Chance, K.; Jacob, D.J.; Kurosu, T.P.; Spurr, R.J.; Bucsela, E.; Gleason, J.F.; Palmer, P.I.; Bey, I.; Fiore, A.M.; et al. An improved retrieval of tropospheric nitrogen dioxide from GOME. *J. Geophys. Res. Atmos.* **2002**, *107*. [CrossRef]

15. Boersma, K.F.; Eskes, H.J.; Brinksma, E.J. Error analysis for tropospheric NO$_2$ retrieval from space. *J. Geophys. Res. Atmos.* **2002**, *109*. [CrossRef]

16. Boersma, K.F.; Eskes, H.J.; Veefkind, J.P.; Brinksma, E.J.; Van Der A, R.J.; Sneep, M.; Van Der Oord, G.H.J.; Levelt, P.F.; Stammes, P.; Gleason, J.F.; et al. Near-real time retrieval of tropospheric NO$_2$ from OMI. *Atmos. Chem. Phys. Discuss.* **2006**, *6*, 12301–12345. [CrossRef]

17. Bucsela, E.J.; Celarier, E.A.; Wenig, M.O.; Gleason, J.F.; Veefkind, J.P.; Boersma, K.F.; Brinksma, E.J. Algorithm for NO$_2$ vertical column retrieval from the Ozone Monitoring Instrument. *IEEE Trans. Geosci. Remote Sens.* **2006**, *44*, 1245–1258. [CrossRef]

18. Ordóñez, C.; Richter, A.; Steinbacher, M.; Zellweger, C.; Nüß, H.; Burrows, J.P.; Prévôt, A.S.H. Comparison of 7 years of satellite-borne and ground-based tropospheric NO$_2$ measurements around Milan, Italy. *J. Geophys. Res. Atmos.* **2006**, *111*. [CrossRef]

19. Kim, N.K.; Kim, Y.P.; Morino, Y.; Kurokawa, J.I.; Ohara, T. Verification of NO$_x$ emission inventory over South Korea using sectoral activity data and satellite observation of NO$_2$ vertical column densities. *Atmos. Environ.* **2013**, *77*, 496–508. [CrossRef]

20. Studies–Seoul, M.A.P. NIER. Available online: https://espo.nasa.gov/home/sites/default/files/documents/MAPS-Seoul_White%20Paper_26%20Feb%202015_Final.pdf (accessed on 17 June 2017).

21. Levelt, P.F.; Noordhoek, R. *OMI Algorithm Theoretical Basis Document Volume I: OMI Instrument, Level 0–1b Processor, Calibration & Operations*; ATBD-OMI-01; Version 1.1; NASA: Washington, DC, USA, 2002. Available online: http://www.sciamachy-validation.org/omi/documents/data/OMI_ATBD_Volume_1_V1d1.pdf (accessed on 17 June 2017).

22. Celarier, E.A.; Brinksma, E.J.; Gleason, J.F.; Veefkind, J.P.; Cede, A.; Herman, J.R.; Ionov, D.; Goutail, F.; Pommereau, J.P.; Lambert, J.C.; et al. Validation of Ozone Monitoring Instrument nitrogen dioxide columns. *J. Geophys. Res. Atmos.* **2008**, *113*. [CrossRef]

23. Levelt, P.F.; Hilsenrath, E.; Leppelmeier, G.W.; van den Oord, G.H.; Bhartia, P.K.; Tamminen, J.; de Haan, J.F.; Veefkind, J.P. Science objectives of the ozone monitoring instrument. *IEEE Trans. Geosci. Remote Sens.* **2006**, *44*, 1199–1208. [CrossRef]

24. Aumann, H.; Gaiser, S.; Ting, D.; Manning, E. *AIRS Algorithm Theoretical Basis Document Level 1B Part 1: Infrared Spectrometer*; EOS Project Science Office, NASA Goddard Space Flight Center: Greenbelt, MD, USA, 1999.

25. Fetzer, E.; McMillin, L.M.; Tobin, D.; Aumann, H.H.; Gunson, M.R.; McMillan, W.W.; Hagan, D.E.; Hofstadter, M.D.; Yoe, J.; Whiteman, D.N.; et al. Airs/amsu/hsb validation. *IEEE Trans. Geosci. Remote Sens.* **2003**, *41*, 418–431. [CrossRef]

26. Aumann, H.H.; Chahine, M.T.; Gautier, C.; Goldberg, M.D.; Kalnay, E.; McMillin, L.M.; Revercomb, H.; Rosenkranz, P.W.; Smith, W.L.; Staelin, D.H.; et al. AIRS/AMSU/HSB on the Aqua mission: Design, science objectives, data products, and processing systems. *IEEE Trans. Geosci. Remote Sens.* **2003**, *41*, 253–264. [CrossRef]

27. Chahine, M.T.; Pagano, T.S.; Aumann, H.H.; Atlas, R.; Barnet, C.; Blaisdell, J.; Chen, L.; Divakarla, M.; Fetzer, E.J.; Goldberg, M.; et al. AIRS: Improving Weather Forecasting and Providing New Data on Greenhouse Gases. *Bull. Am. Meteorol. Soc.* **2006**, *87*, 911–926. [CrossRef]

28. Lee, C.J.; Brook, J.R.; Evans, G.J.; Martin, R.V.; Mihele, C. Novel application of satellite and in-situ measurements to map surface-level NO$_2$ in the Great Lakes region. *Atmos. Chem. Phys.* **2011**, *11*, 11761–11775. [CrossRef]

29. Knepp, T.; Pippin, M.; Crawford, J.; Chen, G.; Szykman, J.; Long, R.; Cowen, L.; Cede, A.; Abuhassan, N.; Herman, J.; et al. Estimating surface NO$_2$ and SO$_2$ mixing ratios from fast-response total column observations and potential application to geostationary missions. *J. Atmos. Chem.* **2015**, *72*, 261–286. [CrossRef] [PubMed]

30. Xue, D.; Yin, J. Meteorological influence on predicting surface SO_2 concentration from satellite remote sensing in Shanghai, China. *Environ. Monit. Assess.* **2014**, *186*, 2895–2906. [CrossRef] [PubMed]

31. Seidel, D.J.; Ao, C.O.; Li, K. Estimating climatological planetary boundary layer heights from radiosonde observations: Comparison of methods and uncertainty analysis. *J. Geophys. Res. Atmos.* **2010**, *115*. [CrossRef]

32. Cui, W.; Yan, X. Adaptive weighted least square support vector machine regression integrated with outlier detection and its application in QSAR. *Chemom. Intell. Lab. Syst.* **2009**, *98*, 130–135. [CrossRef]

33. Kutner, M.H.; Nachtsheim, C.; Neter, J. *Applied linear Regression Models*; McGraw-Hill/Irwin: New York, NY, USA, 2004.

34. Sellke, T.; Bayarri, M.J.; Berger, J.O. Calibration of ρ values for testing precise null hypotheses. *Am. Stat.* **2001**, *55*, 62–71. [CrossRef]

35. Lamsal, L.N.; Martin, R.V.; Van Donkelaar, A.; Steinbacher, M.; Celarier, E.A.; Bucsela, E.; Dunlea, E.J.; Pinto, J.P. Ground-level nitrogen dioxide concentrations inferred from the satellite-borne Ozone Monitoring Instrument. *J. Geophys. Res. Atmos.* **2008**, *113*. [CrossRef]

36. Zeng, X.; Brunke, M.A.; Zhou, M.; Fairall, C.; Bond, N.A.; Lenschow, D.H. Marine atmospheric boundary layer height over the eastern Pacific: Data analysis and model evaluation. *J. Clim.* **2004**, *17*, 4159–4170. [CrossRef]

37. Team, O.M.I. *Ozone Monitoring Instrument (OMI) Data User's Guide*; NASA: Washington, DC, USA, 2009; Volume 7, pp. 6–10.

remote sensing

MDPI

Article

Estimating Ground Level NO₂ Concentrations over Central-Eastern China Using a Satellite-Based Geographically and Temporally Weighted Regression Model

Kai Qin [1,*], Lanlan Rao [1], Jian Xu [2], Yang Bai [3,*], Jiaheng Zou [1], Nan Hao [4], Shenshen Li [5] and Chao Yu [5]

[1] School of Environment Science and Spatial Informatics, China University of Mining and Technology, Xuzhou 221116, China; raolanlan2016@163.com (L.R.); zoujiaheng@cumt.edu.cn (J.Z.)
[2] German Aerospace Center (DLR), Remote Sensing Technology Institute, 82234 Weßling, Germany; jian.xu@dlr.de
[3] College of Environment and Planning, Henan University, Kaifeng 475001, China
[4] European Organization for the Exploitation of Meteorological Satellites, 64283 Darmstadt, Germany; Nan.Hao@eumetsat.int
[5] Institute of Remote Sensing and Digital Earth, Chinese Academy of Sciences, Beijing 100094, China; lishenshen@126.com (S.L.); yuchao@radi.ac.cn (C.Y.)
* Correspondence: qinkai20071014@163.com (K.Q.); baiyang_cumt@163.com (Y.B.); Tel.: +86-159-5066-3287 (K.Q.)

Received: 3 July 2017; Accepted: 9 September 2017; Published: 13 September 2017

Abstract: People in central-eastern China are suffering from severe air pollution of nitrogen oxides. Top-down approaches have been widely applied to estimate the ground concentrations of NO₂ based on satellite data. In this paper, a one-year dataset of tropospheric NO₂ columns from the Ozone Monitoring Instrument (OMI) together with ambient monitoring station measurements and meteorological data from May 2013 to April 2014, are used to estimate the ground level NO₂. The mean values of OMI tropospheric NO₂ columns show significant geographical and seasonal variation when the ambient monitoring stations record a certain range. Hence, a geographically and temporally weighted regression (GTWR) model is introduced to treat the spatio-temporal non-stationarities between tropospheric-columnar and ground level NO₂. Cross-validations demonstrate that the GTWR model outperforms the ordinary least squares (OLS), the geographically weighted regression (GWR), and the temporally weighted regression (TWR), produces the highest R² (0.60) and the lowest values of root mean square error mean (RMSE), absolute difference (MAD), and mean absolute percentage error (MAPE). Our method is better than or comparable to the chemistry transport model method. The satellite-estimated spatial distribution of ground NO₂ shows a reasonable spatial pattern, with high annual mean values (>40 µg/m³), mainly over southern Hebei, northern Henan, central Shandong, and southern Shaanxi. The values of population-weight NO₂ distinguish densely populated areas with high levels of human exposure from others.

Keywords: NO₂; ground level; OMI; GTWR; China

1. Introduction

High ground level nitrogen oxides (NO$_x$ = NO + NO₂) are identified to be deleterious to human health, including decreased lung function and an increased risk of respiratory symptoms [1,2]. In addition, NO$_x$ can also produce other photochemical pollutants like O₃ in photochemical reactions, and acts as a gaseous precursor of aerosols and acid rain. Thus, the NO$_x$ concentration has been included in multi-pollutant health indices [3] and its monitoring with complete spatial coverage is needed for exposure assessment. Since 1995, a series of satellites sensors, e.g., the Global Ozone Monitoring Experiment (GOME), the Scanning Imaging Absorption Spectrometer for Atmospheric Cartography (SCIAMACHY), and the Ozone Monitoring Instrument (OMI) have been successfully used to retrieve vertical NO_2 columns [4–8]. A dramatic increase in tropospheric NO_2 columns was revealed by the GOME and SCIAMACHY observations over China [9–12], the world's largest developing country along with the fastest growing economy.

Given that the existing ambient monitoring stations are sparse and unevenly distributed, there is a growing interest in the top-down satellite approach to obtain timely map of the spatial variations of surface concentrations of NO_2. A close relationship between ground level NO_2 concentrations and satellite-retrieved tropospheric NO_2 columns is expected based on two facts: (1) ground level NO_2 accounts for the majority of tropospheric NO_2 columns since human activities are their main source; and (2) the short lifetime of near-surface NO_2 results in little transport, both vertically and horizontally [13]. Petritoli et al. [14] demonstrated a significant correlation between in situ NO_2 measurements and the GOME tropospheric NO_2 columns. Recently, satellite observations were combined with land use regression models to provide spatio-temporally resolved ambient NO_2 [15–17]. In addition, an approach proposed by Lamsal et al. [18] that combines the vertical profiles of NO_2 generated by the chemical transport model and satellite tropospheric NO_2 columns, has been widely used to estimate ground level NO_2 concentrations [19,20]. However, the emission inventories used for the model simulations are based on outdated statistical data about human activities. These model-based profiles may not capture the actual vertical distribution of NO_2, especially where anthropogenic NO_x emissions are undergoing rapid changes such as in China [21]. Kim et al. [22] estimated the surface NO_2 volume mixing ratio by using multiple regression models with OMI data.

In this study, a geographically and temporally weighted regression (GTWR) model is introduced to estimate the ground level NO_2 concentrations by using the OMI tropospheric NO_2 columns over central-eastern China. The GTWR model is adapted from the geographically weighted regression (GWR) model [23–26] by taking into account spatio-temporal non-stationarity, which has been proven to effectively establish the relation between satellite-retrieved aerosol optical depth and fine particulate matter ($PM_{2.5}$) [27,28]. Furthermore, population-weighted ground level NO_2 concentrations are calculated to evaluate population exposure levels in different regions.

2. Study Area and Data

2.1. Study Area

This study focuses on the central-eastern China with a geographic scope of 20°N–45°N and 105°E–124°E (major populated areas in China, see left panel in Figure 1). The study area covers 20 province-level administrative units in mainland China, including the regions of the North China Plain, Yangtze River Delta, and Pearl River Delta that are most polluted. 715 ambient monitoring stations are located in this study area (see the right panel in Figure 1).

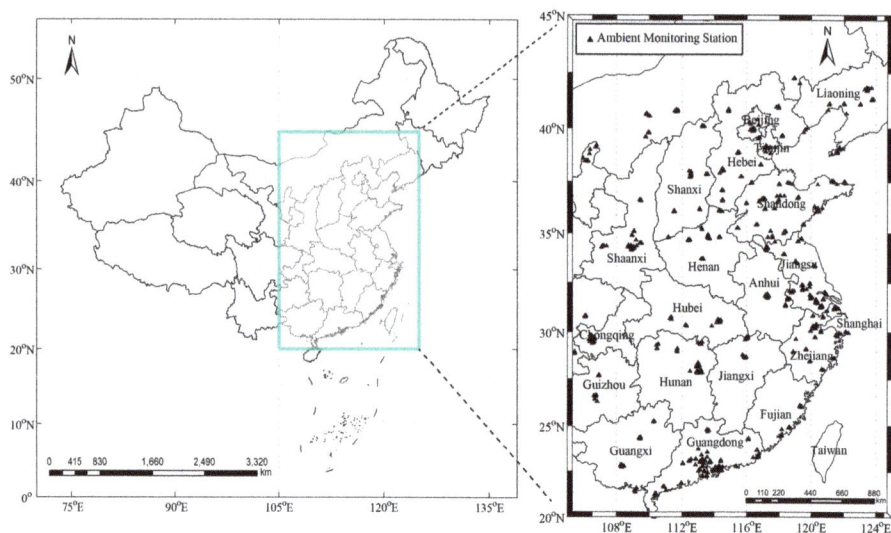

Figure 1. Study area and locations of ambient monitoring stations.

2.2. OMI Tropospheric NO$_2$ Columns

OMI is a Dutch-Finnish nadir-viewing hyperspectral instrument onboard the Earth Observing System Aura satellite in a Sun-synchronous orbit with an equatorial crossing time of approximately 13:45 local time. It measures sunlight backscattered radiances from the Earth in three channels covering a wavelength range of 270 to 500 nm (UV-1: 270 to 310 nm; UV-2: 310 to 365 nm; and, visible: 365 to 500 nm) at a spectral resolution of 0.45 to 0.63 nm [29]. OMI makes simultaneous measurements in a swath of width 2600 km, divided into 60 fields of view (FoVs). The FoVs vary in size from ~13 × 26 km near nadir to ~40 × 250 km at the outermost FoVs. The OMI measurements in the spectral range 402–465 nm are used to retrieve the NO$_2$ columns. First, NO$_2$ slant columns are determined from the OMI calibrated earthshine radiance spectra by using the differential optical absorption spectroscopy (DOAS) algorithm [30]. Second, the slant columns are then converted into the vertical columns using air mass factors (AMFs) calculated from radiative transfer models. Finally, the stratospheric and tropospheric column amounts are derived separately under the assumption that the two quantities are largely independent [31].

Here, we used the Version 3 Aura OMI NO$_2$ Standard Product (OMNO2) available from the NASA Goddard Earth Sciences Data and Information Services Center (http://disc.gsfc.nasa. gov/Aura/OMI/omno2_v003.shtml). The major improvements include: (1) an improved spectral fitting algorithm for retrieving slant column densities, including the use of monthly mean solar spectral irradiances; (2) improved Global Modeling Initiative model-based monthly a priori NO$_2$ and temperature profiles [32]. For further details, please refer to [33]. The main error sources in determining tropospheric NO$_2$ columns are associated with uncertainties in the surface albedo, aerosols, cloud interference, and the NO$_2$ vertical profile [34–37]. Overall, OMI retrievals tend to be lower in urban regions and higher in remote areas, but generally agree with other measurements within ±20% [38].

The data were filtered using a number of criteria [39] to ensure retrieval quality including: (1) cloud radiance fraction <0.3, (2) surface albedo <0.3, (3) solar zenith angles <85°, (4) 10 < cross-track positions < 50, and (5) root mean squared error of fit <0.0003. In addition, the cross track pixels affected by row anomaly (http://www.knmi.nl/omi/research/product/rowanomaly-background.php) were excluded, which was first noticed in the data in June 2007. Then, the NO$_2$ tropospheric column densities from the Level-2 OMNO2 Swath product were binned on to a $0.1 \times 0.1°$ grid by calculating the area-weighted averages at each grid cell.

2.3. Ambient Monitoring Station Data

The Ministry of Environmental Protection of Republic of China has built 1497 ambient monitoring stations over 367 cities in order to assess the air quality in China. Hourly mean concentrations of air pollutants including $PM_{2.5}$, PM_{10}, NO_2, SO_2, and O_3 are available since 2013 in the national air quality publishing platform (http://106.37.208.233:20035/). In this study, hourly mean ground-based NO_2 concentrations of 715 stations in central-eastern China from 1 May 2013 to 30 April 2014 (13:00–15:00 local time) were included. The locations of these stations are shown in Figure 1.

2.4. Meteorological Data

In order to improve the performance of our regression model, a number of meteorological parameters such as air temperature, relative humidity, planetary boundary layer height, wind speed, and air pressure from the Weather Research & Forecasting Model (WRF, version 3.4.1) were used. NCEP FNL Operational Model Global Tropospheric Analyses dataset of $1 \times 1°$ resolution (http://rda.ucar.edu/dsszone/ds083.2/) was adopted in the WRF model. The WRF model is a mesoscale numerical weather prediction system designed for both atmospheric research and operational forecast, and serves as a wide range of meteorological applications across scales from tens of meters to thousands of kilometers. The nested domain scheme with 30 km horizontal grid space of WRF output centered at 115°E, 32.5°N was adopted, and the temporal resolution of WRF outputs was set 1 h intervals. The number of altitude levels is 30 and the top-level pressure is 50 hPa. The physical options used in WRF include the single-moment 3-class (WSM3) microphysics, the Yonsei University (YSU) PBL scheme, the Rapid Radiative Transfer Model (RRTM) longwave and Dudhia shortwave radiation schemes, and Noah land surface model. Then, the hourly mean meteorological data from 13:00 to 15:00 local time with a spatial resolution of 30 km was interpolated to a $0.1 \times 0.1°$ grid same as the NO_2 tropospheric column products.

2.5. Population Data

Worldwide gridded population data are available at 5-year intervals from 1995 to 2020 from the NASA Socioeconomic Data and Applications Center (Gridded Population of the World, v4; http://sedac.ciesin.columbia.edu/). The population data in 2013 was obtained by linearly-interpolating the data in 2010 and 2015 using $0.1 \times 0.1°$ resolution.

3. Methodology

3.1. GTWR Model

The GTWR model for the relationship of ground NO_2 concentrations and satellite tropospheric columns can be expressed as [40]:

$$NO_{2\_ground(i)} = \beta_0(u_i, v_i, t_i) + \beta_1(u_i, v_i, t_i)NO_{2\_Trop(i)} + \varepsilon_i, (i = 1, 2, \dots, n) \tag{1}$$

where (u_i, v_i, t_i) represents the given coordinates of the training sample i in location (u_i, v_i) at time t_i. $NO_{2\_ground(i)}$ is the ground level NO_2 concentration observed by the ambient monitoring station at (u_i, v_i, t_i). $NO_{2\_Trop(i)}$ is the OMI NO_2 column density, $\beta_0(u_i, v_i, t_i)$ indicates the intercept of the GTWR model, $\beta_1(u_i, v_i, t_i)$ is a coefficient describing the unique spatial and temporal relationship between $NO_{2\_ground(i)}$ and $NO_{2\_Trop(i)}$. ε_i is the random error.

We introduced a number of meteorological parameters to the GTWR, i.e., air temperature at 2 m above the ground (T), relative humidity (RH), wind speed at 10 m above the ground (WS), planetary boundary layer height ($PBLH$), dew point temperature at 2 m above the ground (T_d), and the ambient pressure near ground (P). Akaike's information criterion (AIC) [41] was used to judge whether the GTWR performance could be improved with the addition of each specific meteorological parameter. The AIC value for the GTWR model is expressed as:

$$AIC = 2n \ln(\hat{\sigma}) + n \ln(2\pi) + n \left(\frac{n + tr(\mathbf{S})}{n - 2 - tr(\mathbf{S})} \right)$$

(2)

where $\hat{\sigma}$ is the maximum likelihood estimation of the standard deviation for random error $\varepsilon_i (i = 1, 2, \dots, n)$. $\mathbf{S}$ is the hat matrix of the dependent variable. $tr(\mathbf{S})$ is the trace of matrix $\mathbf{S}$. $\mathbf{S}$ and $\hat{\sigma}$ are calculated using Equations (17) and (18), respectively. The smaller AIC is, the better the model performance will be. As indicated in Table 1, the model performance improves substantially when the meteorological parameters of *PBLH, RH, WS, T*, and *P* are included. This is because that: (1) high temperature can increase photochemical reactions and hence reduce the lifetime of NO_2; (2) high relative humidity is related to low NO_2 concentration since it enhances the conversion rate of secondary aerosol from NO_X; (3) high *PBLH* is often related to low NO_2 concentration when it is supposed that NO_2 are well-mixed and confined within the planetary boundary layer; (4) high wind speed is favorable to pollutant dispersion that will result in the decrease of NO_2 concentration; and (5) high pressure increases atmospheric stability, leading to less atmospheric general circulation and thus more NO_2.

Table 1. Akaike's information criterion (AIC) values when satellite, planetary boundary layer height (PBLH), relative humidity (RH), wind speed at 10 m above the ground (WS), air temperature at 2 m above the ground (T), and ambient pressure near ground (P) data are included respectively in the geographically and temporally weighted regression (GTWR) model.

Satellite	PBLH	RH	WS	T	P
373,664	372,215	371,529	370,684	370,049	369,750

The GTWR can be modified as:

$$NO_{2\_ground(i)} = \beta_0(u_i, v_i, t_i) + \beta_1(u_i, v_i, t_i) \times NO_{2\_Trop(i)} + \beta_2(u_i, v_i, t_i) \times RH_{(i)} + \beta_3(u_i, v_i, t_i) \times T_{(i)}$$
$$+ \beta_4(u_i, v_i, t_i) \times PBLH_{(i)} + \beta_5(u_i, v_i, t_i) \times WS_{(i)} + \beta_6(u_i, v_i, t_i) \times P_{(i)} + \varepsilon_i, (i = 1, 2, \dots, n)$$

(3)

$\beta_1(u_i, v_i, t_i), \beta_2(u_i, v_i, t_i), \beta_3(u_i, v_i, t_i), \beta_4(u_i, v_i, t_i), \beta_5(u_i, v_i, t_i)$, and $\beta_6(u_i, v_i, t_i)$ denote the slope of *T, RH, PBLH, WS*, and *P*, respectively. In the GTWR model, a local weighted least squares algorithm is employed to determine the parameters of $\beta(u_i, v_i, t_i)$:

$$\hat{\beta}(u_i, v_i, t_i) = (\mathbf{X}^T \mathbf{W}(u_i, v_i, t_i)\mathbf{X})^{-1} \mathbf{X}^T \mathbf{W}(u_i, v_i, t_i)\mathbf{Y}$$

(4)

where $\mathbf{W}(u_0, v_0, t_0)$ is a square matrix comprising the geographically and temporally weighted values of training datasets for measurement i by the diagonal elements. $\mathbf{X}$ and $\mathbf{Y}$ are, respectively, expressed as:

$$\mathbf{W}(u_0, v_0, t_0) = \begin{pmatrix} w_1(u_0, v_0, t_0) & 0 & \cdots & 0 \\ 0 & w_2(u_0, v_0, t_0) & \cdots & 0 \\ \vdots & \vdots & \ddots & \vdots \\ 0 & 0 & \cdots & w_n(u_0, v_0, t_0) \end{pmatrix}$$

(5)

$$\mathbf{X} = \begin{pmatrix} 1 & NO_{2\_Trop(1)} & RH_{(1)} & T_{(1)} & PBLH_{(1)} & WS_{(1)} & P_{(1)} \\ 1 & NO_{2\_Trop(2)} & RH_{(2)} & T_{(2)} & PBLH_{(2)} & WS_{(1)} & P_{(2)} \\ \vdots & \vdots & \vdots & \vdots & \vdots & \vdots & \vdots \\ 1 & NO_{2\_Trop(n)} & RH_{(n)} & T_{(n)} & PBLH_{(n)} & WS_{(1)} & P_{(n)} \end{pmatrix}$$

(6)

$$\mathbf{Y} = \begin{pmatrix} NO_{2\_ground(1)} \\ NO_{2\_ground(2)} \\ \vdots \\ NO_{2\_ground(n)} \end{pmatrix} \tag{7}$$

The temporal distance d_{i0}^t and the spatial distance d_{i0}^s are given by:

$$d_{i0}^t = |t_i - t_0| \tag{8}$$

$$d_{i0}^s = \sqrt{(u_i - u_0)^2 + (v_i - v_0)^2} \tag{9}$$

By combining the temporal distance d_{i0}^t and the spatial distance d_{i0}^s, the spatio-temporal distance is defined as:

$$d_{i0}^{st} = d_{i0}^s \otimes d_{i0}^t \tag{10}$$

where $\otimes$ denotes different kinds of operators. Here, the "+" operator is adopted, the d_{i0}^{st} is hence computed by:

$$d_{i0}^{st} = \lambda d_{i0}^s + \mu d_{i0}^t \tag{11}$$

where λ and μ stand for the scale factors of temporal and spatial distance, respectively. Furthermore, an ellipsoidal coordinate system is used to calculate the d_{i0}^{st}:

$$\begin{aligned} \left(d_{i0}^{st}\right)^2 &= \lambda \left(d_{i0}^s\right)^2 + \mu \left(d_{i0}^t\right)^2 \\ &= \lambda \left[(u_i - u_0)^2 + (v_i - v_0)^2\right] + \mu(t_i - t_0)^2 \end{aligned} \tag{12}$$

Gaussian distance decay-based functions and Euclidean distance are chosen to construct the spatio-temporal weight matrix $\mathbf{W}(u_0, v_0, t_0)$. The diagonal element $w_i(u_0, v_0, t_0)$ of the $\mathbf{W}(u_0, v_0, t_0)$ can be obtained by:

$$\begin{aligned} w_i(u_0, v_0, t_0) &= \exp\left[-\frac{1}{2}\left(\frac{d_{0i}}{h_{ST}}\right)^2\right], i = 1, 2, 3, \ldots, n \\ &= \exp\left\{ -\frac{1}{2}\left(\frac{\lambda\left[(u_i - u_0)^2 + (v_i - v_0)^2\right] + \mu(t_i - t_0^2)}{h_{ST}^2} \right) \right\} \\ &= \exp\left\{ -\frac{1}{2}\left(\frac{\left(d_{i0}^S\right)^2}{h_S^2} + \frac{\left(d_{i0}^T\right)^2}{h_T^2} \right) \right\} \\ &= \exp\left\{ -\frac{1}{2}\frac{\left(d_{i0}^S\right)^2}{h_S^2} \right\} \times \exp\left\{ -\frac{1}{2}\frac{\left(d_{i0}^T\right)^2}{h_T^2} \right\} \end{aligned} \tag{13}$$

where h_{ST}, h_T and h_S are the parameters of spatio-temporal, spatial, and temporal bandwidths, respectively.

Adaptive spatio-temporal bandwidths are adopted according to the density of sample points around the given point (u_0, v_0, t_0). When many sample points are closely distributed around the given point, the bandwidths are small. On the contrary, if there are not enough sample points near it, the bandwidths are larger when THE sample points are sparsely distributed. In practice, the bandwidths are determined with an optimization technique by cross-validation through minimizing Equation (14).

$$CV(h_{ST}) = \sum_i (y_i - \hat{y}(h_{ST}))^2 \tag{14}$$

where the function $\hat{y}_i(h_{ST})$ denotes the predicted value from the GTWR which is built without sample i.

The ground level NO_2 at (u_i, v_i, t_i) is estimated by:

$$\hat{NO}_{2\_ground(i)} = \mathbf{x}_i^T \left(\mathbf{X}^T \mathbf{W}(u_i, v_i, t_i)\mathbf{X}\right)^{-1}\mathbf{X}^T \mathbf{W}(u_i, v_i, t_i)\mathbf{Y} \tag{15}$$

where $x_i^T = \begin{pmatrix} 1, & NO_{2\_Trop(i)} & ,RH_{(i)} & ,T_{(i)} & ,PBLH_{(i)} & ,WS_{(i)} & ,P_{(i)} \end{pmatrix}$, and

$$\hat{Y} = \begin{pmatrix} \hat{N}O_{2\_ground(1)} \\ \hat{N}O_{2\_ground(2)} \\ \vdots \\ \hat{N}O_{2\_ground(i)} \end{pmatrix} = \begin{pmatrix} x_i^T(X^TW(u_i,v_i,t_i)X)^{-1}X^TW(u_i,v_i,t_i)Y \\ x_i^T(X^TW(u_i,v_i,t_i)X)^{-1}X^TW(u_i,v_i,t_i)Y \\ \vdots \\ x_i^T(X^TW(u_i,v_i,t_i)X)^{-1}X^TW(u_i,v_i,t_i)Y \end{pmatrix} = SY \qquad (16)$$

where S is the hat matrix of Y and is calculated as:

$$S = \begin{pmatrix} x_i^T(X^TW(u_i,v_i,t_i)X)^{-1}X^TW(u_i,v_i,t_i) \\ x_i^T(X^TW(u_i,v_i,t_i)X)^{-1}X^TW(u_i,v_i,t_i) \\ \vdots \\ x_i^T(X^TW(u_i,v_i,t_i)X)^{-1}X^TW(u_i,v_i,t_i) \end{pmatrix} \qquad (17)$$

The maximum likelihood estimation of the standard deviation for rand error is calculated as:

$$\hat{\sigma} = \sqrt{\frac{RSS}{n - tr(S)}} \qquad (18)$$

where RSS is the residual sum of squares between estimated ground level NO_2 concentrations and observed ones:

$$RSS = Y^T(I_n - S)^T(I_n - S)Y \qquad (19)$$

3.2. Population-Weighted NO$_2$

The population data are introduced to calculate the population-weighted NO_2 (PNO_2) for different province-level administrative units:

$$PNO_2^j = \frac{\sum_{k=1}^{m} NO_2^{j,k} \times Population^{j,k}}{\sum_{k=1}^{m} Population^{j,k}} \qquad (20)$$

where PNO_2^j is the population-weighted NO_2 for province j, $NO_2^{j,k}$ and $Population^{j,k}$ are the NO_2 concentration and population data of pixel k in province j respectively.

3.3. Implementation Process and Statistical Indicators

To correlate the ground-based measurements with satellite data, the 715 ambient monitoring stations in the central-eastern China were merged into 509 stations by averaging all of the measurements within a grid of $0.1 \times 0.1°$. For the 509 grid cells, the total numbers of satellites and ambient monitoring observations are 54,867 (Figure 2a) and 110,545 (Figure 2b), respectively. Combining the satellite and ground observations, there are 31,463 valid data pairs (Figure 2c). The spatial distribution of the numbers of filtered satellite observations in Figure 2a shows a north-south difference, which is likely due to a higher cloud fraction over southern China. These 509 stations with total 31,463 dataset were divided randomly into 10 groups. The model fitting and cross-validation process was repeated 10 times, for every time one group was used for the cross-validation and the rest were used to train the fitting model until all groups were entered into the cross-validation once, thereby creating out-of-sample predictions for all the stations [42]. To be more specific, all of the 31,463 datasets were used both in the fitting and the cross-validation.

Some statistical indicators were employed to quantitatively assess the model performances. They are the coefficient of determination (R^2), whose higher value indicating better fitting accuracy, the root mean square error (RSME), that is sensitive to both systematic and random errors, the mean absolute difference (MAD), that measures the mean error magnitude, and the mean absolute percentage error (MAPE), which characterizes the prediction accuracy of a statistical model.

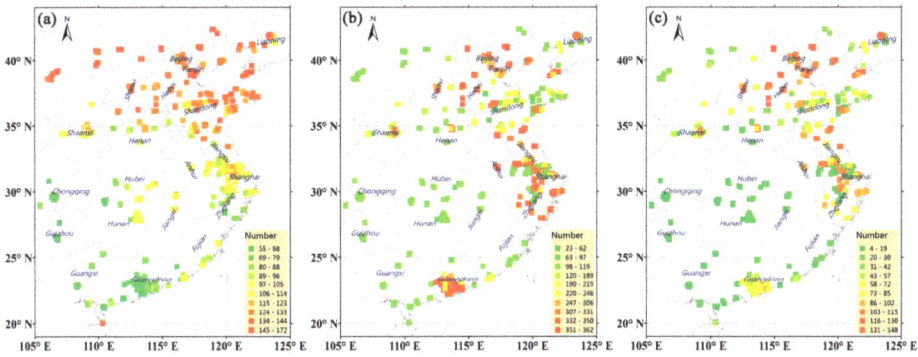

Figure 2. Numbers of satellite observations (**a**), ambient monitoring observations (**b**), and valid satellite and ground observation pairs (**c**) for 593 grid cells.

4. Results and Discussion

4.1. Spatio-Temporal Non-Stationarities between Tropospheric-Columnar and Ground Level NO_2

According to previous studies [43,44], the tropospheric NO_2 profiles show a large spatial-temporal variation. It is necessary to assess the impact of the spatio-temporal non-stationarities on the satellite-estimated ground level NO_2 concentrations. Lamsal et al. [38] showed that OMI retrievals are underestimated in urban regions and overestimated in remote areas about 20%. To isolate the influence of different land covers types, 293 pure urban grids and corresponding ambient stations were picked out from the total 509 grids and stations. As shown in Figure 3a, the mean values of tropospheric-columnar and corresponding ground level NO_2 over three provinces in eastern China (see also Figure 1) i.e., Shandong, Zhejiang, and Hunan, are compared. The mean values of OMI tropospheric NO_2 columns of the three provinces are different when the column data is composited with respect to the ground level NO_2 mass concentrations from ambient monitoring stations. This is related to the spatial difference in tropospheric NO_2 profiles due to different topographies and meteorological conditions. Moreover, the mean values of OMI tropospheric NO_2 columns in summer (May to July 2013), autumn (August to October 2013), winter (November 2013 to January 2014), and spring (February to April 2014) are compared in Figure 3b. The relationship between the NO_2 columns and ground level NO_2 shows a significant seasonal variation. The NO_2 columns in winter and autumn are higher than those in summer and spring when the values of ground level NO_2 are at the same level. This seasonal difference is more notable when ground concentrations increase, which is likely because of the longer lifetime of NO_2 in winter and autumn as compared to that in summer and spring. Consequently, it can exist for a longer time in the upper layer in the case of high ground emissions. The numbers of satellite observations used in Figure 3a,b are given in Tables 2 and 3, respectively. It should be pointed out that the numbers of satellite observations for high ground level NO_2 (>100 $\mu g/m^3$) are less than five in Hunan and in summer.

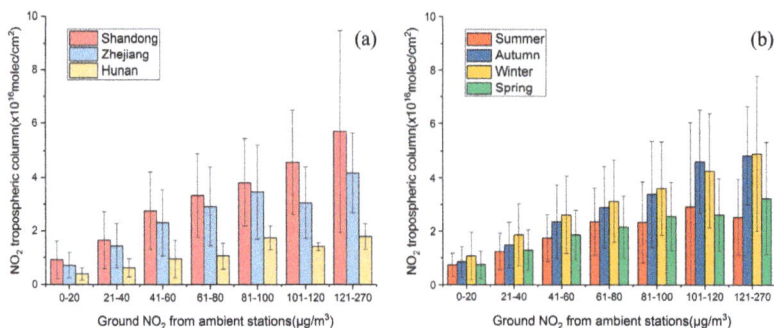

Figure 3. Mean values of tropospheric NO_2 columns ($\times 10^{16}$ molec/cm^2) in different provinces (**a**) and different seasons (**b**) when the column data is composited with respect to the ground level NO_2 mass concentrations observed at 293 pure urban ambient monitoring stations. Error bars stand for one standard deviation.

Table 2. Numbers of satellite observations used in Figure 3a.

Province	Ground Level NO_2 Mass Concentrations						
	0–20	21–40	41–60	61–80	81–100	101–120	121–270
Shandong	1002	1265	611	299	126	68	79
Zhejiang	1371	1240	491	156	57	37	33
Hunan	66	101	57	24	13	3	2

Table 3. Numbers of satellite observations used in Figure 3b.

Season	Ground Level NO_2 Mass Concentrations						
	0–20	21–40	41–60	61–80	81–100	101–120	121–270
Summer	1567	1137	283	94	23	3	4
Autumn	1815	1423	436	169	51	21	13
Winter	1059	1785	1248	653	313	174	175
Spring	2022	2336	870	296	96	37	30

4.2. Comparison between Model Fitted and Ground-Observed NO_2

The ordinary least squares (OLS), GWR, temporally weighted regression (TWR), and GTWR models were tested using the same datasets. As shown in Tables 4 and 5, the OLS performance reveals that the tropospheric NO_2 columns are potentially useful for ground level NO_2 with R^2 of 0.45 and 0.44 for fitting and validation, respectively. The TWR outperforms the GWR with significant increases of R^2 values from 0.55 and 0.49 to 0.61 and 0.55. This suggests that the temporal non-stationarity is more dominant than the spatial non-stationarity between the tropospheric NO_2 columns and ground level NO_2. Among the four models, the GTWR has the best performance in both model-fitting and cross-validation with the highest R^2 and lowest errors (RMSE, MAD, and MAPE). Nevertheless, the GTWR regression shows a slight over-fitting, i.e., the R^2 generated from the cross-validation is 0.09 smaller than that from the model-fitting. In addition, the scatter plots in Figure 4 shows the largest correlation slope and the smallest intercept for the GTWR model. It is worth noting that all of the regression line slopes for the four models are less than 1. Figure 5 is present to assess the impact of the numbers of valid observations on the GTWR performance. The R^2 over Hunan (Figure 5a) is smaller than those over Shandong (Figure 5b) and Zhejiang (Figure 5c), due to less observations.

Table 4. Quantitative assessment of model-fitting through ordinary least squares (OLS), geographically weighted regression (GWR), temporally weighted regression (TWR), and GTWR.

Model	R^2	RMSE ($\mu g/m^3$)	MAD ($\mu g/m^3$)	MAPE (%)
OLS	0.45	0.11	12.54	73.24
GWR	0.55	0.10	11.16	61.10
TWR	0.61	0.09	10.59	60.52
GTWR	0.69	0.08	9.38	52.10

Table 5. Quantitative assessment of cross-validation through OLS, GWR, TWR, and GTWR.

Model	R^2	RMSE ($\mu g/m^3$)	MAD ($\mu g/m^3$)	MAPE (%)
OLS	0.44	0.33	12.57	73.45
GWR	0.49	0.31	12.09	68.83
TWR	0.55	0.29	11.27	64.63
GTWR	0.60	0.28	10.68	60.19

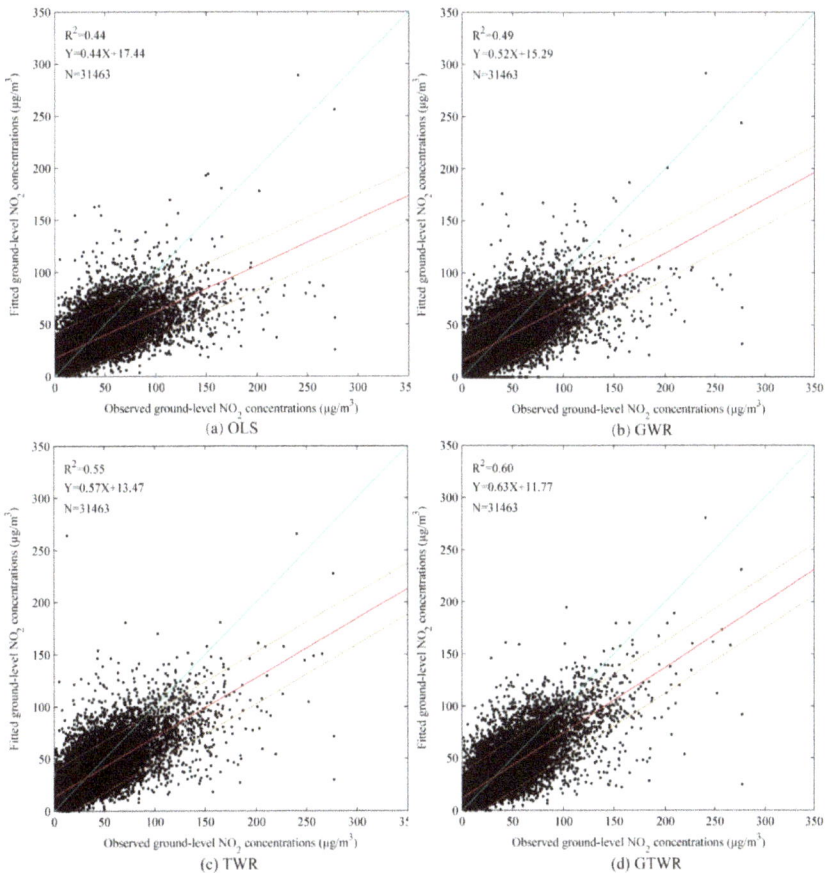

Figure 4. Scatter plots between the observed NO_2 and predicted NO_2 concentrations using OLS (**a**), GWR (**b**), TWR (**c**), and GTWR (**d**) for cross validation over central-eastern China from May 2013 to April 2014.

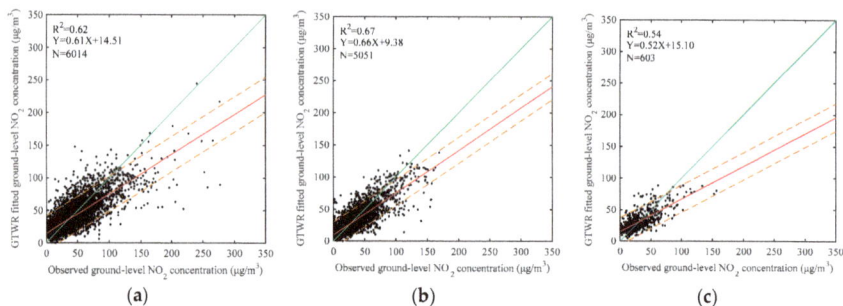

Figure 5. Scatter plots between the observed NO$_2$ and predicted NO$_2$ concentrations for cross validation over Shandong (**a**), Zhejiang (**b**), and Hunan (**c**) from May 2013 to April 2014.

There are three possible errors in the estimation of ground level NO$_2$ concentrations using the satellite-based GTWR model. First, satellite data are collected over an area of hundreds of km^2, while in-situ measurements are point observations. Second, the errors in the retrieval of OMI tropospheric NO$_2$ columns are underestimated in urban regions and overestimated in remote areas by about -20% and 20%, respectively [38]. Third, the uncertainty in the meteorological parameters can affect the vertical distribution of tropospheric NO$_2$. Zhang et al. [45] validated the NCEP FNL data against meteorological station data over Henan, China during 2012, and they found the errors of air temperature and pressure are $-3\sim2$ K and $-10\sim10$ hPa, respectively. We introduced the expected random errors (Gaussian distribution) from the tropospheric NO$_2$ columns, air temperature, and air pressure, to assess their impact on the performance of the GTWR model. As shown in Table 6 and Figure 6, our model uncertainties are relatively low with the expected uncertainties from the model parameters.

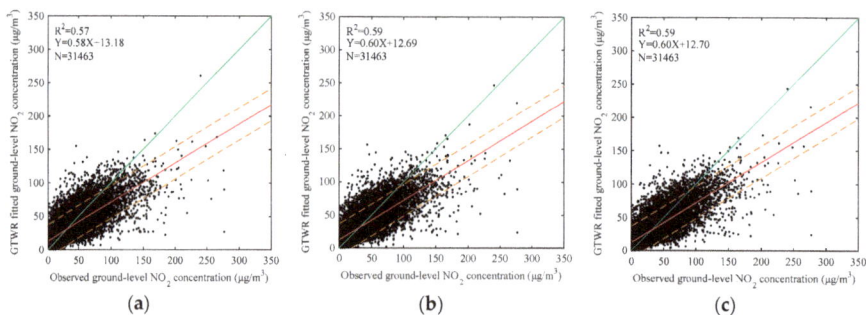

Figure 6. Scatter plots between the observed NO$_2$ and GTWR predicted NO2 concentrations with random errors from tropospheric NO$_2$ columns (20%) (**a**), air temperature (2 K) (**b**), and air pressure (10 hPa) (**c**) over central-eastern China from May 2013 to April 2014.

In the GTWR model, the smaller the spatio-temporal distance between two samples is, the greater weight coefficients are given. As illustrated in Figure 7, the GTWR performs much better than the OLS for the samples whose distances to the ambient monitoring stations are within 100 km, whereas the GTWR performance is worse than the OLS (0.41 versus 0.44 for R^2) for the samples that are more than 100 km away from the ambient monitoring stations. In the regions like Anhui, Jiangxi, and Fujian, where the ambient monitoring stations are very sparse and unevenly distributed, the nearest samples are mostly more than 100 km away. Hence, we used adjustable bandwidths according to the sample distances rather than the fixed ones. As compared to Figure 7, the R^2 in Figure 8 improves from 0.41 to 0.50 for the samples with larger distances (>100 km) after adjusting the bandwidth.

Table 6. Quantitative assessment of GTWR cross-validation with random errors from tropospheric NO_2 columns, air temperature, and air pressure.

Variations	Random Errors	R^2	RMSE ($\mu g/m^3$)	MAD ($\mu g/m^3$)	MAPE (%)
Tropospheric NO_2 columns	20%	0.57	0.29	11.01	62.40
Air temperature	2 K	0.59	0.28	10.75	61.11
Air pressure	10 hPa	0.59	0.28	10.75	60.86

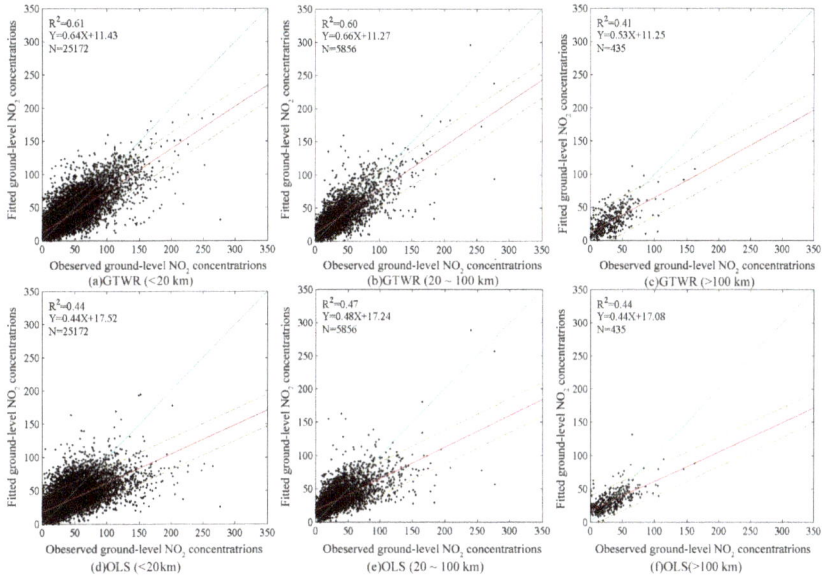

Figure 7. Scatter plots between the observed NO_2 and predicted NO_2 concentrations using GTWR ((**a–c**) and OLS (**d–f**) for samples with different distances (**a,d**) <20 km; (**b,e**) 20–100 km; (**c,f**) >100 km) for cross validation over central-eastern China from May 2013 to April 2014.

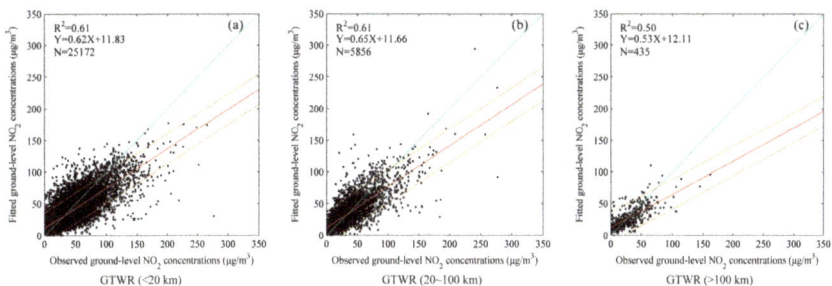

Figure 8. Scatter plots between the observed NO_2 and GTWR predicted NO_2 concentrations for samples with different distances ((**a**) <20 km; (**b**) 20-100 km; (**c**) >100 km) after adjusting the bandwidth for cross validation over central-eastern China from May 2013 to April 2014.

To compare the GTWR method with the chemistry transport model (CTM) approach, we generated the tropospheric NO_2 profiles by using a WRF-Chem model with the monthly MIX Asian anthropogenic emission inventory [46]. This emission inventory has a spatial resolution of $0.25 \times 0.25°$ and involves four emission categories, including industry, power, transport, and residential. The model has 20 vertical levels and the top level pressure is 200 hPa. The RADM2 chemical mechanism is used for the gas-phase chemical reaction calculations. The Modal Aerosol Dynamics Model for Europe-MADE/SORGAM is chosen for the aerosol scheme. Then, we estimated the ground level NO_2 concentrations in January 2014 over central-eastern China using the approach described by Lamsal et al. [18,19]. As shown in Figure 9, the result of the GTWR fitted is much better than the WRF-Chem. Recently, Gu et al. [43] estimated the ground level NO_2 over China using the chemistry transport model approach with the Community Multi-scale Air Quality (CMAQ) model by considering the influence of China's high atmospheric pollution on obtaining the vertical distribution of tropospheric NO_2 profiles. They achieved a correlation coefficient (R) of 0.80 for January 2014, which is comparable to the coefficient of determination (R^2) of 0.60 obtained by the GTWR.

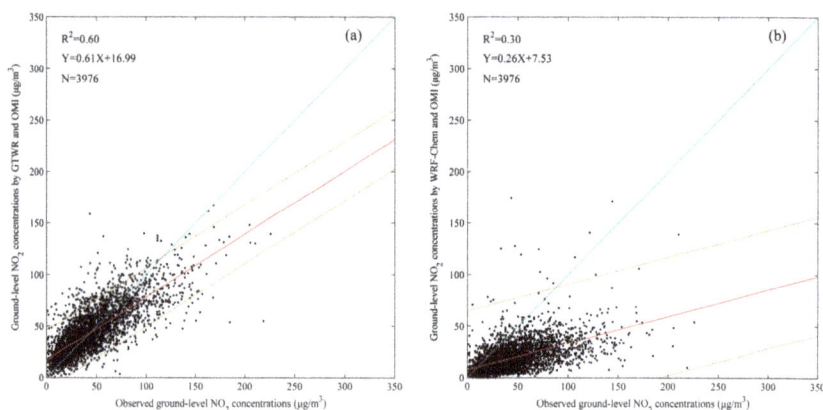

Figure 9. Scatter plots between the observed NO_2 and predicted NO_2 concentrations by GTWR (**a**) and WRF-Chem (**b**) for cross validation over central-eastern China in January 2014.

To further evaluate the performance of the GTWR model, the comparison of the annual mean of NO_2 concentrations between the model-fitted and ground-observed data is given in Figure 10. Overall, the NO_2 concentrations estimated by the GTWR model agree well with the ground-based measurements. More than 90% of the cross-validation stations possess low mean discrepancies of less than 10 $\mu g/m^3$.

4.3. Spatial Distribution of GTWR Fitted Ground-Observed NO_2

The spatial distributions of annual mean NO_2 values are shown in Figure 11. The fitted ground-observed NO_2 concentrations by GTWR in (a) have similar spatial patterns to the satellite tropospheric NO_2 columns in (b). The concentrations are comparable to the interpolated in situ observations using the Kriging method in (c) over the region with high values. Importantly, in the areas without monitoring stations (e.g., southern Jiangxi and northern Fujian), Figure 11a provides more reasonable estimations that are overestimated in Figure 11c. In Figure 11a, high NO_2 concentrations are clustered in the regions of North China Plain, Yangtze River Delta, and Pearl River Delta. Especially, the NO_2 concentrations in southern Hebei, northern Henan, central Shandong, and southern Shaanxi exceeded the Level 2 standard of the Chinese National Ambient Air Quality Standard (40 $\mu g/m^3$). Figure 12 denotes dramatic seasonal changes in the spatial distribution of GTWR fitted ground level NO_2. Unparalleled high values are found in winter, while the lowest values are found in summer.

Remote Sens. **2017**, *9*, 950

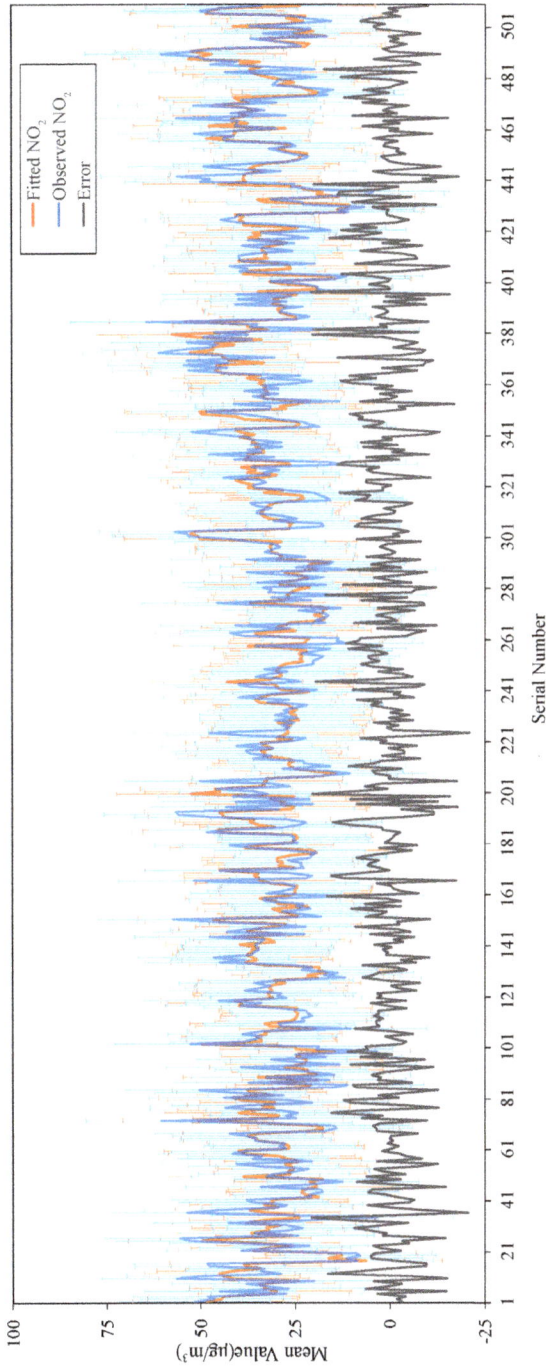

Figure 10. Annual means of observed NO₂ and fitted NO₂ by the GTWR model for all 509 stations.

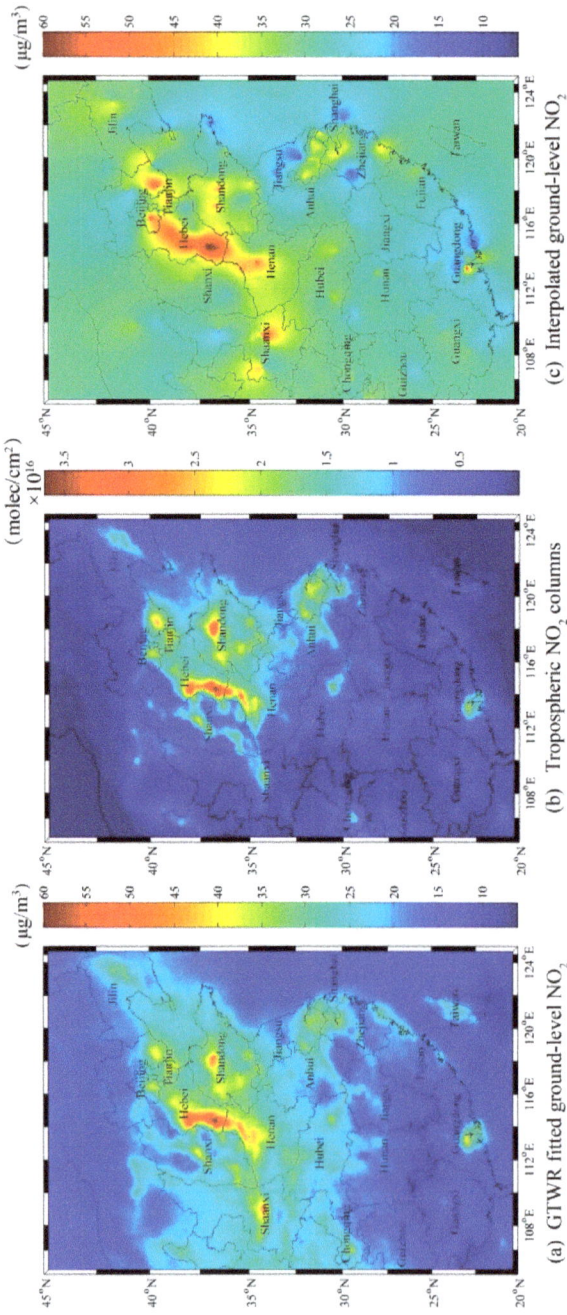

Figure 11. Spatial distribution of annual mean values of (**a**) ground level NO$_2$ concentrations fitted by GTWR model, (**b**) NO$_2$ tropospheric columns, and (**c**) ground level NO$_2$ concentrations interpolated by Kriging method.

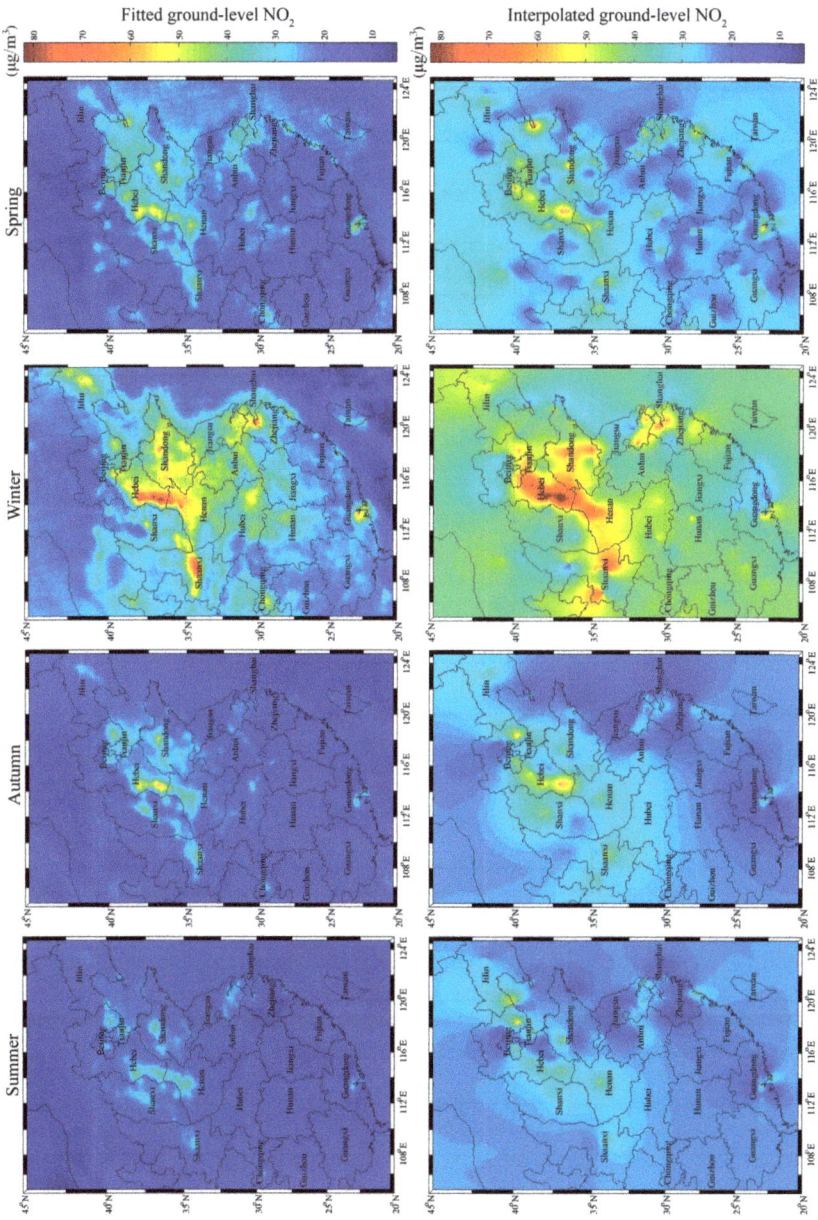

Figure 12. Spatial distribution of seasonal mean values of GTWR fitted (**upper**) and Kriging interpolated (**lower**) ground level NO$_2$ concentrations.

4.4. Population-Weighted Ground Level NO₂ Concentrations

Given the NO_2 toxicity to human health, it is necessary to evaluate population exposure levels over different provinces. Traditionally, the province-level mean NO_2 concentration provided by the Chinese environmental protection agencies are the arithmetic means of all values in administered cities. Here, we calculated the annual mean population-weighted NO_2 (AMPNO2) concentrations by using Equation (20). The annual mean NO_2 concentrations (AMNO2) and AMPNO2 of 17 provinces in central-eastern China are summarized in Table 7. AMPNO2 is higher than AMNO2 for all of the provinces, especially for densely populated provinces, e.g., Hebei, Beijing, and Guangdong. People from these 17 provinces except Anhui, Fujian, Jiangxi, and Hunan, are exposed to high-level NO_2 concentrations (>30 μg/m³). Heibei, Tianjin, and Beijing suffer from the most serious NO_2 pollution, with more than 70% of people affected by high-level NO_2. The satellite-estimated ground level NO_2 concentration is observed at afternoon (13:00–15:00) leading to underestimated annual mean values.

Table 7. Annual mean NO_2 (AMNO2) concentrations and population-weighted NO_2 (AMPNO2) concentrations for 17 provinces in central-eastern China.

Province	AMNO₂ (μg/m³)	AMPNO₂ (μg/m³)	Population (Millions)	Proportion of People Exposed to High-Level NO₂ Concentrations (>30 μg/m³)
Hebei	27.47	35.23	108.74	74.14%
Tianjin	31.67	34.38	23.04	85.84%
Beijing	26.09	33.86	33.73	87.03%
Shaanxi	25.66	32.43	53.35	55.78%
Henan	29.73	32.12	132.31	56.30%
Shandong	30.21	31.3	139.65	58.28%
Shanxi	24.59	28.43	54.76	46.07%
Shanghai	26.02	27.87	33.56	14.61%
Jiangsu	24.86	27.04	113.16	35.40%
Hubei	23.4	25.56	81.48	19.49%
Chongqing	22.13	25.29	38.34	21.16%
Zhejiang	21.17	25.08	76.21	21.03%
Anhui	21.8	23.48	84.41	0%
Guangdong	14.91	21.09	138.15	18.92%
Fujian	16.9	18.76	47.94	0%
Jiangxi	15.76	17.23	62.96	0%
Hunan	15.55	16.9	89.44	0%

5. Conclusions

In this study, a satellite-based GTWR model has been applied to estimate ground level NO_2 concentrations over central-eastern China. OMI tropospheric NO_2 columns, together with ambient monitoring station measurements and meteorological data from May 2013 to April 2014 were considered. The results show that the GTWR model produces the highest cross-validation R^2 (0.60) and the lowest errors (RMSE, MAD, and MAPE), in comparison with other models, i.e., OLS, GWR, and TWR. The model performance is significantly correlated with the meteorological parameters that likely describe the NO_2 vertical profile shapes. Our method is better than or comparable to the CTM method.

The satellite-estimated spatial distribution of annual mean NO_2 shows a similar spatial pattern to the tropospheric NO_2 column and possesses similar value with the in situ observation. High annual mean NO_2 concentrations (>40 μg/m³) are found in southern Hebei, northern Henan, central Shandong, and southern Shaanxi. Seasonal changes in the spatial distribution of ground level NO_2 are easily identifiable with unparalleled high values in winter and the lowest values in summer. The population-weighted NO_2 demonstrates that people who lived in densely populated areas are more likely to be exposed to high NO_2 pollution.

One of the major error sources in the estimation of ground level NO_2 concentrations using OMI data is the spatial gradient and the horizontal inhomogeneity between individual satellite pixels. In September 2017, the TROPOMI/S5P will be launched and measure tropospheric NO_2 columns with a higher spatial resolution (7 km × 7 km) [47], which enables an improved accuracy in the ground level NO_2 estimation.

Acknowledgments: This study was supported by the Fundamental Research Funds for the Central Universities (2014QNA32). We acknowledge the free use of tropospheric NO_2 column data from the OMI sensor.

Author Contributions: Kai Qin, Lanlan Rao, Yang Bai and Jian Xu conceived and designed the experiments; Lanlan Rao, Yang Bai and Kai Qin performed the experiments and analyzed the data; Kai Qin, Lanlan Rao and Yang Bai prepared the manuscript; Jian Xu, Jiaheng Zou, Nan Hao, Shenshen Li and Chao Yu improved the ideas and made revisions.

Conflicts of Interest: The authors declare no conflict of interest.

References

1. Gauderman, W.J.; Gilliland, G.F.; Vora, H.; Avol, E.; Stram, D.; Mcconnell, R.; Thomas, D.; Lurmann, F.; Margolis, H.G.; Rappaport, E.B. Association between air pollution and lung function growth in southern California children. *Am. J. Respir. Crit. Care Med.* **2000**, *162*, 1383–1390. [CrossRef] [PubMed]
2. Chiusolo, M.; Cadum, E.; Stafoggia, M.; Galassi, C.; Berti, G.; Faustini, A.; Bisanti, L.; Vigotti, M.A.; Patrizia, D.M.; Cernigliaro, A. Short-term effects of nitrogen dioxide on mortality and susceptibility factors in 10 Italian cities: The EpiAir study. *Environ. Health Perspect.* **2011**, *119*, 1233. [CrossRef] [PubMed]
3. Stieb, D.M.; Burnett, R.T.; Smithdoiron, M.; Brion, O.; Shin, H.H.; Economou, V. A new multipollutant, no-threshold air quality health index based on short-term associations observed in daily time-series analyses. *J. Air Waste Manag. Assoc.* **2008**, *58*, 435–450. [PubMed]
4. Sioris, C.E.; Kurosu, T.P.; Martin, R.V.; Chance, K. Stratospheric and tropospheric NO_2 observed by SCIAMACHY: First results. *Adv. Space Res.* **2004**, *34*, 780–785. [CrossRef]
5. Boersma, K.F.; Eskes, H.J.; Veefkind, J.P.; Brinksma, E.J.; van der A, R.J.; Sneep, M.; Oord, G.H.J.V.; Levelt, P.F.; Stammes, P.; Gleason, J.F. Near-real time retrieval of tropospheric NO_2 from OMI. *Atmos. Chem. Phys. Discuss.* **2006**, *6*, 12301–12345. [CrossRef]
6. Valks, P.; Pinardi, G.; Richter, A.; Lambert, J.C. Operational total and tropospheric NO_2 column retrieval for GOME-2. *Atmos. Meas. Tech.* **2011**, *4*, 1491. [CrossRef]
7. Irie, H.; Boersma, F.; Kanaya, Y.; Takashima, H.; Xiaole, P.; Wang, Z. Quantitative bias estimates for tropospheric NO_2 columns retrieved from SCIAMACHY, OMI, and GOME-2 using a common standard for East Asia. *Atmos. Meas. Tech.* **2012**, *5*, 2403–2411. [CrossRef]
8. Hassinen, S.; Balis, D.; Bauer, H.; Begoin, M.; Delcloo, A.; Eleftheratos, K.; Gimeno Garcia, S.; Granville, J.; Grossi, M.; Hao, N. Overview of the O3M SAF GOME-2 operational atmospheric composition and UV radiation data products and data availability. *Atmos. Meas. Tech.* **2016**, *9*, 383–407. [CrossRef]
9. Richter, A.; Burrows, J.P.; Nüss, H.; Granier, C.; Niemeier, U. Increase in tropospheric nitrogen dioxide over China observed from space. *Nature* **2005**, *437*, 129–132. [CrossRef] [PubMed]
10. Van Der, A.R.J.; Peters, D.; Eskes, H.; Boersma, K.F.; Roozendael, M.V.; Smedt, I.D.; Kelder, H.M. Detection of the trend and seasonal variation in tropospheric NO_2 over China. *J. Geophys. Res. Atmos.* **2006**, *111*. [CrossRef]
11. Zhang, Q.; Geng, G.N.; Wang, S.W.; Richter, A.; He, K.B. Satellite remote sensing of changes in NOx emissions over China: 1996–2010. *Chin. Sci. Bull.* **2012**, *57*. [CrossRef]
12. Zhang, L.; Lee, C.S.; Zhang, R.; Chen, L. Spatial and temporal evaluation of long term trend (2005–2014) of OMI retrieved NO_2 and SO_2 concentrations in Henan Province, China. *Atmos. Environ.* **2017**, *154*, 151–166. [CrossRef]
13. Liu, F.; Beirle, S.; Zhang, Q.; Dörner, S.; He, K.; Wagner, T. NOx lifetimes and emissions of cities and power plants in polluted background estimated by satellite observations. *Atmos. Chem. Phys.* **2016**, *16*, 5283–5298. [CrossRef]
14. Petritoli, A.; Bonasoni, P.; Giovanelli, G.; Ravegnani, F.; Kostadinov, I.; Bortoli, D.; Weiss, A.; Schaub, D.; Richter, A.; Fortezza, F. First comparison between ground-based and satellite-borne measurements of tropospheric nitrogen dioxide in the Po basin. *J. Geophys. Res. Atmos.* **2004**, *109*. [CrossRef]
15. Vienneau, D.; Hoogh, K.D.; Bechle, M.J.; Beelen, R.; Donkelaar, A.V.; Martin, R.V.; Millet, D.B.; Hoek, G.; Marshall, J.D. Western European land use regression incorporating satellite-and ground-based measurements of NO_2 and PM_{10}. *Environ. Sci. Technol.* **2013**, *47*, 13555–13564. [CrossRef] [PubMed]

16. Lee, H.J.; Koutrakis, P. Daily ambient NO_2 concentration predictions using satellite ozone monitoring instrument NO_2 data and land use regression. *Environ. Sci. Technol.* **2014**, *48*, 2305–2311. [PubMed]
17. Hoek, G.; Eeftens, M.; Beelen, R.; Fischer, P.; Brunekreef, B.; Boersma, K.F.; Veefkind, P. Satellite NO_2 data improve national land use regression models for ambient NO_2 in a small densely populated country. *Atmos. Environ.* **2015**, *105*, 173–180. [CrossRef]
18. Lamsal, L.N.; Martin, R.V.; Donkelaar, A.V.; Steinbacher, M.; Celarier, E.A.; Bucsela, E.; Dunlea, E.J.; Pinto, J.P. Ground level nitrogen dioxide concentrations inferred from the satellite-borne Ozone Monitoring Instrument. *J. Geophys. Res. Atmos.* **2008**, *113*, 280–288. [CrossRef]
19. Lamsal, L.N.; Duncan, B.N.; Yoshida, Y.; Krotkov, N.A.; Pickering, K.E.; Streets, D.G.; Lu, Z. US NO_2 trends (2005–2013): EPA Air Quality System (AQS) data versus improved observations from the Ozone Monitoring Instrument (OMI). *Atmos. Environ.* **2015**, *110*, 0e143. [CrossRef]
20. Kharol, S.K.; Martin, R.V.; Philip, S.; Boys, B.; Lamsal, L.N.; Jerrett, M.; Brauer, M.; Crouse, D.L.; Mclinden, C.; Burnett, R.T. Assessment of the magnitude and recent trends in satellite-derived ground-level nitrogen dioxide over North America. *Atmos. Environ.* **2015**, *118*, 236–245. [CrossRef]
21. Chan, K.L.; Hartl, A.; Lam, Y.F.; Xie, P.H.; Liu, W.Q.; Cheung, H.M.; Lampel, J.; Pöhler, D.; Li, A.; Xu, J. Observations of tropospheric NO_2 using ground based MAX-DOAS and OMI measurements during the Shanghai World Expo 2010. *Atmos. Environ.* **2015**, *119*, 45–58. [CrossRef]
22. Kim, D.; Lee, H.; Hong, H.; Choi, W.; Yun, G.L.; Park, J. Estimation of surface NO_2 volume mixing ratio in four metropolitan cities in Korea using multiple regression models with OMI and AIRS Data. *Remote Sens.* **2017**, *9*, 627. [CrossRef]
23. Song, W.; Jia, H.; Huang, J.; Zhang, Y. A satellite-based geographically weighted regression model for regional $PM_{2.5}$ estimation over the Pearl River Delta region in China. *Remote Sens. Environ.* **2014**, *154*, 1–7. [CrossRef]
24. You, W.; Zang, Z.; Zhang, L.; Li, Z.; Chen, D.; Zhang, G. Estimating ground-level PM_{10} concentration in northwestern China using geographically weighted regression based on satellite AOD combined with CALIPSO and MODIS fire count. *Remote Sens. Environ.* **2015**, *168*, 276–285. [CrossRef]
25. Zou, B.; Pu, Q.; Bilal, M.; Weng, Q.; Zhai, L.; Nichol, J.E. High-resolution satellite mapping of fine particulates based on geographically weighted regression. *IEEE Geosci. Remote Sens. Lett.* **2016**, *13*, 495–499. [CrossRef]
26. You, W.; Zang, Z.; Zhang, L.; Li, Y.; Pan, X.; Wang, W. National-scale estimates of ground-level $PM_{2.5}$ concentration in China using geographically weighted regression based on 3 km resolution MODIS AOD. *Remote Sens.* **2016**, *8*, 184. [CrossRef]
27. Bai, Y.; Wu, L.; Qin, K.; Zhang, Y.; Shen, Y.; Zhou, Y. A geographically and temporally weighted regression model for ground-level $PM_{2.5}$ estimation from satellite-derived 500 m resolution AOD. *Remote Sens.* **2016**, *8*, 262. [CrossRef]
28. Guo, Y.; Tang, Q.; Gong, D.Y.; Zhang, Z. Estimating ground-level $PM_{2.5}$ concentrations in Beijing using a satellite-based geographically and temporally weighted regression model. *Remote Sens. Environ.* **2017**, *198*, 140–149. [CrossRef]
29. Levelt, P.F.; Oord, G.H.J.V.; Dobber, M.R.; Malkki, A.; Visser, H.; Vries, J.D.; Stammes, P.; Lundell, J.O.V.; Saari, H. The ozone monitoring instrument. *IEEE Trans. Geosci. Remote Sens.* **2006**, *44*, 1093–1101. [CrossRef]
30. Platt, U.; Stutz, J. Differential Absorption Spectroscopy. In *Differential Optical Absorption Spectroscopy*; Springer: Berlin/Heidelberg, Germany, 2008; pp. 135–174.
31. Bucsela, E.J.; Krotkov, N.A.; Celarier, E.A.; Lamsal, L.N. A new stratospheric and tropospheric NO2 retrieval algorithm for nadir-viewing satellite instruments: Applications to OMI. *Atmos. Meas. Tech.* **2013**, *6*, 2607. [CrossRef]
32. Douglass, A.R.; Stolarski, R.S.; Strahan, S.E.; Connell, P.S. Radicals and reservoirs in the GMI chemistry and transport model: Comparison to measurements. *J. Geophys. Res. Atmos.* **2004**, *109*. [CrossRef]
33. Bucsela, E.J.; Celarier, E.A.; Gleason, J.L.; Krotkov, N.A.; Lamsal, L.N.; Marchenko, S.V.; Swartz, W.H. OMNO2 Readme Document Data Product Version 3.0. 2016. Available online: https://aura.gesdisc.eosdis.nasa.gov/data//Aura_OMI_Level2/OMNO2.003/doc/README.OMNO2.pdf (accessed on 16 September 2016).
34. Boersma, K.F.; Eskes, H.J.; Brinksma, E.J. Error analysis for tropospheric NO_2 retrieval from space. *J. Geophys. Res. Atmos.* **2004**, *109*. [CrossRef]

35. Boersma, F.; Dirksen, R.; Brunner, D.; Zhou, Y.; Huijnen, V.; Eskes, H.; Veefkind, P.; Kleipool, Q.; Dobber, M.; Stammes, P. An improved retrieval of tropospheric NO2 columns from the Ozone Monitoring Instrument. *Atmos. Meas. Tech.* **2011**, *4*, 1905–1928. [CrossRef]

36. Hong, H.; Lee, H.; Kim, J.; Jeong, U.; Ryu, J.; Lee, D.S. Investigation of simultaneous effects of aerosol properties and aerosol peak height on the air mass factors for space-borne NO_2 retrievals. *Remote Sens.* **2017**, *9*, 208. [CrossRef]

37. Wang, Y.; Beirle, S.; Lampel, J.; Koukouli, M.; De Smedt, I.; Theys, N.; Li, A.; Wu, D.; Xie, P.; Liu, C. Validation of OMI, GOME-2A and GOME-2B tropospheric NO_2, SO_2 and HCHO products using MAX-DOAS observations from 2011 to 2014 in Wuxi, China: Investigation of the effects of priori profiles and aerosols on the satellite products. *Atmos. Chem. Phys.* **2017**, *17*, 5007. [CrossRef]

38. Lamsal, L.N.; Krotkov, N.A.; Celarier, E.A.; Swartz, W.H.; Pickering, K.E.; Bucsela, E.J.; Martin, R.V.; Philip, S.; Irie, H.; Cede, A. Evaluation of OMI operational standard NO_2 column retrievals using in situ and surface-based NO_2 observations. *Atmos. Chem. Phys.* **2014**, *14*, 11587–11609. [CrossRef]

39. Wang, S.; Zhang, Q.; Martin, R.V.; Philip, S.; Liu, F.; Li, M.; Jiang, X.; He, K. Satellite measurements oversee China's sulfur dioxide emission reductions from coal-fired power plants. *Environ. Res. Lett.* **2015**, *10*, 114015. [CrossRef]

40. Huang, B.; Wu, B.; Barry, M. Geographically and temporally weighted regression for modeling spatio-temporal variation in house prices. *Int. J. Geogr. Inf. Sci.* **2010**, *24*, 383–401. [CrossRef]

41. Beelen, R.; Hoek, G.; Pebesma, E.; Vienneau, D.; Hoogh, K.D.; Briggs, D.J. Mapping of background air pollution at a fine spatial scale across the European union. *Sci. Total Environ.* **2009**, *407*, 1852–1867. [CrossRef] [PubMed]

42. Young, M.T.; Bechle, M.J.; Sampson, P.D.; Szpiro, A.A.; Marshall, J.D.; Sheppard, L.; Kaufman, J.D. Satellite-based NO_2 and model validation in a national prediction model based on universal Kriging and land-use regression. *Environ. Sci. Technol.* **2016**, *50*, 3686–3694. [CrossRef] [PubMed]

43. Gu, J.; Chen, L.; Yu, C.; Li, S.; Tao, J.; Fan, M.; Xiong, X.; Wang, Z.; Shang, H.; Su, L. Ground-level NO2 concentrations over China inferred from the Satellite OMI and CMAQ Model simulations. *Remote Sens.* **2017**, *9*, 519. [CrossRef]

44. Ma, J.Z.; Beirle, S.; Jin, J.L.; Shaiganfar, R.; Yan, P.; Wagner, T. Tropospheric NO_2 vertical column densities over Beijing: Results of the first three years of ground-based MAX-DOAS measurements (2008–2011) and satellite validation. *Atmos. Chem. Phys.* **2013**, *13*, 1547–1567. [CrossRef]

45. Zhang, Y.T.; Chen, Y.D. Error comparison analysis between FNL data and observation data of air temperature, air pressure and ground temperature in Henan Province in 2012. *Meteorol. Environ. Sci.* **2014**, *37*, 93–97. (In Chinese)

46. Li, M.; Zhang, Q.; Kurokawa, J.; Woo, J.H.; He, K.; Lu, Z.; Ohara, T.; Song, Y.; Streets, D.G.; Carmichael, G.R. MIX: A mosaic Asian anthropogenic emission inventory under the international collaboration framework of the MICS-Asia and HTAP. *Atmos. Chem. Phys.* **2017**, *17*, 935. [CrossRef]

47. Veefkind, J.P.; Aben, I.; Mcmullan, K.; Förster, H.; Vries, J.D.; Otter, G.; Claas, J.; Eskes, H.J.; Haan, J.F.D.; Kleipool, Q. TROPOMI on the ESA Sentinel-5 Precursor: A GMES mission for global observations of the atmospheric composition for climate, air quality and ozone layer applications. *Remote Sens. Environ.* **2012**, *120*, 70–83. [CrossRef]

remote sensing

MDPI

Article

Linear and Non-Linear Trends for Seasonal NO$_2$ and SO$_2$ Concentrations in the Southern Hemisphere (2004–2016)

Adrián Yuchechen [1,*], Susan Gabriela Lakkis [2,3] and Pablo Canziani [1]

[1] Universidad Tecnológica Nacional (UTN), Facultad Regional Buenos Aires (FRBA), Consejo Nacional de Investigaciones Científicas y Técnicas (CONICET), Unidad de Investigación y Desarrollo de las Ingenierías (UIDI), Buenos Aires C1407IVT, Argentina; pocanziani@frba.utn.edu.ar

[2] Pontificia Universidad Católica Argentina, Facultad de Ingeniería y Ciencias Agrarias, Buenos Aires C1107AAZ, Argentina; gabylakkis@uca.edu.ar

[3] UTN, FRBA, UIDI, Buenos Aires C1407IVT, Argentina

* Correspondence: aeyuchechen@frba.utn.edu.ar; Tel.: +54-11-4567-7268

Received: 30 June 2017; Accepted: 24 August 2017; Published: 28 August 2017

Abstract: In order to address the behaviour of nitrogen dioxide (NO$_2$) and sulphur dioxide (SO$_2$) in the context of a changing climate, linear and non-linear trends for the concentrations of these two trace gases were estimated over their seasonal standardised variables in the Southern Hemisphere—between the Equator and 60° S—using data retrieved by the Ozone Monitoring Instrument, for the period 2004–2016. A rescaling was applied to the calculated linear trends so that they are expressed in Dobson units (DU) per decade. Separately, the existence of monotonic—not necessarily linear—trends was addressed by means of the Mann-Kendall test. Results indicate that the SO$_2$ exhibits significant linear trends in the planetary boundary layer only; they are present in all the analysed seasons but just in a small number of grid cells that are generally located over the landmasses or close to them. The SO$_2$ concentrations in the quarterly time series exhibit, on average, a linear trend that is just below 0.08 DU decade^{-1} when significant and not significant values are considered altogether, but this figure increases to 0.80 DU decade^{-1} when only the significant trends are included. On the other hand, an important number of pixels in the lower troposphere, the middle troposphere, and the lower stratosphere have significant monotonic upward or downward trends. As for the NO$_2$, no significant linear trends were found either in the troposphere or in the stratosphere, yet monotonic upward and downward trends were observed in the former and latter layers, respectively. Unlike the linear trends, semi-linear and non-linear trends were seen over the continents and in remote regions over the oceans. This suggests that pollutants are transported away from their sources by large-scale circulation and redistributed hemispherically. The combination of regional meteorological phenomena with atmospheric chemistry was raised as a possible explanation for the observed trends. If extrapolated, these trends are in an overall contradiction with the projected emissions of both gases for the current century.

Keywords: nitrogen dioxide; sulphur dioxide; concentrations; linear trends; non-linear trends; Mann-Kendall test; Southern Hemisphere

1. Introduction

There is now widespread consensus that changes in the composition of the Earth's atmosphere caused by human activities play a relevant role in the Earth's climate system. Unlike the greenhouse gases that induce a positive radiative forcing, aerosol particles influence the global radiation budget causing a net negative radiative forcing associated with a cooling effect on the atmosphere ([1] and references therein). Broadly speaking, the radiative contribution of the aerosols can be centred in three main categories depending on what they interact with: aerosols-surface, aerosols-radiation and aerosols-clouds, or considering the aerosols' influence on the atmosphere as a direct, indirect, or semi-direct effect [2–5]. In addition to the radiative influence, aerosols are significant contributors to air pollution and they have a direct linkage with the biogeochemical cycles of the atmosphere, the oceans and the surfaces, acting as micronutrients for the marine and terrestrial biosphere. Aerosol deposition can also have detrimental environmental effects (e.g., the acidification of precipitation by sulphurs [6]) with impacts on the aquatic and terrestrial ecosystems [7–9], yet the benefits or the detrimental effects on ecological processes depend upon both the amount and composition of deposition and the underlying ecosystem conditions.

Amidst the aerosols with greater relevance, nitrogen dioxide (NO_2) and sulphur dioxide (SO_2) must be considered since they are reactive short-lived atmospheric trace gases, with both anthropogenic and natural sources that strongly impact on human health and the environmental degradation either directly or through the formation of secondary aerosols [10,11]. The main sources of nitrogen oxide compounds, nitric oxide (NO) and NO_2—collectively referred to as NO_x—include fuel combustion, biomass burning, soil emissions and lightning, and they can impact on climate in a number of interconnected ways [12]. Tropospheric NO_2 is a highly reactive and toxic gas which, in the presence of sunlight, water vapour (H_2O) and carbon monoxide (CO) or volatile organic compounds, drives the production of ozone (O_3) and hydroxyl radicals (OH), the principal tropospheric oxidants [12,13]. In the stratosphere, NO_x contributes to the ozone-loss cycles [14] and may indicate long-term changes in the tropospheric emissions of the long-lived nitrous oxide (N_2O) [13]. On the other hand, SO_2 is a colourless, non-flammable, non-explosive gas, toxic at high concentrations, and its principal contribution to air pollution is related to the acidification of precipitation and subsequent impacts on the receiving ecosystems [6].

SO_2 dissolves in cloud droplets and oxidises to form sulphuric acid (H_2SO_4) [15], which can fall to the Earth as acid rain or snow, or form sulphate aerosol particles in the atmosphere through oxidation [11]. The main contributions of SO_2 are related to anthropogenic emissions (including the combustion of sulphur-containing fuels [15]) and natural phenomena (including biomass burning [16]) and the oxidation of dimethyl sulphide (CH_3SCH_3), emitted from phytoplankton [15], and from the degassing and eruptions of volcanoes [16]. During long-term persistent volcanic eruptions SO_2 can be injected into the stratosphere and converted to sulphate aerosols, reflecting the sunlight and therefore inducing a cooling effect on the Earth's climate. They also have a role in ozone depletion. Volcanic SO_2 is often injected into the atmosphere at altitudes above the planetary boundary layer (PBL), while anthropogenic SO_2 emissions are predominantly in or just above the PBL [11]. Additionally, SO_2 in the atmosphere is associated with adverse health effects, including respiratory and cardiovascular diseases. The United States Environmental Protection Agency (EPA) has estimated that two thirds of SO_2 and a fourth of NO_x found in the atmosphere come from the burning of fossil fuels to generate electricity [17]. World Health Organization guidelines recommended daily SO_2 exposure levels not to exceed 125 µg m^{-3} on average over a 24 h period [18]. NO_2, SO_2, and their oxidised products O_3 and PM2.5 (particulate matter with aerodynamic diameter less than 2.5 µm) are designated as "criteria pollutants" by both the European Commission and the EPA (see e.g., [11] and references therein). PM2.5 have serious health effects, and it also causes acidification of water and the biosphere, with adverse consequences on plants, soils, and weather and climate through direct radiative forcing and indirect modification of cloud formation and optical properties ([11] and references therein).

The preceding lines highlighted the importance of assessing the role of aerosols and their impacts on the Earth's climate system, for which a global understanding of their spatial and temporal behaviour is required. Reliable, up-to-date inventories of emissions and concentrations are the first step when attempting to evaluate these impacts and to address the effects on the climate system over different timescales. Although in situ measurements provide valuable information, they are insufficient to these particular aims for they are scarce spatially, temporally, or both and they must be at least complemented by remotely-sensed data. Following this, the Ozone Monitoring Instrument (OMI) provides the scientific community with a valuable source of information since it is the first space-borne hyperspectral ultraviolet/visible spectrometer that enables a continuous mapping of several trace gases and ozone, including SO_2 and NO_2, globally and on a daily basis [19]. There have been an increasing number of studies related to NO_x and SO_x emissions in the last decade [10,13,20–25]. However, few of them were devoted to analysing linear and non-linear trends in the Southern Hemisphere (SH), and those papers dealing with this particular topic only studied specific locations [26]. In order to fill this gap and to complement the existing studies, this paper presents a comprehensive analysis of NO_2 and SO_2 linear and non-linear trends in the entire SH within the 2004–2016 period using OMI data.

2. Materials and Methods

The NO_2 and SO_2 data used in this research were retrieved by the OMI aboard the Aura spacecraft, which was launched in July 2004 [27,28]. The OMI is a nadir solar backscatter spectrometer that operates in the 270–500 nm spectral range [19]. It began collecting data in August 2004 [27], and data production commenced in October the same year [28]. OMI has the highest spatial resolution, the least degradation, and the longest record among all satellite ultraviolet-visible instruments, which permits an improved space-borne estimation of NO_2 and SO_2 emissions and the study of their temporal behaviours ([11] and references therein).

The datasets for the two gas species were obtained from the Goddard Earth Sciences Data and Information Services Center (GES DISC) through the National Aeronautics and Space Administration's (NASA) Mirador search engine (https://mirador.gsfc.nasa.gov/). The NO_2 daily product corresponds to the OMNO2G version 3 dataset, whose coverage has a resolution of $0.25° \times 0.25°$ on a global basis [29]. NO_2 measurements made by the OMI are performed in the visible spectrum within the 402–465 nm range [30]. Each OMNO2G daily file includes pixel information regarding the column concentrations of this gas in the troposphere (Trop) and the stratosphere (Strat), as well as the total column. All these quantities are expressed in molecules cm^{-2}. A single quality flag (QF) for these three concentrations and a number of ancillary variables are included too. Only those records with a value of "0" for this QF should be used [29]. In a first processing of each daily file only those registries fulfilling this condition were selected. Under these circumstances, negative concentrations, should there be any, were flagged as missing data. Simultaneously, the concentrations were converted to Dobson Units (DU) (1 DU = 2.69×10^{16} molecules cm^{-2} [31]). This was done for the sake of homogeneity since the SO_2 concentrations are given in DU (see below), and also because it is a much more familiar unit that has been traditionally used in research efforts mainly related to ozone [32–34], but also dealing with other atmospheric constituents [35].

On the other hand, the SO_2 daily product used here was the OMSO2G version 3 dataset. It has a resolution of $0.125° \times 0.125°$ [36]. The OMSO2G daily files include the following concentration estimates for this gas: PBL, lower troposphere (TRL), middle troposphere (TRM), and upper troposphere and lower stratosphere (STL), the former two associated with anthropogenic activity and the latter two associated with volcanic activity [27]. SO_2 PBL concentrations are estimated using the Band Residual Difference Algorithm [37], whereas concentrations in the rest of the layers are estimated using the Linear Fit Algorithm [16]. Unlike the NO_2 dataset, all these quantities are directly expressed in DU. Another difference with the NO_2 dataset is that there is an individual QF for each concentration. As with the NO_2, a first processing was carried out for SO_2 but considering each individual QF. In cases where negative concentrations fulfilled the QF condition the values were flagged as missing.

Apart from the fact that both datasets have a different spatial resolution the concentrations for these two species are provided in an irregular grid that varies from day to day. A regridding was therefore carried out in order to overcome these drawbacks. Data was mapped into a common grid having a resolution of 1° and 1.25° in latitude and longitude, respectively, for both constituents. This was done in order to match a standard grid of 180×288 pixels in which different Total Ozone Mapping Spectrometer (TOMS) products—ozone, reflectivity, etc.—were given (e.g., [38,39]). More specifically, the regridding process assigned a daily mean value to the centre of each of the 51,840 possible boxes. This mean value was calculated over all the daily non-missing available data that fell within $-89.5 + (i - 1) \pm 0.5$ for $i = 1, 2, \ldots, 180$ and $-179.375 + 1.25(j - 1) \pm 0.625$ for $j = 1, 2, \ldots, 288$ (latitude and longitude, respectively, both expressed in degrees). The outcome of this regridding is the mapping of daily irregular NO_2 and SO_2 global fields into a common regular grid. They constitute the starting point of this research. The period of analysis is the 2004–2016 period.

Long-term seasonal means and standard deviations (SDs) were calculated at each of the pixels of the TOMS-like grid for southern summer, autumn, winter, and spring, including data from December, January and February (DJF), March, April and May (MAM), June, July and August (JJA), and September, October and November (SON), respectively. A minimum number of three values were required in order not to consider the seasonal mean at any pixel as missing. All the seasons were brought together in order to represent the quarterly (Q) cycle. The long-term means and SDs for this case were calculated too. The seasonal means and SDs show prominent, yet spurious, loci of maximum values for the SO_2 concentrations—but not for the NO_2 concentrations—in southern Brazil (BRA). As an example, Figure 1 shows the Q mean concentrations for both gases. The high prominent SO_2 concentrations in the specified region are attributed to the so-called South Atlantic Anomaly (SAA), a region centred in central South America (SA) where the intensity of the Earth's magnetic field has a minimum, enabling the entrance of high-energy particles from space [40]. Given that satellites are exposed to high levels of radiation when they fly over this region [41], the SAA increases the noise in OMI-retrieved data in a significant fashion [24]. The remarkable difference in Figure 1 can be attributed to the photon energy being proportional to its frequency, which is greater by two orders of magnitude for the ultraviolet wavelengths with respect to the visible ones.

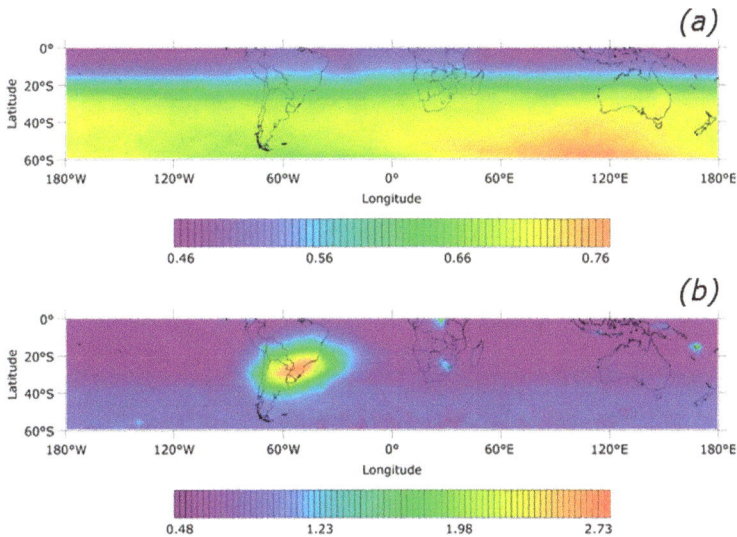

Figure 1. Mean concentrations for the quarterly (Q) time series over the study period for (**a**) NO_2 in the stratospheric (Strat) and (**b**) SO_2 in the planetary boundary layer (PBL). Values expressed in DU. The prominent mean concentrations in southern Brazil (BRA) in (**b**) are related to the South Atlantic Anomaly (SAA).

Standardised anomalies for each of the seasons and for the Q time series were built in order to homogenise the entire study region and to remove (or at least attenuate) the distortion created in the SAA region. Standardised anomalies are a useful tool to make sets of different data comparable to each other. They have been used across a number of applications [42–44]. Figure S1 shows a loop for the DJF SO$_2$ standardised anomalies in the PBL. This is an example that shows how the simple procedure of using standardised anomalies removed the SAA signal, for it is not discernible in each of the figures of this loop. A trend analysis was carried out on the standardised anomalies for each of the seasons individually and for the Q time series. The trends for the standardised anomalies were calculated and statistically tested using a level of significance of 95%. These trends were rescaled by multiplying them by the corresponding SD in order to get a linear trend for the original seasonal concentrations. A rescaling was also applied to the statistic used to test the rescaled trends in order to assess the significance of the trends for the original variables. As pointed out in the Intergovernmental Panel on Climate Change's (IPCC) Fifth Assessment Report there is no physical reason for the time series to have a linear behaviour in time [45]. Apart from the linear trend analysis, the Mann-Kendall (MK) test [46], which evaluates the existence of monotonic upward or downward trends, were implemented on the standardised anomalies. According to [46] the test relies upon the calculation of the following quantity:

$$S = \sum_{j=1}^{n-1} \sum_{i=k+1}^{n} \text{sgn}(x_i - x_j), \tag{1}$$

In (1) n represents the number of points in each of the time series of standardised anomalies, and

$$\text{sgn}(x_i - x_j) = \begin{cases} -1 & \text{if } x_i < x_j \\ 0 & \text{if } x_i = x_j \\ 1 & \text{if } x_i > x_j \end{cases} \tag{2}$$

For $n \geq 10$ (which is the case here) the normal approximation test is used, for which the value of the statistic Z is defined as

$$Z = \begin{cases} \frac{S-1}{\sqrt{VAR(S)}} & \text{if } S > 0 \\ 0 & \text{if } S = 0 \\ \frac{S+1}{\sqrt{VAR(S)}} & \text{if } S < 0 \end{cases} \tag{3}$$

In (3) $VAR(S)$ stands for the variance of S, defined as $VAR(S) = \left[n(n-1)(2n+5) - \sum_{p=1}^{q} t_p(t_p - 1)(2n+5)\right]/18$ with q and t_p being the number of tied groups and the number of values in the p − th group, respectively [46]. Positive and negative values of Z indicate upward and downward trends, respectively. The value of (3) is compared against the critical value Z_c of a normal distribution, which depends upon the choice of the level of significance (e.g., $Z_c = 1.96$ for a level of significance of 95%). The results of the trend analyses are presented in the next section.

3. Results

3.1. SO$_2$ Trends

3.1.1. PBL

Seasonal SO$_2$ linear trends in the PBL are shown in Figure 2. Pixels with a significant trend are cross-hatched. Trend calculations were not carried out for grid cells that had at least one missing value. This is particularly relevant at the higher latitudes of the study region where the number of missing values is remarkable. Skipping the trend calculation in these cases avoids the estimation of potential spurious trends that may arise from time series with missing values at the beginning or the end of the study period. This is the reason why some panels of the figure show a lowermost blank latitudinal band, most notably in the JJA one (and consequently in the Q one) owing to polar night effects. Each of the panels was built using the same scale. Trends in the PBL range from −1.375 to 2.825 DU decade^{-1}, disregarding the season. Trends in the Q time series are smoother, ranging from −0.50 to 1.50 DU decade^{-1} (Figure 1a). Only 42 pixels (out of 10,686 analysed) have a significant linear trend.

Significant positive trends occur in central SA, to the east of the Democratic Republic of the Congo (COD), and over Vanuatu (VUT), and significant negative trends were seen in north-eastern Papua New Guinea (PNG). The significant positive trends in eastern COD and over VUT, approximately of 1 DU decade^{-1} and 0.30 DU decade^{-1}, respectively, are in line with an increase in annual SO_2 emissions in the 2012–2014 period when compared with the 2005–2007 biennium [24]. Eastern COD has the largest concentration of significant adjoining points. Nyiamuragira and Nyiragongo are the two active volcanoes in the region that contribute to the emissions [47], the latter ranking among Africa's most active volcanoes [48] and classified as a "Decade" volcano [49]—i.e., a volcano that was particularly selected for studies due to its proneness to cause natural disasters. The significant negative trends in PNG are also in agreement with the existing literature, with several volcanoes contributing to these figures [50]. Overall, the hemispheric average trend (HAT) is just below 0.08 DU decade^{-1} if all 10,686 analysed points are considered, but it experiences a tenfold increase—to 0.80 DU decade^{-1}—if only the 42 pixels with a significant trend are included. In both cases these figures are in contradiction with the IPCC's predicted decrease in SO_2 concentrations within the century [51].

Linear trends for the DJF quarter are shown in Figure 2b; they range from −0.80 to 2.60 DU decade^{-1}. The number of analysed pixels is 13,982 with only 62 of them showing a significant linear trend. Like in the Q case, the HAT including only these points is much stronger (0.57 DU decade^{-1}) than that obtained by averaging the entire set of grid cells (0.07 DU decade^{-1}). As in Figure 2a, eastern COD shows a significant positive trend (of the order of 0.60 DU decade^{-1} in this case), whereas northern PNG shows a significant negative trend (approximately −0.60 DU decade^{-1}). Two new regions with a significant linear trend that were not present in Figure 2a appear over the eastern Indian Ocean, yet they are negligible when compared with the trends found over the continents. Overall, the region that has the largest increase in significant values occurs in SA, particularly over north-western Argentina (ARG) where the trends are in the order of 0.30 DU decade^{-1}. Trends in MAM (Figure 2c) range from −0.60 and 1.50 DU decade^{-1}. The number of points with a significant trend reduces to 40 (out of 13,030 analysed pixels) with the location of them restricted to eastern COD and SA. Linear trends in the former region are in the order of 1 DU decade^{-1} on average. As to SA, positive and negative significant values occur: the most positive trends were over the south-eastern Brazilian coast (around 1.30 DU decade^{-1}) while the most negative values (of the order of −0.30 DU decade^{-1}) were located approximately over the southernmost portion of the Province of Buenos Aires (ARG). Once again, the HAT estimated only using the pixels that have a significant value (0.90 DU decade^{-1}) is ten times greater than that using the entire set.

The season that has the maximum number of grid cells with a significant trend is JJA, accounting for 75 of them (out of 11,802 points) (Figure 2d). The HAT calculated using the significant points only is −0.14 DU decade^{-1} but it reverses the sign and is approximately seven times weaker when the entire domain is used to estimate it. The two regions with significant values are the same that were present in MAM, but with the one in SA including a largest portion of their pixels with a significant trend in the South Atlantic coasts. Regarding eastern COD, the trend is again positive and in the order of 2 DU decade^{-1}, the strongest one for the season. However, the most dramatic effect occurs in the eastern coasts of SA, where pixels having negative trends account for the majority of the significant points, with an average trend of −0.60 DU decade^{-1}. The number of pixels with a significant trend in SON accounts for 55 (out of 12,130) (Figure 2e). They consist of a single pixel in northern PNG (−0.70 DU decade^{-1}) and the rest of them in central SA, spanning the 20°S–40°S latitudinal band, with negative trends off the coast (approximately −0.60 DU decade^{-1} on average) and positive ones over the landmass. Significance in eastern COD vanishes in this season. The strongest positive values in central SA are in the order of 1 DU decade^{-1}; they take place in central Chile (CHL), where Santiago (SCL), one of Latin America's most polluted cities [52,53] is located. The dispersion of pollutants in central CHL is hindered by the presence of inversion layers associated with anticyclonic conditions, the effect of which can be seen in the trends across all the seasons and even in the Q case. However, if these subsidence-related inversions are a cause of the strongest SON trends there they are not the only cause since these inversions are not more frequent during SON but span the entire year [52,53]. The HAT—considering the significant trends only—is almost 0.60 DU decade^{-1}; this value is more than eleven times greater than that estimated using all the points.

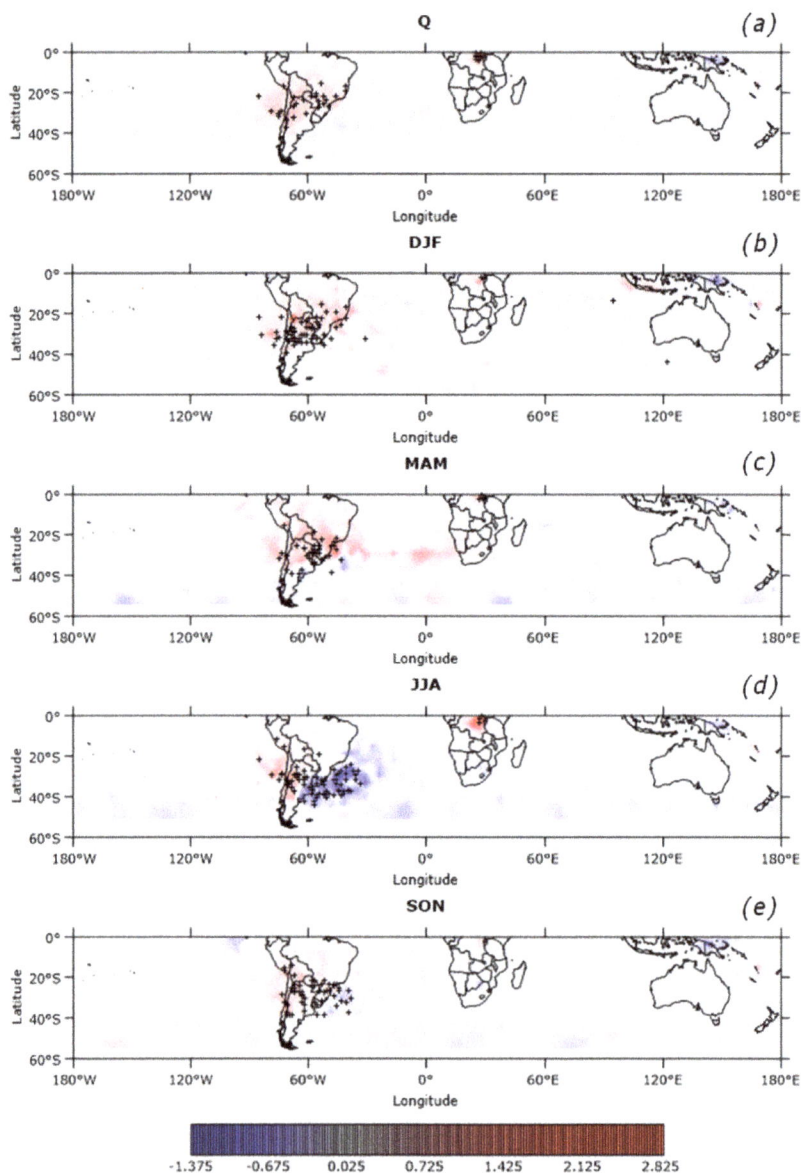

Figure 2. Linear trends (background colours) for the SO$_2$ seasonal concentrations in the planetary boundary layer (PBL) (in DU decade^{-1}) for (**a**) Q, (**b**) December, January, February (DJF), (**c**) March, April, May (MAM), (**d**) June, July, August (JJA), and (**e**) September, October, November (SON). Significant trends are cross-hatched. The level of significance is 95%. Pixels that had at least a seasonal missing value were not included in the analysis.

3.1.2. TRL, TRM and STL

Unlike the SO_2 concentrations in the PBL, the linear trend analysis revealed significance at none of the pixels for the SO_2 seasonal concentrations in the TRL, TRM and the STL. Figures 3–5 show these non-significant linear trends in the SO_2 seasonal concentrations in the TRL, the TRM and the STL, respectively. These trends are expressed in DU century^{-1} in view of their weakness. They range from -3.60 to 8.90, -1.51 to 8.24, and -1.35 and 5.15 DU century^{-1} in the TRL, the TRM and the STL, respectively.

The most extreme linear trends in the TRL occur in JJA both for the positive and the negative values (Figure 3). The former ones occur in eastern COD (around 8 DU century^{-1} on average) and the latter ones take place in the South Atlantic Ocean (SAO) off the Brazilian coasts (in the order of -3 DU century^{-1} on average). In the case of the Q time series (Figure 3a) the number of pixels with a significant monotonic trend is 619 (out of 10,325) split into 555 and 64 for upward and downward trends, respectively (Table 1). These points do not extend beyond 42.5°S, and they are distributed as follows: 388 between 0°S and 20°S (i.e., the tropical band), 227 for the 20°S–40°S band (i.e., the subtropical band), and just 4 in the 40°S–60°S band (i.e., the high-latitude band). Furthermore, there are regions with a relatively small amount of these pixels, most notably northern Amazonia (NAM), the eastern Tropical Pacific off the coasts of Ecuador and Peru (PER), and the SAO. By contrast, there are regions that have a large concentration of such points, most notably the eastern Pacific off the coasts of northern CHL and southern PER, all of them with an upward trend, and northern PNG, with grid cells exhibiting a downward trend.

Table 1. Number of pixels with a significant monotonic trend (not necessarily linear) for the SO_2 seasonal concentrations in the lower troposphere (TRL).

Latitudinal Band	Q	DJF	MAM	JJA	SON
0°S–20°S	388	387	298	379	359
20°S–40°S	227	573	387	454	382
40°S–60°S	4	159	98	56	62
Total upward	555	921	480	314	602
Total downward	64	198	303	575	201
Total analysed pixels	10,325	13,490	12,706	11,551	11,693

The number of grid cells with a monotonic trend for the DJF time series increases to 1119 (out of 13,490) with 387, 573 and 159 of them filling the tropical, the subtropical and the high-latitude bands, respectively (Table 1). Furthermore, 921 (198) of these points are associated to an upward (downward) trend. The south Pacific and the south Atlantic oceans between 45°S and 60°S are the two most notable regions that are empty of these pixels (Figure 3b). On the other hand, the region with a greater number of contiguous points showing a monotonic downward trend is northern PNG; in contrast, the points with an upward trend seem to be evenly distributed across the Atlantic and the Pacific basins. Regarding the MAM time series, the number of pixels that have a trend is 783 (out of 12,706); 298, 387 and 98 occur in the tropical, the subtropical and the high-latitude bands, respectively (Table 1). The number of pixels with an upward (downward) trend is 480 (303). The most notable feature of this season is the dipole that is present in SA, with pixels showing a positive trend occurring in central SA and pixels with a negative trend located off the coasts of ARG. On the other hand, points with a significant monotonic trend at high latitudes are almost void in the south Pacific and the south Atlantic oceans (Figure 3c); these regions coincide with the void areas present in the DJF case.

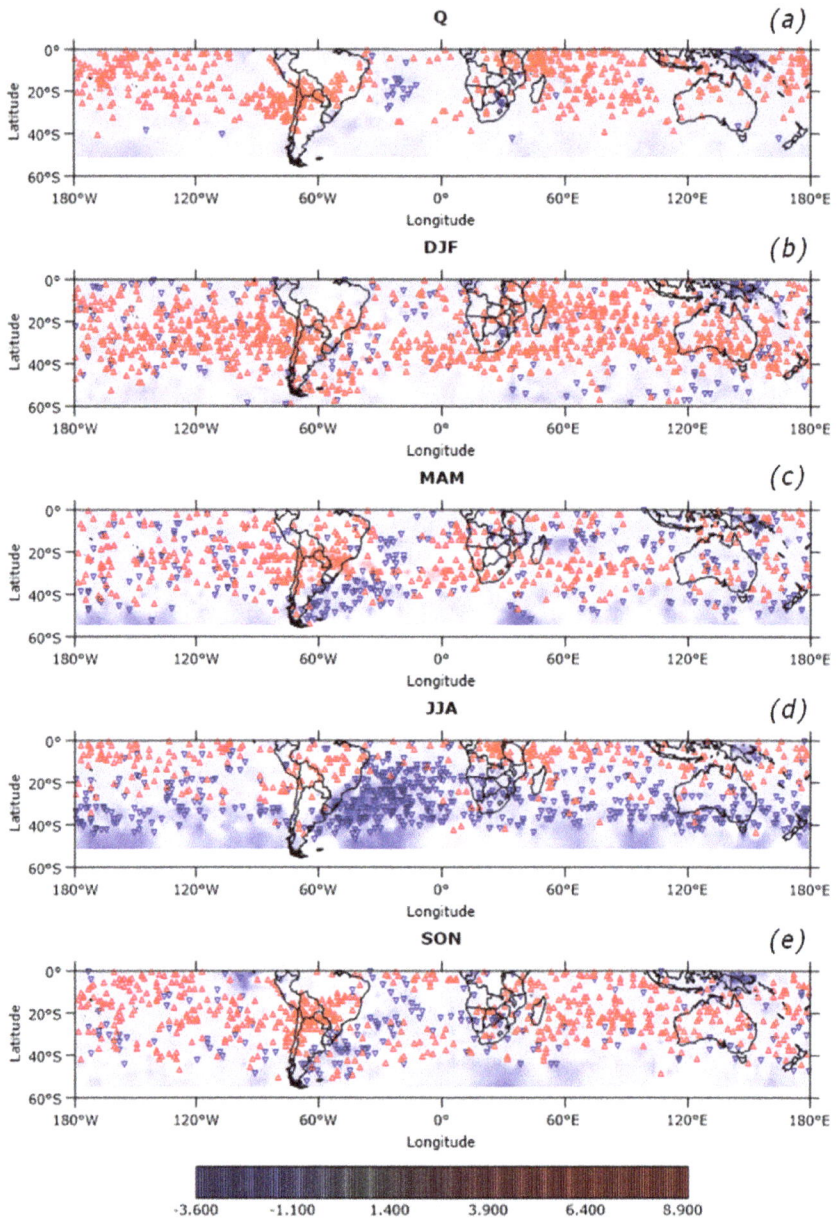

Figure 3. Linear trends (background colours) for the SO$_2$ seasonal concentrations in the lower troposphere (TRL) (in DU century^{-1}) for (**a**) Q, (**b**) DJF, (**c**) MAM, (**d**) JJA, and (**e**) SON. Red and blue arrows mark the pixels that have a significant monotonic upward and downward trend (not necessarily linear), respectively. A level of significance of 95% was set. As in Figure 2, pixels whose time series had at least a seasonal missing value were not included in the analysis.

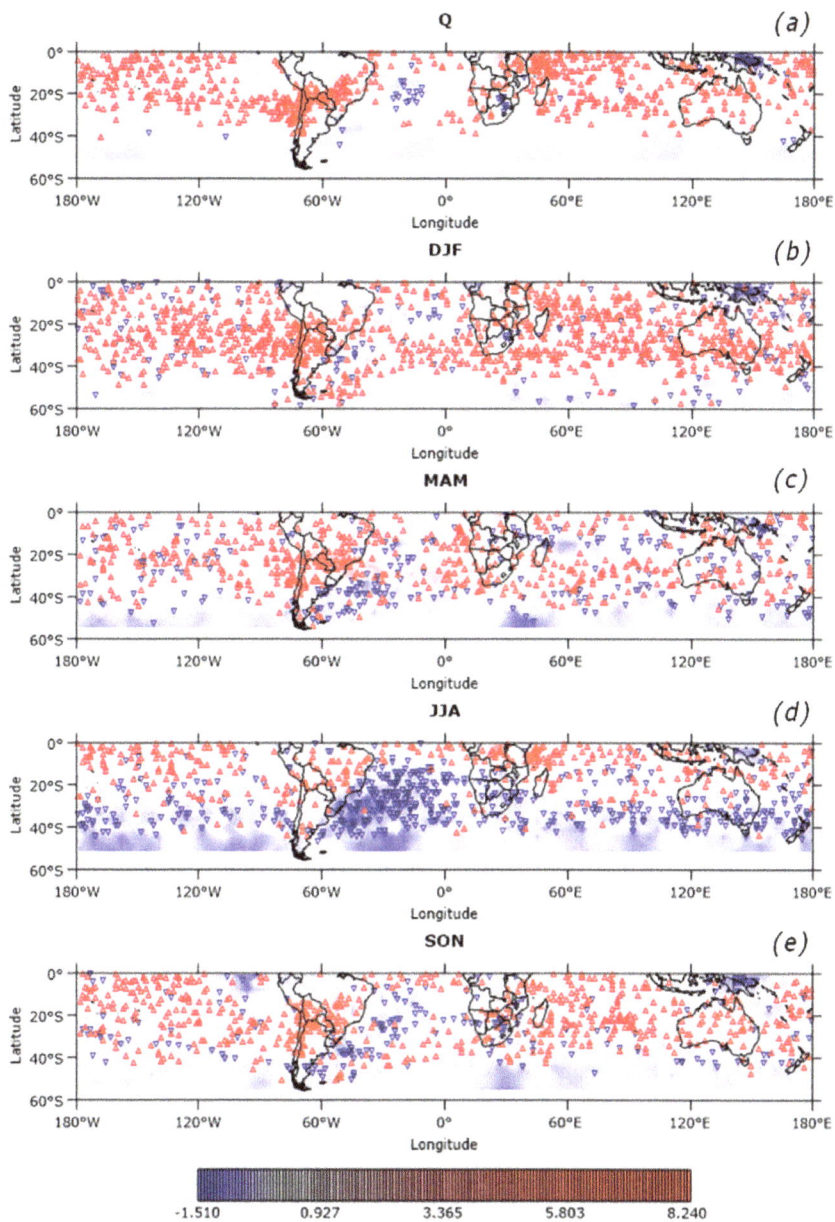

Figure 4. As in Figure 3 but for the SO$_2$ seasonal concentrations in the middle troposphere (TRM). Q, DJF, MAM, JJA and SON figures are shown in panels (**a**–**e**), respectively.

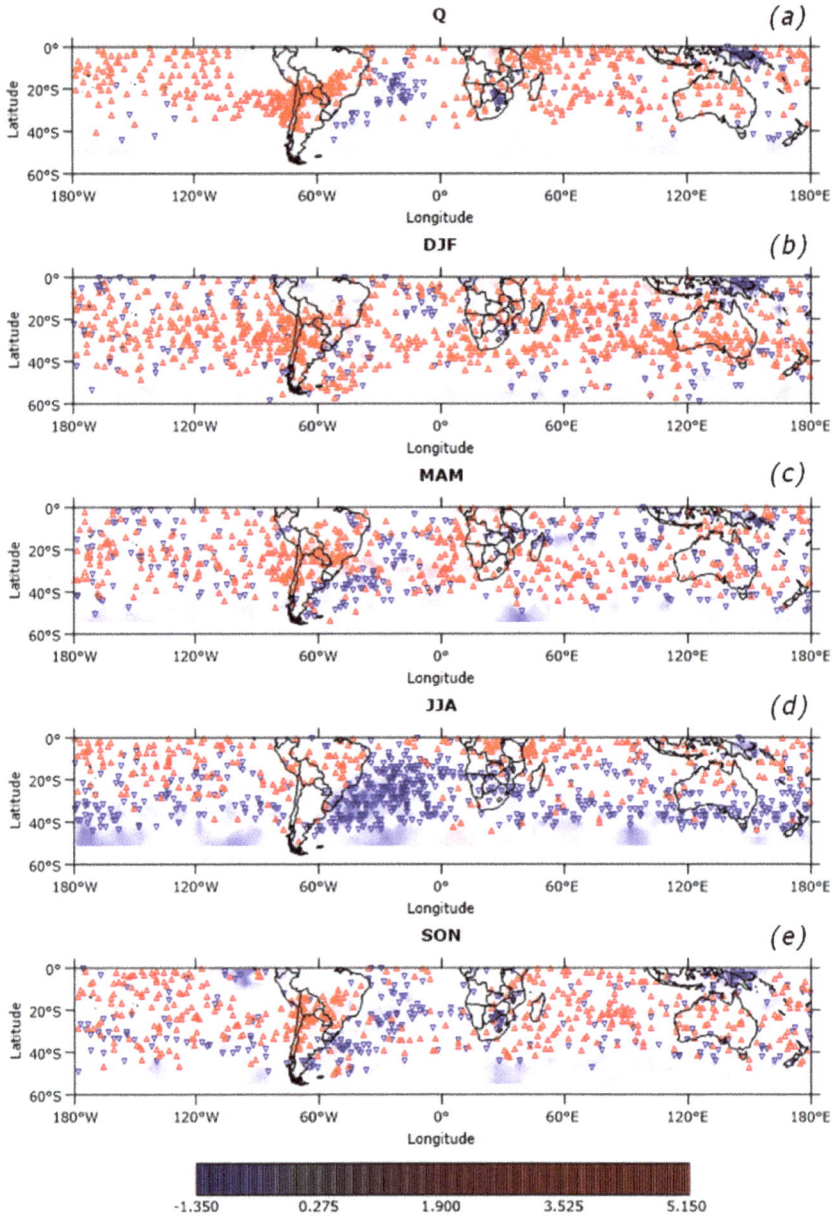

Figure 5. As in Figure 3 but for the SO$_2$ seasonal concentrations in the upper troposphere and the lower stratosphere (STL). Q, DJF, MAM, JJA and SON figures are shown in panels (**a**–**e**), respectively.

The JJA time series record 889 pixels (out of 11,551) with a significant monotonic trend, with the most number of them lying in the subtropical belt (454), followed by the tropical and the high-latitude bands with 379 and 56, respectively (Table 1). There are no pixels in such condition beyond 46.5°S. Even though north of this latitude all regions seem to be fairly populated, these points are particularly dense in the eastern coast of SA and over the Atlantic, where there is a large region with pixels having a downward trend (Figure 3d). Actually, JJA is the only season that is dominated by points with such a characteristic and, in general, the subtropical (tropical) band seems to include the majority of the points with a downward (upward) trend. As for the SON quarter, the number of grid cells with a significant trend is 803 (out of 11,693), with the tropical and the subtropical bands including almost the same number (359 and 382, respectively), and the high-latitude band including the remainder with a count of just 62 (Table 1). In addition, three quarters of these pixels have an upward trend. The largest concentration of points with a monotonic trend takes place over central SA where the trends are upwards (Figure 3e).

The non-significant linear trends for the seasonal SO_2 concentrations in the TRM (Figure 4) are weaker than those in the TRL. The strongest negative values (in the order -1.50 DU century^{-1}) occur in DJF in north-eastern PNG, while the strongest positive ones (approximately 8 DU century^{-1}) take place in SON in central BRA. Even though these trends are not significant, the latter result should be taken with caution as the effect of the SAA might not have been completely removed by the normalisation that was carried out on the variables, yet other effects cannot be ruled out. Notwithstanding, biomass burning in Amazonia traditionally peaks in September [54,55], and hotter fires can create their own convective systems so that the trace gases can be transported well into the higher troposphere ([55] and references therein).

The number of pixels with a significant monotonic trend in the Q time series is 736 (out of 10,359) with 666 representing an upward trend and the rest of them exhibiting the opposite trend (Table 2). The count of grid cells with a trend is greater in the tropical band with more than half of them located there, followed by the subtropical and the high-latitude bands. For the rest of the seasons the latitudinal band that has the greatest population of pixels with a significant trend is the subtropical one, followed by the tropical and the high-latitude bands.

Table 2. As in Table 1 but for the SO_2 concentrations in the TRM.

Latitudinal Band	Q	DJF	MAM	JJA	SON
0°S–20°S	439	399	324	394	333
20°S–40°S	292	592	396	412	395
40°S–60°S	5	148	93	62	66
Total upward	666	956	555	362	592
Total downward	70	183	258	506	202
Total analysed pixels	10,359	13,521	12,725	11,567	11,732

The largest density of pixels with a trend occurs in SA, from the northern coasts of CHL to southern BRA, extending along northern ARG (Figure 4a). They are associated with upward trends with the contribution of at least two volcanoes (Isluga in CHL and Sabancaya in PER) [24]. The 70 points that have a downward trend mainly locate over the Atlantic east of BRA, in eastern South Africa, and in northern PNG. In the last case, the contribution of several volcanoes with decaying activity in the recent years is documented [50]. On the other hand, NAM is another region with no pixels showing any trend at all, apart from those in the high latitudes.

The number of points with a significant trend in DJF is 1139 (out of 13,521) with 956 (183) having an upward (downward) trend. Most of them are distributed across the tropical and the subtropical latitudes. Apart from the high latitudes, NAM is again the region that has the least concentration of pixels in that condition, followed by the SAO (Figure 4b). By contrast, the regions that show the higher density of such pixels coincide with those present in the Q case, i.e., the region between northern CHL and southern BRA for upward trends and northern PNG for downward trends. Regarding MAM, the points that have a significant monotonic trend are 813 (out of 12,725) with 555 (258) of them having an upward (downward) trend. The largest concentration of grid cells showing an upward trend occurs in the same region than in the Q and the DJF cases, i.e., from central CHL to southern BRA across northern ARG. By contrast, the least concentration of points with a trend occurs over the southern oceans. Unlike the Q and the DJF cases NAM does show points with a significant monotonic trend in this season, and their trend is upwards (Figure 4c). As to the JJA quarter, 868 pixels (out of 11,567) show a significant monotonic trend. They are split between 362 (506) points with an upward (downward) trend. As in the TRL case (cf. Table 1), this makes JJA the only season with the number of pixels having a downward trend exceeding their upward counterpart. Downward trends dominate the subtropical latitudes whereas upward trends abound in the tropical latitudes. The region that has the higher density of points with downward trends is the SAO and these pixels are in conjunction with negative linear trends that are the season's strongest negative ones (Figure 4d). In the case of SON, 794 pixels (out of 11,732) have a significant trend, 592 (202) of them exhibiting upward (downward) trends. Most of these points are scattered across the tropical and the subtropical latitudes. The region that has a relative larger concentration of pixels with a significant monotonic trend is the same as in the Q case, i.e., the region that extends from the northern coasts of CHL to southern BRA across northern ARG, conjoined with positive linear trends (Figure 4e).

The non-significant linear trends for the SO_2 seasonal concentrations in the STL are shown in Figure 5. These linear trends are also expressed in DU century^{-1}. They range from -1.35 to 5.15 DU century^{-1} considering individual points. The regions that have the average strongest positive and negative linear trends are Amazonia (approximately 4.40 DU century^{-1}) and northern PNG (around -1.30 DU century^{-1}), respectively, both in the SON quarter.

The count of pixels with a significant monotonic trend in the Q time series is 619 (out of 10,379), split between 495 and 124 for the upward and the downward trends, respectively (Table 3). Most of these points are distributed across the tropical latitudes, followed by the subtropical and the high-latitude bands. As with the previous layers, the higher number of points with a significant trend shifts from the tropical to the subtropical band for the rest of the seasons. The location of the most concentration of points with upward and downward trends in the Q case is similar to that of the TRM's counterpart (Figure 5a, cf. Figure 4a).

Table 3. As in Table 1 but for the SO_2 concentrations in the STL.

Latitudinal Band	Q	DJF	MAM	JJA	SON
0°S–20°S	339	345	304	368	280
20°S–40°S	270	502	393	417	367
40°S–60°S	10	140	84	65	68
Total upward	495	768	487	300	450
Total downward	124	219	294	550	265
Total analysed pixels	10,379	13,554	12,734	11,590	11,747

Regarding the DJF time series, the number of grid cells with a significant trend is 987 (out of 13,554), with 768 (219) of them exhibiting an upward (downward) trend. The highest density of points with an upward monotonic trend occurs over central ARG and CHL; this region is associated with upward linear trends. On the other hand, the location of the highest concentration of pixels with a downward trend is once again northern PNG and they are conjoined to negative linear trends (Figure 5b). As to MAM, the count of points with an upward (downward) trend is 487 (294), out of a total of 12,734. The location that has the most concentration of pixels with an upward trend in this season seems to replicate the DJF's one, yet with a thin zonal line of disruption across approximately 30°S; the higher density of points with a significant downward trend occurs in the SAO east of the Argentine coasts. In both cases these pixels are in compliance with their corresponding linear trends (Figure 5c). The matter concerning the downward trends is more exaggerated in JJA where these pixels span a much larger region than in the MAM's case (Figure 5d). As before (cf. Tables 1 and 2), JJA is the only season with the number of points having a downward trend exceeding that with an upward trend (Table 3). Concerning the SON quarter, the count of points with a significant trend stands at

715 (out of 11,747), split into 450 (265) with an upward (downward) trend. The regions that have the largest concentration of pixels with and upward and downward trends coincide with the ones in the Q counterpart and they are in correspondence with their linear trends (Figure 5e).

3.2. NO₂ Trends

The linear trends for the seasonal time series of Trop and Strat NO_2 concentrations and the pixels that show a monotonic upward or downward trend for the standardised anomalies of the same variables are shown in Figures 6 and 7, respectively. As with the SO_2 seasonal concentrations in most of the layers of the atmosphere all these linear trends are not significant across the analysed seasons both in the Trop and the Strat. In the former layer the linear trends range from -3.25 to 2.50 DU century^{-1} (Figure 6) and they are weaker in the latter layer, ranging from -1.184 to 0.363 DU century^{-1} (Figure 7).

Unlike the SO_2 case the number of pixels with a monotonic upward trend in the Trop is the dominant characteristic in all the seasons (Figure 6), with the maximum number of them (7954 out of 8095) taking place in the Q time series (Table 4)—they are largely concentrated in the subtropical and high-latitude bands over the oceans. On the other hand, MAM is the season that shows the largest number of pixels exhibiting a monotonic downward trend, with these points being mainly distributed across the tropical latitudes but with a large concentration of them also taking place in southern SA (Figure 6c). Despite this, grid cells with an upward trend are by far the most dominant in all the seasons. These results are quite in agreement with an overall increase in the global tropospheric NO_2 concentrations in the recent years [56]. The regions exhibiting positive and negative monotonic trends in all the seasons are in correspondence with the linear trends shown at these locations. According to [57] the months of maximum NO_2 concentrations over central SA/Amazonia, southern Africa and northern Australia (AUS) take place in the SON quarter. There are pixels with a significant monotonic trend in these regions within the specified season, with southern Africa and large portions of central SA/Amazonia exhibiting upward trends and northern AUS exhibiting downward trends; the presence of downward trends in western Amazonia is noteworthy as well (Figure 6e). The upward trends are evidence that the emissions in the corresponding regions increased in the analysed period, and the converse situation occurs for the downward trends, most notably in northern AUS. Unfortunately, no further links of this kind can be established for the rest of the seasons as much of the maximum NO_2 concentrations in the SH take place during SON.

The lack of previous research covering the topics dealt with in our paper does not permit direct comparison at a hemispheric scale. However, trends for tropospheric NO_2 concentrations were presented at major urban agglomerations in the SH [26]. According to Figure 6a the monotonic upward trends in Lima and SCL are in qualitative match with the results presented in [33] as they are in Jakarta and Sydney for a downward trend, whereas they do not match in Buenos Aires, Johannesburg, San Pablo and Rio de Janeiro, as the monotonic trends at these cities are not significant. The case of Sydney is interesting since Figure 6a shows that the only pixel with a significant downward trend in the region is located at this city.

The fact that the standardised seasonal NO_2 Strat concentrations time series have no pixels with a monotonic upward trend for any of the analysed seasons is the most striking feature of this paper. The Q time series have the least number of points exhibiting a downward trend with a count of just 70 (Table 5) that are located in the subtropical latitudes (Figure 7a). This number increases to 6042 for DJF with most of these pixels distributed across the subtropical and the high-latitude bands but with blanks in several zones, most notably in the southern SAO (Figure 7b). MAM is the season that has the largest number of pixels with a monotonic trend with a count of 10,292, most of them beyond 20°S with a noticeable blank region at high latitudes south of AUS (Figure 7c). A feature of the Q, the DJF and the MAM time series is that there are virtually no points with upward linear trends. This is not the case for JJA and SON (Figure 7d,e, respectively) whose count of pixels with a monotonic trend reduces dramatically to 1168 and 372, respectively.

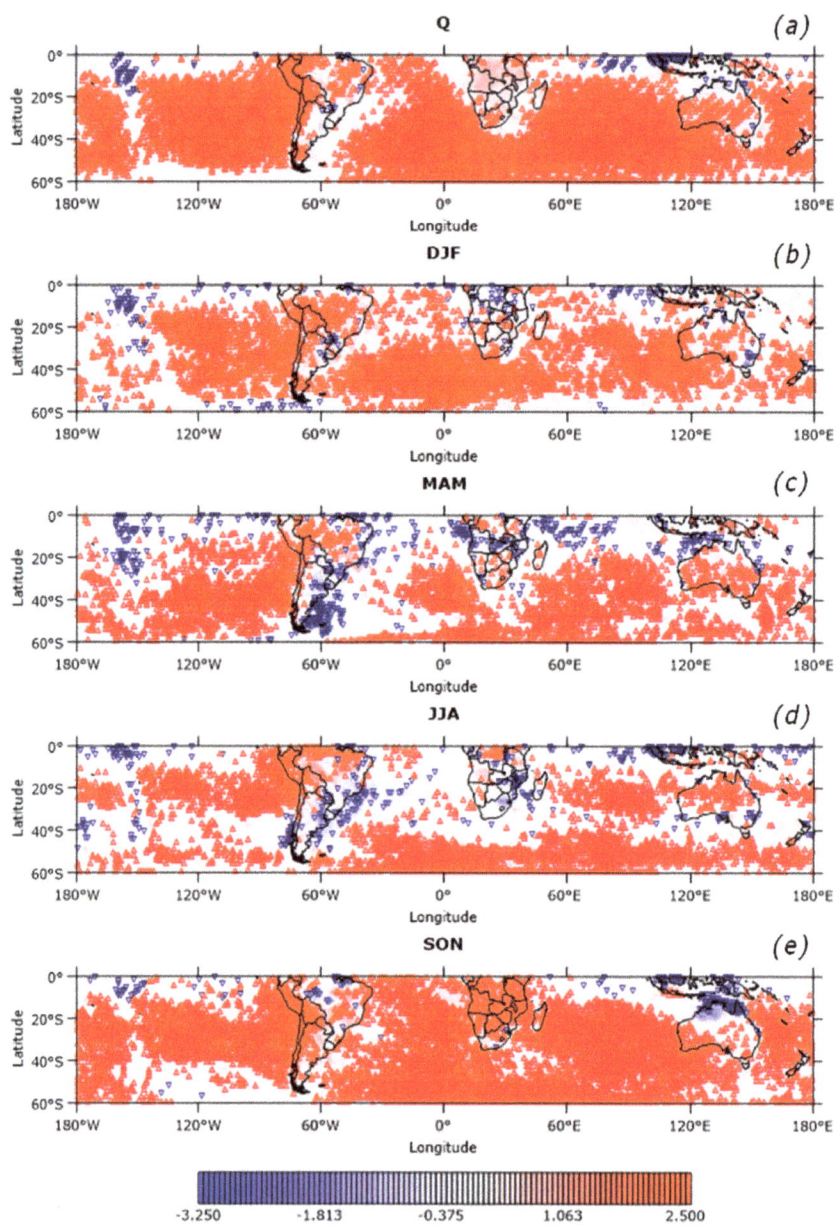

Figure 6. As in Figure 3 but for the seasonal concentrations of NO_2 in the troposphere (Trop). Q, DJF, MAM, JJA and SON figures are shown in panels (**a–e**), respectively.

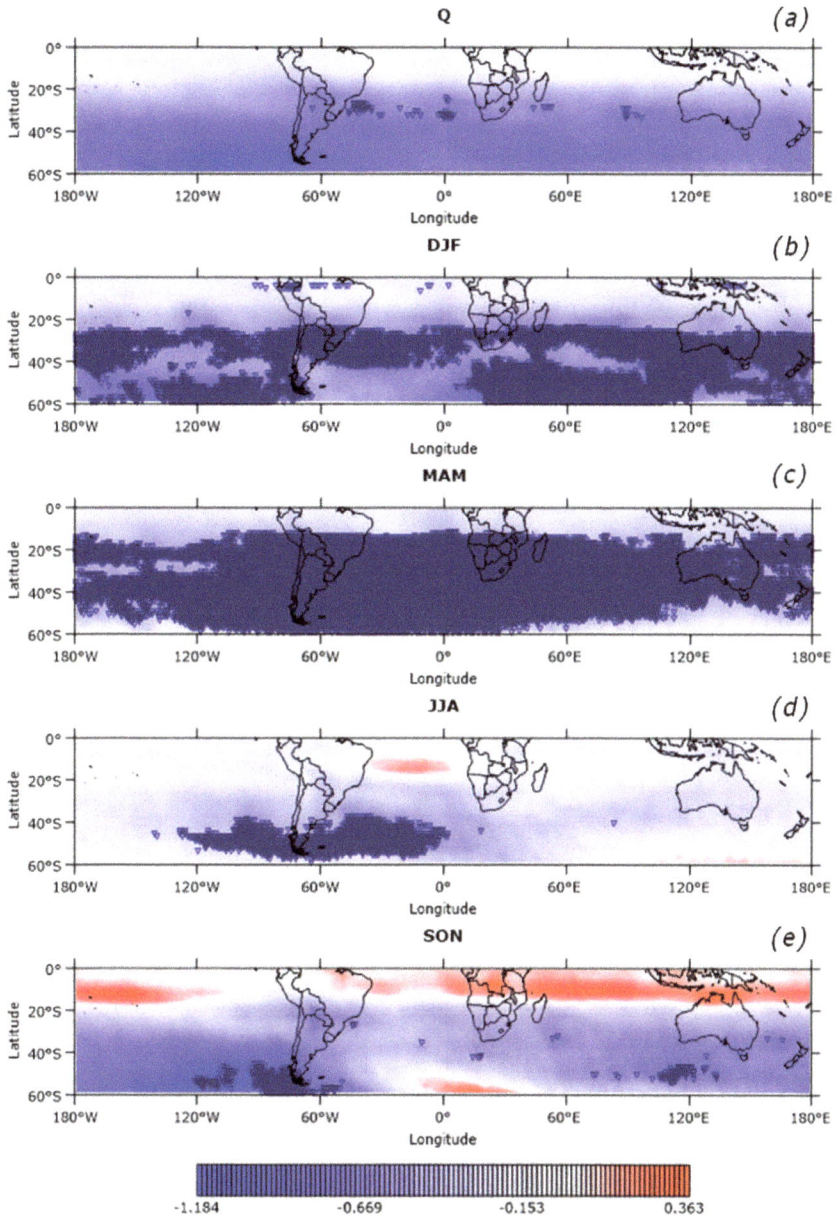

Figure 7. As in Figure 3 but for the seasonal concentrations of NO_2 in the stratosphere (Strat). Q, DJF, MAM, JJA and SON figures are shown in panels (**a–e**), respectively.

Table 4. As in Table 1 but for the NO_2 concentrations in the Trop.

Latitudinal Band	Q	DJF	MAM	JJA	SON
0°S–20°S	1455	792	650	938	1630
20°S–40°S	3382	1883	1575	1031	3144
40°S–60°S	3258	1641	1715	1838	2459
Total upward	7954	4108	3472	3461	7054
Total downward	141	208	468	346	179
Total analysed pixels	17,280	17,280	17,280	17,280	17,280

Table 5. As in Table 1 but for the NO_2 concentrations in the Strat.

Latitudinal Band	Q	DJF	MAM	JJA	SON
0°S–20°S	0	47	1338	0	0
20°S–40°S	70	3179	5172	48	8
40°S–60°S	0	2816	3782	1120	364
Total upward	0	0	0	0	0
Total downward	70	6042	10292	1168	372
Total analysed pixels	17,280	17,280	17,280	17,280	17,280

4. Discussion

The prevailing meteorological conditions were briefly mentioned as a cause of SCL being a pollution-prone city. These conditions, along with other factors (such as topography) may conspire to make other regions of the world as polluted as, or even more polluted than, SCL, either seasonally or on an annual basis. Notwithstanding, the degree of relationship between concentrations and emissions is in general not as straightforward as this example suggests. This can be figured out from the trends observed in remote areas over open waters of the SH where there are no sources and the pollutants owe their existence to atmospheric transport. The number of research efforts devoted to identifying different linkages between the concentration of pollutants and the meteorological conditions is on the rise in both hemispheres, but they are more numerous in the Northern Hemisphere due to a number of factors that include this hemisphere having a large number of megacities, and therefore the highest concentration of regions with strong industrial activity. Even though the areas closer to the sources of pollutants (natural or anthropogenic) are expected to be the most affected by the emissions, the local and regional meteorology, the general circulation [58,59] and the chemistry should be considered altogether in order to establish the spatial extent of the influence and the actual concentrations. Recently, a study carried out for China found that the discrepancy in the relative values of the emissions and the SO_2 concentrations in the lower troposphere owing to a change in local meteorological conditions can represent up to 20–30%. Under the same scenario, the deviations from a linear relationship linking the emissions and the columnar concentrations can be up to 50% [60]. A similar study for the NO_x regarding trends showed that meteorology may account for up to a 30% difference in emission/concentration differences [61]. More generally, the characteristics of the emissions—e.g., increase or decrease over time, seasonality—combined with a global warming scenario that leads to the alteration of the meteorological conditions may have an impact on the evolution of the concentrations and hence on the observed trends. The expansion of the tropical belt, the intensification of the Hadley Cell (HC) and the strengthening of upper tropospheric jets (UTJs) are part of the meteorological aspects of the global warming scenario that may play a role in the distribution of these pollutants, thereby modulating their concentrations across both hemispheres.

This work focused on the SH. The count of analysed pixels across all the seasons covered the entire study region for the NO_2 concentrations both in the troposphere and in the stratosphere. This was not the case for the SO_2 concentrations in the different analysed atmospheric layers owing to missing data. Our results show that there is a marked seasonality in the characteristics of the linear and non-linear trends over the SAO for the SO_2 concentrations, and in the sub-tropics and the mid-latitudes for the NO_2 concentrations. In particular, both the linear and the non-linear trends in the SO_2 concentrations—and to a lesser extent in the NO_2 concentrations—over the SAO exhibit a general reversion in their condition in JJA with respect to the rest of the seasons (including the Q case), from upward to downward.

As mentioned above, the overall features in the global distribution of the seasonal SO_2 concentration trends can be at least partly interpreted by recalling some components of the large-scale circulation and their evolution in the recent years. A noteworthy distinction of these trends is the absence of significance in the linear trends in NAM in Figure 2 and in the monotonic non-linear trends in the same region in Figures 3–5, for the Q case as well as for the DJF and the SON seasons in all these sets of figures. This can be due to the effects of the HC's ascending branch, whose vigorous vertical currents remove the specie from the equatorial latitudes and deposits it in the descending regions, located around 30°S on average, where the concentration of significant trends is considerable. Taking into account the described mechanism, the upward trends found in the sub-tropical latitudes of SA can be attributed to two distinct effects: (a) increasing emissions and (b) intensification of the HC. The intensification of the HC has been proven in [62] and it is present in different reanalysis datasets from 1979 through 2009. It is a plausible explanation for what is observed, provided the intensification of the HC is also valid within our study period. Although not statistically different from zero, the background linear trends in NAM shown in Figures 3–5 are worth mentioning as they are consistent with a removal of this specific pollutant there. Grid cells with a significant positive non-linear trend in SA reach their southernmost position in DJF in concordance with the descending branch of the HC occupying its most poleward position during the summer months [62]. This agreement reinforces the intensification of the HC as a plausible mechanism for the latitudinal distribution of the observed trends. The poleward extent of the HC depends upon a number of land-sea contrast parameters—most notably the meridional temperature gradient—and moisture content [63] and so does the southernmost position of the significant trends. The relationship is complex, however, since the expansion of the tropical belt seems to have a dependence on the state of some modes of variability of the coupled atmosphere/ocean system, which in turn seems to depend on the concentration of anthropogenic aerosols [64]. The study of SO_2 concentrations in the SH is particularly important to this topic considering the effects SO_2 has on ozone depletion and the role played by the polar stratospheric ozone depletion in driving the widening of the tropical belt [65,66].

The importance of the UTJs to synoptic processes is that they drive the location of the storm tracks through baroclinic instability, but they also play a role acting as waveguides or inhibiting the poleward transport of wave activity [67], therefore interfering with the meridional circulation described above. The latitudinal distribution of the trends can be partly ascribed to the semi-horizontal transport of the pollutants away from their sources aided by the UTJs and redistributed by the eddy perturbations they contribute to create. UTJs are stronger in the winter hemisphere [68] thus creating more favourable conditions for baroclinic perturbations to develop. Generally speaking, the widening of the tropical belt implies an intensification (weakening) of the polar (subtropical) UTJ in DJF, and the converse situation takes place for JJA, i.e., the subtropical (polar) UTJ experiences a strengthening (weakening), yet the regions exhibiting significance are much more reduced [68]. Particularly over SA, the subtropical UTJ shows upward linear trends in both seasons [68] and this feature can be used to easily interpret the linear trends in PBL SO_2 concentrations for JJA (Figure 2d). Indeed, a strengthened UTJ has the potential to create more intense storm tracks, and the region in the SAO concentrating the downward trends is located in one of the most cyclogenetic regions of the SH, particularly during JJA [67].

The clockwise rotation of these low pressure systems removes the pollutants from the oceanic regions and accumulates them east of the Andes so that the positive trends there are also in agreement with this mechanism. Regarding the non-linear downward trends, the highest concentration of them in the SAO during JJA (Figures 3d, 4d and 5d) may respond to a chemical process combined with the aforementioned changing atmospheric conditions. The oxidation of SO_2 in the troposphere is at its maximum in the winter months [69]. Furthermore, due to its high aqueous solubility [70], the production of sulphates via in-cloud oxidation of SO_2 has been reported to be important [71,72]. The intensification of the subtropical UTJ in the SAO region from central SA to southern Africa during JJA impact on both the frequency and the strength of the storm tracks there, with both characteristics expected to increase, leading to more cloudiness and hence more proneness to in-cloud oxidation of the SO_2. Upward trends in the column of integrated water vapour content over the oceans in most of the regions of interest [45] are in concordance with the proposed mechanisms. The strengthening of the subtropical UTJ is also seen in the southern Indian Ocean from southern Africa to AUS [68], and the neighbouring regions exhibit non-linear downward trends in SO_2 concentrations as well. A similar situation is also observed over New Zealand. There are no reasons for not to consider the same meteorological mechanisms for the NO_2 trends. Concerning this particular pollutant, significant downward trends are generally closer to the landmasses—i.e., closer to the emission sources when compared with their SO_2 counterparts. This may respond either to an actual reduction in the emissions or to the fact that NO_2 has a lifetime of a few minutes against photolysis, leading to the generation of O_3 [73], another tropospheric pollutant. In general, both NO_2 and SO_2 react with OH in polluted atmospheres, leading to the production of acids—it actually constitutes the dominant loss mechanism for NO_x and the removal is much more efficient for the NO_x species than it is for the SO_2 [70]. The benefits of having a downward trend in both SO_2 and NO_2 concentrations at certain regions may be only apparent, as their reduction may imply these two compounds are oxidising and leading to the potential formation of acid rain.

5. Conclusions

The seasonal standardised anomalies of sulphur dioxide (SO_2) and nitrogen dioxide (NO_2) concentrations in different layers of the atmosphere were analysed for linear and non-linear trends in order to address the behaviour of these two pollutants in the context of a changing climate. The studied region was the Southern Hemisphere (SH) between the Equator and 60°S, the analysed period was 2004–2016, and Ozone Monitoring Instrument (OMI) data was used for both gases. To the best of our knowledge this is the first time the characterisation of non-linear trends was carried out in the entire SH. Standardisation was carried out as a twofold purpose: to remove (or at least attenuate) the known influence of the South Atlantic Anomaly in OMI data and to homogenise the datasets in the entire domain. On the one hand, linear trends were estimated and statistically tested. This procedure was carried out on the standardised anomalies of seasonal SO_2 concentrations in the planetary boundary layer (PBL), the lower troposphere (TRL), the middle troposphere (TRM) and the upper troposphere and the lower stratosphere (STL), and NO_2 concentrations in the troposphere (Trop) and the stratosphere (Strat), for the austral summer (DJF), autumn (MAM), winter (JJA) and spring (SON) time series, as well as for the quarterly (Q) time series. The obtained linear trends for the standardised anomalies were converted so that the trends for the original time series could be informed. On the other hand, non-linear trends were estimated for the series of standardised anomalies by means of the Mann-Kendall test and it was assessed whether the grid cells had a significant monotonic upward or downward trend.

The main findings of this paper can be summarised as follows. The SO_2 concentrations show significant linear trends in the PBL only, but just a few pixels located mainly over the landmasses display such behaviour. The location of such pixels and their trends are in general in agreement with the existing literature. Even though an important number of grid cells in the TRL, the TRM, and the STL do not have significant linear trends, they exhibit significant monotonic upward or downward trends depending upon the region and the seasons considered, both over the landmasses and in remote regions over the oceans. A noteworthy feature is that JJA shows a large number of points with a downward trend in all the layers, while the opposite holds for the rest of the seasons. Concerning the NO_2 concentrations, no significant linear trends were found either in the troposphere or in the stratosphere, but the former (latter) layer shows monotonic upward (downward) trends. Results are in agreement with a general increase in NO_2 and SO_2 emissions in the recent years [74,75] but they are not in accordance with the predicted emissions for these two gases within the different scenarios of climate change for the current century. The presence of trends in remote areas of the hemisphere away from the sources (which are mostly located in the landmasses) suggests that the general circulation combined with local processes—both subject to climate change—and chemistry play an important role in the transport and the spatial distribution of these pollutants. The statistic used to test the significance of the monotonic upward or downward trends revealed different degrees of non-linearities but the exact types of these non-linear evolutions were not addressed. The determination of the most suitable function to fit will likely help in further understanding the way the concentrations of these two gases will behave in the future. This is a matter for future investigation.

Supplementary Materials: The following are available online at www.mdpi.com/2072-4292/9/9/891/s1. Figure S1: Loop of standardised anomalies of PBL SO_2 concentrations in DJF. Significant values are cross-hatched; the number of them is shown in parenthesis. The level of significance was set to 95%.

Acknowledgments: We gratefully acknowledge the academic editor and three anonymous reviewers for their valuable comments and suggestions. PIP 2012–2014 N° 0075, PICT 2012–2927 and PIDDEF 2014 N° 26 grants partly funded this paper. The funds for covering the costs to publish in open access were provided by the Facultad Regional Buenos Aires of the Universidad Tecnológica Nacional.

Author Contributions: A.Y. and S.G.L. conceived the underlying ideas; A.Y. performed the calculations and the figures; A.Y. and S.G.L analysed the data; A.Y. and S.G.L. wrote the paper. All the authors reviewed the manuscript.

Conflicts of Interest: The authors declare no conflict of interest. The founding sponsors had no role in the design of the study; in the collection, analyses, or interpretation of data; in the writing of the manuscript, and in the decision to publish the results.

References

1. Myhre, G.; Shindell, G.D.; Bréon, F.-M.; Collins, W.; Fuglestvedt, J.; Koch, D.; Lamarque, J.-F.; Lee, D.; Mendoza, B.; Nakajima, T.; et al. Anthropogenic and Natural Radiative Forcing. In *Climate Change 2013: The Physical Science Basis. Contribution of Working Group I to the Fifth Assessment Report of the Intergovernmental Panel on Climate Change*; Stocker, T.F., Qin, D., Plattner, G.-K., Tignor, M., Allen, S.K., Boschung, J., Nauels, A., Xia, Y., Bex, V., Midgley, P.M., Eds.; Cambridge University Press: New York, NY, USA, 2013; pp. 659–740.
2. Lohmann, U.; Feichter, J. Can the direct and semi-direct aerosol effect compete with the indirect effect on a global scale? *Geophys. Res. Lett.* **2001**, *28*, 159–161. [CrossRef]
3. Johnson, B.T.; Shine, K.P.; Forster, P.M. The semi-direct aerosol effect: Impact of absorbing aerosols on marine stratocumulus. *Q. J. R. Meteorol. Soc.* **2005**, *130*, 1407–1422. [CrossRef]
4. Lohmann, U.; Feichter, J. Global indirect aerosol effects: A review. *Atmos. Chem. Phys.* **2005**, *5*, 715–737. [CrossRef]
5. Ghan, S.J.; Liu, X.; Easter, C.; Zaveri, R.; Rasch, P.J.; Yoon, J.-H. Toward a Minimal Representation of Aerosols in Climate Models: Comparative Decomposition of Aerosol Direct, Semidirect, and Indirect Radiative Forcing. *J. Clim.* **2012**, *25*, 6461–6476. [CrossRef]
6. Mahowald, N.M.; Scanza, R.; Brahney, J.; Goodale, C.L.; Hess, P.G.; Moore, J.K.; Neff, J. Aerosol Deposition Impacts on Land and Ocean Carbon Cycles. *Curr. Clim. Chang. Rep.* **2017**, *3*, 16–31. [CrossRef]
7. Martin, J.H.; Fitzwater, S.E. Iron deficiency limits phytoplankton growth in the northeast Pacific subarctic. *Nature* **1988**, *331*, 341–343. [CrossRef]
8. Okin, G.; Mahowald, N.; Chadwick, O.; Artaxo, P. Impact of desert dust on the biogeochemistry of phosphorus in terrestrial ecosystems. *Glob. Biogeochem. Cycles* **2004**, *18*, GB2005. [CrossRef]
9. Stier, P.; Feichter, J.; Roeckner, E.; Kloster, S.; Esch, M. The evolution of the global aerosol system in a transient climate simulation from 1860 to 2100. *Atmos. Chem. Phys.* **2006**, *6*, 3059–3076. [CrossRef]
10. Kiros, F.; Shakya, K.M.; Rupakheti, M.; Regmi, R.P.; Maharjan, R.; Byanju, R.M.; Naja, M.; Mahata, K.; Kathayat, B.; Peltier, R.E. Variability of Anthropogenic Gases: Nitrogen Oxides, Sulfur Dioxide, Ozone and Ammonia in Kathmandu Valley, Nepal. *Aerosol Air Qual. Res.* **2016**, *16*, 3088–3101. [CrossRef]
11. Krotkov, N.A.; McLinden, C.A.; Li, C.; Lamsal, L.N.; Celarier, E.A.; Marchenko, S.V.; Swartz, W.H.; Bucsela, E.J.; Joiner, J.; Duncan, B.N.; et al. Aura OMI observations of regional SO_2 and NO_2 pollution changes from 2005 to 2015. *Atmos. Chem. Phys.* **2016**, *16*, 4605–4629. [CrossRef]

12. Jaffe, D.A.; Weiss-Penzias, P.S. Nitrogen Cycle. In *Encyclopedia of Atmospheric Sciences*; Holton, J.R., Curry, J.A., Pyle, J.A., Eds.; Academic Press: London, UK, 2003; Volume 1, pp. 205–213.

13. Bucsela, E.J.; Krotkov, N.A.; Celarier, E.A.; Lamsal, L.N.; Swartz, W.H.; Bhartia, P.K.; Boersma, K.F.; Veefkind, J.P.; Gleason, J.F.; Pickering, K.E. A new stratospheric and tropospheric NO_2 retrieval algorithm for nadir-viewing satellite instruments: Applications to OMI. *Atmos. Meas. Tech.* **2013**, *6*, 2607–2626. [CrossRef]

14. Kondo, Y. Reactive Nitrogen (NO_x and NO_y). In *Encyclopedia of Atmospheric Sciences*; Holton, J.R., Curry, J.A., Pyle, J.A., Eds.; Academic Press: London, UK, 2003; Volume 5, pp. 2193–2202.

15. Seinfeld, J.H. Tropospheric Chemistry and Composition: Aerosols/Particles. In *Encyclopedia of Atmospheric Sciences*; Holton, J.R., Curry, J.A., Pyle, J.A., Eds.; Academic Press: London, UK, 2003; Volume 6, pp. 2349–2354.

16. Yang, K.; Krotkov, N.A.; Krueger, A.J.; Carn, S.A.; Bhartia, P.K.; Levelt, P.F. Retrieval of large volcanic SO_2 columns from the Aura Ozone Monitoring Instrument: Comparisons and limitations. *J. Geophys. Res.* **2007**, *112*. [CrossRef]

17. United States Environmental Protection Agency. What is Acid Rain? Available online: https://www.epa.gov/acidrain/what-acid-rain (accessed on 30 July 2017).

18. World Health Organization (WHO). WHO Air Quality Guidelines for Particular Matter, Ozone, Nitrogen Dioxide and Sulfur Dioxide. Global Update 2005. Available online: http://apps.who.int/iris/bitstream/10665/69477/1/WHO_SDE_PHE_OEH_06.02_eng.pdf (accessed on 29 June 2017).

19. Levelt, P.F.; van den Oord, G.H.J.; Dobber, M.R.; Mälkki, A.; Visser, H.; de Vries, J.; Stammes, P.; Lundell, J.O.V.; Saari, H. The Ozone Monitoring Instrument. *IEEE Trans. Geosci. Remote Sens.* **2006**, *44*, 1093–1101. [CrossRef]

20. Corbett, J.J.; Koehler, H.W. Updated emissions from ocean shipping. *J. Geophys. Res.* **2003**, *108*, 4650. [CrossRef]

21. Endresen, Ø.; Sørgård, E.; Behrens, H.L.; Brett, P.O.; Isaksen, I.S.A. A historical reconstruction of ships' fuel consumption and emissions. *J. Geophys. Res.* **2007**, *112*, D12301. [CrossRef]

22. Beirle, S.; Boersma, K.F.; Platt, U.; Lawrence, M.G.; Wagner, T. Megacity Emissions and Lifetime of Nitrogen Oxides Probed from Space. *Science* **2011**, *333*, 1737–1739. [CrossRef] [PubMed]

23. Janssens-Maenhout, G.; Crippa, M.; Guizzardi, D.; Dentener, F.; Muntean, M.; Pouliot, G.; Keating, T.; Zhang, Q.; Kurokawa, J.; Wankmüller, R.; et al. HTAP_v.2.2: A mosaic of regional and global emission grid maps for 2008 and 2010 to study hemispheric transport of air pollution. *Atmos. Chem. Phys.* **2015**, *15*, 11411–11432. [CrossRef]

24. Fioletov, V.E.; McLinden, C.A.; Krotkov, N.; Li, C.; Joiner, J.; Theys, N.; Carn, S.; Moran, M.D. A global catalogue of large SO_2 sources and emissions derived from the Ozone Monitoring Instrument. *Atmos. Chem. Phys.* **2016**, *16*, 11497–11519. [CrossRef]

25. McLinden, C.A.; Fioletov, V.; Shepard, M.W.; Krotkov, N.; Li, C.; Martin, R.V.; Moran, M.D.; Joiner, J. Space-based detection of missing sulphur dioxide sources of global air pollution. *Nat. Geosci.* **2016**, *9*, 496–500. [CrossRef]

26. Schneider, P.; Lahoz, W.A.; van der A., R. Recent satellite-based trends of tropospheric nitrogen dioxide over large urban agglomerations worldwide. *Atmos. Chem. Phys.* **2015**, *15*, 1205–1220. [CrossRef]

27. OMSO2 README File v1.1.1 Released Feb 26, 2008 Updated: August 18, 2008. Available online: https://acdisc.gesdisc.eosdis.nasa.gov/data/s4pa/Aura_OMI_Level2G/OMSO2G.003/doc/OMSO2G_OSIPS_README_V003.pdf (accessed on 17 March 2017).

28. Schoberl, M.R.; Douglass, A.R.; Hilsenrath, E.; Bhartia, P.K.; Beer, R.; Waters, J.W.; Gunson, M.R.; Froidevaux, L.; Gille, J.C.; Barnett, J.J.; et al. Overview of the EOS Aura Mission. *IEEE Trans. Geosci. Remote Sens.* **2006**, *44*, 1066–1074. [CrossRef]

29. Krotkov, N.A.; Lamsal, L.N.; Celarier, E.A.; Swartz, W.H.; Marchenko, S.V.; Bucsela, E.J.; Chan, K.L.; Wenig, M. The version 3 OMI NO_2 standard product. *Atmos. Meas. Tech. Discuss.* **2017**. [CrossRef]

30. The OMI Nitrogen Dioxide Algorithm Team. OMNO2 README Document Data Product Version 3.0. Document Version 7.0, September 2016. Available online: https://acdisc.gesdisc.eosdis.nasa.gov/data/Aura_OMI_Level2G/OMNO2G.003/doc/README.OMNO2.pdf (accessed on 17 March 2017).

31. Seinfeld, J.H.; Pandis, S.N. Chapter 2: Atmospheric Trace Constituents. In *Atmospheric Chemistry and Physics: From Air Pollution to Climate Change*; John Wiley & Sons: Hoboken, NJ, USA, 2016; pp. 45–46.

32. McElroy, M.B.; Salawitch, R.J.; Wofsy, S.C.; Logan, J.A. Reductions of Antarctic ozone due to synergistic interactions of chlorine and bromine. *Nature* **1986**, *321*, 759–762. [CrossRef]

33. Bowman, K.P. Global Trends in Total Ozone. *Science* **1988**, *239*, 48–50. [CrossRef] [PubMed]

34. Pyle, J.A. Stratospheric chemistry and composition, Overview. In *Encyclopedia of Atmospheric Sciences*; Holton, J.R., Curry, J.A., Pyle, J.A., Eds.; Academic Press: London, UK, 2003; Volume 5, pp. 2150–2154.

35. Douglass, A.R.; Newman, P.A.; Solomon, S. The Antarctic ozone hole: An update. *Phys. Today* **2014**, *67*, 42–48. [CrossRef]

36. Krotkov, N.A.; Li, C.; Leonard, P. *OMI/Aura Sulphur Dioxide (SO2) Total Column Daily L2 Global Gridded 0.125 Degree x 0.125 Degree V3*; Goddard Earth Sciences Data and Information Services Center (GES DISC): Greenbelt, MD, USA, 2014.

37. Kroktov, N.A.; Carn, S.A.; Krueger, A.J.; Bhartia, P.K.; Yang, K. Band Residual Difference Algorithm for Retrieval of SO_2 From the Aura Ozone Monitoring Instrument (OMI). *IEEE Trans. Geosci. Remote Sens.* **2006**, *44*, 1259–1266. [CrossRef]

38. TOMS Science Team. TOMS Nimbus-7 Total Column Ozone Daily L3 Global 1 deg x 1.25 deg Lat/Lon Grid V008, Greenbelt, MD, Goddard Earth Sciences Data and Information Services Center (GES DISC). Available online: http://disc.sci.gsfc.nasa.gov/datacollection/TOMSN7L3dtoz_008.html (accessed on 16 March 2017).

39. TOMS Science Team. TOMS Nimbus-7 UV Reflectivity Daily L3 Global 1 deg x 1.25 deg Lat/Lon Grid V008, Greenbelt, MD, Goddard Earth Sciences Data and Information Services Center (GES DISC). Available online: http://disc.sci.gsfc.nasa.gov/datacollection/TOMSN7L3dref_008.html (accessed on 16 March 2017).

40. Pavón-Carrasco, F.J.; De Santis, A. The South Atlantic Anomaly: The Key for a Possible Geomagnetic Reversal. *Front. Earth Sci.* **2016**. [CrossRef]

41. Dobber, M.R.; Dirksen, R.J.; Levelt, P.F.; van den Oord, G.H.J.; Voors, R.H.M.; Kleipool, Q.; Jaross, G.; Kowalewski, M.; Hilsenrath, E.; Leppelmeier, G.W.; et al. Rozemeijer. Ozone Monitoring Instrument Calibration. *IEEE Trans. Geosci. Remote Sens.* **2006**, *44*, 1209–1238. [CrossRef]

42. Ropelewski, C.F.; Jones, P.D. An Extension of the Tahiti-Darwin Southern Oscillation Index. *Mon. Weather Rev.* **1987**, *115*, 2161–2165. [CrossRef]

43. Gutman, G.; Ignatov, A. Global land monitoring from AVHRR: Potential and limitations. *Int. J. Remote Sens.* **1995**, *16*, 2301–2309. [CrossRef]

44. Grumm, R.H.; Hart, R. Standardized Anomalies Applied to Significant Cold Season Weather Events: Preliminary Findings. *Weather Forecast.* **2001**, *16*, 736–754. [CrossRef]

45. Hartmann, D.L.; Klein Tank, A.M.G.; Rusticucci, M.; Alexander, L.V.; Brönnimann, S.; Charabi, Y.; Dentener, F.J.; Dlugokencky, E.J.; Easterling, D.R.; Kaplan, A.; et al. Observations: Atmosphere and Surface. In *Climate Change 2013: The Physical Science Basis. Contribution of Working Group I to the Fifth Assessment Report of the Intergovernmental Panel on Climate Change*; Stocker, T.F., Qin, D., Plattner, G.-K., Tignor, M., Allen, S.K., Boschung, J., Nauels, A., Xia, Y., Bex, V., Midgley, P.M., Eds.; Cambridge University Press: New York, NY, USA, 2013; pp. 159–254.

46. Salmi, T.; Määttä, A.; Anttila, P.; Ruoho-Airola, T.; Amnell, T. *Detecting Trends of Annual Values of Atmospheric Pollutants by the Mann-Kendall Test and Sen's Slope Estimates-The Excel Template Application Makesens*; Publications on air quality, volume 31; Finish Meteorological Institute: Helsinki, Finland, 2002.

47. Thousands Flee as Volcano Erupts in Congo. Daily Mail, 18 January 2002. Available online: http://www.dailymail.co.uk/news/article-95518/Thousands-flee-volcano-erupts-Congo.html (accessed on 15 June 2017).

48. Petersen, K. One of Africa's Most Active Volcanoes Is Showing New Signs of Life. Public Radio International, 2 May 2016. Available online: https://www.pri.org/stories/2016--05--02/one-africas-most-active-volcanos-showing-new-signs-life (accessed on 16 June 2017).

49. Oppenheimer, C. Satellite Observations of Lava Lake Activity at Nyiragongo Volcano, Ex-Zaire, during the Rwandan Refugee Crisis. *Disasters* **1998**, *22*, 268–281. [CrossRef] [PubMed]

50. Carn, S.A.; Fioletov, V.E.; McLinden, C.A.; Li, C.; Krotkov, N.A. A decade of global volcanic SO_2 emissions measured from space. *Sci. Rep.* **2017**, *7*, 44095. [CrossRef]

51. Core Writing Team. *Climate Change 2014: Synthesis Report. Contribution of Working Groups I, II and III to the Fifth Assessment Report of the Intergovernmental Panel on Climate Change*; Pachauri, R.K., Meyer, L.A., Eds.; IPCC: Geneva, Switzerland, 2014.

52. Rutllant, J.; Garreaud, R. Meteorological air pollution potential for Santiago, Chile: Towards an objective episode forecasting. *Environ. Monit. Assess.* **1995**, *34*, 223–244. [CrossRef] [PubMed]

53. Romero, H.; Ihl, M.; Rivera, A.; Zalazar, P.; Azocar, P. Rapid urban growth, land-use changes and air pollution in Santiago, Chile. *Atmos. Environ.* **1999**, *33*, 4039–4047. [CrossRef]

54. Kirchhoff, V.W.J.H.; Setzer, A.W.; Pereira, M.C. Biomass burning in Amazonia: Seasonal effects on atmospheric O_3 and CO. *Geophys. Res. Lett.* **1989**, *16*, 469–472. [CrossRef]

55. Duncan, B.N.; Martin, R.V.; Staudt, A.C.; Yevich, R.; Logan, J.A. Interannual and seasonal variability of biomass burning emissions constrained by satellite observations. *J. Geophys. Res.* **2003**, *108*, 4100. [CrossRef]

56. Xiao, Z.; Jiang, H.; Song, X.; Zhang, X. Monitoring of atmospheric nitrogen dioxide using Ozone Monitoring Instrument remote sensing data. *J. Appl. Remote Sens.* **2013**, *7*. [CrossRef]

57. Van der A, R.J.; Eskes, H.J.; Boersma, K.F.; van Noije, T.P.C.; Van Roozendael, M.; De Smedt, I.; Peters, D.H.M.U.; Meijer, E.W. Trends, seasonal variability and dominant NO_x source derived from a ten year record of NO_2 measured from space. *J. Geophys. Res.* **2008**, *113*, D04302. [CrossRef]

58. Civil Aviation Authority of New Zealand. Ash Cloud from Chilean Volcano Entering New Zealand Airspace. 11 June 2011. Available online: https://www.caa.govt.nz/publicinfo/med-rel-ash-cloud/ (accessed on 24 July 2017).

59. More Delays Ahead as Ash Cloud Circles. ABC News, 13 June 2011. Available online: http://www.abc.net.au/news/2011-06-14/more-delays-ahead-as-ash-cloud-circles/2757060 (accessed on 24 July 2017).

60. Calkins, C.; Ge, C.; Wang, J.; Anderson, M.; Yang, K. Effects of meteorological conditions on sulphur dioxide air pollution in the North China plain during winters of 2006-2015. *Atmos. Environ.* **2016**, *147*. [CrossRef]

61. Qu, Z.; Henze, D.K.; Capps, S.L.; Wang, Y.; Xu, X.; Wang, J.; Keller, M. Monthly top-down NO_x emissions for China (2005−2012): A hybrid inversion method and trend analysis. *J. Geophys. Res.* **2017**, *122*, 4600–4625. [CrossRef]

62. Nguyen, H.; Evans, A.; Lucas, C.; Smith, I.; Timbal, B. The Hadley Circulation in Reanalyses: Climatology, Variability, and Change. *J. Clim.* **2013**, *26*, 3357–3376. [CrossRef]

63. Vallis, G.K. The Overturning Circulation: Hadley and Ferrel Cells. In *Atmospheric and Oceanic Fluid Dynamics: Fundamentals and Large-Scale Circulation*; Cambridge University Press: Cambridge, UK, 2006; pp. 451–484.

64. Allen, R.J.; Norris, J.R.; Kovilakam, M. Influence of anthropogenic aerosols and the Pacific Decadal Oscillation on tropical belt width. *Nat. Geosci.* **2014**, *7*, 270–274. [CrossRef]

65. Polvani, L.M.; Waugh, D.W.; Correa, G.J.P.; Son, S.-W. Stratospheric Ozone Depletion: The Main Driver of Twentieth-Century Atmospheric Circulation Changes in the Southern Hemisphere. *J. Clim.* **2011**, *24*, 795–812. [CrossRef]

66. Garfinkel, C.I.; Waugh, D.W.; Polvani, L.M. Recent Hadley cell expansion: The role of internal atmospheric variability in reconciling modelled and observed trends. *Geophys. Res. Lett.* **2015**, *42*, 10824–10831. [CrossRef]

67. Hoskins, B.J.; Hodges, K.I. A New Perspective on Southern Hemisphere Storm Tracks. *J. Clim.* **2005**, *20*, 4108–4129. [CrossRef]

68. Archer, C.L.; Caldeira, K. Historical trends in the jet streams. *Geophys. Res. Lett.* **2008**, *35*, L08803. [CrossRef]

69. Calvert, J.G.; Su, F.; Bottenheim, J.W.; Strausz, O.P. Mechanism of the homogeneous oxidation of sulfur dioxide in the troposphere. *Atmos. Environ.* **1978**, *12*, 197–226. [CrossRef]

70. Clemitshaw, K.C.; Weiss-Penzias, P.S. Tropospheric chemistry and composition: Hydroxyl Radical. In *Encyclopedia of Atmospheric Sciences*; Holton, J.R., Curry, J.A., Pyle, J.A., Eds.; Academic Press: London, UK, 2003; Volume 6, pp. 2403–2411.

71. Hegg, D.A.; Hobbs, P.V. Oxidation of sulphur dioxide in aqueous systems with particular reference to the atmosphere. *Atmos. Environ.* **1978**, *12*, 241–253. [CrossRef]

72. Berresheim, H.; Andreae, M.O.; Ayers, G.P.; Gillett, R.W.; Merrill, J.T.; Davis, V.J.; Chameides, W.L. Airborne measurements of dimethylsulfide, sulfur dioxide, and aerosol ions over the Southern Ocean South of Australia. *J. Atmos. Chem.* **1990**, *10*, 341–370. [CrossRef]

73. Ehhalt, D.H.; Wahner, A. Tropospheric chemistry and composition: Oxidizing Capacity. In *Encyclopedia of Atmospheric Sciences*; Holton, J.R., Curry, J.A., Pyle, J.A., Eds.; Academic Press: London, UK, 2424; Volume 6, pp. 2415–2424.

74. Duncan, B.N.; Lamsal, L.N.; Thompson, A.M.; Yoshida, Y.; Lu, Z.; Streets, D.G.; Hurwitz, M.M.; Pickering, K.E. A space-based, hih-resolution view of notable changes in urban NO_x pollution around the world (2005–2014). *J. Geophys. Res.* **2016**, *121*, 976–996. [CrossRef]
75. Smith, S.J.; van Aardenne, J.; Klimont, Z.; Andres, R.J.; Volke, A.; Delgado Arias, S. Anthropogenic sulfur dioxide emissions: 1850–2005. *Atmos. Chem. Phys.* **2011**, *11*, 1101–1116. [CrossRef]

remote sensing

MDPI

Article

Plume Segmentation from UV Camera Images for SO$_2$ Emission Rate Quantification on Cloud Days

Matías Osorio [1,*], Nicolás Casaballe [1], Gastón Belsterli [1], Miguel Barreto [2], Álvaro Gómez [2], José A. Ferrari [1] and Erna Frins [1]

[1] Instituto de Física, Facultad de Ingeniería, Universidad de la República, J. Herrera y Reissig 565, Montevideo 11200, Uruguay; ncasabal@fing.edu.uy (N.C.); gbelster@fing.edu.uy (G.B.); jferrari@fing.edu.uy (J.A.F.); efrins@fing.edu.uy (E.F.)

[2] Instituto de Ingeniería Eléctrica, Facultad de Ingeniería, Universidad de la República, J. Herrera y Reissig 565, Montevideo 11200, Uruguay; barretomiguel@gmail.com (M.B.); agomez@fing.edu.uy (Á.G.)

* Correspondence: mosorio@fing.edu.uy; Tel.: +598-2711-5444

Academic Editors: Yang Liu, Jun Wang, Omar Torres, Xiaofeng Li and Prasad S. Thenkabail
Received: 27 March 2017; Accepted: 21 May 2017; Published: 24 May 2017

Abstract: We performed measurements of SO$_2$ emissions with a high UV sensitive dual-camera optical system. Generally, in order to retrieve the two-dimensional SO$_2$ emission rates of a source, e.g., the slant column density of a plume emitted by a stack, one needs to acquire four images with UV cameras: two images including the emitting stack at wavelengths with high and negligible absorption features ($\lambda_{on/off}$), and two additional images of the background intensity behind the plume, at the same wavelengths as before. However, the true background intensity behind a plume is impossible to obtain from a remote measurement site at rest, and thus, one needs to find a way to approximate the background intensity. Some authors have presented methods to estimate the background behind the plume from two emission images. However, those works are restricted to dealing with clear sky, or almost homogeneously illuminated days. The purpose of this work is to present a new approach using background images constructed from emission images by an automatic plume segmentation and interpolation procedure, in order to estimate the light intensity behind the plume. We compare the performance of the proposed approach with the four images method, which uses, as background, sky images acquired at a different viewing direction. The first step of the proposed approach involves the segmentation of the SO$_2$ plume from the background. In clear sky days, we found similar results from both methods. However, when the illumination of the sky is non homogeneous, e.g., due to lateral sun illumination or clouds, there are appreciable differences between the results obtained by both methods. We present results obtained in a series of measurements of SO$_2$ emissions performed on a cloudy day from a stack of an oil refinery in Montevideo City, Uruguay. The results obtained with the UV cameras were compared with scanning DOAS measurements, yielding a good agreement.

Keywords: UV cameras; SO$_2$ emissions rates; DOAS; plume segmentation

---

1. Introduction

The development of remote sensing systems based on ultraviolet (UV) cameras for gas detection in the atmosphere has been growing in recent years. Focus has been primarily on the quantification of volcanic sulphuric dioxide (SO$_2$) emissions, and more recently, on emissions from industrial sources. Since the pioneering work of Mori and Burton [1] and Bluth et al. [2], several studies (Burton et al. [3]) and intercomparisons (Kern et al. [4]) have been performed on volcanic emissions and fumarolic fields (Tamburello et al. [5]). The use of UV cameras has the advantage of providing images with high spatial and temporal resolution, providing more information in a single reading than other techniques, such as scanning methods with similar outputs (two dimensional SO_2 map) like Imaging Differential Optical Absorption Spectroscopy (IDOAS) [6]. Some issues concerning UV cameras have been discussed in the literature, for example, light dilution effects (Mori et al. [7], Campion et al. [8]), calibration issues (Lübcke et al. [9]), image corrections (e.g., vignetting and others effects), and the selection of different system parameters (Kantzas et al. [10], Kern et al. [11]). Further, with a similar approach using images acquired at different wavelengths, water fluxes from volcanic plumes can be quantified (Pering et al. [12]).

This technique was more recently employed for monitoring anthropogenic SO_2 emissions to the atmosphere from sources such as power plants, oil refineries or, more broadly, any industrial stack, provided these are strong enough to overcome the detection limit. A low-cost design (Wilkes et al. [13]) of this kind of systems helps to reach that objective. Those emissions have been largely studied with scanning methods based on Differential Optical Absorption Spectroscopy (DOAS) (Platt and Stutz [14]), including stationary ground based measurements with multi axis—DOAS (MAX-DOAS) (Rivera et al. [15], Frins et al. [16]), and mobile measurements (e.g., Rivera et al. [17], Frins et al. [18]). Consistent results are obtained between the emission rates found by the application of those methods and data obtained from in situ devices, validating these techniques.

In comparison with the large amount of work on volcanic plumes, few measurements of anthropogenic sources have been performed with UV cameras. Early studies were carried out by McElhoe and Conner [19], Dalton et al. [20]. More recently, Smekens et al. [21] have remotely measured emissions from a coal-burning power plant using an UV SO_2 camera system, obtaining good agreement with data provided by sensors located within the stacks. Further, a SO_2 camera system to measure ship emissions was used by Prata [22].

Some specific problems may arise when UV cameras are used to quantify emissions of anthropogenic sources. For example, typical industrial emissions have lower concentration of SO_2 than volcanoes, so the detection limit plays an important role. Also, the acquisition of a SO_2-free (clean) reference image for an accurate retrieval can sometimes be complicated, due to the landscape surrounding the stacks, both in industrial environments or volcanoes. To solve this, a great effort to obtain the background images is required, as discussed below. Furthermore, in systems consisting of two cameras, pixels with the same coordinates in both cameras do not necessarily correspond to the same spatial point. Thus, it is necessary to find a correlation between the two images, prior to applying any evaluation method.

A method to obtain background images is by pointing the UV cameras to an emission-free sky region, which is especially troublesome on cloud days. Some authors (see [4,10,21]) have also presented methods to estimate the background behind the plume from two emission images. However, most of those works are restricted to deal with clear sky or almost homogeneously illuminated days. The method proposed by Smekens et al. [21] deserves special attention. They apply a polynomial fit to the sky images that include the SO_2 plume. Then, a transmittance image is calculated by taking the ratio between the original image and the polynomial approximation. Finally, applying an iterative thresholding procedure, they intend to extract (segment) the plume region. On cloudy days, however, the sky image could present spurious structures, similar to the ones appearing in the plume image, so the described procedure does not always allow distinguishing between plume and clouds.

In the present paper, we describe a new (non-iterative) procedure to estimate the intensity background behind the plume, before light has traversed it. Thus, instead of four images (two images of the SO_2-plume and two images of the background), only images of the plume are necessary. The first step of the proposed method involves the segmentation of the SO_2-plume from the emission images, by taking the ratio between the (raw) sky images at two specific wavelengths, λ_{on} and λ_{off}, where the SO_2-absorption cross-section is significantly different (see below). This approach is efficient in locating the plume region still under cloudy conditions. The second step consists of a polynomial interpolation procedure to estimate the light intensity behind the plume.

The purpose of this work is to present our new two-image method (2-IM) that uses background images constructed from the emission images, and to compare its performance with the four-images method (4-IM), which uses as background sky images acquired at a different viewing direction relative to the emission images.

We tested the 4-IM and 2-IM approaches by performing measurements of SO_2-emissions from a stack of an oil refinery placed in Montevideo City, Uruguay. The optical system used consists of two ultraviolet sensitive cameras provided with narrow band UV filters, which can simultaneously acquire two images at different wavelengths. Considerations on image processing and system detection limits are discussed. We also compared the results with MAX-DOAS measurements.

In the next section, the basics of the measurement techniques are presented. The experimental setup and results are discussed in Sections 3 and 4, respectively. Conclusions are presented in Section 5.

2. SO_2-Emission Retrieval with UV-Cameras

2.1. Radiative Transfer Considerations

Based on the Lambert-Beer law, the light reaching a detector after passing through a certain air mass (e.g., a plume containing several trace gases and aerosols) can be expressed as:

$$I(\lambda) = I_0(\lambda) \exp\left\{ -\left[\sum_k \sigma_k(\lambda) \int c_k(l) \, dl \right] - \int \varepsilon_s(\lambda, l) \, dl \right\} \quad (1)$$

where λ is the wavelength, $I(\lambda)$ and $I_0(\lambda)$ are the light intensities after and before traversing the air mass, respectively, $\sigma_k(\lambda)$ is the absorption cross-section of the k-th gas species present in the air mass, $c_k(l)$ its concentration, and l denotes the optical path inside the air mass. ε_s is the scattering extinction coefficient due to aerosols present in the air mass. Equation (1) is valid as long as the aerosol load is low enough to ensure single scattering processes.

When using UV cameras, $I(\lambda)$ and $I_0(\lambda)$ are represented by images filtered by the instrument function of the optical system. These images will be denoted as $I(\lambda,i,j)$ and $I_0(\lambda,i,j)$, respectively, where (i,j) are pixel coordinates. Roughly, we can say that $I(\lambda,i,j)$ is an image (at wavelength λ) of the gas emission, while $I_0(\lambda,i,j)$ is an image of the background intensity before light traverses the plume.

In the following, we will assume that there is only one absorbing trace gas species, specifically SO_2, and we will consider two specific wavelengths: λ_{on} is chosen in the spectral range where the SO_2-absorption cross-section is high, and λ_{off} lies as close as possible to λ_{on}, but the SO_2-absorption cross-section is almost negligible. In this work, λ_{on} is ~310 nm and λ_{off} is ~330 nm (the full detail can be seen in Section 3.2).

From (1), the optical depths at these wavelengths are:

$$\ln\left(\frac{I_0(\lambda_{on}, i, j)}{I(\lambda_{on}, i, j)} \right) = \sigma_{SO2}(\lambda_{on}) \int c_{SO2}(l) \, dl + \int \varepsilon_s(\lambda_{on}, l) \, dl \quad (2)$$

and

$$\ln\left(\frac{I_0(\lambda_{off}, i, j)}{I(\lambda_{off}, i, j)} \right) = \int \varepsilon_s(\lambda_{off}, l) \, dl \quad (3)$$

To describe Mie scattering at $\lambda_{on,off}$, it is assumed (see e.g., [14], Section 4.2):

$$\frac{\varepsilon_s(\lambda_{on}, l)}{\varepsilon_s(\lambda_{off}, l)} = \left(\frac{\lambda_{off}}{\lambda_{on}} \right)^{\alpha} \quad (4)$$

where α is the Angström exponent.

Then, from (2) and (4), the cumulative optical depth due to the SO_2 absorption along the optical path reaching the pixel (i,j) will be:

$$
\begin{aligned}
\tau_{SO2}(i,j) &= \sigma_{SO2}(\lambda_{on}) \int c_{SO2}(l)\, dl \\
&= \ln\left(\frac{I_0(\lambda_{on},i,j)}{I(\lambda_{on},i,j)}\right) - \int \varepsilon_s(\lambda_{on},l)\, dl \\
&= \ln\left(\frac{I_0(\lambda_{on},i,j)}{I(\lambda_{on},i,j)}\right) - \left(\frac{\lambda_{off}}{\lambda_{on}}\right)^{\alpha} \int \varepsilon_s(\lambda_{off},l)\, dl .
\end{aligned}
\tag{5}
$$

Substituting (3) into (5), we obtain an expression for the SO_2 optical depth:

$$
\tau_{SO2}(i,j) = \ln\left(\frac{I_0(\lambda_{on},i,j)}{I(\lambda_{on},i,j)}\right) + \left(\frac{\lambda_{off}}{\lambda_{on}}\right)^{\alpha} \ln\left(\frac{I(\lambda_{off},i,j)}{I_0(\lambda_{off},i,j)}\right)
\tag{6}
$$

In the particular case when one assumes that in the volume of interest (e.g., a plume) no aerosols are present, one has $\varepsilon_s(\lambda_{off},l) = 0$, and then:

$$
I(\lambda_{off},i,j) = I_0(\lambda_{off},i,j)
\tag{7}
$$

And thus, expression (6) reduces to:

$$
\tau_{SO2}(i,j) = \ln\left(\frac{I_0(\lambda_{on},i,j)}{I(\lambda_{on},i,j)}\right)
\tag{8}
$$

In general, however, four images are necessary for retrieving a 2D-map of the cumulative SO_2 optical depth (τ_{SO2}), as shown in Equation (6). Two of them, $I(\lambda_{on/off},i,j)$, are images of the emission of a certain source, e.g., images of the plume of a SO_2-emitting stack, and the other two, $I_0(\lambda_{on/off},i,j)$, are images of the background.

In the example of a plume emitted by a stack, the background images should be images of the light intensity behind the plume, which are impossible to acquire from a remote site in the presence of the plume. The typical way (see e.g., [9,11]) of acquiring background images is by changing the viewing direction to obtain plume-free images of the sky, as schematically illustrated in Figure 1. In practice, this could mean a 90° change in looking direction.

Figure 1 depicts a broad panorama of the region to be considered. The region delimited by the dashed red lines allows obtaining the emission images, while the region delimited by the dashed green lines allows obtaining the emission-free background images. Henceforth, this imaging procedure will be called the four-image method (4-IM).

The procedure described above works well if the illumination in the background viewing direction is approximately equal to that in the plume direction, since the light reaching the cameras depends on the solar zenith and azimuth angle, assuming no clouds are present. Additional posterior corrections could be necessary, according to user criterion, for example, subtracting a constant offset from the image as proposed by Mori and Burton [1].

In order to reduce the number of images needed to retrieve the SO_2 map (τ_{SO2}) from Equation (6), in the following section we propose constructing plume-free background images ($I_0(\lambda_{on/off},i,j)$) through an interpolation from a portion of emission free sky available in the two emission images $I(\lambda_{on/off},i,j)$. Henceforth, this approach will be called the two-image method (2-IM).

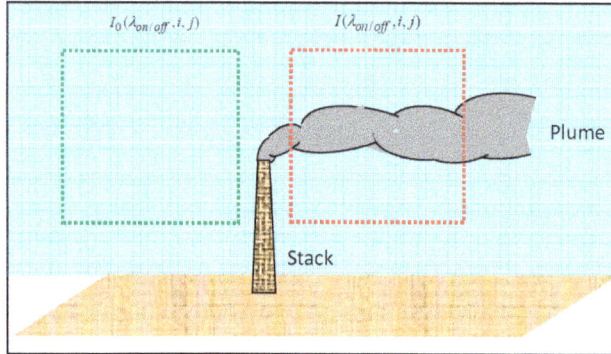

Figure 1. Typical procedure for obtaining the required four images in the four-image method (4-IM).

2.2. 2-IM Approach: Artificial Background Generation

Since the optical system consists of two cameras for acquiring images at different wavelengths, pixels with the same coordinates in both cameras do not necessarily correspond to the same spatial point. Thus, prior to applying any evaluation method (2-IM or 4-IM), the process starts with an image pre-processing for establishing a correspondence between pixels of different cameras. This starts with the subtraction of the dark current images. After that, a binary segmentation is performed in order to keep the same shapes at both wavelengths, i.e., stacks. Then, a 2D cross correlation is made to obtain the correspondence between pixels in the images acquired by both cameras. All these steps are performed in an automatic way.

For the sake of simplicity, to build the artificial background images ($I_0(\lambda_{on/off},i,j)$), we will assume that the plume moves almost horizontally and the light intensity on each side of the plume (immediately above and below) is similar to that directly behind the plume, and that the plume cross section is small in comparison with the whole image. We will model the background as:

$$I_0(\lambda, x, y) = p(\lambda, x, y) \tag{9}$$

where (x,y) is a Cartesian coordinate system (with x and y along the horizontal and vertical direction, respectively). p is a low degree (e.g., third to fifth degree) polynomial matching the sky intensity on each side of the plume and filling the gap in the region inside it, as schematically illustrated in Figure 2 (the procedure described above could be applied when the plume moves in any arbitrary direction.)

The procedure to build artificial background images through polynomial fitting requires processing the emission images $I(\lambda_{on/off},i,j)$. The first step is to know the position of the plume in the images. To do this, we compute the quotient $I(\lambda_{on},i,j)/I(\lambda_{off},i,j)$. The intensity quotient on the plume is smaller than in the surrounding sky, due to the SO_2 absorption at λ_{on}. Thus, this operation does not alter the contrast between plume and background, and eliminates clouds structures because they are present in both wavelengths. After that, a global thresholding is performed, and the result is labelled by connecting neighbour pixels with 8-connectivity, i.e., a pixel z with coordinates (i,j) and its 8 adjacent neighbors that fulfil the condition to have the same image values as z. These two conditions, location and value, determine if the pixel belongs to the "neighbors vicinity with 8-connectivity" or not, defining and labelling each simply connected region (Gonzalez and Woods [23]). Thus, the plume is obtained by selecting the region with the larger number of labels.

Our automatic (and non-iterative) plume segmentation approach could be useful when large numbers of images have to be processed.

Once the plume is segmented from the images $I(\lambda_{on/off},i,j)$, we take vertical profiles at horizontal separations of one pixel. For every vertical profile without plume, a low-order polynomial fit (filling the gap) is applied, as illustrated in the inlet of Figure 2. This generates a sheet that approximates the background at the position of the plume. A more detailed description of the procedure is given in Section 4.

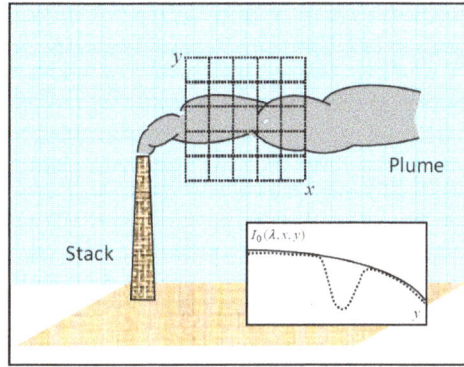

Figure 2. Scheme of the procedure for generating artificial background images in the two-image method (2-IM).

The dashed curve in the inlet of Figure 2 depicts a vertical intensity cut across the plume, while the solid curve illustrates the polynomial fit needed to generate the artificial background images. The fitting procedure allows estimating the light intensity behind the plume.

3. Materials and Methods

3.1. Site Description

We tested our approach by observing SO_2 emissions as a by-product of fossil fuel combustion of an oil refinery located north of the Montevideo Bay (34°52′10″S, 56°13′21″W). The complex has an area of approximately 1 km^2, as shown in Figure 3. From outside the plant, one can observe two stacks from which SO_2-emissions were identified (green marker in Figure 3).

Figure 3. Measurement site (yellow marker), stack (green marker) and oil refinery facility, marked with green outline.

The measurement site (yellow marker in Figure 3) was selected by taking into account the (southern) wind direction. Its location (approximately 850–900 m east from the source) was chosen to be as close as possible to the stack, in order to reduce light dilution effects. The camera system was oriented looking to western direction, so the viewing direction was perpendicular to the plume. Trees and buildings present in the camera's field of view were filtered out before starting the evaluation process. During measurement time, only scattered radiation (not direct sun light) reached the cameras.

3.2. UV Cameras: Image Acquisition and Pre-Processing

We used a pair of Alta U6 cameras developed by Apogee Instruments. The cameras have a CCD (model KAF-1001E, 1024 × 1024 pixels, 24 × 24 μm pixel size), with response in the spectral range of 300–1100 nm, and a quantum efficiency of approximately 10–30% at 310 nm and 330 nm, respectively. Two Coastal Optics telescopic quartz lenses of 105 mm focal length, and two 25 mm diameter interferometric filters from Asahi Spectra, were mounted in front of each camera. The measured central wavelengths were (λ_{on}) 310 nm and (λ_{off}) 331 nm, with Full Width at Half Maximum (FWHM) of 10 nm and 9 nm, respectively.

The filter transmission curves and the SO_2 cross section (Vandaele et al. [24]) are shown in Figure 4. In order to avoid light from lateral directions that could reach the CCD at different wavelengths compared with the central one, each filter was positioned between the lens and the camera detector [11] using custom adapters to fit the filter and the F-mount lens. The focal length of the zoom lenses and their apertures were manually set to obtain the desired image exposure.

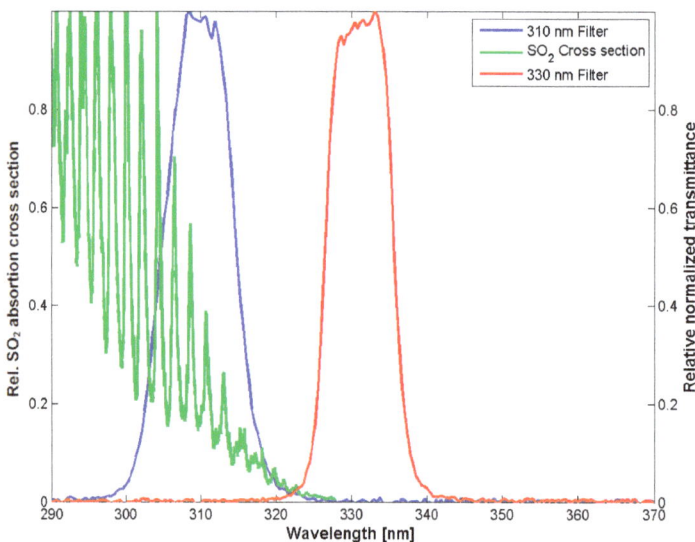

Figure 4. Normalized SO_2 absorption cross-section (green) (from [24]), and normalized filter transmittances obtained in laboratory measurements.

The cameras were fixed side by side to a solid aluminum track, both looking in the same direction. Thus, the acquisition system (two cameras) simultaneously captures the current state of the emission without mechanical movement of the filters. In order to fully control the acquisition process and ensure raw images, a software in C++ language was written to allow communication between the cameras and PC, setting and controlling the internal temperature of the system and acquiring the images.

Since the acquisition system used in the present work consists of two cameras, we developed an algorithm to establish the correspondence between the pixels of one image and the other. The first step is a binary segmentation of the images acquired by the two cameras. In this way, we obtained new images with the stack position without the plume. Then, we proceed to make a 2D cross correlation in order to find the pixels shifts, and construct a transformation matrix with the information to match the pixels of both cameras. Objects with approximately the same pixel coordinates at different distances from the observation site may have a pixels mismatch. If this distance is of the order of kilometers, the pixel mismatches could be low and it would not need the matching procedure at all.

4. Results and Discussion

On 26 March 2015, between 10:46 and 12:04 local time, we acquired several hundred pairs of images of the plume emitted from a stack of the oil refinery with the UV cameras, as described in Section 3.1. We adapted the exposure times in order to acquire images under different illumination levels due to changes in the solar zenith angle (SZA). The exposure times were selected in the range of 400 to 600 ms, and 200 to 450 ms, for 310 and 330 nm, respectively. This was done in order to obtain a good SO_2 signal, which corresponds to 60–80% of the intensity saturation level. Longer exposure times are required for the camera with the 310 nm band pass filter, due to the low intensity of the solar radiation and low quantum efficiency of the CCD at this spectral region. The internal temperature of the cameras was set to 10 °C. After the emission observations, the entrance optics was blocked in order to acquire dark current images and correct them before the evaluation process. Figure 5 shows an example of a pair of raw images $I(\lambda_{310},i,j)$ and $I(\lambda_{330},i,j)$.

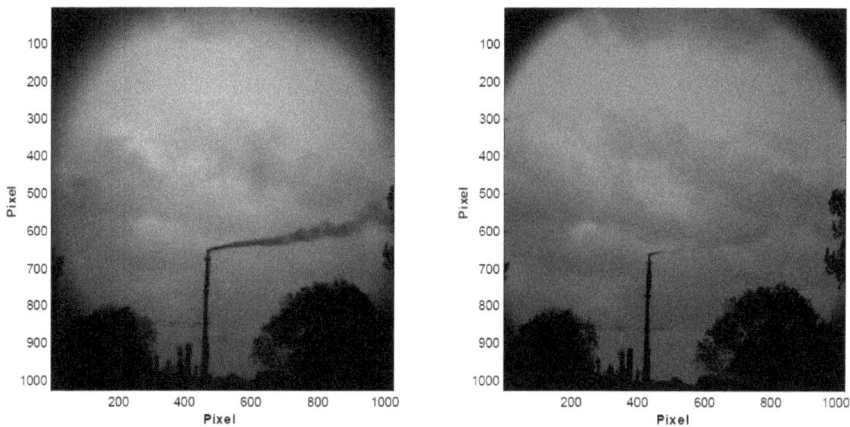

Figure 5. Example of raw images $I(\lambda_{on/off},i,j)$ at λ_{on} = 310 nm (left image) and at λ_{off} = 330 nm (right image) of a plume observed on 26 March 2015. The dark overlaps in the corners of the images are caused by the filter position in the optical systems. The cameras used are designed for 50 mm diameter lens, so we adapted the system to place the 25 mm diameter filters, causing a small reduction in the effective area of the CCD.

Then, we established correspondence between the pixels of the raw images $I(\lambda_{on/off},i,j)$, as described in Section 3.2, and a spatial low-pass filtering was performed to remove some small artifacts from the images. Afterwards, we performed the quotient $I(\lambda_{on},i,j)/I(\lambda_{off},i,j)$ and selected a working region enclosing plume and sky to both sides of the plume, disregarding trees and image borders (Figure 6a). By thresholding and labelling (Gonzalez and Woods [23]), we located the plume spatial distribution (Figure 6b). Due to the low-pass filtering and the posterior thresholding operation, the region occupied by the plume looks a little broadened, but this involves only a few pixels which do not affect the final results (emission rate calculation).

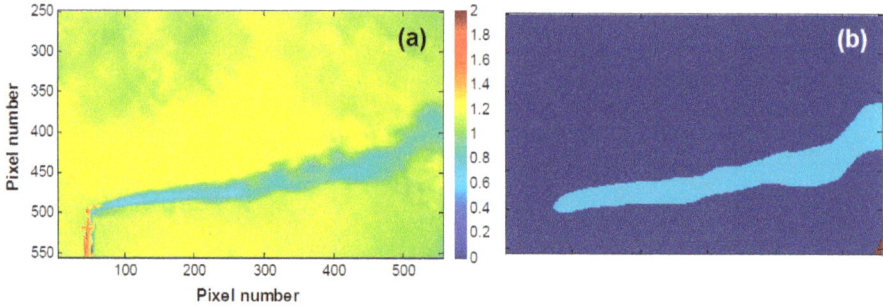

Figure 6. (a) Quotient $I(\lambda_{on},i,j)/I(\lambda_{off},i,j)$ expressed in arbitrary units; **(b)** Image obtained after the process of thresholding and labelling. The region marked with light blue correspond to the plume, obtained by selecting the region with the major number of labels.

As a next step, the region occupied by the plume was removed (segmented) from the images $I(\lambda_{on/off},i,j)$ (see Figure 7a,b). Then, we took vertical profiles at horizontal separations of one pixel, For every vertical profile without plume, a fifth-order polynomial fit (filling the gap) was applied that generated a sheet that approximates the background $I_0(\lambda_{on/off},i,j)$, as mentioned in Section 2.2 (see inlet in Figure 2). This procedure is shown in Figure 8.

Figure 9a,b shows the generated artificial background images $I_0(\lambda_{310},i,j)$ and $I_0(\lambda_{330},i,j)$ resulting from the procedure described above. It is important to mention that the artificial background does not need to be recalculated each time. We want to retrieve an emission map, as long as the illumination conditions (or clouds) do not change. If this is not the case, i.e., clouds behind the plume move rapidly, in order to minimize the error in the retrieval process, the artificial background should be generated for each set of images.

On the other hand, Figure 10a,b shows background images acquired by changing the viewing direction (~70° to the left of the stack), as required by the 4-IM method. Clearly, the background images acquired by varying the viewing direction do not match with the artificial background generated by the 2-IM procedure.

Figure 11 shows the subtraction between the background acquired changing the viewing direction and the background derived from the fitting process for each wavelength. Clearly, it can be seen different structures due to the clouds.

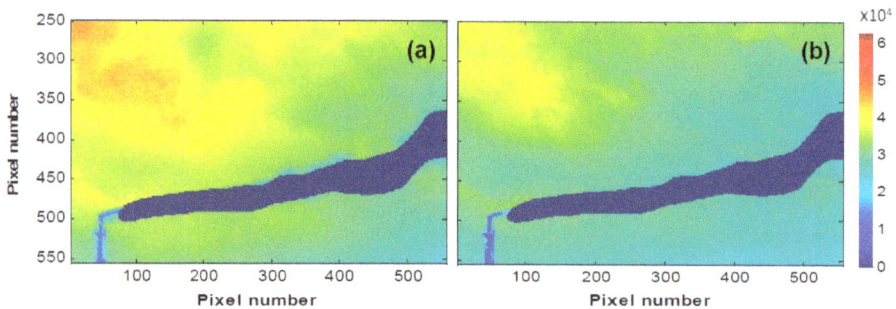

Figure 7. Images $I(\lambda_{on/off},i,j)$ with the plume removed. **(a)** $\lambda_{on} = 310$ nm; **(b)** $\lambda_{off} = 330$ nm.

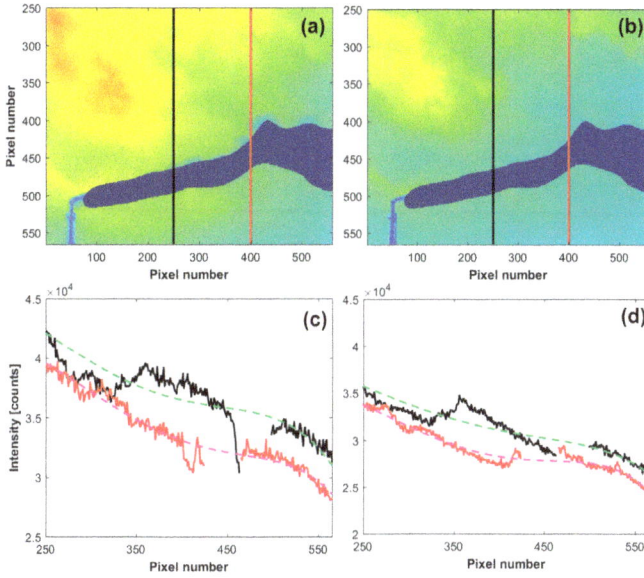

Figure 8. Segmented plume and vertical profiles resulted for images acquired at: (**a**) 310 nm; (**b**) 330 nm. In (**c**,**d**), the marked vertical profiles without the plume are shown, together with the results of the applied polynomial fit.

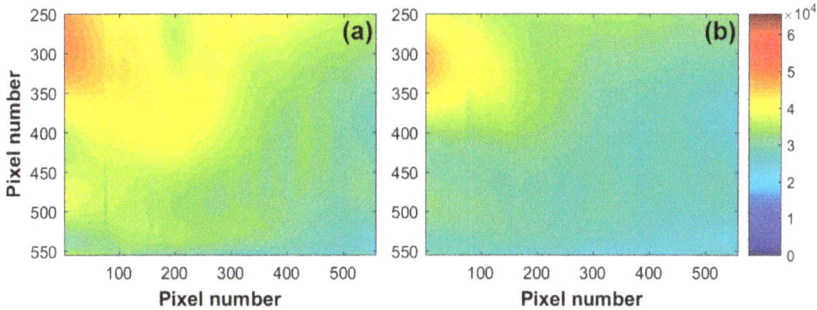

Figure 9. Artificial background sky. (**a**) $I_0(\lambda_{310}, i, j)$; (**b**) $I_0(\lambda_{330}, i, j)$.

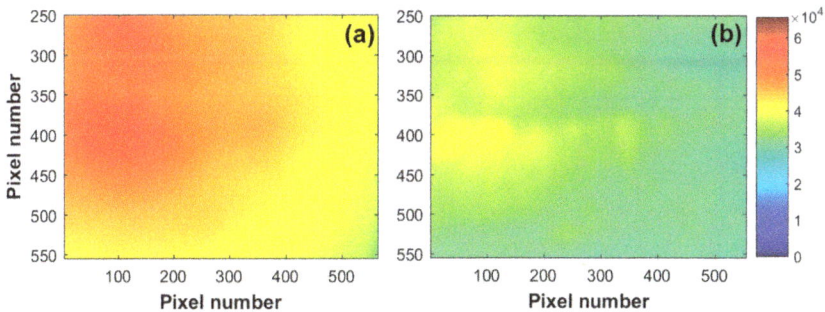

Figure 10. Background images acquired on 26 March 2015 by changing the viewing direction (4-IM method) to the left of the stack, as illustrated in Figure 1, acquired at: (**a**) 310 nm; (**b**) 330 nm.

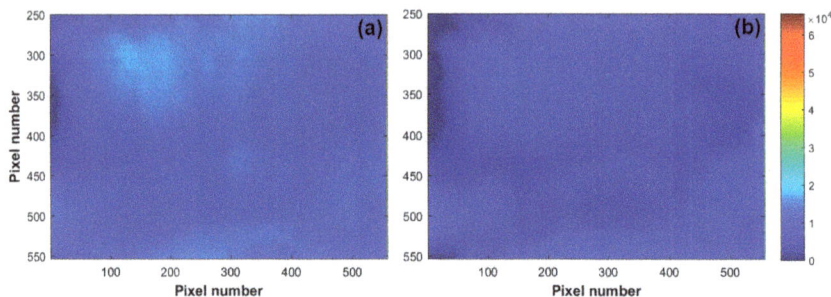

Figure 11. Result of the subtraction of background obtained by the 4-IM and 2-IM procedures: (**a**) 310 nm; (**b**) 330 nm.

In order to retrieve a two-dimensional map of optical depth $\tau_{SO2}(i,j)$ (alternatively, the differential slant column density, $dSCD$, derived after a calibration procedure), in a first approximation, we will set $(\lambda_{off}/\lambda_{on})^{\alpha} \approx 1$ in Equation (6).

Illustrative examples of the two-dimensional emission maps obtained from Equation (6) by using the 2-IM approach are presented in Figure 12a in false colour—SO$_2$-free sky (i.e., zero SO$_2$-concentration) is depicted in dark blue and high SO$_2$-concentration is depicted in red as shown in the scale on the right. The plots in Figure 12b show vertical intensity cuts (optical depth $\tau_{SO2}(i,j)$) across the plume at different horizontal positions.

Figure 12c shows the result obtained by the 4-IM method starting from the same raw images as before (Figure 5), but now by using the background images shown in Figure 10a,b. Figure 13 shows other examples of two-dimensional maps of SO$_2$ optical depths obtained from Equation (6) using the 2-IM and 4-IM approaches.

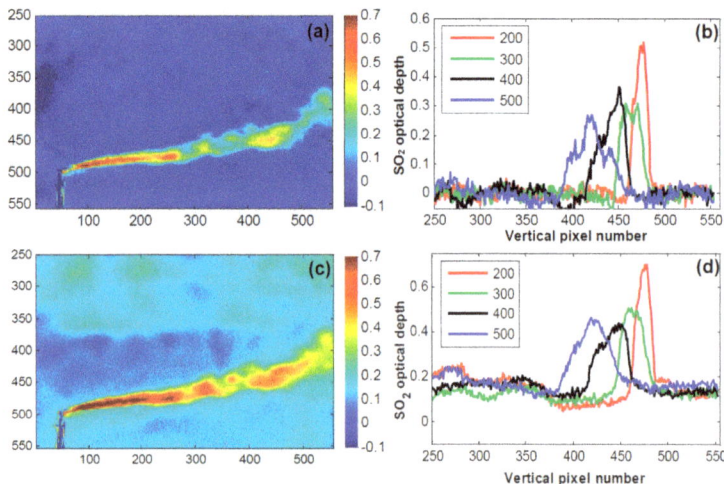

Figure 12. (**a**) Result of SO$_2$ differential optical depth retrieval by the 2-IM method: two-dimensional map of SO$_2$ optical depth obtained from Equation (6) using the artificial background shown in Figure 8a,b. (**b**) Vertical cuts of the plume at different horizontal positions (e.g., the red line indicates a vertical cut at the position of the horizontal pixel number 200). (**c**) Result of SO$_2$ differential optical depth retrieval by the 4-IM method: two-dimensional map of SO$_2$ optical depth obtained from Equation (6) using the background shown in Figure 9a,b. (**d**) Vertical cuts of the plume presented in (**c**) at different horizontal positions.

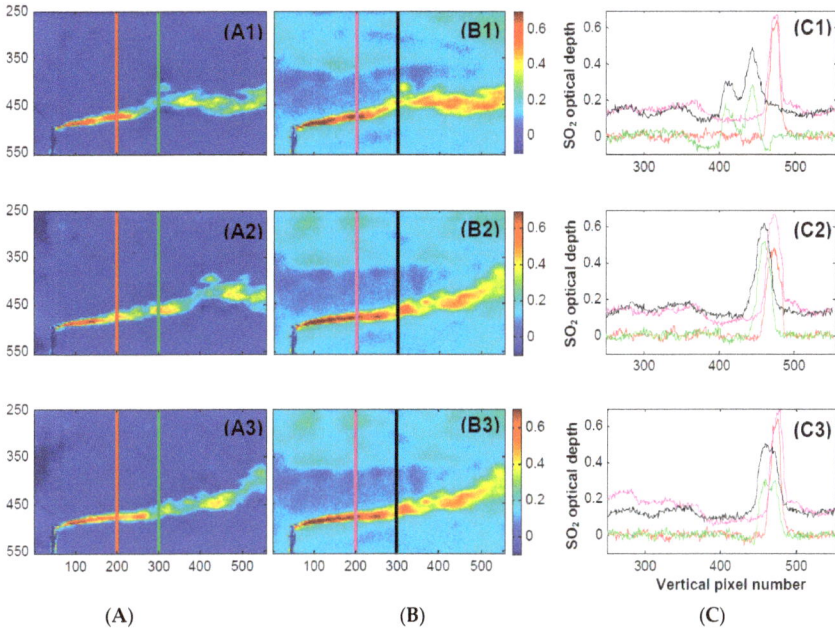

Figure 13. Examples of two-dimensional maps of SO_2 optical depths obtained from Equation (6) for a cloudy day (26 March 2015). (**A**) results obtained by applying the 2-IM (red and green profiles); (**B**) results obtained by applying the 4-IM (magenta and black cuts). (**C**) vertical cuts at different horizontal positions of the plumes shown in (**A**) and (**B**) plotted in the same color as the cuts.

The comparison of Figures 12 and 13A with Figure 13B clearly show that the retrieved SO_2 optical depths obtained from both methods are quite different when clouds are present. For example, the artificial sky derived from applying the 4-IM procedure appears inhomogeneous and shows a large amount of spurious SO_2 optical depth outside the emitting plume. Meanwhile, the artificial sky obtained using the 2-IM procedure appears homogeneous and shows an optical depth near to zero as expected. Also, the peak values of the SO_2 optical depth at the plume are overestimated by the 4-IM method in a magnitude of 25–30%, with respect to the 2-IM ones.

4.1. Calibration and Detection Limit

In order to calculate the actual value of the SO_2-emission rate of the stack, the two-camera system requires calibration. For the purposes of the present work, however, calibration plays a secondary role, since the relative performance comparison between 2-IM and 4-IM approaches can be done in terms of SO_2-optical depth.

In order to estimate the SO_2-density [ppm.m] from the measured optical depth ($\tau_{SO2}(i,j)$), we performed calibration measurements using SO_2-calibration cells containing 94, 480, 985 and 1740 ppm.m. After a linear fitting procedure, we estimate that a SO_2-density of the order of 800 ppm.m corresponds to a value of $\tau_{SO2}(i,j) \approx 0.3$, which results from the linear calibration curve $\tau_{SO2}(i,j) = (3.7299 \times 10^{-4})\ \tau_{SO2}(i,j)$ [ppm.m]. The calibration was performed on a clear day (with a cloud-free sky).

The calibration uncertainty can be estimated by performing a characterization during different days and sky conditions (e.g., SZA and presence of clouds). We found that under different conditions, around midday, the calibration changes at most 20%.

The resulting optical depth (applying the 2-IM method) at image areas outside the plume will be considered as noise, for example, the rapid fluctuations outside the plume region in the plots in Figures 12 and 13C. We estimated the detection limit of our measurements through the standard deviation of the noise, as discussed in [4]. For the measurements performed on 26 March, the estimated detection limit was 47.5 ppm.m. This confirms that our system works well in conditions of relatively low emissions of industrial stacks.

4.2. Emission Rate Calculation

Assuming that the trace gases of the plume are transported at wind speed, the SO_2-flux can be calculated through the expression (Frins et al. [25]):

$$\Phi_{SO_2} = \hat{n} \cdot \vec{v} \, R \sum_i S_{SO_2}(\alpha_i)\Delta\alpha_i \tag{10}$$

where $\vec{v}$ is wind speed, R the distance to the plume in the viewing direction, $\hat{n}$ is the normal to the plume cross-section, and S_{SO2} is the SO_2 differential slant column density at the elevation angle α_i (in the interval of width $\Delta\alpha_i$).

Error sources in the emission rate calculation are due to: (1) uncertainties in the wind speed; (2) uncertainties in the determination of differential slant column densities, and (3) uncertainties in the determination of distance to the plume.

The wind direction was estimated by visual inspection of the plume, while the wind speed was estimated, on average, at 24.6 km·h^{-1}, derived from the movement of plume features in successively captured images. This, and other methods to determine the plume speed, has been used in similar systems [21], and by ground based measurements with a single [26] or multiple spectrometers [27]. It should be mentioned that optical flow methods are accurate speed estimation tools, especially for turbulent plumes [28]. We estimate an uncertainty of approximately 10% due to the wind speed. The error in the slant column density is partially due to the calibration process, and its estimated value is in the order of 4%. Potential errors derived from radiative transfer effects like light dilution were ignored, since we are dealing with distances to the plume in the order of 850–900 m (Kern et al. [29]).

Finally, the distance between plume and observation site was estimated with the help of Google Earth, and its uncertainty was approximately 10% (estimated considering small changes by the wind direction). Thus, taking into account all the aforementioned uncertainties, we found an estimated error in the order of 20% in the emission rate. Certainly, the total error estimation will depend on the method utilized for generating the background, but it is difficult to quantify a priori its order of magnitude.

It is important to mention that for the calculation of emission rates, the crude SCDs obtained by the 4-IM procedure could not be utilized without—at least—a previous convenient compensation of the non-uniform intensity distribution in the picture. Thus, the results of the estimations of emission rates by the 4-IM procedure shown in Figure 12c,d and Figure 13B require some caution.

This problem practically does not occur in the 2-IM procedure, because the artificial background images are extracted from the proper emission images by applying a polynomial fit. Therefore, when some cloudiness is present in the sky, the construction of the background images utilizing the 2-IM method could produce a better retrieval of emissions, than by the 4-IM-method of pointing the camera in a different direction.

For the 2-IM procedure, emission rates between 162 ± 36 kg·h^{-1} and 626 ± 140 kg·h^{-1} were obtained. On average, an emission rate of 329 ± 74 kg·h^{-1} was obtained. Figure 14 shows the results for the emission rates quantified by both methods. It is clear that the flux derived by 4-IM method is overestimated (the emission rate for this method was computed after a manually correction of the offset that is shown in Figure 13C), showing a dynamic behavior along the measurement time. This is due to the changing of background conditions due to the presence of clouds. On the other hand, the 2-IM method shows an approximately constant behavior, coherent with the assumption of a constant work regime of the facility.

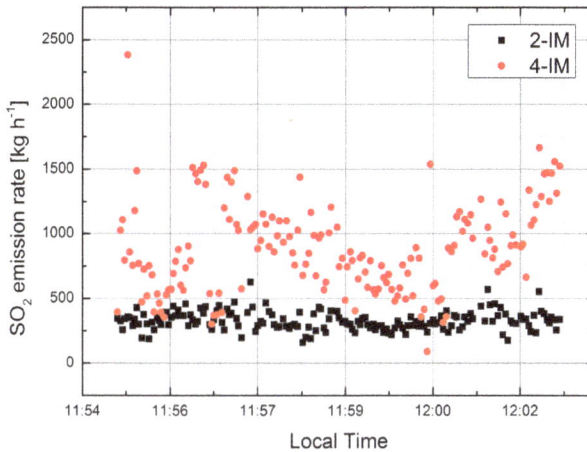

Figure 14. Comparison of the SO$_2$ emission rates obtained by the 2-IM (black squares) and 4-IM (red dots) methods.

4.3. Comparison with MAX-DOAS Measurements

In order to validate the SO$_2$ fluxes obtained using the 2-IM approach, we simultaneously performed measurements with a MAX-DOAS instrument (Hoffmann Messtechnik GmbH) from the same location where the cameras were placed. The instrument includes an Ocean Optics USB2000+ spectrometer, with a spectral range between 317 and 465 nm. The resolution varies in the range of 0.45–0.65 nm, and has a full field of view of approximately 0.8 degrees. The internal temperature of the instrument was set to 10 °C before starting the measurements, and we acquired a dark current and offset spectrum for correction at the end of the day.

To perform these measurements, the MAX-DOAS instrument was set to acquire spectra with an acquisition time of 30 s in a loop routine of elevation angles from 0° to 20° in 1° steps, 45°, and finally a zenith measurement, which was later used as reference in the DOAS evaluation. The whole plume cross section was completely scanned three times. During the measurements, the wind speed was 24.2 km·h^{-1} (derived from the images) and the sky was covered with clouds.

For the calculation of slant column densities, the WINDOAS software (Fayt and van Roozendael [30]) was used. For the SO$_2$ analysis, the SO$_2$-cross section at 294 K (Vandaele et al. [24]), O$_3$ at 243 K (Bogumil et al. [31]), HCHO (Meller et al. [32]), NO$_2$ (Vandaele et al. [33]), a synthetic Ring spectrum calculated using DOASIS software (Kraus [34]), and a fourth degree polynomial were included. The analysis window was set between 317 nm and 335 nm. Once the SO$_2$ slant column densities at different elevation angles crossing the plume were obtained, we proceeded to obtain the corresponding SO$_2$ emission rates.

With the MAX-DOAS instrument, were measured SO$_2$-emission rates between 265 ± 59 kg·h^{-1} and 382 ± 85 kg·h^{-1}. On average, we obtained an emission rate of the order of 315 ± 41 kg·h^{-1}, which is consistent, within the experimental error, with the SO$_2$-emission rate of 352 ± 77 kg·h^{-1} obtained by the 2-IM procedure.

5. Conclusions

In this work, we presented a new approach that uses background images constructed from emission images by an automatic segmentation and interpolation procedure to estimate the light intensity behind plumes. We also compared its performance with the four images method, which uses as background sky images acquired at a different viewing direction. Unlike other methods published in the literature (e.g., [21]), we focus our efforts to present a non-iterative approach that locates the plume region in an acceptable way, trying to minimize user intervention.

We demonstrated that, even when clouds are present, the SO_2 emission rate derived by the 2-IM approach agrees, within the experimental error, with the results obtained by the MAX-DOAS measurements. This is not the case for the 4-IM approach due to spurious non-homogeneities in the images, which resulted from the comparison of different cloud structures. As can be observed in Figure 14, the 2-IM method provides an emission rate that varies in the range of 160–626 kg·h^{-1}, while the 4-IM method primarily varies between 400 and 1500 kg·h^{-1}, in the same period of observation. The results obtained by 2-IM, agrees with the MAX-DOAS observation, which was on average in the order of 352 kg·h^{-1}.

It is important to note that, for evaluating emission rates by using the 2-IM procedure, two images less are needed than by the 4-IM method, which requires moving the cameras from their original position pointing in a new direction. As this direction selection is not always representative of the background, and also depends on the personal discretion of the user, it could lead to different results.

The 2-IM approach has the advantage that the cameras are always at rest pointing in direction of the SO_2-emitting source, which is a necessary condition for acquiring an image sequence (i.e., a video) with the purpose of studying its temporal evolution. On the contrary, in the 4-IM procedure, the need of mechanical movement of the cameras may be a drawback in dynamical situations (e.g., background illumination changes due to the variation of the solar zenith angle, displacements of clouds behind the plume) in which the emissions and lighting conditions are changing with time.

Under stationary clear sky conditions, appreciable differences in the results obtained by both methods should not be expected. However, the change of viewing direction in the 4-IM approach also changes the relative sun position, which may lead to a kind of spatially variant offset in the emission results that should be corrected.

The 2-IM procedure requires certain additional software tasks, in comparison with the 4-IM. However, the computational cost is not relevant if we perform the image processing by, for example, using a GPU. The main assumptions of the 2-IM approach is that the light intensity on each side of the plume (right and left in case of a vertical plume, and above and below in the case presented here) is similar to that directly behind the plume, and that the plume cross section is small in comparison with the whole image.

Acknowledgments: This work was funded by the Comisión Sectorial de Investigación Científica (CSIC, UdelaR, Uruguay), Programa de Apoyo a las Ciencias Básicas (PEDECIBA) and Agencia Nacional de Investigación e Innovación (ANII). Erna Frins acknowledges funding by the L'Oreal National Award for Women in Science. Also, we would like to thank Tom. D. Pering and two anonymous referees for their comments and improving the manuscript.

Author Contributions: Matías Osorio, Erna Frins and José A. Ferrari conceived the ideas, performed the experiments, and wrote the manuscript. Álvaro Gómez provided image processing expertise knowledge. Nicolás Casaballe and Gastón Belsterli helps in the field measurements and data evaluation. Miguel Barreto helps to design the experiment. All authors contributed to the analysis and discussion of the data.

Conflicts of Interest: The authors declare no conflict of interest.

References

1. Mori, T.; Burton, M. The SO_2 camera: A simple, fast and cheap method for ground-based imaging of SO_2 in volcanic plumes. *Geophys. Res. Lett.* **2006**, *33*. [CrossRef]
2. Bluth, G.J.S.; Shannon, J.M.; Watson, I.M.; Prata, A.J.; Realmuto, V.J. Development of an ultra-violet digital camera for volcanic SO_2 imaging. *J. Volcanol. Geotherm. Res.* **2007**, *161*, 47–56. [CrossRef]
3. Burton, M.; Prata, F.; Platt, U. Volcanological applications of SO_2 cameras. *J. Volcanol. Geoterm. Res.* **2015**, *300*. [CrossRef]
4. Kern, C.; Lübcke, P.; Bobrowski, N.; Campion, R.; Mori, T.; Smekens, J.F.; Stebel, K.; Tamburello, G.; Burton, M.; Platt, U.; et al. Intercomparison of SO_2 camera systems for imaging volcanic gas plumes. *J. Volcanol. Geotherm. Res.* **2015**, *300*, 22–36. [CrossRef]

5. Tamburello, G.; Kantzas, E.P.; McGonigle, A.J.S.; Aiuppa, A.; Giudice, G. UV camera measurements of fumarole field degassing (La Fossa crater, Vulcano Island). *J. Volcanol. Geotherm. Res.* **2010**, *199*. [CrossRef]
6. Bobrowski, B.; Hönninger, G.; Lohberger, F.; Platt, U. IDOAS: A new monitoring technique to study the 2D distribution of volcanic gas emissions. *J. Volcanol. Geotherm. Res.* **2006**, *150*, 329–338. [CrossRef]
7. Mori, T.; Kazahaya, K.; Ohwada, M.; Hirabayashi, J.; Yoshikawa, S. Effect of UV scattering on SO_2 emission rate measurements. *Geophys. Res. Lett.* **2006**, *33*. [CrossRef]
8. Campion, R.; Delgado Granados, H.; Mori, T. Image-based correction of the light dilution effect for SO_2 camera measurements. *J. Volcanol. Geotherm. Res.* **2015**, *300*. [CrossRef]
9. Lübcke, P.; Bobrowski, N.; Illing, S.; Kern, C.; Alvarez Nieves, J.M.; Vogel, L.; Zielcke, J.; Delgado Granados, H.; Platt, U. On the absolute calibration of SO_2 cameras. *Atmos. Meas. Tech.* **2013**, *6*, 677–696. [CrossRef]
10. Kantzas, E.P.; McGonigle, A.J.S.; Tamburello, G.; Aiuppa, A.; Bryant, R.G. Protocols for UV camera volcanic SO_2 measurements. *J. Volcanol. Geotherm. Res.* **2010**, *194*, 55–60. [CrossRef]
11. Kern, C.; Kick, F.; Lübcke, P.; Vogel, L.; Wöhrbach, M.; Platt, U. Theoretical description of functionality, applications, and limitations of SO_2 cameras for the remote sensing of volcanic plumes. *Atmos. Meas. Tech.* **2010**, *3*, 733–749. [CrossRef]
12. Pering, T.D.; McGonigle, A.J.S.; Tamburello, G.; Aiuppa, A.; Bitetto, M.; Rubino, C.; Wilkes, T.C. A novel and inexpensive method for measuring volcanic plume water fluxes at high temporal resolution. *Remote Sens.* **2017**, *9*, 146. [CrossRef]
13. Wilkes, T.C.; Pering, T.D.; McGonigle, A.J.; Tamburello, G.; Willmott, J.R. A Low-Cost smartphone sensor-based UV camera for volcanic SO_2 emission measurements. *Remote Sens.* **2017**, *9*, 27. [CrossRef]
14. Platt, U.; Stutz, J. Differential Optical Absorption Spectroscopy—Principles and Applications. In *Physics of Earth and Space Environments*; Springer: Berlin/Heidelberg, Germany, 2008.
15. Rivera, C.; Mellqvis, J.; Samuelsson, J.; Lefer, B.; Alvarez, S.; Patel, M.R. Quantification of NO_2 and SO_2 emissions from the Houston Ship Channel and Texas City industrial areas during the 2006 Texas Air Quality Study. *J. Geophys. Res.* **2010**, *115*. [CrossRef]
16. Frins, E.; Ibrahim, O.; Casaballe, N.; Osorio, M.; Arismendi, F.; Wagner, T.; Platt, U. Ground based measurements of SO_2 and NO_2 emissions from the oil refinery "la Teja" in Montevideo city. *J. Phys. Conf. Ser.* **2011**, *174*. [CrossRef]
17. Rivera, C.; Sosa, G.; Wöhrnschimmel, H.; de Foy, B.; Johansson, M.; Galle, B. Tula industrial complex (Mexico) emissions of SO_2 and NO_2 during the MCMA 2006 field campaign using a mobile mini-DOAS system. *Atmos. Chem. Phys.* **2009**, *9*, 6351–6361. [CrossRef]
18. Frins, E.; Bobrowski, N.; Osorio, M.; Casaballe, N.; Belsterli, G.; Wagner, T.; Platt, U. Scanning and mobile multi-axis DOAS measurements of SO_2 and NO_2 emissions from an electric power plant in Montevideo, Uruguay. *Atmos. Environ.* **2014**, *98*, 347–356. [CrossRef]
19. McElhoe, H.B.; Conner, W.D. Remote Measurement of Sulfur Dioxide Emissions Using an Ultraviolet Light Sensitive Video System. *J. Air Pollut. Control Assoc.* **1986**, *36*, 42–47. [CrossRef]
20. Dalton, M.P.; Watson, I.M.; Nadeau, P.A.; Werner, C.; Morrow, W.; Shannon, J.M. Assessment of the UV camera sulfur dioxide retrieval for point source plumes. *Atmos. Environ.* **2009**, *188*, 358–366. [CrossRef]
21. Smekens, J.F.; Burton, M.; Clarke, A. Validation of the SO_2 camera for high temporal and spatial resolution monitoring of SO_2 emissions. *J. Volcanol. Geotherm. Res.* **2015**, *300*, 37–47. [CrossRef]
22. Prata, A.J. Measuring SO_2 ship emissions with an ultraviolet imaging camera. *Atmos. Meas. Tech.* **2014**, *7*, 1213–1229. [CrossRef]
23. Gonzalez, R.; Woods, R. *Digital Image Processing*, 3rd ed.; Pearson: Upper Saddle River, NJ, USA, 2007.
24. Vandaele, A.C.; Simon, T.C.; Goilmont, J.M.; Carleer, C.M.; Colin, R. SO_2 absorption cross section measurement in the UV using a Fourier transform spectrometer. *J. Geophys. Res.* **1994**, *99*, 25599–25605. [CrossRef]
25. Frins, E.; Osorio, M.; Casaballe, N.; Belsterli, G.; Wagner, T.; Platt, U. DOAS-measurement of the NO_2 formation rate from NOx emissions into the atmosphere. *Atmos. Meas. Tech.* **2012**, *5*, 1165–1172. [CrossRef]
26. McGonigle, A.J.S.; Inguaggiato, S.; Aiuppa, A.; Hayes, A.R.; Oppenheimer, C. Accurate measurement of volcanic SO_2 flux: Determination of plume transport speed and integrated SO_2 concentration with a single device. *Geochem. Geophys. Geosyst.* **2005**, *6*. [CrossRef]

27. Williams-Jones, G.; Horton, K.A.; Elias, T.; Garbeil, H.; Mouginis-Mark, P.J.; Sutton, A.J.; Harris, A.J.L. Accurately measuring volcanic plume velocity with multiple UV spectrometers. *Bull. Volcanol.* **2006**, *68*, 328–332. [CrossRef]

28. Peters, N.; Hoffmann, A.; Barnie, T.; Herzog, M.; Oppenheimer, C. Use of motion estimation algorithms for improved flux measurements using SO2 cameras. *J. Volcanol. Geotherm. Res.* **2015**, *300*, 58–69. [CrossRef]

29. Kern, C.; Deutschmann, T.; Vogel, L.; Wöhrbach, M.; Wagner, T.; Platt, U. Radiative transfer corrections for accurate spectroscopic measurements of volcanic gas emissions. *Bull. Volcanol.* **2009**, *72*, 233–247. [CrossRef]

30. Fayt, C.; van Roozendael, M. WinDOAS 2.1, Software User Manual, Belgian Institute for Space Aeronomy, Brussels, Belgium. Available online: http://bro.aeronomie.be/WinDOAS-SUM-210b.pdf (accessed on 22 May 2017).

31. Bogumil, K.; Orphal, J.; Homann, T.; Voigt, S.; Spietz, P.; Fleischmann, O.C.; Vogel, A.; Hartmann, M.; Bovensmann, H.; Frerick, J.; et al. Measurements of molecular absorption spectra with the SCIAMACHY pre-flight model: Instrument characterization and reference data for atmospheric remote sensing in the 230–2380 nm region. *J. Photochem. Photobiol. A Chem.* **2003**, *157*, 167–184. [CrossRef]

32. Meller, R.; Moortgat, G.K. Temperature dependence of the absorption cross sections of formaldehyde between 223 and 323 K in the wavelength range 225–375 nm. *J. Geophys. Res.* **2000**, *105*, 7089–7101. [CrossRef]

33. Vandaele, A.C.; Hermans, C.; Simon, P.C.; Carleer, M.; Colins, R.; Fally, S.; Mérienne, M.F.; Jenouvrier, A.; Coquart, B. Measurements of the NO_2 absorption cross-sections from 42000 cm^{-1} to 10000 cm^{-1} (238–1000 nm) at 220 K and 294 K. *J. Quant. Spectrosc. Radiat. Transf.* **1998**, *59*, 171–184. [CrossRef]

34. Kraus, S.G. DOASIS—A Framework Design for DOAS. Ph.D. Thesis, University of Mannheim, Mannheim, Germany, 2006.

![remote sensing logo] *remote sensing*

MDPI

Article

Attributing Accelerated Summertime Warming in the Southeast United States to Recent Reductions in Aerosol Burden: Indications from Vertically-Resolved Observations

Mika G. Tosca [1,2,*,†], James Campbell [3], Michael Garay [1], Simone Lolli [4], Felix C. Seidel [1], Jared Marquis [5] and Olga Kalashnikova [1]

[1] Jet Propulsion Laboratory and California Institute of Technology, Pasadena, CA 91109, USA; michael.j.garay@jpl.nasa.gov (M.G.); felix.c.seidel@gmail.com (F.S.); olga.kalashnikova@jpl.nasa.gov (O.K.)

[2] School of the Art Institute of Chicago (SAIC), Chicago, IL 60603, USA

[3] Naval Research Laboratory, Monterey, CA 93943, USA; james.campbell@nrlmry.navy.mil

[4] NASA-JCET, University of Maryland, Baltimore Country and NASA Goddard Space Flight Center, Greenbelt, MD 20771, USA; slolli74@gmail.com

[5] University of North Dakota, Grand Forks, ND 58202, USA; jared.marquis@und.edu

* Correspondence: mtosca1@artic.edu

† Current address: School of the Art Institute of Chicago (SAIC), Chicago, IL, USA.

Received: 10 April 2017; Accepted: 26 June 2017; Published: 1 July 2017

Abstract: During the twentieth century, the southeast United States cooled, in direct contrast with widespread global and hemispheric warming. While the existing literature is divided on the cause of this so-called "warming hole," anthropogenic aerosols have been hypothesized as playing a primary role in its occurrence. In this study, unique satellite-based observations of aerosol vertical profiles are combined with a one-dimensional radiative transfer model and surface temperature observations to diagnose how major reductions in summertime aerosol burden since 2001 have impacted surface temperatures in the southeast US. We show that a significant improvement in air quality likely contributed to the elimination of the warming hole and acceleration of the positive temperature trend observed in recent years. These reductions coincide with a new EPA rule that was implemented between 2006 and 2010 that revised the fine particulate matter standard downward. Similar to the southeast US in the twentieth century, other regions of the globe may experience masking of long-term warming due to greenhouse gases, especially those with particularly poor air quality.

Keywords: warming hole; air quality; southeast US; global warming; climate change; aerosols

---

1. Introduction

During the latter half of the twentieth century, while the globally-averaged surface temperature increased [1], the southeastern United States (SEUS) experienced cooling [2–4]. Annually-averaged surface temperature observations calculated from three widely recognized datasets (Figure 1a, described in Section 2.1) show a warming trend of $+0.54 \pm 0.30\,°\text{C century}^{-1}$ between 1900 and 2008 for the continental United States, for instance, but a minimal cooling trend of $-0.02 \pm 0.39\,°\text{C century}^{-1}$ in the SEUS (30–35° N; 95–80° W) (Figure 1b). While this so-called "warming hole" is noted in the literature, its origin is still unknown, though several recent studies link it to changes in large-scale convective precipitation [5], low-level circulation [6], decadal swings of the Pacific Decadal Oscillation (PDO) [7], or interannual variations in tropical Pacific sea surface temperatures [8]. However, recent well-supported hypotheses attribute the warming hole instead to high regional aerosol emissions [9–13]. In support of this hypothesis, the widespread cooling is thought to have persisted most strongly between 1970 and 1990 when the surface concentration of particulate matter peaked in the SEUS [11,14]. Indeed, observations of net surface solar radiation over the United States show a corresponding multi-decadal decrease from 1961 through 1990 [15].

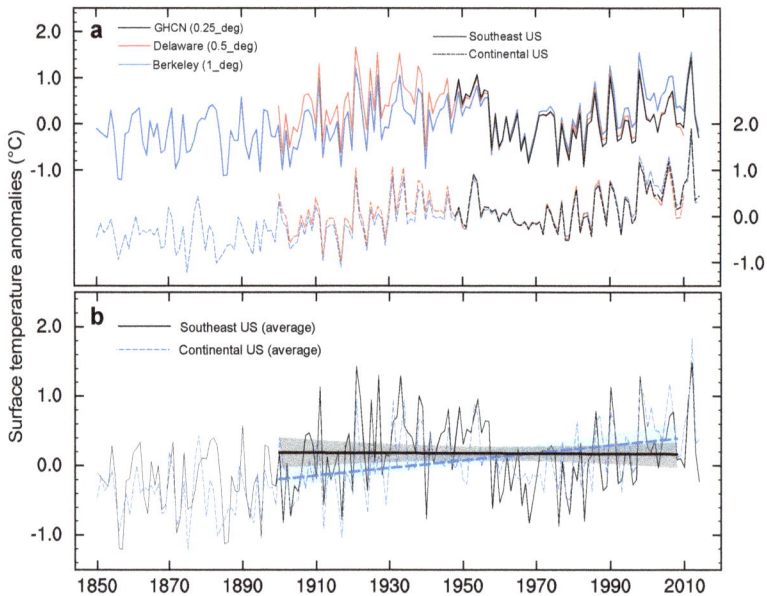

Figure 1. Time series of mean annual temperature anomalies for the southeast US (solid lines; 30–35° N, 95–80° W) and the continental US (dashed lines; 25–50° N, 125–65° W). For panel (**a**), the unique colors represent three independent gridded temperature datasets–red = Delaware, black = GHCN and blue = Berkeley; For panel (**b**), the lines are each respective region's mean temperature anomaly, averaged across all three datasets in (**a**). The trend lines in (**b**) are linear trends of temperature anomalies from 1900–2008, with error estimates, calculated using bootstrapping techniques (n = 1000), shown as the solid shaded regions.

Anthropogenic aerosols can locally cool the climate both directly through their effect on incoming solar radiation [16,17] and indirectly through their influence on clouds [18–20]. Reflective aerosols increase the planetary albedo and cool the surface [16], while their indirect modification of cloud reflectivity and lifetime accomplishes the same effect [21,22]. In general, high aerosol emissions in the twentieth century are thought to have masked up to a third of continental global warming [23]. Though evidence suggests that anthropogenic emissions of reflective sulfate aerosols in the SEUS produce a direct local cooling effect [24,25], other research shows that the effect may be more limited [26], and may induce mesoscale changes to cloud and wind fields that significantly limit and modify the overall surface temperature effect of aerosols [27,28]. Shindell et al. [29] also suggest that while inhomogeneous radiative forcing from aerosols produces a wide range of surface temperature responses in climate models, the strongest sensitivity to this inhomogeneous forcing is in the extratropical northern hemisphere, including the SEUS. Finally, there is evidence that anthropogenic global warming also indirectly increases emissions of biogenic aerosol precursors, the presence of which is likely to further enhance the cooling effect [30]. The sum of these effects suggests that aerosols have indeed likely played an outsized role in modulating global warming in the SEUS.

Recent work by Attwood et al. [31] and Kim et al. [32] indicates that recent decreases (since the 1990s) in aerosol concentration in the SEUS have led to increased solar radiation at the surface.

In seeming contradiction, other studies link increased surface radiation in SEUS to changes in cloud cover and not aerosols [33]. However, a related study suggests that trends in precipitation, and not cloud cover, are responsible for recent trends in surface temperatures [34]. Moreover, another study suggests that neither clouds nor precipitation can fully account for the observed temperature trends in the SEUS, and that aerosols need to be considered [2]. We also note that Zhang et al. [35] report an increase in absorbing aerosol optical depth over the continental US during the 21st century, which occurred in tandem with an overall decrease in total optical depth, suggesting an overall shift in the species composition of aerosol emissions over the past two decades or so. Here, we present evidence, using a unique vertically-resolved dataset of aerosol extinction, that aerosols have contributed to accelerating positive temperature trends in the SEUS since 2001, though they are likely only one piece of the puzzle. We note that all post-2001 aerosol and temperature trends reported in the text are statistically significant.

2. Materials and Methods

2.1. Surface Temperature Trend Calculation

We analyzed regional temperature trends using three reliable, high-resolution ($\leq 1°$) gridded temperature datasets at varying temporal and spatial resolutions: the University of Delaware terrestrial air temperature time series (http://climate.geog.udel.edu/~climate/html_pages/Global2011/), available from 1900–2010 at $0.5° \times 0.5°$ spatial resolution; the Berkeley Earth Land + Ocean dataset (http://berkeleyearth.org/data/) available from 1750-present at $1° \times 1°$ spatial resolution; and the Global Historical Climatology Network (GHCN) CAMS (Climate Anomaly Monitoring System) land air temperature time series, available from 1948-present at $0.5° \times 0.5°$ spatial resolution [36]. The three datasets were chosen because they fit four criteria: high-resolution, gridded, globally-resolved and publicly available. Additionally, the datasets are widely respected [37] and were all constructed completely independent from each other. We calculated both least-squares and Theil–Sen linear trends and estimated the significance using a bootstrapping technique ($n = 1000$).

2.2. CALIOP Aerosol Retrievals for 2006–2014

We used version 3.01, 3.02, and 3.30 NASA Cloud Aerosol Lidar with Orthogonal Polarization (CALIOP) Level 2 5-km 532 nm daytime aerosol extinction coefficient data to create summertime-mean (May–September) vertical profiles at 75 m vertical resolution from data collected for 2006–2014. As described in Section 3.1, we restricted our study to the summertime because that was when the warming hole was strongest and when aerosol optical depth was highest. The data were screened for quality assurance using the rubric described by Campbell et al. [38], which very closely mirrors the method described by Winker et al. [39] that is formally endorsed by the CALIPSO (Cloud-Aerosol Lidar and Infrared Pathfinder Satellite Observations project development team. However, the following screening metrics in Campbell et al. [38] improve the robustness of the retrieval method for the purposes of this investigation. First, no profiles were considered if cloud was observed, so as to minimize the corresponding signal attenuation impacting the retrieval of aerosol extinction coefficient in any way. Second, no profiles were considered if the corresponding retrieval failed to resolve aerosol within 250 m of the surface. Finally, profiles were removed if the retrieval failed to observe any aerosol (i.e., aerosol optical depth; AOD = 0). This choice was based on personal communication with Travis Toth (University of North Dakota), and will discussed in a forthcoming publication (which is currently still in preparation). These restrictions as a whole, however, limit the effects of signal attenuation that we believe could compromise the averaged solutions off extinction coefficient profile derived for this study. Approximately 2500 qualifying profiles were averaged each year to construct these profiles. Averages were solved using a Gaussian weighting function to increase the significance of retrievals relative to the center of the $5° \times 15°$ study domain. We further constructed domain-average CALIOP 532 nm aerosol extinction profiles at 75 m vertical resolution corresponding with each of the native

Level 2 species resolved for use in a radiative transfer model experiment described below. Campbell et al. [38], among others, characterize CALIOP AOD and its relative accuracy.

We calculated the total aerosol optical depth (AOD) from the CALIOP data by integrating the extinction coefficient by height. We used atmospheric boundary layer (ABL) data from the ERA (ECMWF Reanalysis) interim [40] to estimate the mean monthly mid-day (1:00 p.m. local time) ABL. We calculated above- (and below) ABL AOD by integrating the extinction coefficient for only that portion of atmosphere that was above (or below) the ERA-estimated noontime ABL height for each month. We also used satellite AOD measurements from the Multi-angle Imaging SpectroRadiometer (MISR) instrument [41] (Version 22, Level 3) for comparison and validation.

Though we note that each seasonally-averaged vertical profile contains approximately 2500 unique CALIOP profiles, we acknowledge that the lower spatial (and temporal) coverage may affect the overall robustness of our analyses, though we anticipate that the overall conclusion to this work is largely unaffected by these limitations.

2.3. Fu–Liou–Gu Radiative Transfer Model

The aerosol direct effect on radiative forcing is investigated using the Fu–Liou–Gu (FLG) model [42–45]. The vertical profiles used to force the model were constructed from the CALIOP data described above and were taken as a mean extinction at each 75 m layer (the mean of approximately 2500 qualifying profiles per year). An individual profile was created for each aerosol species defined by CALIOP and then matched with the corresponding species in the FLG model. To calculate the net radiative forcing and heating rates for each single species extinction profile, the partial contribution to the total AOD and the value of the species optical depth at each altitude level is required as input for the model

FLG parameterization uses eighteen different types of aerosols, with single scattering aerosol properties parameterized through the Optical Properties of Aerosol and Clouds (OPAC) catalog [46,47]. However, for our study, four main aerosol types (as defined by CALIOP) contributed more than 95% to the total AOD; these were dust ("transported dust" in FLG), polluted continental ("urban" in FLG), polluted dust ("half urban, half dust" in FLG), and smoke ("black carbon" in FLG). We note the concerns raised by Burton et al. [48] in resolving polluted dust in the Version 3 Level 2 CALIOP algorithms, and concede some measurable offset is introduced here when applying the speciated terms in the manner described. Since we are comparing profiles derived from separate years, however, and our concern focuses on the relative differencing of forcing calculations between the two, our belief is that the uncertainty is self-contained overall and the analysis is ultimately reasonable.

According to Gu et al. [46], the FLG also takes into account the effect of water vapor (available from the standard atmosphere atmospheric profile) on aerosols. We performed two simulations: "pristine", where the aerosols were excluded (totally clear conditions) and "total sky", where the individual aerosol species were included. We combined these results into two experiments: FULL, which is simply the "total sky" simulation for each year, and AERO, which is the "total sky" simulation minus the "pristine" simulation for each year. For both experiments, the US1976 standard summer mid-latitude thermodynamics profile was used. Use of the standard atmosphere implies that the water vapor profile was invariant year-to-year. We acknowledge that this is a limitation as water vapor does absorb in the near infrared and that this could impact our direct radiative forcing measurements. However, while we recognize this could cause a small error in our calculations, the data suggest that it does not approach the magnitude of biases derived and discussed or affects the nature of our conclusions.

3. Results and Discussion

3.1. Summertime Warmth Linked to Improved Air Quality

The twentieth century warming hole was especially notable during the summer (Figure 2a) when anthropogenic aerosol concentrations in the SEUS are highest [49]. Summertime temperatures

(May–September) in the SEUS decreased by a negligible -0.01 ± 0.35 °C century^{-1} between 1900 and 2008, compared with a substantial $+0.99 \pm 0.35$ °C century^{-1} increase in Western US (35–48° N; 125–110° W) temperatures during the same time period (Figure 2c). Time series analysis using LOESS (locally weighted scatterplot smoothing) curve smoothing shows that much of the 20th century decrease in SEUS summertime surface temperature occurred before 1975, plateauing between 1975 and 1990, and with a slight rebound beginning afterward; indeed, the 1900–1975 trend in SEUS surface temperature was a robust -0.27 ± 0.25 °C century^{-1} (Figure 3). This period roughly coincided with the peak in SEUS tropospheric aerosol concentration [14].

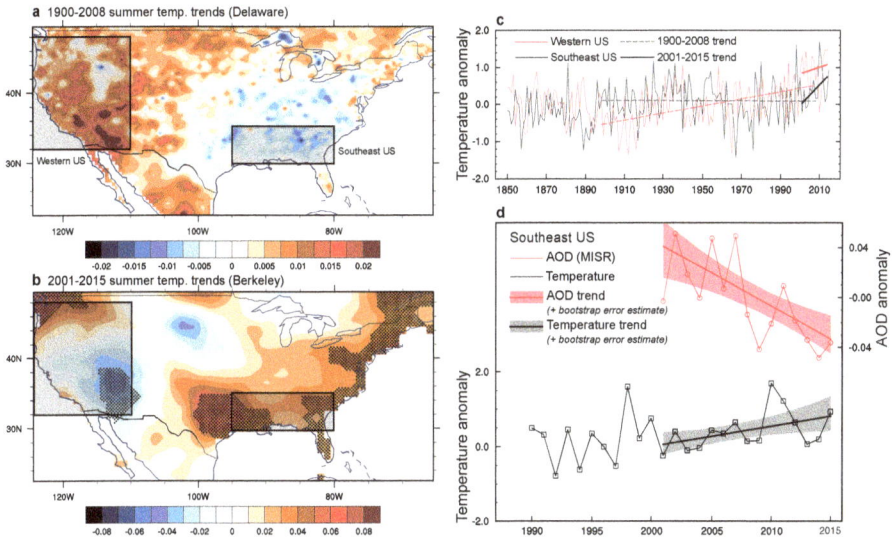

Figure 2. (**a**) linear trends in summer (May–September) surface temperatures from the Delaware dataset for 1900–2008 (0.5 degree resolution); (**b**) linear trends in summer surface temperatures from the Berkeley dataset for 2001–2015 (one-degree resolution). Stipling indicates statistically significant trends at 95% confidence; (**c**) regionally averaged summer surface temperature anomalies for 1850–2015 for the Southeast US (black) and the Western US (red) (regions defined by boxes in (**a**,**b**)). Linear trends for 1900–2008 (dashed lines) and 2001–2015 (solid lines) are noted; (**d**) summertime surface temperature and aerosol optical depth (AOD) anomalies for the southeast US for 1990–2015. The 2001–2015 linear trend is noted by solid lines. Error estimates calculated using the bootstrapping method (*n* = 1000) are shown as shaded solids.

In recent years, the summertime warming hole has not only disappeared but reversed (Figure 2b). The 2001–2015 temperature trend in the SEUS was $+0.54 \pm 0.52$ °C decade^{-1}, compared with $+0.18 \pm 0.62$ °C decade^{-1} in the Western US over the same period. Using the alternative Theil–Sen linear trend-fitting algorithm [50], the reversal is still notable and significant, with a positive $+0.50$ °C decade^{-1} trend in SEUS versus a negative -0.18 °C decade^{-1} trend in the Western US. When only the most recent decade (2001–2010) is considered, ensuring that the average contains input from all three primary surface temperature records, the contrast is more striking: $+0.69 \pm 0.68$ °C decade^{-1} in SEUS and -0.48 ± 0.74 °C decade^{-1} in the Western US.

Contemporary (2001–2015) satellite AOD measurements from the Multi-angle Imaging SpectroRadiometer (MISR) instrument [41] (Version 22, Level 3) show high summertime aerosol loading in the SEUS, especially when compared with the Western US where population densities are lower (Figure 4a). Summertime AOD (from MISR) in the SEUS was, on average, 0.05 to 0.10 greater than the long-term annual mean. In contrast, summertime AOD in the Western US was only 0.00 to 0.04 greater during the summer months.

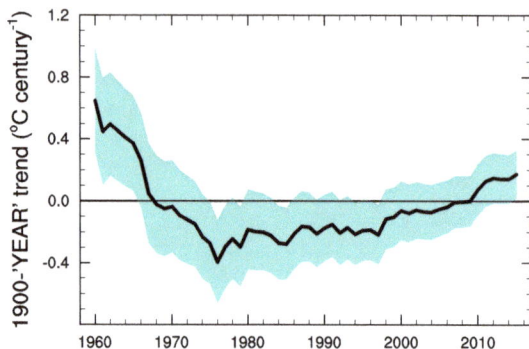

Figure 3. Time series of the 1900 to "year" linear trend (least-squares) for each "year" from 1960 to 2015. The green shaded region represents the error estimate (95% confidence) of the linear trend. Note the negative trends for the time period ranging from 1900 to 1970 through the time period ranging from 1900 to 2010, and the recent flip to positive trends after 2010.

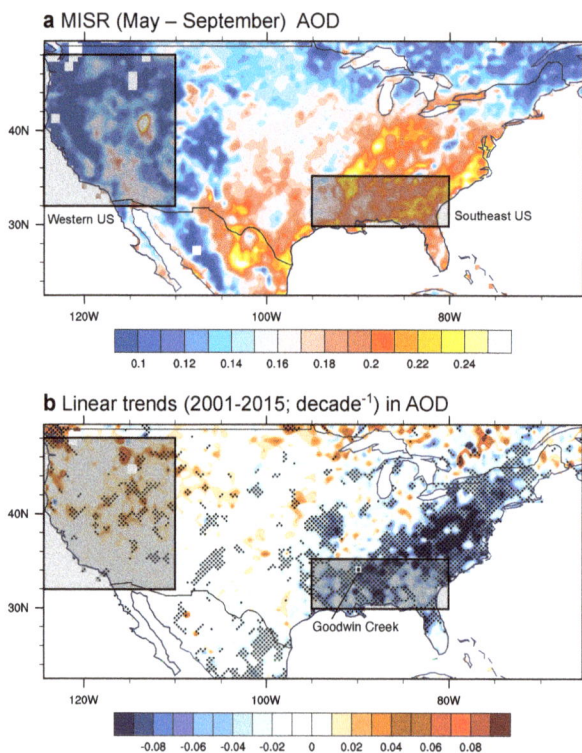

Figure 4. *Cont.*

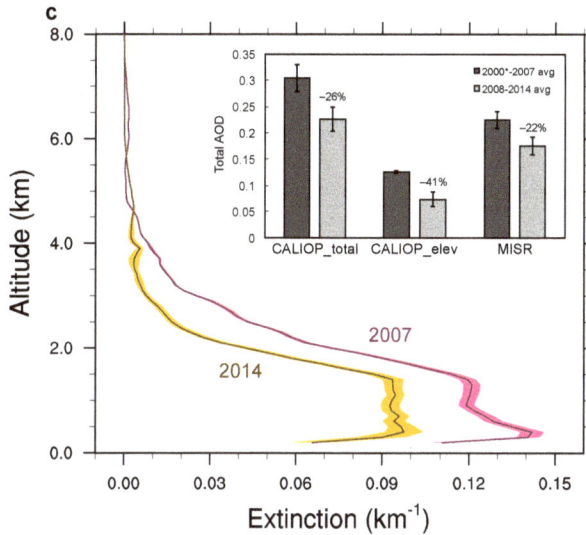

Figure 4. (**a**) summertime (May–September) aerosol optical depth (AOD) averaged for 2001–2015; (**b**) the linear trend (2001 to 2015) in summertime AOD. Aerosol data were taken from MISR (Multi-Angle Imaging Spectroradiometer) Stipling indicates statistically significant trends at 95% confidence; (**c**) *main:* vertical profiles of average summertime (May–September) aerosol extinction coefficients (km^{-1}) in the southeast US (SEUS) for both 2007 and 2014 calculated using CALIOP (Cloud Aerosol Lidar Orthogonal Polarization) lidar data. *inset:* summertime SEUS aerosol optical depth (AOD) averages for two periods (2000*–2007 and 2008–2014). CALIOP_total = total column AOD from the CALIOP instrument; CALIOP_elev = only elevated AOD (above the boundary layer) from the CALIOP instrument and MISR = MISR total column AOD. * For the two CALIOP variables: 2006–2007.

The trend in regionally averaged summertime AOD in the SEUS was −0.05 decade^{-1} between 2001 and 2015 (Figure 2d, Figure 4b). In contrast, summertime reductions in Western US AOD were minimal, where they existed at all (Figure 4b). Using a bootstrapping technique for error calculation ($n = 1,000$), we estimate that the 14-year reduction in SEUS AOD was statistically significant (−0.05 ± 0.03). However, a decomposition of the linear trend reveals a sharp drop in AOD between 2007 and 2008. The linear trend from 2001–2007 was near-zero (+0.01 decade^{-1}), and the linear trend from 2008 to 2015 was insignificantly negative (−0.03 decade^{-1}), suggesting that the majority of the decrease occurred between 2006 and 2009; indeed, the trend from 2005 to 2010 was −0.16 decade^{-1}. The smoothed curve, fit using LOESS smoothing techniques ($n = 5$ years), further illustrates this sharp transition. We note that an EPA rules change to the Clean Air Act (CAA) National Ambient Air Quality Standards (NAAQS) for Fine Particulate Matter (PM$_{2.5}$), implemented beginning in 2006 (Environmental Protection Agency, 40 C.F.R. § 50, 71 FR 2620) corresponds well with the aerosol changes resolved in the data. The rule lowered the 24-h PM$_{2.5}$ standard from 65 µg m^{-3} to 35 µg m^{-3}. That the timing of this rule and its subsequent implementation matches well with the sudden trend reversal represents a circumstantial link that may reconcile the observed relationship [51], though we acknowledge that we have only limited evidence indicating a steep drop in surface-level particulate matter concentration from 2007 to 2009 for several southeast US stations.

The 2001–2015 AOD decrease in the eastern United States (Figure 4b) coincided with a marked increase in average surface temperature (Figure 2d). The spatial correlation of AOD decrease and temperature increase implies that warmer regional temperatures were driven in some manner by improved air quality. To reconcile this relationship, we used vertical profiles of aerosol extinction combined with a one-dimensional radiative transfer model (Sections 2.2 and 2.3) to assess whether the link could be corroborated quantitatively or was mere coincidental correlation.

Summertime monthly mean profiles of SEUS aerosol light extinction coefficient were derived for 2006–2014 using data from the CALIOP instrument [38] (see Section 2). Similar to our results using MISR, CALIOP data show significant reductions in aerosol burden even over the shorter 2006–2014 time series (Figure 4c). More importantly, CALIOP data show that reductions in aerosol burden were not restricted to the surface layer but persisted as high as 5 km above the surface (though not much higher). The mean summertime (May–September) aerosol extinction coefficient decreased by an average of 35% from 2007 (the first full year of data) to 2014 (Figure 4c). In fact, the mean integrated AOD above the ABL was 41% lower (from 0.13 ± 0.00 to 0.07 ± 0.01) during 2008–2014 when compared with 2006–2007. This contrasts with a 26% reduction for the same time period for the entire tropospheric column (from 0.30 ± 0.02 to 0.23 ± 0.02) (Figure 4c). Furthermore, this decrease was not the result of a positive trend in ABL height, which can enhance the near-surface contribution to total aerosol loading. The ERA interim data show that, while the mean summertime regional ABL experiences interannual fluctuations, the overall trend was neither positive or negative.

Seasonal increases in the amount of secondary aerosols present above the boundary layer in the summer have been proposed to explain the apparent discrepancies noted between the total column aerosol burden from satellites and the measured near-surface pollution reported in EPA pollution measurements [49]. Our data support these findings, but, more notably, the CALIOP data show no apparent trend in aerosol composition from 2007 to 2014, from the surface through the mid-troposphere, indicating a continuity in relative airmass physical properties approaching the surface despite an overall reduction in their relative magnitudes.

3.2. Modeling Results Corroborate Observations

The FLG radiative transfer model was used to estimate the direct effect of these reductions in aerosol burden on the surface energy budget and column radiative heating profile, and to examine how that effect compares with the observed temperature trends previously discussed. We forced the model (see Section 2.3) with observed changes in vertically-resolved, CALIOP-derived aerosol extinction, broadband surface reflectance from MISR, and standard meteorological conditions, and estimated the direct radiative forcing of the reduced aerosol burden from 2007 to 2014 at solar noon. We performed two experiments, which are described in Section 2.3: FULL (direct forcing effect of aerosols) and AERO (direct forcing effect of aerosols minus a control simulation with no aerosols). Here, we present results from both, but focus on the FULL simulation.

The observed 35% decrease in CALIOP-measured aerosol extinction between 2007 and 2014 resulted in a 29 W m^{-2} (30 W m^{-2} for AERO) increase in solar noon net surface energy flux ($R_{n,srf}$) (Figure 5, inset). When we consider the respective CALIOP aerosol profiles for every year (2007 to 2014), the linear trend, which factors in all the intermediate years (2008–2013; see Table 1) in $R_{n,srf}$ was 2.3 W m^{-2} year^{-1} (2.5 W m^{-2} year^{-1} for AERO), which, when multiplied by seven (the number of years from 2007 to 2014), was less than the 29 W m^{-2} absolute difference between 2014 and 2007, but was nevertheless similar. These values are similar to what we derived using an empirical model generated for a Rayleigh-only atmosphere with a direct and diffuse component and gaseous absorption and the observed total column change in AOD (-0.09) [52]. Despite being the average for the entire SEUS, the positive surface energy fluxes also correspond remarkably well with observations from the Goodwin Creek, MS, USA SURFRAD (Surface Radiation Budget Network) 34.25° N, 89.87° W; https://www.esrl.noaa.gov/gmd/grad/surfrad/goodwin.html) surface radiation monitoring station (Figure 5, inset, Figure 6), which is located in a region of particularly large aerosol reductions (e.g., Figure 4b), is located outside of urbanized area in an area of representative vegetation, and is the only SURFRAD site within our region of interest.

Table 1. Fu-Liou-Gu (FLG) radiative transfer model reported net surface forcing values for each simulation.

Year	Net, Noontime Forcing (W m^{-2})	
	FULL	AERO
2007	−50.3	460.1
2008	−29.4	480.9
2009	−29.0	483.9
2010	−35.4	477.3
2011	−30.2	483.7
2012	−28.1	484.2
2013	−32.8	478.1
2014	−21.3	490.4
Trend (year^{-1})	+2.30	+2.45

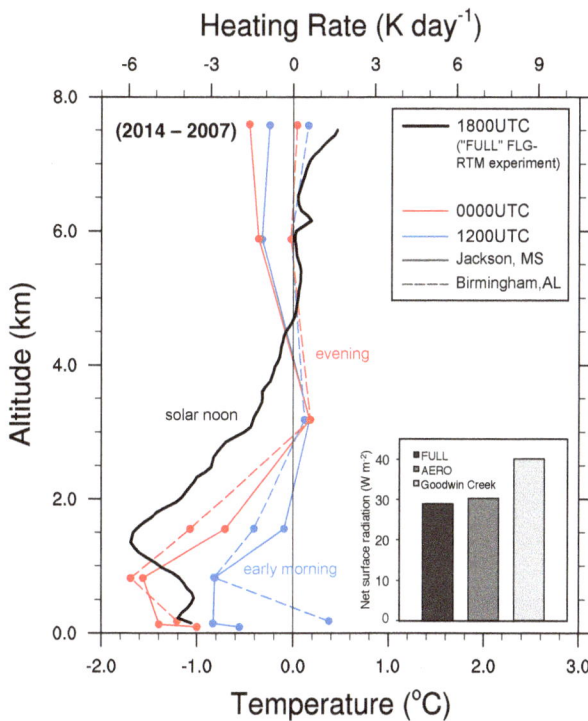

Figure 5. *main:* vertical profiles of the difference (2014–2007) in heating rates (thick black line) and atmospheric temperatures (red and blue lines) in SEUS from the 'FULL' FLG-RTM (Fu-Lio-Gu Radiative Transfer Model) model experiment (black line) at solar noon (1800 UTC) and atmospheric sounding data from Birmingham, AL, USA (dashed lines) and Jackson, MS, USA (solid lines) at 8:00 a.m. local time (1200 UTC) and 8:00 p.m. local time (0000 UTC). *inset:* net surface radiation differences (2014 minus 2007) at solar noon from the 'FULL' and 'AERO' (see Section 2.3) FLG-RTM experiments and the integrated 1700–1900 UTC average from the Goodwin Creek, MS, USA SURFRAD (surface radiation) measuring station. Goodwin Creek values were computed by pairing the number of days and the times so that the intrinsic sampling was identical.

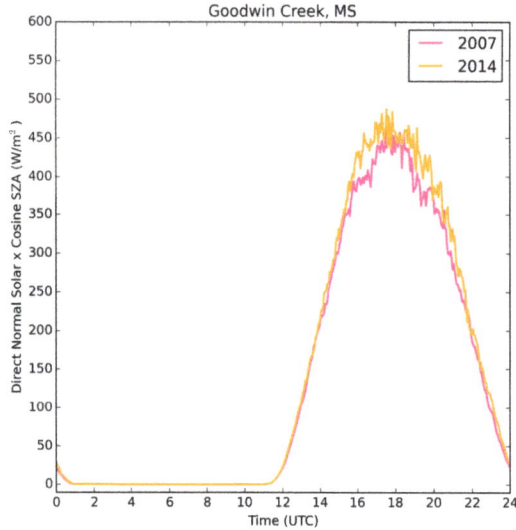

Figure 6. A diurnal time series of average summertime (May–September) direct minus normal surface radiation received at Goodwin Creek, MS, USA for 2007 (pink) and 2014 (yellow).

Curiously, while surface energy fluxes increased in the radiative transfer model over the temporal study period, total top of atmosphere (TOA) fluxes decreased between 2007 and 2014: -53.3 W m^{-2} in the FULL simulation and -50.5 W m^{-2} in the AERO simulation. As described below, this was probably a result of decreased aerosol absorption in the middle troposphere. In fact, while the trend in TOA radiative forcing was negative, the magnitude was still positive (e.g., 114.8 W m^{-2} in 2007 to 61.6 W m^{-2} in 2014 in the FULL simulations).

The FLG-estimated positive change in surface energy flux is a result of less solar absorption and scattering by aerosols in the atmospheric column. In the FULL simulation, lowered aerosol burdens from 2007 to 2014 resulted in a -3.4 K day^{-1} average decrease in solar noon heating rates between the surface and 4 km (Figure 5). The maximum decrease in heating rates occurred between 0.5 and 2.5 km, corresponding with large observed temperature reductions (2014–2007) at 0.7 km in the temperature sounding data from Birmingham, AL and Jackson, MS (Figure 5). The slight misalignment of the modeled and observed peaks is likely due to differences in the vertical resolution of the sounding data and the model as well as temporal differences; the soundings were taken at 7:00 a.m. (1200 UTC) and 7:00 p.m. (0000 UTC next day) local time, while the model simulations were performed for solar noon (1:00 p.m. local time). Atmospheric heating rates (and, subsequently, temperatures) throughout the lower troposphere were lower in 2014 because the atmospheric aerosol burden was lower. The model results indicate that a decrease in aerosol burden allowed more radiation to reach the surface and atmospheric absorption and semi-direct heating were suppressed. The results are consistent with the observed surface temperature trends, indicating that aerosols played a substantial role in the increasing temperatures observed in this region.

4. Conclusions

In summary, our results show that reduced aerosol direct radiative effects were of sufficient magnitude to contribute to the reversal of negative trends in twentieth century SEUS surface temperature. We present observations from both MISR and CALIOP showing statistically significant decreasing aerosol concentrations both at the surface and at elevated layers from 2007 to 2014 (supplementing results from Attwood et al. [31]). We demonstrate, using reliable profile data and a physically-based model that total column aerosol reductions increase the surface energy flux via direct effects, decrease the column atmospheric heating rate, and act as a probable contribution to an increase in regional surface temperatures since 2001. The consistency of our radiative transfer model results forced with changes in the observed aerosol budget with both surface radiation and atmospheric temperature profile measurements indicates that our results are robust and that the direct influence of aerosols played a significant role in the reversal of the warming hole in the SEUS. We note that Yu et al. [13] find evidence that a substantial portion of the trend reversal may be due to indirect effects. Specifically, they suggest that aerosol effects on shortwave cloud forcing combined with offset from the greenhouse warming of increased water vapor, may explain the temperature trend in SEUS. Our results neither support or contradict this hypothesis and suggest that direct and indirect effects may both play a role in SEUS temperature trends.

The local, direct effect from aerosols is unlikely to explain the entirety of the observed temperature trend reversal, and we acknowledge that the combination of the aerosol forcing with changes in clouds may also play a role. For example, we note significant noise, and relatively low temporal correlation, in both the surface temperature and aerosol trends, suggesting the existence of external mechanisms. Furthermore, Ruckstuhl et al. [53] and Philipona et al. [54] describe a similar rebound in surface temperature trends over mainland Europe in the 1980s as a response to changes in both aerosol burden and cloud distribution—though they note that the change due to direct aerosol forcing greatly outweighed changes from clouds—and our study seemingly validates their results in a different region. Using 2007–2014 cloud fraction data from CALIOP, we find that there is no trend in both total summer cloud fraction ($-0.7\% \pm 1.7\%$) or liquid cloud fraction ($+1.4\% \pm 1.7$) (Figure 7). Similarly, there is no evidence of a trend in monthly cloud fraction or cloud top height anomalies in the MISR data for the same region and a slightly longer time period (Figure 8). While these results do not immediately prove that clouds play no role in the observed temperature trend reversal, they do lend evidence to our analysis that the trend reversal may be driven more directly by changes to atmospheric aerosol burden.

However, while clouds may not have played a role in the reversal of the warming hole, we note that our work only addresses the relative role of aerosol trends. We acknowledge that other factors may also have contributed to positive temperature trends in the SEUS in the twenty-first century (e.g., [55]). In particular, we note that while multi-year trends in AOD and temperature appeared consistent with a direct influence of aerosols on surface temperature, the lack of covariance on interannual timescales suggests that other factors may have been substantial contributors. Our work indicates that aerosols were likely to have played a role in modulating surface temperatures but that other factors—which we did not explicitly account for in this study—may outweigh the aerosol effect. Further investigation of the relative roles of these other factors using a general circulation climate model is a logical next step.

We note that our work does not resolve the larger question of why surface temperatures in the SEUS were particularly sensitive to aerosol trends. While our analysis was restricted to the deep South (where the warming hole was largest), we acknowledge that equally substantial negative AOD trends along the eastern seaboard (Figure 4b) roughly corresponded to positive temperature trends (Figure 2b). Furthermore, extreme warming in eastern Texas did not correspond to equally notable aerosol trends (Figure 2b), suggesting eminent contributions from other meteorological and climatic factors. We also note that while temperature and aerosol trends did co-vary at the regionally-averaged scale, they did not do so at smaller spatial scales; further investigation is necessary to determine why the aerosol effect was not more geographically targeted, though other regional aerosol-specific studies may help elucidate why [56,57]. Going forward, we feel that this work would benefit from the integration of several ground-based observations, including, but not limited to, lidar, ceilometers and photometers. The Aerosol Robotic Network (AERONET), for example, can provide a more direct measure of AOD with potential to characterize aerosol properties, which, when combined with surface radiation measurements, can help elucidate more regionally targeted anomalies [58]. Futhermore, in this work, vertically resolved measurements were critical toward explaining the atmospheric warming and surface radiative response to changing aerosol burden. While the network of ground-based lidar equipment (MPLNET: https://mplnet.gsfc.nasa.gov/) does not currently include data within our study domain, these data are valuable for similar or expanded studies of this problem and further work should attempt to utilize any vertically-resolved aerosol data that may be available now or in the future.

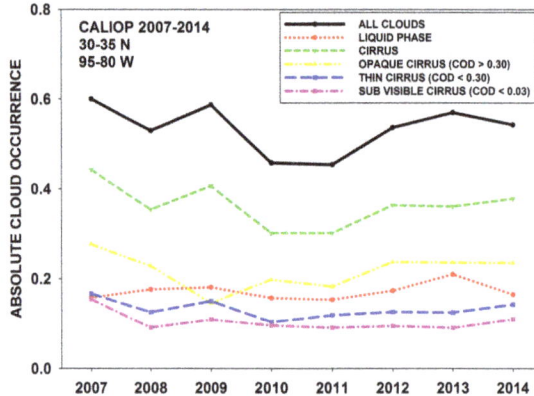

Figure 7. CALIOP-derived cloud fraction over the Southeast US (SEUS) for 2007 through 2014 showing no evident trend during the period.

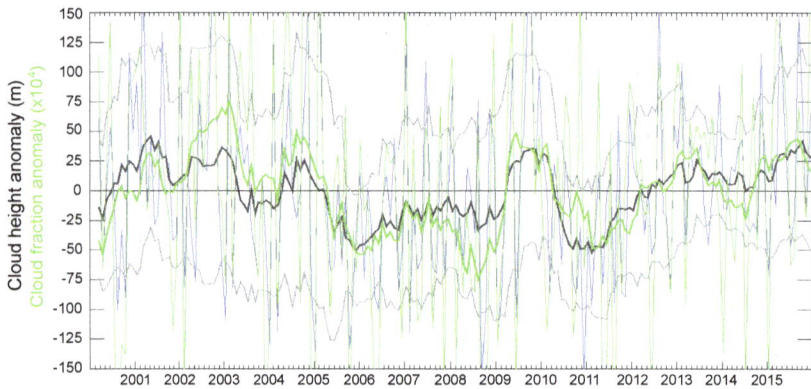

Figure 8. MISR-derived monthly cloud fraction over the Southeast US (SEUS) for March 2000 through 2015 showing no evident trend during the period. The thick gray line represents the anomaly in the mean cloud height smoothed with a 12-month running mean. The actual data are in blue and the gray lines indicate the 1-sigma bounds. The thick green line is the cloud fraction anomaly.

Despite many caveats, this study presents evidence that aerosols played a first-order role in accelerating positive surface temperature trends in the SEUS in recent years. It follows, then, that degraded air quality and aerosol forcing in other regions may similarly be masking climate warming effects within surface temperature records. This is especially true for regions with extremely poor air quality like northern India and eastern China [59–62]. This study provides contextual evidence for considering aerosol radiative effects and regional cooling when interpreting corresponding surface temperature trends. The logical next step includes characterizing the relative contribution of air quality improvements in the SEUS to increased warming using a regional climate model (such as the Weather Research and Forecasting model); we anticipate running these simulations in the near term.

Acknowledgments: The research included in this manuscript was supported by an ACCDAM grant from the NASA Earth Sciences Division, Radiation Sciences program. We thank the MISR team for providing facilities and useful discussions. We also thank the Naval Research Laboratory for computing facilities. The MISR and CALIOP data were obtained from the NASA Langley Research Center Atmospheric Science Data Center. The links to the temperature datasets are found in Section 2. We also thank Yu Gu (UCLA) for productive conversations and help with the FLG radiative transfer model and Jeffrey Reid (NRL) for productive conversations and suggestions over the past year.

Author Contributions: All authors contributed substantially to this work. All authors performed the research and helped produce the figures. Kalashnikova funded the research and conceived the hypothesis with Tosca, Garay and Campbell. Tosca assembled the paper.

Conflicts of Interest: The authors declare no conflict of interest.

References

1. Hansen, J.; Ruedy, R.; Sato, M.; Lo, K. Global surface temperature change. *Rev. Geophys.* **2010**, *48*, RG4004.
2. Portmann, R.W.; Solomon, S.; Hegerl, G.C. Spatial and seasonal patterns in climate change, temperatures, and precipitation across the United States. *Proc. Natl. Acad. Sci. USA* **2009**, *106*, 7324–7329.
3. Capparelli, V.; Franzke, C.; Vecchio, A.; Freeman, M.P.; Watkins, N.W.; Carbone, V. A spatiotemporal analysis of U.S. station temperature trends over the last century. *J. Geophys. Res.* **2013**, *118*, 7427–7434.
4. Melillo, J.M.; Richmond, T.C.; Yohe, G.W. *Climate Change Impacts in the United States: The Third National Climate Assessment*; Technical Report; U.S. Global Change Research Program: Washington, DC, USA, 2014.
5. Liang, X.Z.; Pan, J.; Zhu, J.; Kunkel, K.E.; Wang, J.X.L.; Dai, A. Regional climate model downscaling of the U.S. summer climate and future change. *J. Geophys. Res.* **2006**, *111*, D10108.
6. Pan, Z.; Arritt, R.W.; Takle, E.S.; Jr., W.J.G.; Anderson, C.J.; Segal, M. Altered hydrologic feedback in a warming climate introduces a "warming hole". *Geophs. Res. Lett.* **2004**, *31*, L17109.
7. Meehl, G.A.; Arblaster, J.M.; Branstator, G. Mechanisms contributing to the warming hole and the consequent U.S. East-West differential of heat extremes. *J. Climate* **2012**, *25*, 6394–6408.
8. Robinson, W.A.; Reudy, R.; Hansen, J.E. General circulation model simulations of recent cooling in the east-central United States. *J. Geophys. Res.* **2002**, *107*, 4748.
9. Yu, S.; Saxena, V.K.; Zhao, Z. A comparison of signals of regional aerosol-induced forcing in eastern China and the southeastern United States. *Geophs. Res. Lett.* **2001**, *28*.
10. Shindell, D.; Faluvegi, G. Climate respnose to regional radiative forcing during the twentieth century. *Nature Geosci.* **2009**, *2*, 294–300.
11. Leibensperger, E.M.; Mickley, L.J.; Jacob, D.J.; Chen, W.T.; Seinfeld, J.H.; Nenes, A.; Adams, P.J.; Streets, D.G.; Kumar, N.; Rind, D. Climatic effects of 1950–2050 changes in US anthropogenic aerosols—Part 2: Climate response. *Atmos. Chem. Phys.* **2012**, *12*, 3349–3362.
12. Mickley, L.J.; Leibensperger, E.M.; Jacob, D.J.; Rind, D. Regional warming from aerosol removal over the United States: Results from a transient 2010–2050 climate simulation. *Atmos. Environ.* **2012**, *46*, 545–553.
13. Yu, S.; Alapaty, K.; Mathur, R.; Pleim, J.; Zhang, Y.; Nolte, C.; Eder, B.; Foley, K.; Nagashima, T. Attribution of the United States "warming hole": Aerosol indirect effect and precipitable water vapor. *Sci. Rep.* **2014**, *4*, 6929.
14. Leibensperger, E.M.; Mickley, L.J.; Jacob, D.J.; Chen, W.T.; Seinfeld, J.H.; Nenes, A.; Adams, P.J.; Streets, D.G.; Kumar, N.; Rind, D. Climatic effects of 1950–2050 changes in US anthropogenic aerosols—Part 1: Aerosol trends and radiative forcing. *Atmos. Chem. Phys.* **2012**, *12*, 3333–3348.
15. Liepert, B.G. Observed reductions of surface solar radiation at sites in the United States and worldwide from 1961 to 1990. *Geophys. Res. Lett.* **2002**, *29*, 1421.
16. Charlson, R.J.; Schwartz, S.E.; Hales, J.M.; Cess, R.D.; Coakley, J.A., Jr.; Hansen, J.E.; Hofman, D.J. Climate forcing by anthropogenic aerosols. *Science* **1992**, *255*, 423–430.
17. Hansen, J.; Sato, M.; Ruedy, R. Radiative forcing and climate response. *J. Geophys. Res.* **1997**, *102*, 6831–6864.
18. Ramanathan, V.; Crutzen, P.J.; Kiehl, J.T.; Rosenfeld, D. Aerosols, climate, and the hydrological cycle. *Science* **2001**, *294*, 2119–2124.
19. Lohmann, U.; Lesins, G. Stronger constraints on the anthropogenic indirect aerosol effect. *Science* **2002**, *298*, 1012–1015.
20. Rosenfeld, D. Aerosols, clouds, and climate. *Science* **2006**, *312*, 1323–1324.
21. Albrecht, B.A. Aerosols, cloud microphysics, and fractional cloudiness. *Science* **1989**, *245*, 1227–1230.

22. Rotstayn, L.D. Indirect forcing by anthropogenic aerosols: A global climate model calculation of the effective-radius and cloud-lifetime effects. *J. Geophys. Res.* **1999**, *104*, 9369–9380.

23. Storelvmo, T.; Leirvik, T.; Lohmann, U.; Phillips, P.; Wild, M. Disentangling greenhouse warming and aerosol cooling to reveal Earth's climate sensitivity. *Nature Geoscience* **2016**, *9*, 206–289.

24. Mitchell, J.F.B.; Johns, T.C. On modification of global warming by sulfate aerosols. *J. Climate* **1997**, *10*, 245–267.

25. Andreae, M.O.; Jones, C.D.; Cox, p.m. Strong present-day aerosol cooling implies a hot future. *Nature* **2001**, *435*, 1187–1190.

26. Levy, H.; Schwarzkopf, M.D.; Horowitz, L.; Ramaswamy, V.; Findell, K.L. Strong sensitivity of late 21st century climate to projected changes in short-lived air pollutants. *J. Geophys. Res.* **2008**, *113*, D06102.

27. Ming, Y.; Ramaswamy, V. Nonlinear climate and hydrological responses to aerosol effects. *J. Clim.* **2009**, *22*, 1329–1339.

28. Rotstayn, L.D.; Cai, W.; Dix, M.R.; Farquhar, G.D.; Feng, Y.; Ginoux, P.; Herzog, M.; Ito, A.; Penner, J.E.; Roderick, M.L.; et al. Have Australian rainfall and cloudiness increased due to the remote effects of Asian anthropogenic aerosols? *J. Geophys. Res.* **2007**, *112*, D09202.

29. Shindell, D.T.; Faluvegi, G.; Rotstayn, L.; Milly, G. Spatial patterns of radiative forcing and surface temperature response. *J. Geophys. Res.* **2015**, *120*, 5385–5403.

30. Paasonen, P.; Asmi, A.; Petaja, T.; Kajos, M.K.; Aijala, M.; Junninen, H.; Holst, T.; Abbatt, J.P.D.; Arneth, A.; Birmili, W.; et al. Warning-induced increase in aerosol number concentration likely to moderate climate change. *Nat. Geosci.* **2013**, *6*, 438–442.

31. Attwood, A.R.; Washenfelder, R.A.; Brock, C.A.; Hu, W.; Baumann, K.; Campuzano-Jost, P.; Day, D.A.; Edgerton, E.S.; Murphy, D.M.; Palm, B.B.; et al. Trends in sulfate and organic aerosol mass in the Southeast US: Impact on aerosol optical depth and radiative forcing. *Geophys. Res. Lett.* **2014**, *41*, 7701–7709.

32. Kim, P.S.; Jacob, D.J.; Fisher, J.A.; Travis, K.; Yu, K.; Zhu, L.; Yantosca, R.M.; Sulprizio, M.P.; Jimenez, J.L.; Campuzano-Jost, P.; et al. Sources, seasonality, and trends of southeast US aerosol: an integrated analysis of surface, aircraft, and satellite observations with the GEOS-Chem chemical transport model. *Atmos. Chem. Phys.* **2015**, *15*, 10411–10433.

33. Augustine, J.A.; Dutton, E.G. Variability of the surface radiation budget over the United States from 1996 through 2011 from high-quality measurements. *J. Geophys. Res.* **2013**, *118*, 43–53.

34. Tang, Q.; Leng, G. Changes in cloud cover, precipitation, and summer temperature in North America from 1982 to 2009. *J. Clim.* **2013**, *26*, 1733–1744.

35. Zhang, L.; Henze, D.K.; Grell, G.A.; Torres, O.; Jethva, H.; Lamsal, L.K. What factors control the trend of increasing AAOD over the United States in the last decade? *J. Geophys. Res.* **2017**, *122*, 1797–1810.

36. Fan, Y.; van den Dool, H. A global monthly land surface air temperature analysis for 1948-present. *J. Geophys. Res.* **2008**, *113*, D01103.

37. Levi, B.G. Earth's land surface temperature trends: A new approach confirms previous results. *Phys. Today* **2013**, *66*.

38. Campbell, J.R.; Tackett, J.L.; Reid, J.S.; Zhang, J.; Curtis, C.A.; Hyer, E.J.; Sessions, W.R.; Wetphal, D.L.; Prospero, J.M.; Welton, E.J.; et al. Evaluating nighttime CALIOP 0.532 um aerosol optical depth and extinction coefficient retrievals. *Atmos. Meas. Tech.* **2012**, *5*, 2143–2160.

39. Winker, D.M.; Tackett, J.L.; Getzewich, B.J.; Liu, Z.; Vaughan, M.A.; Rogers, R.R. The global 3-D distribution of tropospheric aerosols as characterized by CALIOP. *Atmos. Chem. Phys.* **2013**, *13*, 3345–3361.

40. Dee, D.P.; Uppala, S.M.; Simmons, A.J.; Berrisford, P.; Poli, P.; Kobayashi, S.; Andrae, U.; Balmaseda, M.A.; Balsamo, G.; Bauer, P.; et al. The ERA-Interim reanalysis: configuration and performance of the data assimilation system. *Q. J. R. Meteorol. Soc.* **2011**, *137*, 553–597.

41. Martonchik, J.V.; Diner, D.J.; Crean, K.A.; Bull, M.A. Regional aerosol retrieval results from MISR. *IEEE Trans. Geosci. Remote Sens.* **2002**, *40*, 1520–1531.

42. Fu, Q.; Liou, K.N. On the correlated k-distribution method for radiative transfer in nonhomogeneous atmospheres. *J. Atmos. Sci.* **1992**, *49*, 2139–2156.

43. Fu, Q.; Liou, K.N. Parametrization of the radiative properties of cirrus clouds. *J. Atmos. Sci.* **1993**, *50*, 2008–2025.

44. Gu, Y.; Farrara, J.; Liou, K.N.; Mechoso, C.R. Parametrization of cloud-radiative processes in the UCLA general circulation model. *J. Climate* **2003**, *16*, 3357–3370.

45. Gu, Y.; Liou, K.N.; Ou, S.C.; Fovell, R. Cirrus cloud simulations using WRF with improved radiation parametrization and increased vertical resolution. *J. Geophys. Res.* **2011**, *116*, D06119.
46. Gu, Y.; Liou, K.N.; Jiang, J.H.; Su, H.; Liu, X. Dust aerosol impact on North African climate: A GCM investigation of aerosol-cloud-radiation interactions using A-Train satellite data. *Atmos. Chem. Phys.* **2012**, *12*, 1667–1679.
47. Hess, M.; Koepke, P.; Schult, I. Optical properties of aerosols and clouds: The software package OPAC. *Bull. Am. Meteor. Soc.* **1998**, *79*, 831–844.
48. Burton, S.P.; Ferrare, M.A.; Vaughan, A.H.; Omar, A.H.; Rogers, R.R.; Hostetler, C.A.; Hair, J.W. Aerosol classification from airborne HSRL and comparisons with the CALIPSO vertical feature mask. *Atmos. Meas. Tech.* **2013**, *6*, 1397–1412.
49. Ford, B.; Heald, C.L. Aerosol loading in the Southeastern United States: Reconciling surface and satellite observations. *Atmos. Chem. Phys.* **2013**, *13*, 9269–9283.
50. Theil, H. A rank-invariant method of linear and polynomial regression analysis. I. *Nederl. Akad. Wetensch. Proc.* **1950**, *53*, 386–392.
51. Hand, J.L.; Schichtel, B.A.; Malm, W.C.; Pitchford, M.L. Particulate sulfate ion concentration and SO_2 emissions trends in the United States from the early 1990s through 2010. *Atmos. Chem. Phys.* **2012**, *12*, 10353–10365.
52. Gregg, W.W.; Carder, K.L. A simple spectral solar irradiance model for cloudless maritime atmospheres. *Limnol. Oceanog.* **1990**, *35*, 1657–1675.
53. Ruckstuhl, C.; Philipona, R.; Behrens, K.; Coen, M.C.; Durr, B.; Heimo, A.; Matzler, C.; Nyeki, S.; Ohmura, A.; Vuilleumier, L.; et al. Aerosol and cloud effects on solar brightening and the recent rapid warming. *Geophys. Res. Lett.* **2008**, *35*, L12708.
54. Philipona, R.; Behrens, K.; Ruckstuhl, C. How declining aerosols and rising greenhouse gases forced rapid warming in Europe since the 1980s. *Geophys. Res. Lett.* **2009**, *36*, L02806.
55. Meehl, G.A.; Arblaster, J.M.; Chung, C.T.Y. Disappearance of the southeast U.S. "warming hole" with the late 1990s transition of the Interdecadal Pacific Oscillation. *Geophys. Res. Lett.* **2015**, *42*, 5564–5570.
56. Che, H.; Xia, X.; Zhu, J.; Dubovnik, O.; Holben, B.; Goloub, P.; Chen, H.; Estelles, V.; Cuebas-Agullo, E.; Blarel, L.; et al. Column aerosol optical properties and aerosol radiative forcing during a serious haze-fog month over North China Plain in 2013 based on ground-based sunphotometer measurements. *Atmos. Chem. Phys.* **2014**, *14*, 2125–2138.
57. Prats, N.; Cachorro, V.E.; Berjon, A.; Toledano, C.; De Frutos, A.M. Column-integrated aerosol microphysical properties from AERONET Sun photometer over southwestern Spain. *Atmos. Chem. Phys.* **2011**, *11*, 12535–12547.
58. Holben, B.N.; Eck, T.F.; Slutsker, I.; Tanre, D.; Buis, J.P.; Setzer, A.; Vermote, E.; Reagan, J.A.; Kaufman, Y.; Nakajima, T.; et al. AERONET-A federated instrument network and data archive for aerosol characterization. *Remote Sens. Environ.* **1998**, *66*, 1–16.
59. Donkelaar, A.V.; Martin, R.V.; Brauer, M.; Kahn, R.; Levy, R.; Verduzco, C.; Villeneuve, P.K. Global estimates of ambient fine particulate matter concentrations from satellite-based aerosol optical depth: Development and application. *Environ. Health. Perspect.* **2010**, *118*, 847–855.
60. Li, P.; Yan, R.; Yu, S.; Wang, S.; Liu, W.; Bao, H. Reinstate regional transport of PM2.5 as a major cause of severe haze in Beijing. *Proc. Natl. Acad. Sci. USA* **2015**, *112*, E2739–E2740.
61. Yan, R.; Yu, S.; Zhang, Q.; Li, P.; Wang, S.; Chen, B.; Liu, W. A heavy haze episode in Beijing in February of 2014: Characteristics, origins and implications. *Atmos. Pollut. Res.* **2015**, *6*, 867–876.
62. Yu, S.; Li, P.; Wang, L.; Wang, P.; Wang, S.; Chang, S.; Liu, W.; Alapaty, K. Anthropogenic aerosols are a potential cause for migration of the summer monsoon rain belt in China. *Proc. Natl. Acad. Sci. USA* **2016**, *11*, E2209–E2210.

MDPI AG

St. Alban-Anlage 66

4052 Basel, Switzerland

Tel. +41 61 683 77 34

Fax +41 61 302 89 18

http://www.mdpi.com

Remote Sensing Editorial Office

E-mail: remotesensing@mdpi.com

http://www.mdpi.com/journal/remotesensing

www.ingramcontent.com/pod-product-compliance
Lightning Source LLC
Chambersburg PA
CBHW051711210326
41597CB00032B/5443

* 9 7 8 3 0 3 8 4 2 6 4 0 0 *